Corwin M. Johnson

HORTICULTURAL SCIENCE

Overleaf, Apple blossoms. [Courtesy International Harvester Corporation.]

Third Edition

HORTICULTURAL SCIENCE

Jules Janick
Purdue University

W. H. Freeman and Company
San Francisco

Cover: An eggplant (*Solanum melongena*) in flower and fruit. From a linoleum-block print by Henry Evans.

Sponsoring Editor: Gunder Hefta
Project Editor: Dick Johnson
Designers: Perry Smith and Brenn Lea Pearson
Production Coordinator: Chuck Pendergast
Illustration Coordinator: Cheryl Nufer
Artists: Evan Gillespie and Donna Salmon
Compositor: York Graphic Services
Printer and Binder: The Maple-Vail Book Manufacturing Group

Library of Congress Cataloging in Publication Data

Janick, Jules, 1931–
  Horticultural science.

  Includes bibliographies and index.
  1. Horticulture.   I. Title.
SB318.J35   1979   635   78-13053
ISBN 0-7167-1031-5

Printed in the United States of America

9 8 7 6 5 4 3 2 1

To Shirley and Peter and Robin

Contents

Contents

III

THE INDUSTRY OF HORTICULTURE

Preface

Horticulture is concerned with those plants whose cultivation brings rewards, whether monetary or personal pleasure, sufficient to warrant the expenditure of intensive effort. This art—which entails judicious timing and many skills—has an ancient tradition. But modern horticulture involves the integration of many natural phenomena with synthetic effects and so is a scientific discipline in its own right. The primary purpose of this textbook is to examine the scientific concepts on which horticulture is based. A comprehension of the science gives meaning and scope to the art and makes possible the improvement of centuries-old practices.

The third edition of this book accommodates various changes in horticultural science that have taken place since the second edition was published in 1972. Some innovations and techniques are covered for the first time in this edition, while others are given expanded treatment. Graphs and tables have been updated as appropriate, and many figures have been added. The sequence of presentation is still sound, however, and so remains unchanged.

Part I introduces the biology of horticulture; horticultural problems are biological problems. The plant, the basis of all horticultural activities, must first be considered as a living entity. A knowledge of plant relationships, structure, growth, and development is necessary if the technology and industry of horticulture are to be understood.

Part II deals with the technology of horticulture. Rather than being considered as they relate to specific crops, techniques have been treated in terms of their broader horticultural implications. Thus it is hoped that the information will become more meaningful and transferable. Specific practices should be discussed by the instructor, since they will vary with geographical location.

Part III describes the industry of horticulture, which is analyzed on the basis of location and specialty. The distinguishing characteristics and special problems of the industry are emphasized. The book concludes with a discussion of the esthetic aspects of horticulture.

The third edition, like the first and second, has been designed primarily for the beginning horticulture student as well as for those whose interests may be only incidentally associated with horticulture. The text assumes no great familiarity with botany or plant science. It is divided into fifteen chapters, which can be covered adequately in a semester's time. The skills associated with horticulture can be reviewed in a laboratory in a sequence similar to that used in this text. Key references are provided at the end of each chapter to facilitate further study.

It is a pleasure to acknowledge my colleagues who have been generous with their time, their information, and their support. Among the many are K. M. Brink, N. W. Desrosier, H. C. Dostal, Dominic Durkin, F. H. Emerson, A. T. Guard, P. M. Hasegawa, C. E. Hoxsie, Jerome Hull, Jr., K. W. Johnson, A. C. Leopold, N. W. Marty, C. A. Mitchell, C. L. Pfeiffer, E. C. Stevenson, R. B. Tukey, G. F. Warren, G. E. Wilcox, and M. Workman. C. E. Hess and A. H. Westing, each of whom has contributed a chapter, have been of inestimable help. I would like to thank J. R. Shay, W. H. Gabelman, M. N. Dana, and I. J. Johnson for their critical reading of the original manuscript.

December 1978   JULES JANICK

HORTICULTURAL SCIENCE

1

The Impact of Horticulture

Virtue? a fig! 'tis in ourselves that we are thus or thus. Our bodies are
our gardens, to the which our wills are gardeners; so that if we will
plant nettles or sow lettuce, set hyssop and weed up tine, supply it with
one gender of herbs or distract it with many, either to have it sterile
with idleness or manur'd with industry—why, the power and corrigible
authority of this lies in our wills.

SHAKESPEARE, *Othello* [I. 3]

The origins of horticulture are intimately associated with the history of mankind.
The term *horticulture*, which is probably of relatively recent origin, first appeared
in written language in the seventeenth century.[1] The word is derived from the
Latin *hortus*, garden, and *colere*, to cultivate. The concept of the culture of
gardens (Anglo-Saxon *gyrdan*, to enclose) as distinct from the culture of fields—
that is, **agriculture**—is a medieval concept, indicative of the practices of that
period. Agriculture now refers broadly to the technology of raising plants and
animals. **Horticulture** is that part of plant agriculture concerned with so-called
"garden crops," as contrasted with **agronomy** (field crops, mainly grains and
forages) and **forestry** (forest trees and products). The relation of these disciplines
to the rest of science and technology is portrayed in Figure 1-1.

Horticulture deals with an enormous number of plants. Garden crops traditionally include fruits, vegetables, and all the plants grown for ornamental
purposes, as well as spices and medicinals. Many horticultural products are
utilized in the living state and are thus highly perishable; constituent water is
essential to their quality. In contrast, the products of agronomy and forestry are
often utilized in the nonliving state and are usually high in dry matter. Custom has
delineated the boundary line for some crops; for example, tobacco and, in some
locations, potatoes are considered agronomic crops in the United States. In the
main, however, horticulture deals with crops that are intensively cultivated—that

---

[1] The first known use of the word *horticultura* is in Peter Lauremberg's treatise of that name,
written in 1631. *Horticulture* is first mentioned in English by E. Phillips in *The New World of English
Words*, London, 1678.

1

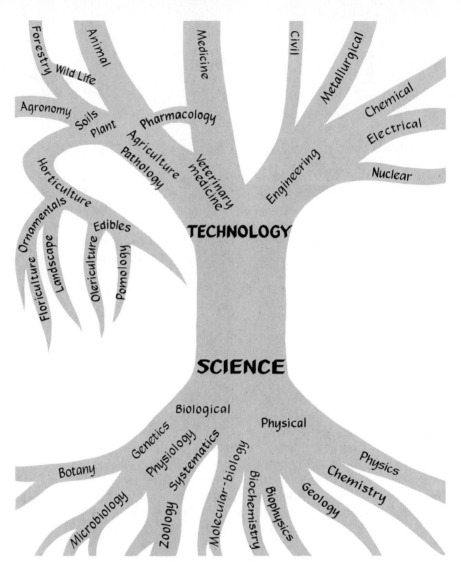

FIGURE 1-1. A tree of the natural sciences and their technologies.

is, plants that are of high enough value to warrant a large input of capital, labor, and technology per unit area of land. Pine trees grown for timber or pulp are an example of extensive agriculture, rather than horticulture: although the value per acre of pine grove may be large if the grove is harvested all at once, the yearly increment of value is relatively small (30–100 dollars per year). Therefore, the economic importance of our forest industries is largely a function of the tremendous acreage involved. When pine trees are grown more intensively, as are Christmas trees (Fig. 1-2), they are often considered a horticultural crop. Pine trees grown as nursery stock for use in ornamental plantings assume sufficient

A

B

C

FIGURE 1-2. Pines grown for Christmas trees in a horticultural operation require intensive management, such as pest control (A) and shearing (B). The result is a well-formed, attractive tree (C).

value to justify large expenditures for fertilization, pruning, harvesting, and marketing, and are therefore a true horticultural crop. (The value of good nursery land may be equivalent to that of suburban real estate!) Similarly, the presence of a sugary gene in maize, an agronomic crop, increases its value enough to warrant using more intensive cultural methods (the use of better seed—single-cross versus double-cross hybrid—spray programs, and expensive harvesting procedures) and transforms it into a horticultural crop, sweet corn.

Horticulture thus can be defined as the branch of agriculture concerned with intensively cultured plants directly used by people for food, for medicinal purposes, or for esthetic gratification. The industry is usually subdivided according to the kinds of products and the uses to which they are put. The production of edibles is represented by **pomology** (fruit crops) and **olericulture** (vegetable crops); the production of ornamentals is represented by **floriculture** and **landscape horticulture.** These terms are not mutually exclusive. For example, many edible plants (apples) are used as ornamentals, and many plants often classed as ornamentals (poppy, pyrethrum) have pharmacological and industrial uses (Fig. 1-3).

The esthetic use of plants is a unique feature of horticulture, distinguishing it from other agricultural activities. It is this aspect of horticulture that has led to its universal popularity. In the United States ornamental horticulture is undergoing a renaissance brought about by an increased standard of living coincidental with the development of suburban living. The satisfaction of this bent in the American

FIGURE 1-3. The dried flowers of pyrethrum (*Chrysanthemum cinerariaefolium*) are the source of a natural insecticide. Ornamental forms (*C. coccineum*) are known as painted daisies. [Photograph courtesy E. R. Honeywell.]

family has created an expanding industry out of ornamental horticulture, whose practice had formerly been confined to well-to-do fanciers.

Horticulture is an ancient art, and many of its practices have been empirically derived. However, modern horticulture, as agriculture, has become intimately associated with science, which has served not only to provide the methods and resources to explain the art but has also become the guiding force for its improvement and refinement. Horticulture will never become wholly a science, nor is this particularly desirable. Its curious mixture of science (botany to physics), technology, and esthetics makes horticulture a refreshing discipline that has continually absorbed people's interest and challenged their ingenuity. (Indeed, the rise of the world environmental movement in the 1960s and 1970s resulted in a virtual explosion of interest in plants and gardening, an interest that shows no sign of waning in the coming decade.) The science of horticulture remains the dynamic influence in the proper use and understanding of the horticultural art, and it is with this phase of horticulture that this book is largely concerned.

ECONOMIC POSITION OF HORTICULTURE

It is difficult to ascertain the precise position of any large and diverse industry in our economy. This is particularly true of horticulture, which involves not only the many facets of production, but the added increments of processing, service, and maintenance. For example, ornamentals such as woody perennials are not consumed but are invested in plantings, which increase in value with the passage of time. The value of this wealth is ordinarily not taken into consideration until we become painfully aware of it through the tolls taken by severe weather or through the encroachment of concrete and steel. The replacement of large trees and shrubs is usually economically prohibitive and is often horticulturally impossible.

Commercial horticulture represents a significant portion of American agricultural wealth. Agriculture, the country's biggest industry, has a farm value of close to 100 billion dollars, half of which originates directly in agriculture, with the rest representing the contributions of other industries—for example, machinery, fertilizer, and chemicals (Fig. 1-4). These production inputs are a reflection of the increased technology inherent in present-day agriculture. Agriculture has increased in production to keep up with a growing population's demand for a high level of nutrition and a high standard of living. Horticulture's percentage share in this expanding industry has been relatively stable over the past 60 years (Fig. 1-5). In 1970, 13.3% of the annual farm receipts, representing six billion dollars, were attributable to horticulture. The retail value of horticultural products, after being processed, transported, and marketed, increased to about 15 billion dollars. In the United States about 40% by weight of the food consumed consists of horticultural products. In view of this fact alone, one may expect horticulture to maintain its increasing importance in our lives and in our economy.

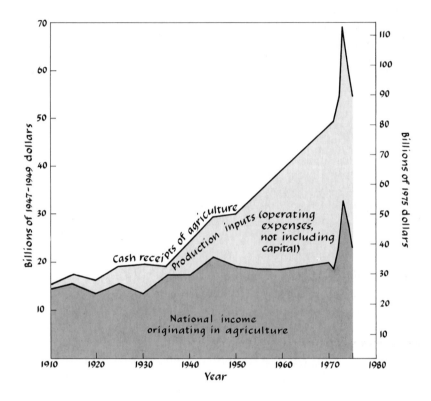

FIGURE 1-4. Cash receipts compared with national income originating in agriculture, 1910–1975. [Data from Economic Research Service, USDA.]

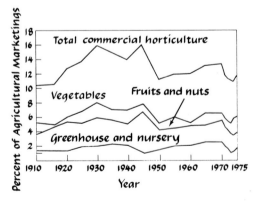

FIGURE 1-5. Horticulture's share of the total agricultural market in the United States, 1910–1975. [Data from Economic Research Service, USDA.]

These figures should not imply that each facet of the horticultural industry has or will share equally in this increase. The fortunes of individual crops in the United States (Fig. 1-6) reflect the changing habits and preferences of the American people and the technological changes in the food industry. For example, the trend in fresh vegetables has been toward an increase in the per capita consumption of those used in salads (lettuce, celery) and a decrease in the consumption of potatoes and starchy root crops. Fruit consumption per capita has gradually increased, with citrus fruits gaining at the expense of apples. The consumption of processed fruits and vegetables has shown a marked increase over the consumption of fresh products, and, on a fresh-equivalent basis, now accounts for more than half of the total consumed.

A HISTORICAL PERSPECTIVE

Along with the discovery of fire, the "invention" of agriculture represents the most significant achievement in human civilization. In primitive societies based on food gathering or hunting, each individual must be totally involved with the urgencies of securing sustenance. Abundance proves to be temporary and exceptional. Notwithstanding the systematic and efficient organization of certain food-gathering societies, each adult is pressed into continual activity. The limiting factor in the development of primitive societies becomes the availability and dependability of a food supply.

Man has been a food collector for the great part of his existence. Food production by the cultivation of edible plants and the domestication of animals is of relatively recent origin, dating back 7000 to 10,000 years to what is known as the Neolithic Age. Only through the gradual development of a system of agriculture could the accumulation of surpluses become a regular occurrence. The immediate reward of a surplus is the release from food production of people who can contribute to society in other ways: The gradual development of agriculture, with its increased efficiency and dependability, encouraged the development of new classes of specialists—artisans, clerks, priests. The standard of living of a people increases as the need for specialists increases. Table 1-1 broadly dates the progress of civilization in terms of advances in agriculture and horticulture.

Selection of Edible Plants

The origin of civilization can be traced to the discovery that it is possible to assure oneself of a plentiful food supply by planting seed. Rapidly growing vegetables and cereals, which produce a crop within a season, must have been the first plants cultivated. The technology involved in cultivating nut or fruit trees, for example, is considerably intricate and time consuming; as a result, these edibles were

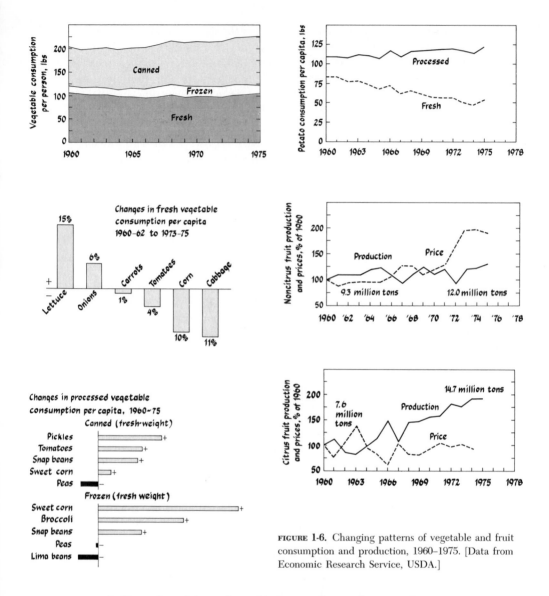

FIGURE 1-6. Changing patterns of vegetable and fruit consumption and production, 1960–1975. [Data from Economic Research Service, USDA.]

probably gathered from the wild. Even today in the United States some food is gathered from indigenous uncultivated plants. For example, Maine's blueberry industry depends upon such a source. The cultivation of cereals and the domestication of animals led to a permanent agriculture, which provided the medium for the growth of an advanced civilization. Although it is not known when the cultivation of plants first took place, it is known that the bulk of our present-day food plants were selected by the people of many lands long before recorded history. The ways in which wild plants were transformed to their present culti-

TABLE 1-1. Dating the past.

Developments in horticulture and plant agriculture	Years ago	Historical era
Gene synthesis		
Genetic code deciphered		Space exploration
Structure of the gene (DNA) discovered		
Isolation of phytochrome		
Mechanical harvesting		
Plastic films		
Radiation preservation		
Polyploid and mutation breeding		
Organic phosphate pesticides		
Chemical weed control		
Tissue culture		Controlled release of atomic
Auxin research		energy
Respiration cycle discovered		
Plant virus studies		
Hybrid corn		
Photoperiodism discovered		
Gasoline powered tractor		
Concept of essential elements		Successful air flight
Bailey's *Cyclopedia of Horticulture*		
Plant nutrition investigations		
Beginnings of agricultural chemistry	100	
U.S. Agricultural Experiment Stations		
Morrill (Land Grant College) Act		
Mendel discovers laws of heredity		American Civil War
Origins of plant pathology		
Modern plow		
Reaper		
Canning of food		American Revolution
Gardens of Versailles		
Discovery of microscope		
Importation of plant species		Discovery of America
Rebirth of botanical sciences		Beginnings of modern science
Monastery gardens	1000	Norman conquest of England
Herbals		Dark Ages
Roman gardens		Roman civilization
Legume rotation		Birth of Christ
Botanical works of Theophrastus		Golden Age of Greece
Fruit cultivars		
Hanging gardens of Babylon		
Grafting		
Irrigation		Egyptian civilization
Domestication of crop plants		
Beginnings of agriculture	10,000	Neolithic Age

vated forms are often obscure, and the original ancestors of many of our crop plants cannot be traced. The same, of course, is true for our domestic animals. Our debt to primitive man is enormous.

Somewhere in the now dry highlands of the Indus, Tigris, Euphrates, or Nile Rivers, the technology we call agriculture was conceived. In about 3000 years, the primitive existence of Stone Age man was transformed to the full-fledged urban cultures of Egypt and Sumeria. By this time the date, fig, olive, onion, and grape, the backbone of ancient horticulture, had been brought under cultivation and the technological base necessary to insure a productive agriculture—land preparation, irrigation, and pruning—had been discovered. By 3000 B.C., field cultivation with a plow pulled by oxen had replaced cultivation with the hoe; by the time of the flowering of Egyptian culture, **agriculture** and **horticulture** were established disciplines (Fig. 1-7).

FIGURE 1-7. Egyptian horticulture as depicted in a tomb at Beni Hasan (about 1900 B.C.). *Top,* picking figs. [From Singer, *History of Technology,* vol. 1, Oxford University Press, 1954.] *Middle,* a round vine arbor; *bottom,* irrigating a vegetable garden. [Both from Gothein, *History of Garden Art,* vol. 1, J. M. Dent & Sons, 1928.]

FIGURE 1-8. Plants and seeds brought back from Syria by Thothmes III, as carved on the temple walls at Karnak, Egypt (about 1450 B.C.). [From Singer, *History of Technology*, vol. 1, Oxford University Press, 1954.]

Egypt and the Fertile Crescent

Roughly 7000 to 8000 years ago, the people living in the Nile valley, which had been inhabited for at least 20,000 years, developed the beginnings of agriculture. In Egypt, man's rendezvous with civilization began. By 3500 B.C., Egypt had established a centralized government with Memphis as its capital, and, by 2800 B.C., it had developed to a high enough level of civilization to support such monumental engineering projects as the pyramids.

The great accomplishment of Egyptian agriculture was systematic irrigation through hydraulic engineering. Among the notable horticultural achievements of the Egyptians were their formal gardens, which were complete with pools and were cared for continuously by gardeners; the creation of a spice and perfume industry; and the development of a **pharmacopoeia**—a collection of drug and medicinal plants (Fig. 1-8).

The Egyptians cultivated a great number of fruits, including date, grape, olive, fig, banana, lemon, and pomegranate, as well as such vegetables as cucumber, artichoke, lentil, garlic, leek, onion, lettuce, mint, endive, chicory, radish, and various melons. In addition, many other plants (such as papyrus, castor bean, and date palm) were cultivated for fiber, oil, and other "industrial" uses. The Egyptians also developed the various technologies associated with the food industry—baking, fermenting, drying, and the making of pottery.

Egyptian influence lasted an incredible 35 centuries. The Egyptians produced, in addition to their great technology, a magnificent art and a complex, if bewildering, theology. By the time Egypt had become a Roman province (3 B.C.), its influence had greatly permeated the ancient world—an influence that is still felt today.

East of Egypt, the ancient cultures of Mesopotamia—Babylonia and Assyria—added to the technology of horticulture such innovations as irrigated terraces, gardens, and parks (Figs. 1-9 and 1-10). Irrigation canals lined with burnt brick with asphalt-sealed joints helped keep 10,000 square miles under cultiva-

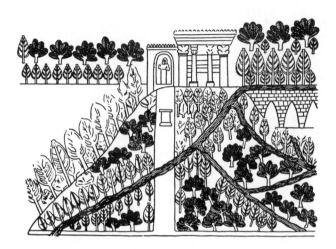

FIGURE 1-9. A royal Assyrian park, watered by streams from an arched aqueduct. From a relief in the palace at Nineveh (seventh century B.C.). [From Singer, *History of Technology*, vol. 1, Oxford University Press, 1954.]

FIGURE 1-10. Darius hunting in a grove of palms (fourth century B.C.?). [From Gothein, *History of Garden Art*, vol. 1, J. M. Dent & Sons, 1928.]

tion, which in 1800 B.C. fed more than 15,000,000 people. By 700 B.C., an Assyrian herbal was compiled that contained the names of more than 900 plants, including 250 vegetable, drug, and oil crops. By the second century B.C., a treatise on agriculture consisting of 28 volumes was compiled by Mago, who lived in the ancient city-state of Carthage in North Africa. The Old Testament prophecy that all nations "shall beat their swords into plowshares and their spears into pruning hooks" (Isaiah 2:4, *ca.* 800 B.C.) attests also to the importance of agriculture and horticulture in that period.

Greece

The Greeks, while adding only modestly to practical agriculture (Fig. 1-11), did devote considerable attention to botany. Although the botanical writings of Aristotle (384–322 B.C.) are lost, a portion of the writings of his most famous student, Theophrastus of Eresos, still exists. These writings cover various botanical subjects, including taxonomy, physiology, and natural history. His books, *History*

FIGURE 1-11. Harvesting olives. From a Greek black-figured vase (sixth century B.C.). [From Singer, *History of Technology*, vol. 2, Oxford University Press, 1958.]

of Plants and *Causes of Plants*, treat such topics as classification, propagation, geographic botany, forestry, horticulture, pharmacology, viticulture, plant pests, and flavors and odors. Theophrastus, now referred to as the Father of Botany, influenced botanical thinking until the seventeenth century—which reflects both the genius of the Greeks and the stagnation of science that was to come about in the Middle Ages.

Later Hellenistic botanists included Attalos III of Pergamon (king from 138–133 B.C.); and, in the following century, Mithridates VI of Pontos, who was much interested in toxicology and poisonous plants, Cratevas (Krateus), physician to Mithridates and author of a treatise on roots as well as an illustrated herbal, and Nicholas of Damascus, author of the great botanical treatise *De Plantis*. Although justifiably best known for their botanical studies, the Greeks also wrote on agriculture. The Roman soldier-encyclopedist Varro (116–27 B.C.) lists more than fifty Greek authorities for his agricultural treatise *Res Rusticae*. The first-century herbal of Pedanius Dioscorides (*De Materia Medica*) remained the standard botanical-medical reference up to medieval times (see Fig. 2-14). Every botanist up to the sixteenth century was either a translator of, or a commentator on, Dioscorides.

There are some who trace the downfall of Greece not primarily to wars or internal decay but to a gradual erosion brought about by the effects of increasing population on declining resources. The loss of a sound agricultural base was due to a combination of factors, among which were shallow soil, poor conservation practices, and perhaps the reluctance of the Greek mind to seriously consider the mundane problems of agriculture. With this loss, the urban centers of Greece, weakened by the intercity struggles and rivalries that culminated in the 27-year Peloponnesian Wars, eventually crumbled before the armies of Macedonia.

Rome

The enigma of the thousand-year history of Rome (roughly 500 B.C. to A.D. 500) is not that Rome fell but that it held together as long as it did, in spite of its being a conglomeration of diverse peoples and lands. Unlike the Greeks, the Romans were extremely interested in practical agriculture. It was a vital part of the economy; the largest single group of producers was agricultural. On this firm foundation, Rome rose to glory, and, according to some, declined along with its eroded soils.

Roman agriculture had many chroniclers. The earliest was Cato (234–149 b.c.), who wrote of farming and useful gardening in his *De Agri Cultura*. Varro (116–27 b.c.) produced a longer commentary on agriculture, which was followed by Columella's twelve-book treatise (*ca.* a.d. 50). It is from these sources, and from the *Georgics* of Virgil (70–19 b.c.), the writings of Pliny the Younger (a.d. 62–116), and especially the *Natural History* and *Palladius* of Pliny the Elder (a.d. 23–79), that the agrarian history of Rome has been reconstructed.

The Romans were great borrowers. Although they produced little that was really new, they did make great improvements upon what they borrowed. The horticultural technology used by the Romans (much of which can be traced to earlier sources, especially Greece and Egypt) was refined, codified, and made more practical. Their agricultural writings mention grafting and budding, the use of many kinds of fruits and vegetables, legume rotation, fertility appraisals, and even cold storage of fruit. Mention can be found of a prototype greenhouse, or *specularium*, constructed of mica and used for vegetable forcing.

In many respects, the well-to-do Roman was a modern type, civilized and urbane, yet bound to the land through business. His problems were largely managerial ones: the care and handling of slaves, the management of income properties, the vagaries of profit and loss. The typical Roman—solidier, farmer, voluptuary—is strikingly similar to a recent counterpart—namely, the aristocrat of the antebellum South.

It was in Rome that the portion of horticulture we now call ornamental horticulture was first developed to a high level. From the beginnings of Roman history hereditary estates ranging in size from one to four acres were referred to as gardens (*hortus*) rather than farms (*fundus*). Early Rome has been described as a market place serving a hamlet of truck gardeners. With the conquest of new lands came the development of large slave plantations, which eventually led to free tenancy and estates, and finally to a manorial system. The great fortunes of Rome were invested in farm land. The good life was that of a gentleman farmer; the sign of wealth was the country estate.

The dwelling on an estate reflected the wealth of its owner. The prosperous Roman had a little place in the country, a *suburbanum*. It included fruit orchards, in which grew apples, pears, figs, olives, pomegranates, and flower gardens, with lilies, roses, violets, pansies, poppies, iris, marigolds, snapdragons, and asters. The mansions of the wealthy were quite splendid. Formal gardens were enclosed by frescoed walls and were amply endowed with statuary and fountains, trellises, flower boxes, shaded walks, terraces, topiary ("bush sculpture"), and even heated swimming pools. The rule was luxury; the desired effect was extravagance.

Rome was largely a parasitic empire based on borrowed culture, slave labor, and stolen goods. This was not destined to last, however, for in the middle of the first millennium A.D., the Roman Empire disintegrated, and Europe took a giant step backward to the village.

Horticulture and Classical Antiquity

Our cultural heritage in art, literature, and ethics is largely traceable to Greek and Roman influences. According to the great historian of science Charles Singer, this has resulted in an overemphasis on the importance of the technology of these cultures, which tended to be lower than the more ancient cultures of Egypt and Mesopotamia, from which they were derived. The rise of both Greece and Rome was similar in some respects to the rise of the Huns and the Goths—a victory of barbarism over worn-out but advanced cultures. The story of civilization and technology is not the steady upward climb of the past 600 years. The technology of horticulture is a good example of this. One is hard pressed for examples of progress made by the Greeks or Romans that are comparable to those of ancient Egypt. Significant advance was to await the Renaissance.

A specific example of this delay concerns the state of knowledge on the role of sex in plants. The cultivation of the date palm and fig in Mesopotamia clearly shows that Mesopotamian horticulturists understood the function of the non-bearing staminate plants of the date palm (Fig. 1-12) and that they recognized the principle underlying caprification of the fig (the use of the wild caprifig, which shelters a parasitic wasp that carries pollen from the caprifig to the pistillate blossoms of the edible fig). Theophrastus was aware of this ancient concept, but

FIGURE 1-12. Winged guardian spirit of the Assyrians pollinating blossoms of the date palm. From the Palace of King Ashur-Nasir-Pal II at Nimrud, Iraq (ninth century B.C.). [Courtesy Museum of Fine Arts, Boston.]

this information was virtually lost until the Dutch botanist Camerarius (1694) experimentally proved the sexual nature of plants. Similarly, the ancient arts of graftage and irrigation, part of the basic technology of horticulture, were not improved until very recently. Advances in technology that came about during the early medieval period are largely traceable to Eastern sources—China, Islam, and the Byzantine Empire.

Medieval Horticulture

After the fall of Rome, during the so-called Dark Ages, some horticultural art and technology survived in monastic gardens. Gardening became an integral part of monastic life, providing food, ornament, and medicines. Many fruit and vegetable strains were preserved, and some of them were even improved. The gardener (*hortulanus* or *gardinarius*) became a regular officer of the monastery. The few botanical and horticultural writings of this period were, for the most part, compilations, and most were based largely on Pliny's *Natural History*. Many centuries passed before the horticultural technology of the Romans was equaled.

Revived interest in the art of horticulture began in Italy with the Renaissance. As feudalism gave way to trade, producing a real rise in the standard of living, garden culture again began to be practiced widely. By the thirteenth and four-teenth centuries, orchards and gardens were common outside of monasteries (Fig. 1-13). As meat became more important in the medieval diet, gardens became

FIGURE 1-13. Representation of a medieval garden.

important as a source of spices and condiments. The horticultural revival spread from Italy to France and then to England. The *Maison Rustique*, published by Charles Estienne (1504–1564) and John Liebault, is a delightful sourcebook of late medieval horticulture. The section on the apple, which is quoted on the next pages,[2] illustrates such practices as fertilization, graftage, pruning, breeding, dwarfing, transplanting, insect "control," girdling to promote flowering, harvesting, processing, and culinary and medicinal utilization.

OF THE APPLE-TREE

The Apple-tree

The Apple-tree which is most in request, and the most precious of all others, and therefore called of *Homer*, the Tree with the goodly fruit, groweth any where, and in as much as it loveth to have the inward part of his wood moist and sweatie, you must give him his lodging in a fat, blacke, and moist ground; and therefore if it be planted in a gravelly and sandie ground, it must be helped with watering, and batling with dung and smal mould in the time of Autumne. It liveth and continueth in all desireable good estate in the hills and mountaines where it may have fresh moisture, being the thing that it searcheth after, but even there it must stand in the open face of the South. Some make nurceries of the pippins sowne, but and if they be not afterward removed and grafted, they hold not their former excellencie: it thriveth somewhat more when it is set of braunches or shoots: but then also the fruit proveth late and of small value: the best is to graft them upon wild Apple-trees, Plum-trees, Peach-trees, Peare-trees, Peare-plum-trees, Quince-trees, and especially upon Peare-trees, whereupon grow the Apples, called Peare-maines, which is a mixture of two sorts of fruits: as also, when it is grafted upon Quince-trees, it bringeth forth the Apples, called Apples of Paradise, as it were sent from heaven in respect of the delicatenesse of their cote, and great sweetnesse, and they are a kind of dwarffe Apples, because of their stocke the Quince-tree, which is but of a small stature.

The Apple loveth to be digged twice, especially the first yeare, but it needeth no dung, and yet notwithstanding dung and ashes cause it prosper better, especially the dung of Sheepe, or for lesse charges sake, the dust which in Sommer is gathered up in the high waies. You must many times set at libertie the boughes which intangle themselves one within another; for it is nothing else but aboundance of Wood, wherewith it being so replenished and bepestred, it becometh mossie, and bearing lesse fruit. It is verie subject to be eaten and spoyled of Pismires and little wormes, but the remedie is to set neere unto it the Sea-onion: or else if you lay

---

[2] From *Maison Rustique, or, The Countrey Farme; Compiled in the French Tongue by Charles Stevens (Estienne), and John Liebault, Doctors of Physicke* (1616 edition, augmented by Gervase Markham).

swines dung at the roots, mingled with mans urine, in as much as the Apple-tree doth rejoyce much to be watered with urine. And to the end it may beare fruit aboundantly, before it begin to blossome, compasse his stocke about, and tie unto it some peece of lead taken from some spout, but when it beginneth to blossome, take it away. If it seeme to be sicke, water it diligently with urine, and to put to his root Asses dung tempered with water. Likewise, if you will have sweet Apples, lay to the roots Goats dung mingled with mans water. If you desire to have red Apples, graft an Apple-tree upon a blacke Mulberrie-tree. If the Apple-tree will not hold and beare his fruit till it be ripe, compasse the stocke of the Apple-tree a good foot from the roots upward, about with a ring of a lead, before it begin to blossome, and when the apples shall begin to grow great, then take it away.

Apples must be gathered when the moone is at the full, in faire weather, and about the fifteenth of September, and that by hand without any pole or pealing downe: because otherwise the fruit would be much martred, and the young siences broken or bruised, and so the Apple-tree by that meanes should be spoyled of his young wood which would cause the losse of the Tree. See more of the manner of gathering of them in the Chapter next following of the Peare-tree: and as for the manner of keeping of them, it must be in such sort as is delivered hereafter. *Gathering of Apples*

You shall thaw frozen Apples if you dip them in cold water, and so restore them to their naturall goodnesse. There is a kind of wild Apple, called a Choake-apple, because they are verie harsh in eating, and these will serve well for hogges to eat. Of these apples likewise you may make verjuice if you presse them in a Cyder-presse, or if you squeese them under a verjuice milstone.

Vinegar is also made after this manner: You must cut these Apples into gobbets, and leave them in their peeces for the space of three dayes, then afterward cast them into a barrell with sufficient quantitie of raine water, or fountaine water, and after that stop the vessell, and so let it stand thirtie daies without touching of it. And then at the terme of those daies you shall draw out vinegar, and put into them againe as much water as you have drawne out vinegar. There is likewise made with this sort of Apples a kind of drinke, called of the Picardines, Piquette, and this they use in steed of Wine. Of other sorts of Apples, there is likewise drinke made, which is called Cyder, as we shall declare hereafter. *Vinegar*

An Apple cast into a hogshead full of Wine, if it swim, it sheweth that the Wine is neat: but and if it sinke to the bottome, it shewes that there is Water mixt with the Wine. *Neat wine Mingled wine*

Infinit are the sorts and so the names of Apples comming as well of natures owne accord without the helpe of man, as of the skill of man, not being of the race of the former: in every one of which there is found some speciall qualitie, which others have not: but the best of all the rest, is the short shanked apple, which is marked with spottings, as tasting and smelling more excellently than any of all the other sorts. And the smell of it is so excellent, as that in the time of the plague there is nothing better to cast upon the coales, and to make sweet perfumes of, than the rinde thereof. The short stalked Apple hath yet furthermore one notable qualitie: for the kernells being taken out of it, and the place fill up with Frankincense, and the hole joyned and fast closed together, and so rosted under hot embers as that it burne not, bringeth an after medicine or remedie to serve when all other fayle, to such as are sicke of the pleurisie, they having it given to eat: sweet apples doe much good against melancholicke affects and diseases, but especially against the pleurisie: for if you roast a sweet apple under the ashes, and season it with the juice of licorice, starch and sugar, and after give it to eat evening and morning two houres before meat unto one sicke of the pleurisie, you shall helpe him exceedingly.

The rise of landscape architecture is one indication of the values of the Renaissance. Gardening became formalized. The design of gardens became as important as the design of the structures they surrounded (Fig. 1-14). The peak of Renaissance horticulture is to be found in the magnificent gardens of LeNôtre

FIGURE 1-14. Astronomical observatory (Arcis Uraniburgi) of Tycho Brahe (1546–1601) at the Danish island of Hveen. [Courtesy Oliver Dunn, from Joan Blaeu, *Grooten Atlas*, 1664–1665.]

(1613–1700), the most notable being those at the Palace of Versailles, which was built for Louis XIV. Building this prodigious chateau and landscaping the countryside around it took approximately 25 years and employed thousands of workmen. The gardens required the engineering of tremendous irrigation projects capable of supplying as many as 1400 jets of water. In one year (1688) 25,000 trees were purchased. Ancient gardens had been more than equaled; modern gardens have not surpassed it.

The New World

The discovery of the New World in 1492—a convenient if inaccurate date to assign to the beginning of the modern age—was inspired by a search for a new route to the spice-rich Orient. The early conquistadores found, in the mixture of advanced and Stone Age cultures that they encountered in the New World, an agricultural technology that was to have a profound influence on the history of the rest of the world. The horticultural contributions of the New World include many new vegetables (maize, potato, tomato, sweetpotato, squash, pumpkin, peanut, kidney bean, and lima bean); fruits and nuts (cranberry, avocado, Brazil nut, cashew, black walnut, pecan, and pineapple); and other important crops (chocolate, vanilla, wild rice, chili, quinine, cocaine, and tobacco). Primitive people in the New World had brought under cultivation practically all of the indigenous plants we now use.

The discovery of America was the most spectacular result of the era of exploration. The broadening of trade routes greatly stimulated horticultural progress. The transplantation of plant species from the Old World to the New World, and vice versa, marks the beginning of our great horticultural industries. The bulb industry in Holland, the cacao industry of Africa, and the banana and coffee industries of Central and South America can all be traced to those importations of plant species.

The Beginnings of Experimental Science

Science is a method of inquiry whose aim is the organization of information. In the ancient world, technology was the parent of science; only recently has science become the parent of technology. The speculative use of information built up through observation and experimentation became a powerful force during the Middle Ages. The discoveries of daVinci (1452–1519), Galileo (1564–1642), and Newton (1642–1727) in astronomy and the physical sciences represent the flowering of this "new" development.

As a result of this approach, the seventeenth century saw a rebirth of botanical studies. Fundamental studies in plant anatomy and morphology were initiated by Marcello Malpighi (1628–1694) and Nehemiah Grew (1641–1712). The discovery of "cells" in cork by Robert Hooke (1635–1703) initiated the study of cytology, which was destined to reunite botany and zoology. The roots of genetics can be traced to the experimental studies of the Dutch botanist Rudolph Jacob Camerarius (1665–1721), who demonstrated sexuality in plants, and to the later hybridization experiments of J. G. Koelreuter (1733–1806). Interest in plant classification was revived, and, as the number of known plants increased, a series of attempts were made to formulate a workable system of classification. It remained for Linnaeus (1707–1778) to develop a workable method based on the structure of reproductive parts. The beginnings of plant physiology were stimulated by Harvey's discovery of blood circulation in animals in 1628. Not until the eighteenth century, however, were fundamental studies in physiology undertaken, such as the investigations by Stephen Hale (1677–1761) of the movement of sap, and the work of Joseph Priestley (1733–1804) on the production of oxygen by plants.

The history of the plant sciences becomes meaningful only when the significance of the fundamental discoveries is understood. The history of botany and experimental horticulture in the last 150 years is in a sense the subject of the following chapters. Similarly, the modern history of the horticultural industry, which represents the accumulated technology and science of many lands, cannot be stated briefly. For the study of this, the reader must investigate particular crops and particular countries. This subject will be discussed briefly, however, in Part III, "The Industry of Horticulture."

Influence of Twentieth-Century Technology

The most remarkable feature of present-day agriculture in the United States is that the increased production of the last 60 years has taken place in spite of a decreasing acreage and a shrinking farm population. In 1910, each farmer produced for himself and 7 others; in 1973 he "supported" 56 people (Fig. 1-15). This increase in efficiency is the result of improved technology. The magnitude of this increase in efficiency has been such that many of the recent agricultural problems in the United States are a result of overproduction.

Technological change can be defined as the change in production resulting directly from the use of new knowledge. In general, technological change is measured by the change in output (production) per unit input of land, labor, and capital within a given period of time. The rate of technological change is not constant, because scientific progress does not ordinarily occur as a steady flow.

In agriculture, technological change may produce savings in either labor or capital. The classic agricultural inventions—the plow, the reaper, and the tractor—have reduced labor. They substitute capital for labor, which does not necessarily increase yield *per se*. The technological improvement of labor-saving devices results in capital savings. Technological changes as a result of genetic gain, improved nutrition, or irrigation bring about an increase in yield per unit of land and become capital improvements, saving land and reducing expenses.

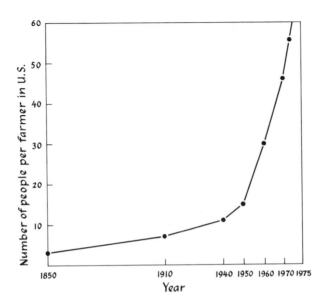

FIGURE 1-15. The number of people fed by each farm in the U.S. has shown a dramatic increase since 1940.

The increase in United States agricultural production from 1880–1920 was largely a reflection of the increase in expenditures—more land and more labor. Returns per acre remained relatively constant. The replacement of horse power by the gasoline engine in the 1920s was the first great step in the twentieth-century scientific revolution. Hundreds of thousands of acres formerly used for feed were made available for other purposes. As a result, farm acreage and the farm labor force, which had been expanding before that time, were stabilized during the decade after World War I.

The proliferation of genetic investigations in the quarter century following the rediscovery in 1900 of the revolutionary paper on inheritance by Gregor Mendel (1822–1884) began to yield technological advances in the form of improved plant cultivars. The development of hybrid corn was the most spectacular of these achievements. These improved genetic stocks in combination with the increased use of inorganic fertilizers accounted for a large part of the tremendous increase in production necessitated by World War II.

In the late 1940s agriculture was stimulated by a whole new set of technological advances made possible by basic research in the preceding decades. Agricultural chemicals in the form of weed killers, organic fungicides, and organic insecticides quickly followed the spectacular commerical success of the broad-leaf weed killer 2,4-D. The effects of these improvements were to increase yields per unit area as well as to conserve labor (Fig. 1-16). Mechanization increased in the 1950s to include even "chore" jobs; and recently, supplemental irrigation in the eastern United States (which, of course, has long been used in the West) has

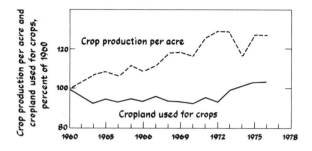

FIGURE 1-16. Crop production per acre has shown a steady increase since the 1960s, with a decrease in the amount of cropland used for crops up to 1970. The increase in cropland used for crops after 1970 has been in response to releases of "soil bank" lands. The releases were motivated by temporarily high grain prices during the middle 1970s. [Data from Economic Research Service, USDA.]

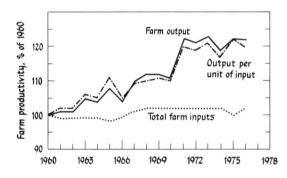

FIGURE 1-17. Farm productivity and efficiency, as measured in terms of output per unit of input, has continued to increase since 1960. [Data from Economic Research Service, USDA.]

permitted the use of additional fertilizer. The net result has been a steady increase in agricultural efficiency. This trend does not appear to be changing (Fig. 1-17).

The horticultural industry has followed this trend in agriculture very closely. The average acre yields of the California tomato processing industry provide a striking example of this pattern. Technological improvements in the form of genetic gain in addition to improved fertilization and irrigation practices have quadrupled the average yield per acre within four decades, as shown in Figure 1-18.

It is clear that technological improvements will continue in the wealthy or "developed" countries. Probably the main increase in efficiency will come from continued reduction of labor. During the 1960s commercial tomato and cucumber harvesters, fruit and nut shakers, and prototype strawberry pickers were devised to eliminate hand harvesting. By the 1970s the mechanical harvesting of grapes had become a commercial reality and strawberry and blackberry harvesters were in the advanced stages of development and testing. Other mechanical devices may soon end hand harvesting for all crops. Hand cultivation of most crops has already

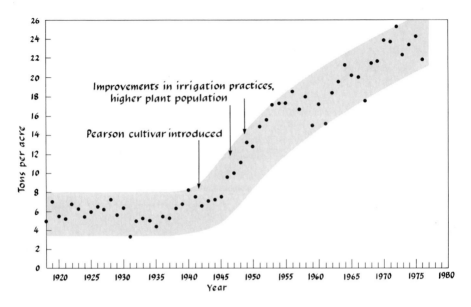

FIGURE 1-18. Yield per acre of processing tomatoes in California. [Data from Agricultural Marketing Service, USDA.]

been eliminated as the result of advances in chemical weed control, and new systems of land preparation and planting may be the next applications of these advances. The recent success of growth regulators in horticulture may mark the beginning of the long-anticipated revolution in the chemical control of growth and development. Highly specific, less hazardous pesticides are under development. Genetic gains will continue to bring forth improved cultivars. Thus, although it is difficult to predict the exact course of technological advance, it is certainly safe to say that, if technology advances at the same rate it has for the past 30 years, this rate will be more than sufficient to provide food for our increasing population. That this can be so during a time when farm acreage and farm population continue to decline surely indicates that the Malthusian law is not yet in operation!

Many of the improvements of recent years have undoubtedly been made possible by the organization of research essential to agriculture. The system of agricultural experiment stations, authorized by the Morrill Act of 1862, has provided a steady supply of basic and applied research. The results are evidenced by bountiful harvests of food and fiber.

It is most unfortunate that these comforting predictions are not applicable to all parts of the world. Many areas with low rates of agricultural productivity are experiencing rapid population increases. A combination of population control and the extension of advanced systems of agriculture to these areas must be accomplished to fulfill our hopes for a better world for all people.

Selected References

Arber, A., 1938. *Herbals, Their Origin and Evolution: A Chapter in the History of Botany, 1470–1670*, 2nd ed. Cambridge: The University Press. (The definitive work on the subject.)

Bailey, L. H., 1947. *The Standard Cyclopedia of Horticulture*. New York: Macmillan. (Probably the best work on horticulture ever assembled. The first edition in 1900 was to compile a "complete record of North American horticulture as it [existed] at the close of the nineteenth century." The work was revised in 1914. Its most valuable use today is for its treatment of plant materials.)

Berrall, J. S., 1966. *The Garden: An Illustrated History*. New York: Viking. (A pictorial romp through garden history.)

Gothein, M. L., 1928. *A History of Garden Art*. New York: Dutton. (Authoritative work on the subject, in two volumes.)

Gras, N. S. B., 1940. *A History of Agriculture*, 2nd ed. New York: F. S. Crofts. (An authoritative treatment.)

Hendrick, U. P., 1950. *A History of Horticulture in America to 1860*. New York: Oxford University Press. (A chronicle of early United States horticulture.)

Henrey, B., 1975. *British Botanical and Horticultural Literature before 1800*. New York: Oxford University Press. (A history and bibliography of botanical and horticultural books from the sixteenth, seventeenth, and eighteenth centuries, in three volumes.

Huxley, A., 1978. *An Illustrated History of Gardening*. New York: Paddington Press. (The newest entry in the world of garden history; profusely illustrated.)

Hyams, E., 1971. *A History of Gardens and Gardening*. New York: Praeger. (The garden, especially in the landscape sense, from stone age to the present.)

Leonard, J. N., 1973. *The First Farmers*. New York: Time-Life Books. (A popular account of the origins of agriculture, profusely illustrated.)

Reed, H. S., 1942. *A Short History of the Plant Sciences*. Waltham, Mass.: Chronica Botanica Co. (An excellent review.)

Singer, C., E. J. Holmyard, and A. R. Hall (editors), 1954–1958. *A History of Technology*. London: Oxford University Press. (A monumental work, in five volumes, on science and technology up to 1900. Agriculture and related fields are well documented.)

Wright, R., 1938. *The Story of Gardening*. Garden City, New York: Garden City Pub. Co. (A popular history of horticulture and gardening.)

I

THE BIOLOGY
OF HORTICULTURE

Corn seedling. [Courtesy International Harvester Corporation.]

2

The Classification
of Horticultural Plants

by Arthur H. Westing

What's in a name? That which
we call a rose
By any other word would
smell as sweet . . . .

Shakespeare, *Romeo and Juliet* [II. 2]

Since ancient times people have named and categorized the many plants that surround them and upon which they are dependent for their very existence. One can readily surmise that the earliest classifications simply divided plants into the harmful ones and the useful ones (a division of value to this day!). Additionally, plants were probably divided according to their uses. Such classifications thus met the need of organizing what must otherwise have been a bewildering array of objects. Practical systems of this sort are, of course, perfectly valid, provided they are logically conceived and consistent, and therefore capable of predictive use.

Often superimposed on the practical systems of classification are those based upon growth habit or other gross physiological characteristics. Thus plants can be characterized as being **succulent (herbaceous)** or **woody.** Succulent seed plants possessing self-supporting stems are known as **herbs.** A plant whose stem requires support for upright growth may be a climbing or trailing plant; if nonwoody, such a plant is known as a **vine,** whereas a woody plant of such habit is correctly called a **liana.** The self-supporting woody plants are either **shrubs** or **trees,** the trees being characterized by a single central axis and the shrubs by several more or less upright stems. Trees are usually taller than shrubs. Occasionally, the distinction between trees and shrubs may be obscured by environmental conditions or by horticultural "training."

Plants that are leafless during a portion of the year (usually the winter) are referred to as **deciduous;** those whose leaves persist the year round are known as **evergreens.** Evergreens actually may lose their leaves annually, but not until a new set of leaves has developed. The deciduous habit is often associated with temperate regions; the persistent, with tropical regions.

Another classification of obvious importance to the horticulturist is based on life span, and divides plants into annuals, biennials, and perennials. The **annual** plant is one that normally completes its entire life cycle during a single growing season. Spinach, lettuce, and petunia are examples of annuals.

The **biennial** plant normally completes its life cycle during a period of two growing seasons. During the first summer its growth is entirely vegetative, the plant often being low in form—a so-called rosette. The winter following the first growing season provides the low temperatures necessary to stimulate this type of plant to "bolt"—to send up a seed stalk during the second growing season, and then to flower and set fruit. Celery, parsnip, and evening primrose are among the biennials. Because certain biennials such as carrots or beets are grown for their overwintering storage organs, they are harvested as annuals at the end of the first growing season. Similarly, if the climate is mild enough, an annual such as spinach can be planted in the fall and harvested the following spring, and thus, although not requiring a period of low temperature, it can be grown in much the same way that biennials are grown.

The **perennial** plant grows year after year, often taking many years to mature. Unlike annuals and biennials, the perennial does not necessarily die after flowering. Although herbaceous plants are found in all three categories, woody plants are usually perennial. The many fruit trees, as well as the ornamental shrubs and trees, are examples known to all (Fig. 2-1). Asparagus, rhubarb, and our various bulb crops are among the herbaceous perennials, whose above-ground parts are killed each year in temperate regions but whose roots remain alive to send up new shoots in the spring. When a subtropical perennial (such as tomato, eggplant, or coleus) is grown in a temperate region, it cannot survive the relatively cold winters, and under such conditions it is treated as an annual. An interesting situation exists with respect to the genus *Rubus* (the raspberries and other brambles) in which the roots are perennial and the shoots are biennial (Fig. 2-2).

Plants can also be variously classified according to their temperature tolerances. Thus, when a horticulturist speaks of **tender** plants and **hardy** plants, he is referring to their ability to withstand low winter temperatures. For woody plants the additional distinction is sometimes made between so-called **wood hardiness** and **flower-bud hardiness.** The former refers to the winter cold-resistance of the plant as a whole; the latter, to the ability of the flower buds to survive low winter temperatures. For example, even though apricot trees could survive in many parts of the United States, their culture is restricted to California because of their limited flower-bud hardiness. Similarly, the ginkgo can be grown as an ornamental in central Canada but cannot "flower" (that is, produce strobili) and set fruit there. (It must be borne in mind that temperate-zone plants "harden off" in the fall to become far more cold-resistant in winter than in summer.)

FIGURE 2-1. Blue spruce (*Picea pungens;* family Pinaceae, order Coniferales, class Gymnospermae). This majestic tree is found scattered primarily on the middle and upper slopes of the Rocky Mountains. Mature trees are often 100 feet tall and live for more than 500 years. The several available cultivars are prized highly as ornamentals, owing not only to their beautiful habit and foliage but also to their ability to withstand drought and extremes of temperature. [Photograph by J. C. Allen & Son.]

FIGURE 2-2. Black raspberry (*Rubus occidentalis;* family Rosaceae, subclass Dicotyledoneae). Raspberries are noted for their fruits, equally delicious raw or as preserves. There are numerous cultivars of this native North American shrub and its various related species. Although the roots of the raspberry are perennial, its shoots are biennial. Members of this genus are often referred to as brambles, particularly in England. [Photograph by J. C. Allen & Son.]

Plants are sometimes also classified according to their temperature requirements during the growing season. For example, peas are a typical cool-season crop, whereas tomatoes are a typical warm-season crop. This characteristic is sometimes related to seed-germination requirements.

The landscape architect may wish to classify plants according to their habitat or site preferences. In addition to recognizing the obvious desert and aquatic types, he must know whether an ornamental plant is best suited to moist or dry sites, to sunny or shady conditions, or to acid or alkaline soils.

A HORTICULTURAL PLANT CLASSIFICATION

Elaborations of the ancient plant classifications, which were based upon the uses to which plants were put, are still the most important ones to the horticulturist. Plants are conveniently separated into those that are edible, those that serve as sources of drugs or spices, those that are of ornamental value, and so forth. Although almost any intensively cultured plant might be considered within the province of horticulture, the primary focus is on the various traditional "garden" plants. The grains, for example, are traditionally excluded from horticulture and are considered field crops.

The horticulturist divides the **edible** garden plants into **vegetables** and **fruits.** Generally considered as **vegetables** are those herbaceous plants of which some portion is eaten, either cooked or raw, during the principal part of the meal (Figs. 2-3 and 2-4). Common examples are spinach (edible leaf), asparagus (edible

FIGURE 2-3. Artichoke (*Cynara scolymus;* family Compositae, subclass Dicotyledoneae). The fleshy flower head of the artichoke is cooked as a vegetable and is considered a delicacy by many. This perennial herb is native to the Mediterranean region and is grown as a field crop in parts of Europe and the United States. The artichoke illustrated is not to be confused with the Jerusalem artichoke (*Helianthus tuberosus*), another vegetable plant in the same family grown for its edible tuber. [Photograph by J. C. Allen & Son.]

FIGURE 2-4. Sweetpotato (*Ipomoea batatas;* family Convolvulaceae, subclass Dicotyledoneae). The edible portion of this vine is the tuberlike root, which is cooked and eaten as a vegetable. Sweetpotatoes originated in tropical America but are now a common starchy food throughout the warmer regions of the world and are a truck crop of some importance in the southern United States. The many cultivars fall into two general types, those having dry, mealy flesh and those having soft, moist flesh. Sweetpotatoes are often known as yams in the southern United States but are not to be confused with the true yam (*Dioscorea* spp., family Dioscoreaceae, subclass Monocotyledoneae), another very important starchy, tuberlike vegetable grown in the tropics. [Photograph by J. C. Allen & Son.]

stem), beet (edible root), cauliflower (edible flower), eggplant (edible fruit), and pea (edible seed). **Fruits,** on the other hand, are the plants from which a more or less succulent fruit or closely related structure is commonly eaten as a dessert or snack. Fruit plants are most often perennial and are usually woody. Whereas those of the temperate zones are primarily deciduous, the tropical and subtropical plants are usually evergreen. Fruits borne on trees are termed **tree fruits,** among which are the pear, cherry (Fig. 2-5), orange, papaya (Fig. 2-6), and date (Fig. 2-7). Fruits borne on low-growing plants, such as shrubs, lianas, and some herbs, are known as the **small fruits.** Examples are the raspberry, cranberry, grape, and strawberry. **Nuts,** which may be considered as a special subcategory of the fruits, are characterized by a hard shell that is separable from a firm inner kernel—the meat. Familiar examples are the pecan, walnut, and cashew.

It must be stressed that no precise distinction can be made between the terms "fruit" and "vegetable." Although the definitions given above hold true for most edible plants, especially those grown in the north temperate regions, the ancient and popular origins of these terms have resulted in certain anomalies. Thus, when the edible portion of a plant is stem, leaf, or root, there is seldom any question that one is dealing with a vegetable. However, rhubarb, with its edible petiole, is considered a fruit in some parts of the world because of its use as a dessert. When the edible portion of an herbaceous plant corresponds to the botanical fruit, the

FIGURE 2-5. Sour cherry (*Prunus cerasus;* family Rosaceae, subclass Dicotyledoneae). Cherries have been an important fruit crop in Europe and in Asia since the beginnings of agriculture. Hundreds of cultivars of this species and the closely related sweet cherry (*P. avium*) are cultivated throughout the temperate portions of the world. Because of their blossoms, cherry trees also rank among our most valued ornamentals. Some species of cherry provide an important cabinet wood that is known for its high silky luster and great beauty. [Photograph by J. C. Allen & Son.]

FIGURE 2-6. Papaya (*Carica papaya;* family Caricaceae, subclass Dicotyledoneae). The tree-like semiherb that bears this delicious fruit is a native of tropical America that is now cultivated pantropically. Papayas have become exceedingly popular in Hawaii and elsewhere. They are mostly eaten raw but are also either boiled as vegetables or pickled. The seeds are a commercial source of the enzyme papain. [Photograph by J. C. Allen & Son.]

FIGURE 2-7. Date palm (*Phoenix dactylifera;* family Palmae, subclass Monocotyledoneae). Date palms have been cultivated for their fruits along the Tigris and Euphrates rivers for more than 4000 years. Apparently originally native to the Near East, date palms are now widely cultivated throughout the warmer regions of the world, including parts of California and Arizona. Dates are the principal food crop from Iran to the Arabian Peninsula and all throughout North Africa. The trees, often 100 feet tall, are also locally important for their wood as well as for their sugary sap. [Photograph by J. C. Allen & Son.]

situation is often more confusing. Thus, the banana and the pineapple are considered as fruits, whereas the melon, tomato, cucumber, squash, pepper, and plantain are variously regarded as vegetables or fruits, depending upon national or even local tradition. It might be mentioned that as a result of a question of import duties the tomato was legally established in this country as a vegetable by a U.S. Supreme Court decision of 1893.

Plants used for their **ornamental** value are commonly separated into **nursery plants** and floricultural plants, or **flowers.** The flowers (often herbaceous) are primarily grown for their blossoms (Fig. 2-8), but are occasionally grown for their showy leaves (Fig. 2-9). They are often separated according to life span: the petunia (an annual), the evening primrose (a biennial), and the tulip (a perennial) are examples. The nursery plants (primarily woody and perennial) are most often raised for landscaping purposes. They are thus often classified according to their form or growth habit. The horticulturist recognizes **lawn** or **turf plants** (herbaceous perennials), **ground covers** (either herbaceous or woody perennials), **vines** (herbaceous or woody, and most often perennial), **shrubs** (usually restricted by the nurseryman to deciduous shrubs), **evergreens** (woody, evergreen shrubs or trees), and **trees** (commonly limited by the nurseryman to deciduous trees). Common examples of each of these categories are bluegrass, periwinkle (Fig. 2-10), English ivy, viburnum (Fig. 2-11), juniper (Fig. 2-12), and birch (Fig. 2-13), respectively.

Of course, the various categories here described are not all mutually exclusive.

FIGURE 2-8. Lady slipper (*Calceolaria herbeohybrida;* family Scrophulariaceae, subclass Dicotyledoneae). Lady slippers or calceolarias are admired for their showy pouch-shaped flowers, which come in a variety of colors. These herbaceous flowers are native to tropical America. The lady slipper illustrated is not to be confused with the orchids of the same name (*Cypripedium* spp.; family Orchidaceae, subclass Monocotyledoneae). [Photograph by J. C. Allen & Son.]

FIGURE 2-9. Caladium (*Caladium bicolor;* family Araceae, subclass Monocotyledoneae). Caladium is a deciduous herbaceous perennial native to tropical America. It is grown as a foliage plant because of its beautifully and variously patterned leaves. [Photograph by J. C. Allen & Son.]

FIGURE 2-10. Periwinkle (*Vinca minor;* family Apocynaceae, subclass Dicotyledoneae). Periwinkle, also known as myrtle, is an often cultivated, trailing, evergreen herb. It makes a hardy ground cover that is especially suited to moist shady areas. The plant illustrated is a particularly attractive variegated cultivar that has solitary blue flowers. [Photograph by J. C. Allen & Son.]

FIGURE 2-11. Arrowwood viburnum (*Viburnum dentatum;* family Caprifoliaceae, subclass Dicotyledoneae). These handsome woody perennials are among our most important ornamental shrubs. They are noted for their attractive clusters of flowers and fruits, as well as for their foliage. There are about two dozen viburnum species of horticultural importance. The arrowwood illustrated here is a native of the United States and prefers moist sites. The tough, pliant shoots were once used to make arrows. [Photograph by J. C. Allen & Son.]

FIGURE 2-12. Pfitzer juniper (*Juniperus chinensis* "Pfitzeriana"; family Cupressaceae, order Coniferales, class Gymnospermae). This beautiful evergreen shrub from China forms a broad pyramid with its spreading branches and nodding branchlets. The grayish-green leaves are needle-shaped on juvenile plants and scale-like on mature ones, a widespread phenomenon in this family. This shrub which may be an interspecific hybrid, is one of the rare gymnospermous polyploids. The blue berry-like fruits of junipers are used to give gin its characteristic flavor. Seen climbing the wall in the background is Boston ivy *Parthenocissus tricuspidata,* a deciduous liana. [Photograph by J. C. Allen & Son.]

FIGURE 2-13. Weeping birch (*Betula pendula;* family Betulaceae, subclass Dicotyledoneae). The slender, gracefully drooping branches and the white papery bark explain the extensive use of the weeping birch as a lawn tree. Numerous cultivars are available. Tannins for the leather industry have been extracted from the bark in Europe, where it is native. [Photograph by J. C. Allen & Son.]

Thus a grape vine, when cultured for its edible berry, is considered to be a small fruit, but when grown in a garden to cover a trellis or arbor it is considered to be an ornamental. A walnut tree can be grown for its nuts, or as an ornamental, or for its highly prized wood. A hazel shrub may be considered to be a crop plant (filberts) by the commercial grower, may be considered an ornamental by the gardener, or may be considered an undesirable or weed species by the forester.

The horticultural plant classification described here (and outlined on the next page) is an operational one based upon the use to which we put the plants. As such it is not only an overlapping classification but may often be an empirical and arbitrary one as well. Moreover, it is a classification that is subject to change over the years as we discover new uses for plants and abandon old ones. The fundamental value of this classification, of course, is that it groups economic plants according to the similarities of their cultural requirements as well as according to the similarities of their utilization and marketing.

A HORTICULTURAL PLANT CLASSIFICATION

EDIBLE PLANTS

Vegetables

PLANTS GROWN FOR AERIAL PORTIONS
Cole crops (cabbage, cauliflower, broccoli)
Legumes or pulse crops (pea, bean, soybean)
Solanaceous fruit crops (tomato, eggplant, pepper)
Vine crops or cucurbits (cucumber, squash, melon)
Greens or pot herbs (spinach, chard, dandelion)
Salad crops (lettuce, celery, parsley, endive)
Miscellaneous (corn, asparagus, okra, mushroom)

PLANTS GROWN FOR UNDERGROUND PORTIONS
Root crops (beet, carrot, radish, turnip, sweetpotato)
Tubers and rootstocks (potato, Jerusalem artichoke, taro, cassava, yam)
Bulb (and corm) crops (onion, garlic, shallot)

Fruits

TEMPERATE (DECIDUOUS) FRUITS
Small fruits (cranberry, grape, raspberry, strawberry)
Tree fruits
Pomes (apple, pear, quince)
Stone fruits or drupes (cherry, peach, plum, apricot)
Nuts (pecan, filbert, walnut)

TROPICAL AND SUBTROPICAL (EVERGREEN) FRUITS
Herbaceous perennials (pineapple, banana)
Tree fruits
Citrus fruits (orange, lemon, grapefruit)
Miscellaneous (fig, date, mango, papaya, avocado)
Nuts (cashew, Brazil nut, macadamia)

ORNAMENTAL PLANTS

Flowers and Foliage Plants

Annuals (petunia, zinnia, snapdragon)
Biennials (evening primrose, hollyhock, sweet william)
Perennials (tulip, peony, chrysanthemum, philodendron)

Nursery Plants

Lawn (turf) plants (bluegrass, Bermuda grass)
Ground cover (periwinkle, sedum)
Vines, both herbaceous and woody (Virginia creeper, grape, English ivy)
Shrubs, commonly restricted to deciduous shrubs (forsythia, lilac)
Evergreens, both shrubs and trees (spreading juniper, rhododendron, white pine)
Trees, commonly restricted to deciduous trees (pin oak, sugar maple, larch)

MISCELLANEOUS PLANTS

Herbs, spices, drugs (dill, nutmeg, spearmint, quinine, digitalis)
Beverage plants (coffee, tea, cacao, maté)
Oil-yielding plants (tung, sunflower)
Rubber plants (Pará rubber tree)
Plants yielding gums or resins (sweetgum, slash pine)
Christmas trees (balsam fir, Scotch pine)

A SCIENTIFIC PLANT CLASSIFICATION

People are by nature methodical creatures, and they have always attempted to discover natural order in the universe. Their physical environment they early classified into its "elements"—air, earth, water, and fire. All living things were classed as either animal or vegetable, and the vegetable or plant kingdom was subdivided in many ways. As long ago as three centuries before Christ, the Greek philosopher Theophrastus classified all plants into annuals, biennials, and perennials according to their life spans, and into herbs, shrubs, and trees according to their growth habits. He further divided the trees according to their deciduous or evergreen natures and their various branching habits.

During the twenty centuries following the time of Theophrastus, a multitude of classifications and innumerable lists of useful plants ("herbals") were accepted and discarded (Fig. 2-14). It was not until the middle of the eighteenth century that a Swedish physician named Carl von Linné (but better known by the Latinized version of his name, Linnaeus) revolutionized the fields of plant and animal classification, or taxonomy. His labors earned him the title "father of taxonomy." He established groups of organisms, large and small, that depended upon structural or morphological similarities and differences. He recognized the value of basing the taxonomic criteria for plants on the morphology of their sexual or reproductive parts—the plant organs least likely to be influenced by environmental conditions. Linnaeus singlehandedly described and assigned names to more than 1300 different plants (and to as many or more animals) from the far reaches of the earth. It was he who brought order out of chaos by standardizing a

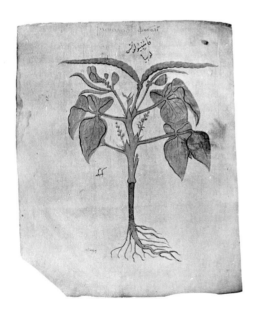

FIGURE 2-14. An illustration of the cowpea (*Vigna sinensis*) taken from a famous early herbal written by the physician Pedanius Dioscorides in the first century A.D. It remained the standard materia medica for centuries. Actually, Dioscorides appears to have taken his illustrations from an even earlier Greek text written in the first century B.C. by Cratevas (Krateuas), physician to Mithridates VI. [Courtesy Austrian National Library, Vienna.]

worldwide system of nomenclature. Linnaeus has been described rightfully as the greatest cataloguer and classifier of all time.

The next important advance in plant classification came in the middle of the nineteenth century, with the publication of Charles Darwin's *Origin of Species*. The principles of evolution propounded in this book had perhaps nowhere a more profound impact than on the fundamentals of taxonomy. Darwin's concept of evolution finally provided a natural framework upon which to hang a scheme of classification. According to this concept, all plants on earth today are more or less closely related taxonomically according to their proximity on the family tree of plant evolution. Thus the classification is based upon the notion that genetic relationships exist between all plants, and that present-day plants are, through successive generations, the offspring of ancient ancestral plants. Of course, the possibility exists that the plant kingdom owes its origin to more than one primordial organism. It is further assumed that there has occurred throughout the history of the earth, and is still occurring, an evolution of plant characteristics that has brought about increasing complexity in structure and genetic organization.

To the taxonomist or systematist falls the task of reconstructing these evolutionary connections—a task that is simple in theory, but that is formidable (and in large part impossible) in actual practice. The task is an immense one because of the countless numbers of now extinct and largely unknown plants. Paradoxically, however, it is the very discontinuities resulting from these extinctions that make it possible for the taxonomist to establish and delimit his various categories. Without them there would be an essentially unclassifiable continuum stretching from the lowest plant to the highest. The work of classification is further complicated by the occasional hybridizations that reticulate the family tree and by different patterns of evolution. The approach the systematist takes when he attempts to determine evolutionary lines and group relationships can perhaps best be described as scientific sleuthing. He bases much of his case on morphological and anatomical comparisons and gleans as much information as possible from distributional or geographic evidence. Of fundamental importance, however, is fossil (or paleobotanical) evidence: each new fossil found fills another small gap in the family tree. The recently developed methods of determining the ages of sedimentary rocks and the plant remains preserved in them have been of inestimable help in this regard.

Particularly among closely allied groups, the modern taxonomist also leans heavily on cytological, genetic, ecological, and even physiological and biochemical evidence. For example, some decades ago the German botanist Karl Mez demonstrated the feasibility of determining plant-group affinities by comparing their constituent proteins. Borrowing standard serological techniques, he would sensitize a rabbit by injecting into its bloodstream the proteins from one plant. He then determined the degree to which this plant was related to another from the degree to which the newly developed antibodies in the rabbit's blood reacted to the proteins of the second plant. The less pronounced the reaction, the more distantly related the plants were assumed to be. More recently the taxonomy of the pines has been verified, and to some degree clarified, on the basis of their oleoresin chemistry. Chromatography, a powerful chemical technique whereby very complex mixtures are separated into their component parts by passing them through selectively adsorbant materials, has become an indispensable adjunct to biochemical taxonomy. Additionally, the electronic computer permits the taxonomist to compare simultaneously and with relative ease several dozen morphological or other characteristics of allied plant groups in order to assist him in determining their taxonomic affinities. Imaginative research in a variety of related fields and a subsequent correlation of the information constitute the difficult task that confronts the modern taxonomist.

One important attempt at a natural classification (although it was based primarily on floral morphology) was made toward the close of the nineteenth century by the German botanist August Eichler. This was elaborated by Adolph Engler, who with his collaborator Karl Prantl published a twenty-volume classification of all the plants then known. Although much of the information that has since come to light indicates that Engler's arrangement of the plants is phylogenetically incorrect in many respects, it is still used by most botanists in the United States and elsewhere in the world. A widespread conversion to more recent systems is unlikely at this time, for none of them is sufficiently improved to warrant such a major effort. The classification of the plant kingdom that follows, however, has been revised to conform to the latest information available.

THE PLANT KINGDOM

The plant **kingdom** is, first of all, separated into about a dozen major phyla or **divisions.** The most advanced (that is, most recently evolved) division is the one with which the horticulturist is directly concerned. This is the division that contains the so-called higher plants—those with roots, stems, leaves, and a vascular or tracheary system (the source of the name Tracheophyta). The remaining divisions contain almost no horticultural crop plants as such (edible mushrooms are an exception) and are therefore primarily of indirect interest to us. They are responsible, however, for many of our crop diseases. Included among these lower divisions are such diverse plants as the mosses, algae, fungi, bacteria, and slime molds. (In a recently developed system of classification, many of the lower plants are separated from the plant kingdom on the basis of their cellular organization. In this system, the fungi are given a kingdom of their own, the blue-green algae and bacteria are put together in a separate kingdom, and many of the other algae are placed in yet another kingdom with the protozoans and the slime molds.)

A synopsis of the traditional plant categories, or **taxa,** is given below. Each taxon is subdivided by the one below it. All categories need not be used, but the sequence must not be altered. Intermediate subdivisions are frequently made and designated by the prefix *sub-*. The categories from kingdom to family are called the major taxa; those below family are called the minor taxa.

Kingdom
Division
Class
Order
Family
Genus (*pl.* genera)
Species (*abbr.* sp., *pl.* spp.)
Variety (*abbr.* var.)
Form (*abbr.* f.)
Individual

The division Tracheophyta is divided by many into about a dozen **classes,** including the horticulturally important Filicinae (or ferns), Gymnospermae (or cycads, ginkgoes, taxads, and conifers), and Angiospermae (the many flowering plants). Figure 2-15, constructed from current information, represents a possible family tree of the living vascular plants. The evolutionary lines or phylogenetic histories within the Filicinae and Gymnospermae have been defined comparatively well, primarily because a relatively complete fossil record exists for these classes. The phylogeny of the Angiospermae, on the other hand, is unfortunately still much more of a mystery, partially owing to the great scarcity of paleobotanical information.

The gymnosperms, represented by less than 700 living species, are primarily evergreen trees of the temperate zones of the world. They characteristically have naked seeds, which are usually borne on cones, and often have narrow or needle-like leaves. Gymnosperms provide lumber, pulp, turpentine, rosin, and some edible seeds, and are often highly prized as ornamental plants.

The angiosperms, numbering some 250,000 species, are worldwide in their distribution and are found in almost all conceivable habitats. Economically, they are the most important class of plants. They are our primary source of food and beverage, shelter and clothing, paper and rubber, oils and spices. The angiosperms characteristically have seeds that are fully enclosed in a fruit, and they often have broad leaves.

Although most classes are divided directly into **orders** (the Cycadales, Ginkgoales, Taxales, and Coniferales previously mentioned are orders of gymnosperms), the class Angiospermae is first divided into two major subclasses, the Dicotyledoneae (or dicots) and the Monocotyledoneae (or monocots). The dicots are characterized by two cotyledons ("seed leaves") in their seedling stage, usually by flower parts in fours or fives or multiples of these numbers, and often by reticulate leaf venation; the monocots are characterized by one cotyledon, by flower parts in threes or multiples thereof, and often by parallel leaf venation. The dicots, which number about 200,000, include most of our broadleafed herbs, shrubs, and trees. The 50,000 or so monocots are grouped into such orders as the

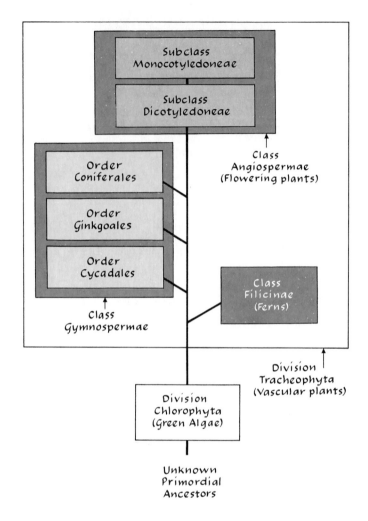

FIGURE 2-15. An abridged family tree of the living vascular plants. The solid lines indicate direct lines of descent. The higher the relative position of a group on the family tree, the more recent its emergence in the evolution of plants.

Liliales (or lilies), the Palmales (or palms), and the Graminales (or grasses and sedges). The grasses and palms together provide the main source of food for most peoples on earth, and the main source of shelter for many.

Orders are split in turn into groups known as **families.** The family is frequently encountered in taxonomic studies because it is often small enough to permit relatively easy study of the natural relationships among its members. The horticulturist will often note structural and cultural similarities among the genera within a family. Well-known families among the monocots are the Orchidaceae (or orchids) and the Gramineae (or grasses). Notable families among the dicots are the Leguminosae (or legumes), the Cucurbitaceae (which includes the gourds and melons), the Umbelliferae (which includes carrots and celery), and the Rosaceae (roses, pome fruits, stone fruits, and their relatives).

SOME WELL-KNOWN FAMILIES OF HORTICULTURAL INTEREST

GYMNOSPERMS

PINACEAE—PINE FAMILY
The pine family consists of nine genera and 210 species of conifers, mostly evergreen trees and shrubs, widely distributed but largely in temperate regions of the Northern Hemisphere. Cultivated genera include *Abies* (fir), *Picea* (spruce), *Pinus* (pine), and *Tsuga* (hemlock).

ANGIOSPERMS

Monocotyledons

AMARYLLIDACEAE—AMARYLLIS FAMILY
Closely related to the Liliaceae, the amaryllis family consists of about 90 genera and 1200 species found principally in the temperate and warm regions of South America, South Africa, and the Mediterranean region. Plants have tunicate bulbs or corms. Horticultural food crops include onion, garlic, and chives—all species of *Allium*. Other cultivated genera include *Amaryllis, Clivia,* and *Narcissus.*

GRAMINEAE (POACEAE)—GRASS FAMILY
The grass family, one of the largest in the plant kingdom, is subdivided into as many as 27 tribes with about 700 genera and 7000 species distributed worldwide. Most are herbaceous plants of low stature, but a few are woody (notably bamboos). The family includes both annuals (many cereal grains) and perennials (turfgrasses). Root systems are fine and fibrous, and stems (culms) are cylindrical and generally hollow. The leaves, which are composed of a sheath and a blade, are parallel-veined; they grow from the base, allowing regrowth after plants are clipped or mowed. The inflorescence is composed of groups of flowers called spikelets, which are variously arranged (in spikes, panicles, or racemes). Flowers are usually perfect, small, and inconspicuous. The fruit is usually a seedlike grain (caryopsis). The grass family is of paramount importance to agriculture, containing all of the cereal grains (rice, wheat, barley, oats, corn, and sorghum) and many forage plants. Horticultural plants include sweet corn (*Zea mays*), bluegrass (*Poa*), and fescues (*Festuca*).

LILIACEAE—LILY FAMILY
With about 240 genera and 3000 species distributed worldwide, the lily family consists mainly of herbaceous plants with specialized food storage organs (rhizomes, bulbs, or fleshy roots). The only horticultural food crop of importance is asparagus (*Asparagus officinalis*). There are many ornamentals, including about 90 *Lilium* species (lilies), *Aloe, Aspidistra, Hemerocallis* (daylily), *Trillium,* and *Tulipa.*

ORCHIDACEAE—ORCHID FAMILY
Among the largest of the families of flowering plants, the orchid family contains 600–800 genera and 17,000–30,000 species. Orchids are found mainly in tropical forests, where they are often epiphytes living high in the branches of large trees. Some species are saprophytic, containing no chlorophyll but obtaining their nourishment from dead organic matter. Terrestrial forms have coarse

roots with sparse branching. The flowers, which are insect pollinated, are usually large and gaudy, and many are bizarrely shaped. *Vanilla planifolia* is economically important as a source of the flavoring vanilla and many other orchids are important as a source of cut flowers (*Cattleya, Dendrobium, Vanda, Cymbidium, Phalaenopsis*) or as "conservatory" plants raised by fanciers and hobbyists.

PALMAE (ARACACEAE)—PALM FAMILY

The palm family is subdivided into 8 tribes, about 210 genera, and more than 4000 species, which are distributed in the tropics and subtropics. Most palms are unbranched woody shrubs or trees with large persistent leaves in a terminal tuft. Flowers are small, usually in a long branched panicle; fruits are berry-like, drupe-like (a fleshy fruit with a single pit), or nut-like. Some palms are cultivated for their fruits or nuts, including the date palm (*Phoenix dactylifera*), the coconut palm (*Cocos nucifera*), and the oil palm (*Elaeis guineensis*). Many are popular landscape ornamentals, such as the royal palm (*Roystonia regia*).

Dicotyledons

CACTACEAE—CACTUS FAMILY

The cactus family comprises 50–220 genera and 800–2000 species of fleshy, succulent herbs, shrubs, and trees, most of them native to the drier regions of North and South America. The fleshy stems, often spiny, may function as leaves. Flowers are often large and brightly colored. Many cacti are cultivated as ornamentals, and a few are of minor significance as food crops.

CHENOPODIACEAE—GOOSEFOOT FAMILY

A small group of about 75 genera and 500 species, the goosefoot family consists of annual or perennial herbs or shrubs often found growing in saline soils. It includes common spinach (*Spinacia oleracea*) and beet (*Beta vulgaris*), which includes both the red garden type and the large white sugar beet. The saltbush (*Atriplex*) is sometimes grown for forage in desert regions.

COMPOSITAE (ASTERACEAE)—SUNFLOWER FAMILY

Perhaps the largest plant family, with 1000 genera and more than 23,000 species, the sunflower family comprises about a tenth of all species of flowering plants. Most of the members of the family are herbaceous; and, although the family is widely distributed, most of its members are best adapted to temperate and cool climates. The inflorescence is typically a head composed of florets on a short or broadened axis. Edible plants in this family include lettuce (*Lactuca sativa*), Jerusalem-artichoke (*Helianthus tuberosus*), sunflower (*Helianthus annuus*), endive (*Cichorium intybus*), salsify (*Tragopogon porrifolius*), globe artichoke (*Cynara scolymus*), and dandelion (*Taraxacum officinale*). The family is rich in ornamentals, including such well-known genera as *Aster, Chrysanthemum, Calendula, Dahlia, Gerbera, Tagetes* (marigold), and *Zinnia*.

CRUCIFERAE (BRASSICACEAE)—MUSTARD FAMILY

The mustard family consists of about 350 genera and 3200 species of pungent herbs indigenous to temperate and cold climates. Garden vegetables in this family include cabbage, broccoli, Brussels sprouts, cauliflower, Chinese cabbage, kohlrabi, turnip, and mustard (all variant forms and species of *Brassica*), and radish (*Raphanus sativa*). Ornamentals include *Alyssum*, sweet allyssum (*Lobularia*), stocks (*Matthiola*), and candytuft (*Iberis*). Shepherd's purse (*Capsella bursa-pastoris*) and wild mustard (*Brassica* sp.) are common temperate-zone weeds.

CUCURBITACEAE—GOURD FAMILY

The gourd family consists of about 100 genera and 500–750 species. The plants in this family have tendrils, leaves that are often rough to the touch, and large fleshy fruits with many seeds. Food plants include watermelon (*Citrullus lan-*

atus), melons (*Cucumis melo*), cucumber (*Cucumis sativus*), pumpkin, summer squash and marrow (*Cucurbita pepo*), and winter squash and pumpkin (*Cucurbita maxima.*)

ERICACEAE—HEATH FAMILY

The heath family consists of about 70 genera and 1900 species, usually shrubs, widely distributed in acid soils. Cultivated genera include *Rhododendron* (rhododendron and azalea) and *Vaccinium* (blueberry and cranberry).

EUPHORBIACEAE—SPURGE FAMILY

The tropical and temperate herbs, trees, and shrubs that constitute the spurge family nearly all contain milky sap, or latex, which will exude from cut surfaces. Many tropical species are adapted to dry areas. There are about 280 genera and 7300 species. The largest genus, *Euphorbia*, includes the poinsettia (*E. pulcherrima*) and the "crown of thorns" (*E. milii*). Economically important crops include the Pará rubber tree (*Hevea brasiliensis*), cassava (*Manihot esculenta*), and castor bean (*Ricinus communis*).

LAURACEAE—LAUREL FAMILY

The laurel family consists of 47 genera and 2000–2500 species of aromatic woody plants, most of which are tropical. Well-known examples are the laurel tree (*Laurus nobilis*), cinnamon (*Cinnamomum zeylanicum*), camphor (*C. camphora*), and avocado (*Persea americana*).

LEGUMINOSAE (FABACEAE)—PEA FAMILY

The pea family, a widely distributed family of about 600 genera and 12,000 species, is one of the three largest in the plant kingdom. It is divided into three subfamilies: Faboideae, with butterfly-like flowers; Caesalpinoideae, with irregular showy flowers; and Mimosoideae, with dense flower heads with many stamens. Herbaceous plants are most common in temperate climates, but trees, shrubs, and vines predominate in the tropics. The leaves are commonly pinnate (feather-like, with leaflets arranged in two rows along a common axis), and the fruit is a flattened, dry, dehiscent pod called a legume, which gives the family its name. Many members of the family accumulate large amounts of nitrogenous substances in their tissues as a result of a symbiotic association with nitrogen-fixing bacteria that live in nodules on their roots. The high protein content makes many legumes important as sources of food and forage. Beans (*Phaseolus*, *Vicia*, and *Vigna*), peas (*Pisum sativum*), soybean (*Glycine max*), and peanuts (*Arachis hypogaea*) are all important as human food, while alfalfa (*Medicago sativa*) and clover (*Trifolium* spp.) are important forage plants. Other members of the family yield gums, such as the famous gum arabic (*Acacia senegal*), and dyes (*Cassia* and *Haematoxylon* spp.). Various tree legumes are used for lumber or as ornamentals: redbud (*Cercis canadensis*), Kentucky coffee-tree (*Gymnocladus dioica*), and honeylocust (*Gleditsia tricanthos*).

ROSACEAE—ROSE FAMILY

The rose family consists of about 100 genera and 3000 species, most of which are perennial herbs, shrubs, and trees. Flower parts (petals, sepals, and stamens) are commonly in multiples of five. The family is sometimes divided into three subfamilies: Rosoideae (rose subfamily), which contains *Rosa* spp. (roses), *Rubus* spp. (raspberries and blackberries), and *Fragaria* spp. (strawberries); Amygdaloideae (peach subfamily), which contains the genus *Prunus* (peach, cherry, apricot, almond, and plum); and Maloideae (apple subfamily), which includes apple (*Malus*), pear (*Pyrus*), quince (*Cydonia*), and hawthorn (*Crataegus*).

RUTACEAE—RUE FAMILY

The rue family comprises about 150 genera and 1500 species, most of them tropical and subtropical aromatic trees and shrubs. Notable are the *Citrus* species (sweet orange, mandarin, lemon, lime, grapefruit, citron, and related hybrids) and rue (*Ruta graveolens*), a pungent, shrubby herb of temperate climates.

SOLANACEAE—NIGHTSHADE FAMILY

There are about 90 genera and 2200 species in the nightshade family, most of them native to South America. *Solanum* is the largest genus with about 1700 species. Many species contain poisonous alkaloids, such as solanine, nicotine, and atropine. Economically important plants include potato (*Solanum tuberosum*), tomato (*Lycopersicon esculentum*), eggplant (*Solanum melongena*), chili peppers (*Capsicum* spp.), tobacco (*Nicotiana tabacum*), and petunia (*Petunia hybrida*).

UMBELLIFERAE (APIACEAE)—CARROT FAMILY

The carrot family consists mostly of north-temperate aromatic herbs, nearly all of them annual or biennial. The 250 genera and 2500 species are characterized by clusters of many small flowers borne in flat-topped umbels or heads. Many are important food or spice plants, including carrot (*Daucus carota*), parsnip (*Pastinaca sativa*), celery (*Apium graveolens*), parsley (*Petroselinum crispum*), anise (*Pimpinella anisum*), caraway (*Carum carvi*), coriander (*Coriandrum sativum*), fennel (*Foeniculum vulgare*), and dill (*Anethum graveolens*). The family also contains various poisonous plants, such as the infamous poison hemlock (*Conium maculatum*).

The Minor Plant Categories

The categories mentioned thus far—the divisions, classes, orders, and families—are known as the major taxa. They are of great evolutionary or phylogenetic importance, but it is the somewhat less arbitrary categories below the family—the minor categories—with which the horticulturist is most concerned. They are, in decreasing order of magnitude (or genetic diversity), the **genus, species, variety, form,** and **individual.** Each of these categories will now be discussed in turn.

The **genus** (the category below family) is a group (taxon) in which the member plants have much in common morphologically. It is usually a small enough group that all of its members can be brought together, and their genetic, cytological, and other relationships studied. It is for this reason the most intensively studied taxon. The generic concept is an old one. Even the ancients were familiar with the oaks (*Quercus*), the roses (*Rosa*), the tulips (*Tulipa*), the pines (*Pinus*), and the maples (*Acer*). Since many of our genera were established long before the time of Darwin, they were based primarily on readily apparent morphological characteristics. As a result of their diverse and frequently popular (that is, nonscientific) origins, these genera are often most annoyingly of different levels of genetic complexity. Modern taxonomists strive to base their systematic studies not merely on the traditional morphological criteria but on genetic and other experimental evidence as well. As a matter of fact, when possible, a taxonomist's main criterion for including or not including a population of plants within a particular genus is a genetic one that can be experimentally verified. The members of such a genus are, by their definition, at least somewhat capable of crossing among themselves, and are absolutely incapable of crossing with the plants of any other genus. A genus of this sort, bounded by total genetic incompatibility, is referred to as a **comparium.**

The white oak and red oak groups within the genus *Quercus* (a traditional genus) could each be considered a comparium. A synopsis of such modern taxonomic categories as the comparium is presented below, each category being subdivided by the one underneath it.

Comparium: Integrity maintained entirely by genetic barriers
(Often equivalent to a traditional genus)

Cenospecies: Integrity maintained by genetic barriers reinforced by ecological barriers
(Often equivalent to a traditional genus or species)

Ecospecies: Integrity maintained by ecological barriers reinforced by genetic barriers
(Often equivalent to a traditional species)

Ecotype: Integrity maintained entirely by ecological barriers
(Often equivalent to a traditional subspecies, variety, or form)

The **species** (the category below genus) is made up of plants that often exhibit many more morphological similarities than do the members of the genus. The members of a species have been referred to as coming from like parents and producing like progeny. They often constitute an exclusive interbreeding population. It was long held that all species had been created originally just as they look today.

The species has always been the basic unit of all taxonomic work. It is the fundamental category upon which Linnaeus based his system of nomenclature. He and his students established that each species be identified by two names, the generic and the specific. Examples of this binomial system of nomenclature are *Quercus alba*, the scientific name for white oak, and *Cornus florida*, that of flowering dogwood. It should be emphasized that a specific epithet is invariably preceded by a generic name.

Throughout the world the genus and species are given in Latin using the Roman alphabet (the genus is capitalized; the species should not be; and both are italicized). The name (or abbreviation of a name) that often follows such a binomial designation is that of the person who first named and described the species—for example, *Spinacia oleracea* Linnaeus (spinach). The naming of plants is governed by the International Code of Botanical Nomenclature, now strictly adhered to by the botanists of every nation. The details of the Code were worked out in Vienna at the turn of the century by botanists from all over the world. Further international congresses have been held from time to time to resolve minor conflicts and to revise the Code where necessary. Here is an example of worldwide cooperation to be envied by our statesmen!

The modern taxonomist (whose interest is to identify evolutionary units and whose system of classification of the minor taxa is based on genetics and ecology to a far greater extent than was the system of his traditional counterpart) has sometimes found it necessary to make use of the concept of the **cenospecies** and that of its subdivision, the **ecospecies.** He thus divides the genetically isolated comparium into one or more cenospecies. Each cenospecies encompasses a group of similar plants separated from other cenospecies almost entirely by genetic barriers. Cenospecies do not cross under natural conditions, but can be made to cross. The offspring of such a cross are sterile, and are often unable even to survive.

The ecospecies maintains its integrity through a combination of ecological and genetic mechanisms. The genetic barriers between ecospecies are somewhat less effective than those found between cenospecies, but these barriers are reinforced by spatial separation that results from the differing ecological requirements of related ecospecies.

Upon investigation, some traditional genera, and even some traditional species, turn out to be equivalent to cenospecies. The one or more ecospecies that make up a cenospecies may often be found the equivalent of the traditional species.

A subclassification of the traditional species is the **variety** (or **botanical variety**—not to be confused with the cultivated variety, or cultivar, which will be discussed in the next section). When one or more of the populations of plants making up a species is sufficiently different in appearance from the remaining members of the species, it is often given varietal status. A variety is designated by a trinomial, such as *Juniperus communis* var. *depressa* (prostrate common juniper). As soon as a variety is established, the remaining typical members of that species are designated by a trinomial formed by repeating the specific name—*Juniperus communis* var. *communis*, for the example just given. A population deviating not quite enough to be called a variety is sometimes called a race or, more properly, a **form** (or **forma**). Two forms of the attractive ornamental *Taxus cuspidata* (or Japanese yew) are *Taxus cuspidata* f. *densa* (an erect form) and *Taxus cuspidata* f. *nana* (a spreading form) (Fig. 2-16). On the other hand, a group deviating perhaps somewhat more from the species norm than a variety may be ranked as a **subspecies.** (It must be added that in practice the fine distinctions among subspecies, varieties, and forms are sometimes lost.)

Two varieties of a single species, through isolation and subsequent divergent evolution, may eventually differ enough to be considered as two species. If, however, the barrier between the two groups breaks down, they may again converge. The taxonomist is often hard pressed to decide whether to combine or split two such groups.

The modern systematist who prefers to revise the classification of plant populations on the basis of experimentation with particular emphasis on genetic and ecological considerations finds it necessary, as we have learned, to introduce such new concepts as the comparium, cenospecies, and ecospecies. Each ecospecies may be further divided into **ecotypes.** The ecotype is thus his substitute for the traditional subspecies, variety, or form. It is fully capable of crossing with other ecotypes within its ecospecies, but is separated from them by differing habitat preferences, which make for different geographic distributions. Ecotypic varia-

FIGURE 2-16. Japanese Yew (*Taxus cuspidata;* family Taxaceae, order Taxales, class Gymnospermae). The dark green foliage and small scarlet fruits make this hardy evergreen shrub from the Orient a favorite ornamental. *Left,* an erect form (*T. cuspidata f. densa*); *right,* a spreading form (*T. cuspidata* f. *nana*). There are numerous other yews, some of which attain tree proportions. Yew wood makes good archery bows. [Photographs by J. C. Allen & Son.]

tion can often be expected to occur in species of wide latitudinal extent. If, on the other hand, the ecospecies exhibits a continuous gradient of characters from one end of its range to the other, this is referred to as clinal or **ecoclinal** variation.

Normally self-pollinating plants such as tomatoes or peas maintain inbred lines that become stabilized after several generations and thenceforth breed true. Such populations of plants, even more genetically homogeneous than ecotypes, are referred to as **biotypes.**

A special situation exists in *Citrus, Mangifera* (mango), some species of *Rubus* (brambles), *Poa* (bluegrass), and several other plants. These plants outwardly seem to depend upon sexual propagation. In reality, although pollination may occur and may trigger fruit production, fertilization occurs only occasionally. Most new plants are produced from seeds that developed asexually from diploid ovular tissue, thereby bypassing the usual meiotic division. This process is called **apomixis,** and the genetically similar lines thus maintained in these plants are referred to as **apomicts.**

The horticulturist is continuously on the lookout for an individual horticultural plant that, by some quirk of genetic recombination or chance mutation, is especially desirable. To perpetuate this valuable selection or **"sport,"** which might not breed true if left to its own sexual devices, he may propagate it vegetatively—that is, create a **clone.** The clone is a group of plants all of which arose from a single individual (the **ortet**) through some means of vegetative or asexual propagation. Each member of a clone, which is technically known as a **ramet,** has a genetic makeup identical with the ortet or clonal progenitor. The 'Redhaven' peach, the 'Delicious' apple, the 'Russet Burbank' potato, and the 'Thompson Seedless' grape are all examples of clones.

It should perhaps be re-emphasized that of the traditional taxa in general use today most are based strictly on floral morphology, some on profound ignorance, and only a handful on sound principles and experimental evidence. *Prunus* (cherry, plum), *Rubus* (raspberries and other brambles), *Vitis* (grape), *Rosa* (rose), and *Crataegus* (hawthorn) are, for example, just a few of the horticulturally important genera requiring extensive revision on the basis of modern taxonomic experimentation. The field is wide open for those with imagination and broad scientific interests.

Cultivar: The Cultivated Variety

The horticulturist is often particularly interested in the botanical variety, form, biotype, and clone. When one of these is intentionally in cultivation it is referred to as a **cultivar,** a contraction of *cultivated variety* that is commonly abbreviated cv. In agricultural terminology, cultivar refers to a named group of plants within a particular cultivated species that is distinguished by a character or a group of characters. Examples of cultivars are the 'Jonathan' apple, the 'Tendercrop' snapbean, the 'Danish Ballhead' cabbage, and the 'Golden Cross' sweet corn. Note that the name of the cultivar is capitalized and set off by single quotation marks.

Cultivar has replaced the older term *variety* to avoid confusion with the taxonomic term. In its taxonomic sense, *variety* means a "botanical variety," and in its horticultural and agricultural sense it means a "cultivar" or "horticultural variety." (The use of the term in its horticultural sense considerably antedates the taxonomic usage defined in the previous section.)

The cultivar is the taxon that is most often of relevance to horticulturists. In fact, cultivars can in a sense be considered the very keystone of horticulture. Cultivars have been known since antiquity. Some present day clonal cultivars, such as the 'Bartlett' pear and the 'Yellow Newtown' apple, have been known for centuries.

The cultivar, biologically speaking, can be one of several entities, often distinguished by the method of reproduction of a particular crop. Thus with reference to crops that are generally asexually propagated, such as the potato and various fruits, the term means a particular clone, the product of a single individual. For example, the names 'Anjou' pear, 'Russet Burbank' potato, and 'Senga Sengana' strawberry refer to unique genotypes perpetuated by vegetative propagation. They are also referred to as **clonal varieties.** Many improved clonal varieties are the result of mutations (known as **sports**) that produced some advantageous change, such as the red pigment in the skin of sports of the 'Delicious' apple. In a few plants clonal propagation can be made by seed that is produced by apomixis. 'Kentucky' bluegrass (*Poa pratensis*), a favorite for lawns as well as a prime pasture species, is largely apomictic.

Cultivars of such self-pollinated plants as the pea or tomato usually constitute **inbred lines** or **pure lines** that breed true naturally. These cultivars are maintained by avoiding contamination and mixing. In such cross-pollinated crops as cabbage, a cultivar may constitute a population of plants distinguished on some morphological or physiological basis and maintained by selection and isolation. These are often referred to as **open-pollinated cultivars.** Finally, the term **hybrid cultivars** may refer to a particular combination of inbred lines. 'Jade Cross' is a recent hybrid cultivar of Brussels sprouts. Thus it can be seen that the precise meaning of *cultivar* in the horticultural or agricultural sense depends on the particular crop in question. The improvement of cultivars is a major goal of plant breeding.

Some of our horticultural plants are known only in cultivation. They have been bred and selected by man for so many centuries that their wild origins have become obscure. Such plants are referred to as **cultigens.** Cabbage and maize are two examples.

THE IDENTIFICATION OF PLANTS

When people are confronted with plants unknown to them they may turn for assistance to an appropriate book of plant descriptions. Horticulturists living in the continental United States are fortunate in having at their disposal several regional manuals, as well as two that span the continent. These latter two manuals (see Selected References at the end of this chapter) describe virtually all of the horticultural species capable of surviving under our environmental conditions (including greenhouse conditions). One was compiled by the renowned American horticulturist Liberty Hyde Bailey, the other by the well-known dendrologist Alfred Rehder.

Should the plant in question be an exotic not known to be cultivated in the continental United States, it will not be covered by either Bailey or Rehder. One can then often learn which manual to turn to by consulting the list compiled by Sidney F. Blake and Alice C. Atwood (see Selected References).

The use of manuals enables the horticulturist to identify most plants he is likely to encounter. The possibility always exists, however, that the unknown plant may have been considered to be too rare or too unimportant horticulturally for inclusion, or even that it is a plant new to science. The safest recourse under such circumstances is to send the plant to some botanical institute for identification (or naming!) by a professional taxonomist.

The actual identification of an unknown plant is usually accomplished through the use of the **analytical keys** that are a part of most manuals. A key contains the diagnostic features of all plants listed in the manual arranged in such a manner that all but the correct plant can be rejected. A dichotomous key, the kind most commonly used, is constructed as a series of couplets, each containing a pair of contrasting statements. One examines the first couplet and chooses the statement that fits the unknown plant; this leads to the next correct couplet to use, and that to the next, and so on until the plant is "keyed out." The following is an example of such an analytical key.

AN ANALYTICAL KEY TO THE WALNUTS (*Juglans*)

1A Leaf scars with a hairy upper fringe;
pith not light brown..2

1B Leaf scars with a glabrous upper edge;
pith light brown...3

2A Pith dark brown;
fruits solitary or in clusters
of 2 to 5.....................................Butternut or white walnut (*J. cinerea*)

2B Pith violet-brown;
fruits in clusters of 12 to 20......................Japanese walnut (*J. sieboldiana*)

3A Leaves glabrous;
bark smooth and
silvery gray ....Persian or English walnut (*J. regia*)

3B Leaves pubescent;
bark rough and dark brown ....Black walnut (*J. nigra*)

Some keys may be ambiguous with respect to species, especially if an attempt has been made to separate the species by characters other than those upon which the species were originally based. If desired, final verification can be made by comparison with a known example of that species. Such examples can be found as dried and pressed specimens (known as herbarium specimens) in many botanical institutes. Finally, it must be pointed out that in order to get full information on varieties one frequently must consult monographs dealing with the particular plant group in question.

CONCLUSION

The subject of plant classification, although perhaps the most elementary branch of horticulture, is also its most inclusive one. Nothing in horticulture can be discussed in a scientific way without some taxonomic knowledge; furthermore, a classification serves to integrate and to summarize all that we know about our plants, whether morphological, genetic, ecological, or physiological. A knowledge of classification often permits the horticulturist to predict the cultural requirements of a plant. It also helps him predict its graft compatibilities and the other plants with which it will hybridize. Finally, it aids him in his search for and development of new plants of horticultural importance.

Selected References

Bailey, L. H., 1949. *Manual of Cultivated Plants Most Commonly Grown in the Continental United States and Canada,* rev. ed. New York: Macmillan. (This manual provides a means for the identification of 5347, or almost all, cultured tracheophyte species covered by the title. Varieties are not described. Identification is done through keys, descriptions, and some illustrations. There is a glossary.)

Bailey, L. H., E. Z. Bailey, and the Staff of the L. H. Bailey Hortorium, 1976. *Hortus Third: A Concise Dictionary of Plants Cultivated in the United States and Canada.* New York: Macmillan. (This monumental work, whose earlier editions were called *Hortus* [1930] and *Hortus Second* [1941], covers 34,305 families, genera, and species of cultivated plants.)

Blake, S. F., and A. C. Atwood, 1942, 1961. *Geographical Guide to Floras of the World; Annotated List with Special Reference to Useful Plants and Common Plant Names. I: Africa, Australia, North America, South America, and Islands of the Atlantic, Pacific, and Indian Oceans* (USDA Misc. Pub. 401; 1942). *II: Western Europe: Finland, Sweden, Norway, Denmark, Iceland, Great Britain with Ireland, Netherlands, Belgium, Luxembourg, France, Spain, Portugal, Andorra, Monaco, Italy, San Marino, and Switzerland* (USDA Misc. Pub. 797; 1961). Washington, D.C.: U.S. Government Printing Office. (This exhaustive bibliography, containing 9866 titles, is the only one of its kind. It covers over half the world, but does not include Germany, central and eastern Europe, or Asia and associated islands.)

Lawrence, G. H. M., 1951. *Taxonomy of Vascular Plants.* New York: Macmillan. (A standard text in systematic botany. Part I covers both the theoretical and practical aspects of classification, nomenclature, and identification. Part II describes all tracheophyte families known to be native to, or introduced into, the United States. The appendix contains a detailed glossary of botanical terms.)

Porter, C. L., 1967. *Taxonomy of Flowering Plants,* 2nd ed. San Francisco: W. H. Freeman and Company. (An excellent elementary text.)

Rehder, A., 1940. *Manual of Cultivated Trees and Shrubs Hardy in North America Exclusive of the Subtropical and Warmer Temperate Regions,* 2nd ed. New York: Macmillan. (This manual includes 2550 species with about 2900 varieties and about 580 hybrids, or almost all the cultured trees, shrubs, lianas, and partially woody plants covered by the title. There are keys to the species, a glossary, and species descriptions that include indications of hardiness.)

Sneath, P. H. A., and R. R. Sokal, 1973. *Numerical Taxonomy: The Principles and Practice of Numerical Classification.* San Francisco: W. H. Freeman and Company. (An advanced text in the quantitative approach to taxonomic classification.)

Swain, T. (editor), 1963. *Chemical Plant Taxonomy.* London: Academic Press. (Chemical and allied techniques for determining relationships among plants.)

3

The Structure
of Horticultural Plants

This is the state of man: to-day he puts forth
The tender leaves of hopes, to-morrow blossoms,
And bears his blushing honors thick upon him,
The third day comes a frost, a killing frost,
And when he thinks, good easy man, full surely
His greatness is a-ripening, nips his root,
And then he falls as I do.

SHAKESPEARE, *Henry VIII* [III. 2]

THE PLANT BODY

Flowering plants, which make up almost the entire range of horticultural interest, show extreme diversity in size and structure. Nevertheless, there are essential similarities: many structures (for example, the air-borne root of the orchid and the swollen root of the sweetpotato), although superficially very different, are functionally and morphologically related. The flowering plant consists of two basic parts—the **root,** the portion that is normally underground, and the **shoot,** the portion that is normally above ground. The shoot is made up of stems and leaves. The leaves grow from enlarged portions of the stem called **nodes.** The shoot shows several significant modifications. The **flower** may be thought of as a specialized stem with leaves adapted for reproductive functions. **Buds** are miniature leafy or flowering stems. Figure 3-1 illustrates the fundamental plant parts.

The plant body is made up ultimately of microscopic components called **cells,** which, although they are essentially similar, may be structurally and functionally very different. Masses of cells in various combinations and arrangements build up the various morphological structures of the plant.

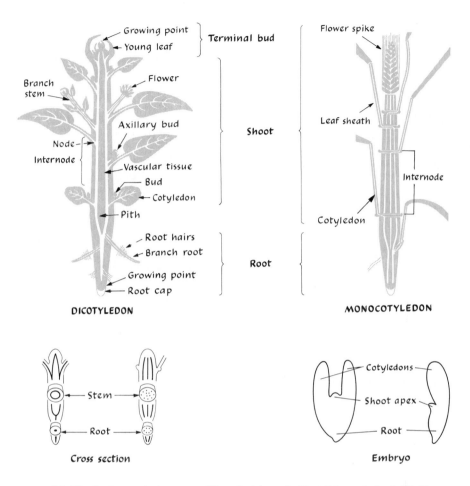

FIGURE 3-1. The fundamental plant parts. [From Janick et al., *Plant Science*, 2nd ed., W. H. Freeman and Company, copyright © 1974.]

THE CELL AND ITS COMPONENTS

The structural unit of plants, as well as of animals, is the cell. **Cytology,** the study of cells, is concerned with their organization, structure, and function. The concept that the cell is the basic unit of all living things is still accepted by biologists as dogma. The complex multicellular organism is an integrated collection of living and nonliving cells. The high degree of synchronization and coordination in the total organism creates an entity that is in effect greater than the sum of its parts.

Plant cells vary in shape from spherical, through polyhedral and ameboidal, to tubular. Most cells are between 0.025 mm and 0.25 mm in size (0.001–0.1 inch), but some cells (long, tubular fibers) are as much as 2 feet long. The concept of the typical or generalized plant cell, such as the one in Figure 3-2, is a useful one. The most prominent components visible under the light microscope are the densely staining nucleus and the more or less rigid cell wall that encloses the cytoplasm. The cytoplasm contains a number of structural bodies, or organelles, such as

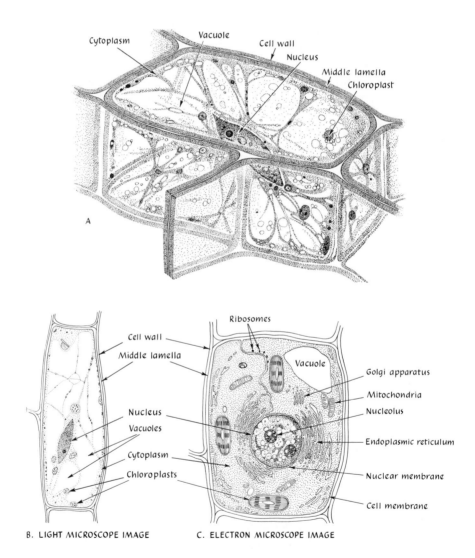

FIGURE 3-2. Three views of a plant cell and its component parts. Recent electron microscope studies have produced a new concept of the ultrastructure of the cell. [Parts A and B after Esau, *Plant Anatomy*, Wiley, 1953, Part C after Brachet, "The Living Cell," copyright © 1961 by Scientific American, Inc., all rights reserved.]

plastids, mitochondria, and vacuoles, and various other entities such as crystals, starch grains, and oil droplets. A distinction is often made between living and nonliving substances in the cell, but such distinctions have little meaning.

The **cytoplasm** (less precisely termed protoplasm) is an extremely complex substance, both physically and chemically. It is composed (by fresh weight) of 85 to 90 percent water; the remaining 10 to 15 percent consists of organic and inorganic substances that are either dissolved (salts and carbohydrates) or in the colloidal state (proteins and fats). The physiological activity of the cell occurs in the cytoplasm. Although the living state appears to be dependent on the cytoplasm within the structural confines of a cell, various so-called vital processes (such as photosynthesis) have recently been found to be possible outside of the cell. The concept of what constitutes a vital process changes as biological techniques increase in sophistication.

Surrounding the cytoplasm is the **plasma membrane,** which lies against the inside of the cell wall. The cell wall is to some degree permeable to all solutes and solvents; the plasma membrane is semipermeable. The plasma membrane is composed of lipoproteins, which account for its elasticity and high permeability to fatty substances. Although it is ultramicroscopic in thickness, the plasma membrane has been viewed in plasmolyzed cells: the electron microscope has revealed that it is not a mere boundary to the cell but has structural definition. The cell appears permeated by internal membranes that, along with the external membrane, are thought to constitute a system called the **endoplasmic reticulum.**

The **nucleus,** a dense, usually spheroidal body, is located within the cytoplasm. The close affinity of the nuclear material for many dyes makes it the most conspicuous feature of a stained cell. The nucleus is, in effect, the control center of the cell, since it contains the **chromosomes** (*chromo,* colored; *soma,* body). The chromosomes contain deoxyribonucleic acid (DNA) in association with protein. The arrangement of the DNA provides genetic information for the cell, in somewhat the same way that punch cards provide information for an electronic computer. The information is relayed to the cytoplasm in the form of RNA (ribonucleic acid), a substance similar to DNA, and it affects the machinery of the cell through the control of protein synthesis. The actual site of protein synthesis is not the nucleus but small particles in the cytoplasm called **ribosomes.** The information relayed by DNA provides the basis for the physiological functioning of the cell and thus determines the metabolic and morphological features of the organism. It is also a major part of the hereditary bridge between generations. Just how DNA does all of this will be discussed in Chapter 9, where we focus our attention on plant reproduction.

The plant cell wall, one of the structures that distinguishes plant cells from animal cells, is usually thought of as a deposit, or secretion, of the cytoplasm. It is composed of three basic groups of compounds: polysaccharide materials (which are long-chain units formed from such simple sugars as glucose), lignin, and pectin.

The principal polysaccharide in the cell wall is generally **cellulose,** which is an *unbranched* polymer of a few thousand glucose molecules. Other cell-wall polysaccharides are frequently referred to as hemicelluloses, which is a vague, broad term for *branched* chain polysaccharides containing a variety of monosaccharide units, only one of which is glucose, as well as such nonsugar components as protein.

Cellulose and its derivatives are the most abundant materials in the structure of plants. They are highly combustible materials. They are indigestible directly by mammals, including humans, because mammals do not possess an enzyme that is capable of breaking the bonds between the glucose units in cellulose. However, various bacteria do secrete an appropriate enzyme, and those mammals (such as ruminants) that sustain themselves on a diet rich in cellulose are able to do so because these bacteria grow in their digestive tracts. Through fermentation, these bacteria begin the decomposition of the cellulose, with the remaining digestion completed by the host animal.

Lignin consists of a complex mixture of chemically related compounds—specifically, polymers of phenolic acid. The deposition of lignin (lignification) hardens the cellulose walls into an inelastic and enduring material resistant to microbial decomposition. Because lignin causes yellowing in paper, it must be dissolved from any wood pulp used in the manufacture of high-quality papers.

Pectins are important components of the cell wall. They are acidic polysaccharides, specifically water-soluble polymers of galacturonic acid that form sols and gels with water. The most familiar kind of pectin is that used as a solidifying agent for jams and jellies.

The cell wall is laid down in distinct layers, and its thickness varies greatly with the age and type of cell (Fig. 3-3). Three regions of the cell wall are generally distinguished:

1. The **middle lamella** is a pectinacious material associated with the intercellular substance. The slimy nature of rotted fruit is due to the dissolving of the middle lamella by fungal organisms.

2. The **primary cell wall** is the first wall formed in the developing cell. It is composed largely of cellulose and pectic compounds, but closely related substances and noncellulose compounds may also be present, and it may become lignified. The primary cell wall is the wall of dividing and growing cells; in many cells it is the only wall.

3. The **secondary cell wall** is laid down inside the primary cell wall after the cell has ceased to enlarge. It appears to have a mechanical function, and is similar in structure to the primary wall, although it is higher in cellulose. It often contains some lignin.

The cell wall in plants in not continuous. It appears to be pierced by cytoplasmic strands (plasmodesmata) that provide a living connection between cells. Furthermore, thin areas called pits occur in the walls of some cells (Fig. 3-4).

Plastids are specialized disc-shaped bodies in the cytoplasm, and they are

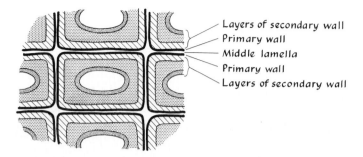

Layers of secondary wall
Primary wall
Middle lamella
Primary wall
Layers of secondary wall

FIGURE 3-3. Diagrammatic cross section of a tracheid, showing the structure of the cell wall. The secondary cell wall deposited inside the primary wall may consist of three layers.

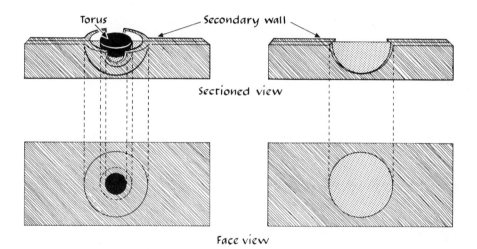

Torus

Secondary wall

Sectioned view

Face view

FIGURE 3-4. Pits are thin areas in the cell wall. *Left*, bordered pit; *right*, simple pit. [Adapted from Eames and MacDaniels, *An Introduction to Plant Anatomy*, McGraw-Hill, 1947.]

peculiar to plant cells. They are classified, on the basis of the presence or absence of pigment, into leucoplasts (colorless) or chromoplasts (colored). Leucoplasts occur in mature cells that are not exposed to light, and some types are involved in the storage of starch. Of the colored plastids, those containing chlorophyll (**chloroplasts**) are the most significant, for they are the complete structural and functional unit of photosynthesis, the process by which carbon dioxide and water are transformed to carbon-containing compounds. The reactions result in the formation of starch grains within the chloroplast.

There are about 20 to 100 chloroplasts in each chlorophyllous cell of a typical

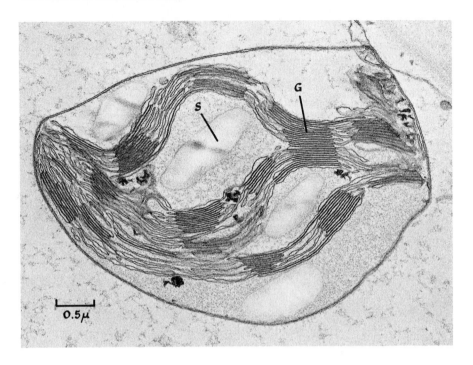

FIGURE 3-5. An electron micrograph of a chloroplast from a tobacco leaf. Note granum (G) and starch grain (S). [Courtesy W. M. Laetsch.]

green leaf. (Mature leaf cells of spinach, however, may contain up to 500 chloroplasts.) Electron microscope photographs have revealed the internal structure of chloroplasts (Fig. 3-5). Structural units called **grana,** which resemble stacked coins, contain the chlorophyll and are the receptors of light. The actual transformation of carbon dioxide to carbon-containing compounds occurs in the surrounding material, called **stroma.**

Chloroplasts have a kind of independent existence in the cell. Although they are under the influence of nuclear genes, the chloroplasts are self-replicating and apparently arise only from pre-existing plastids.

Mitochondria are the cell's power centers. They appear as small, dense granules under the light microscope, but electron microscopy has revealed that they have a complex involuted internal structure. The mitochondria are made up of proteins and phospholipids. All of the known functions of mitochondria are related to enzymatic activity connected with oxidative metabolism. This activity occurs through the formation of the energy-carrying substance called adenosine triphosphate (ATP).

Vacuoles are membrane-lined cavities located within the cytoplasm. They are filled with a watery substance known as the cell sap, which contains a number of dissolved materials—salts, pigments, and various organic metabolic constituents. In actively dividing cells the numerous vacuoles are small; in mature cells they coalesce into one large vacuole that occupies the center of the cell, pushing the cytoplasm and the nucleus next to the cell wall.

A number of complex materials may be found in the cytoplasm; among these are crystals, starch grains, oil droplets, silica, resins, gums, alkaloids, and many organic substances. Many of these compounds are reserve or waste products of the cell.

TISSUES AND TISSUE SYSTEMS

Although the plant ultimately originates from a single cell (the fertilized egg), the marvels of cell division and differentiation produce an organism composed of many kinds of cells that are structurally and physiologically diverse. It is this difference in cell morphology and cell arrangement that results in the complex variation between plants and within an individual plant.

Plants can be shown to be made up of groups of similar types of cells that are organized in a definite pattern. Continuous, organized masses of similar cells are known as **tissues.** Tissues have been classified in several ways. No universally accepted system exists. The following system is a logical one that retains the customary botanical terms.

> **Meristematic tissue:** actively dividing undifferentiated cells
>
> **Permanent tissues:** nondividing differentiated cells
>> **Simple tissues:** composed of one type of cell
>>> **Parenchyma:** simple thin-walled cells
>>> **Collenchyma:** thicker-walled "parenchyma"
>>> **Sclerenchyma:** thick-walled supporting cells
>> **Complex tissue:** composed of more than one type of cell
>>> **Xylem:** water-conducting tissue
>>> **Phloem:** food-conducting tissue

Meristematic Tissue

Meristematic tissue is composed of cells actively or potentially involved in cell division and growth. The meristem not only perpetuates the formation of new tissue but also perpetuates itself. Since many so-called permanent tissues may, under proper stimulation, assume meristematic activity, no strict line of demarcation exists between meristematic and permanent tissue.

Meristematic tissues are located in various portions of the plant (Fig. 3-6). Those at the tips of shoots and roots are known as **apical meristems.** The shoot

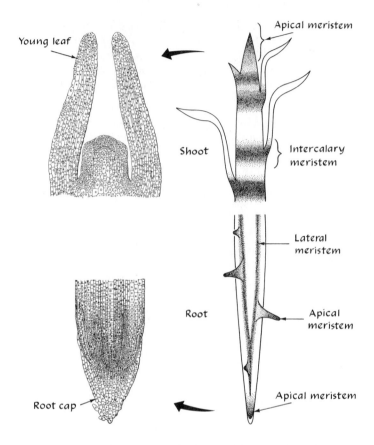

Young leaf

Apical meristem

Shoot

Intercalary meristem

Lateral meristem

Root

Apical meristem

Apical meristem

Root cap

FIGURE 3-6. Diagrammatic longitudinal section of a grass plant, showing the location of the meristems. These shaded areas are the youngest parts of the plant. [Adapted from Eames and MacDaniels, *An Introduction to Plant Anatomy*, McGraw-Hill, 1947.]

apical meristem is known horticulturally as the growing point. The increase in girth of woody stems results from the growth of lateral meristems, specifically referred to as the **cambium.** The meristematic regions of grasses become "isolated" near the nodes, and are called **intercalary meristems.** Thus, the mowing of lawns does not interfere with the growth of the grass plant because the growing points are not damaged by mowing. Tissues differentiated from apical meristems are referred to as **primary tissues.** Others, especially tissues formed from the cambium, are **secondary tissues.**

Although there are many exceptions, meristematic cells are usually small, often roughly spherical to brick-shaped, and have thin walls with inconspicuous vacuoles. In sections prepared for microscopic examination they appear darkly stained, owing to the small amount of cytoplasm in relation to the nucleus.

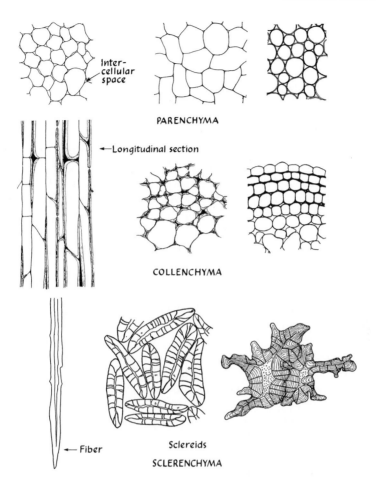

FIGURE 3-7. Simple tissues. [Adapted from Eames and MacDaniels, *An Introduction to Plant Anatomy*, McGraw-Hill, 1947.]

Permanent Tissues

Permanent tissues are made up of nondividing differentiated cells derived directly from meristems. They are referred to as **simple tissues** when they are composed of one type of cell (Fig. 3-7), and as **complex tissues** when they are mixtures of cell types.

Simple Tissues

Simple tissues are of three types: parenchyma, collenchyma, and sclerenchyma.
   Parenchyma is relatively undifferentiated, unspecialized vegetative tissue. It

makes up a large portion of many plants, such as the fleshy portion of fruits, roots, and tubers.

Collenchyma is tissue characterized by elongated cells with thickened primary walls composed of cellulose and pectic compounds. (It may be thought of as thick-walled parenchyma.) This tissue functions largely as mechanical support in early growth. The strands at the outer edge of a celery stalk are collenchyma (Fig. 3-8).

Sclerenchyma is tissue composed of especially thick-walled cells that are often lignified. When these cells are long and tapered they are usually referred to as fibers. Others are referred to as sclereids. Clusters of these sclereids, or "stone cells," are responsible for the gritty texture of pears. In masses, sclereids are responsible for the hardness of walnut shells and of peach and cherry pits. Unlike parenchyma and collenchyma, sclerenchyma cells are nonliving when mature.

Complex Tissues

Combinations of simple tissue and specialized tissue form complex tissues, the two major types of which are xylem and phloem.

Xylem, the principal water-conducting tissue, is an enduring tissue that consists of living and nonliving cells. Wood is largely xylem. Herbaceous plants also contain xylem, although the amount is very much less than that contained in "woody" plants. The water-conducting function in xylem is accomplished through specialized nonliving cells called **tracheids.** These are elongated tapered cells with walls that are hard and usually lignified, although not especially thick. The water, in the form of cell sap, moves readily through the empty tracheid, flowing from cell to cell through the numerous pits between them. The cell walls of tracheids

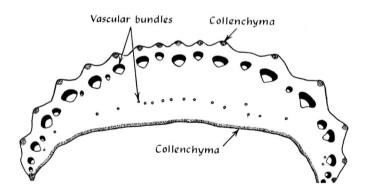

FIGURE 3-8. Cross section of a celery petiole, showing the distribution of collenchyma and vascular bundles. [Adapted from Esau, *Plant Anatomy,* Wiley, 1953.]

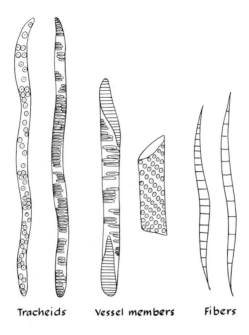

Tracheids   Vessel members   Fibers

FIGURE 3-9. Fibers, tracheids, and vessels of the xylem. [Adapted from Esau, *Plant Anatomy*, Wiley, 1953.]

are often unevenly thickened or sculptured, which makes these cells easy to distinguish in longitudinal section.

In addition to tracheids, a specialized series of cell members called vessels also functions in water conduction. These are formed from meristematic cells from which the cell contents and end walls have been dissolved. Such cells are lined up end to end, and the series may be many feet long. Xylem typically includes fibers and parenchyma cells (Fig. 3-9).

Xylem is formed by differentiation of the apical meristems of root and shoot. In perennial woody plants, secondary xylem is also formed from familiar annual rings. The spring wood consists of larger cells with thinner walls and appears lighter, or less dense, than the summer wood.

The principal food-conducting tissue is the **phloem.** Its basic components are series of specialized cells called **sieve elements.** It is through these elongated living cells with thin cellulose walls that the food is conducted from one part of the plant to another. Upon maturity the nucleus of the sieve cell disappears. Specialized cells called **companion cells** are in intimate association with the sieve cells (Fig. 3-10). In addition to these, fibers and sclereids may be present. The fibers of hemp and flax are derived from the phloem tissue.

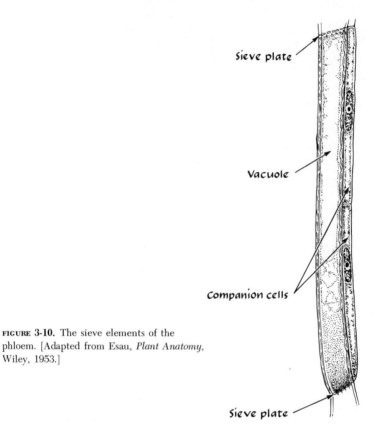

FIGURE 3-10. The sieve elements of the phloem. [Adapted from Esau, *Plant Anatomy*, Wiley, 1953.]

The phloem is formed, as is the xylem, both by the apical meristem and by the cambium. The phloem, however, is not enduring, and the old phloem disintegrates in woody stems. It is protected by special meristematic tissues (**cork cambium**) that produce parenchymatous tissue. The phloem, the corky tissue, and the other incidental tissues constitute **bark.**

ANATOMICAL REGIONS

The tissues that form the various regions of the plant can be classified in terms of structure and function. In much the same way, the bricks, boards, pipe, and wire used in the construction of a house (figuratively speaking, its cells and tissues) can be classified, on the basis of structure and function, as masonry, frame, plumbing system, and wiring system. Since these are interrelated, it is sometimes hard to decide where one region ends and another begins. (Is an electric hot water heater part of the electrical system or the plumbing system?)

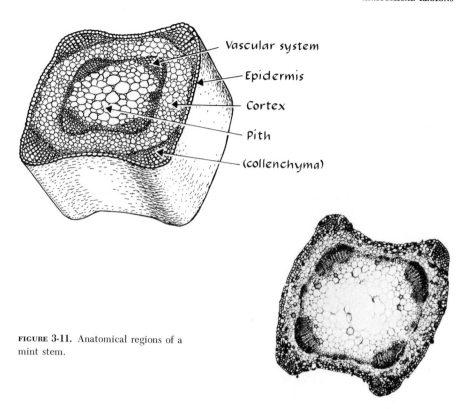

Vascular system

Epidermis

Cortex

Pith

(collenchyma)

FIGURE 3-11. Anatomical regions of a mint stem.

Horticultural plants are grossly divided into the vascular system (plumbing), the cortex (frame and insulation), and the epidermis (siding, floor, and roof) (Fig. 3-11). The pith, pericycle, endodermis, and secretory glands, however, are components of one or more of these regions.

The Vascular System

The vascular system consists principally of the xylem and phloem tissues. The vascular system serves as the conduction system of the plant, but because it also serves in support it may be compared to both the circulatory and skeletal systems of animals. There are differences between the structural relationship of xylem and of phloem. Typically, the vascular system forms a continuous ring in the stem, in which the inner portion is xylem, surrounding an area of parenchymatous tissue known as the **pith.** The vascular system, however, may be discontinuous, and may appear as a series of strands in longitudinal section and as bundles in cross section (Fig. 3-12). This is usual in monocotyledonous plants.

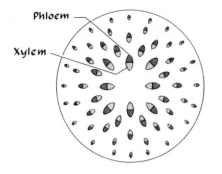

Phloem

Xylem

Phloem

Cambium

Xylem

Pith

Discontinuous vascular system
of a monocotyledonous stem.
Note lack of distinct pith.

Continuous vascular system
of a dicotyledonous stem.
Vascular system surrounds pith

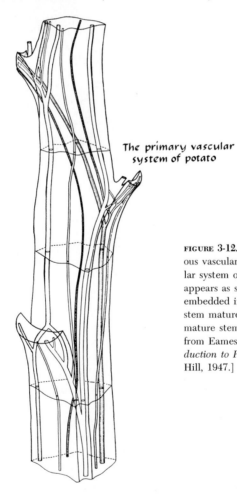

The primary vascular
system of potato

FIGURE 3-12. Discontinuous and continuous vascular systems. The primary vascular system of the potato stem initially appears as separate bundles but becomes embedded in secondary tissue as the stem matures. The vascular system of the mature stem is continuous. [Adapted from Eames and MacDaniels, *An Introduction to Plant Anatomy*, McGraw-Hill, 1947.]

In roots, the vascular system makes up the core, and the pith is absent. It is separated from the cortex by specialized tissues called the **pericycle** and the **endodermis.** The pericycle encircles the vascular system. It is composed of parenchymatous tissue and is the source of the branch roots and stems that arise from the root. The endodermis is commonly a single sheet of cells separating the vascular system from the cortex, and it is not absolutely clear whether it is part of the vascular system or of the cortex. It appears to have a protective function. The pericycle and endodermis are usually absent in the stem.

Cortex

The cortex is the region between the vascular system and the epidermis. It is made up of primary tissues, predominantly parenchyma. In older woody stems, the formation of cork in the cortex, with the subsequent disintegration of the outer areas, tends to obliterate the cortex as a distinct area. Cork is formed when mature tissue is infiltrated by a waxy substance known as **suberin,** which essentially waterproofs the cell walls. This process is known as **suberization,** and the corky protective sheath that it produces is called the **periderm.** Callus tissue (the "scar tissue" that is formed in response to wounding) may also become corky. The cork industry is based on the large amounts of this tissue produced by wounding of the cork oak (*Quercus suber*).

Differentiated portions of the cortex that form ruptured, rough areas on woody stems and in bark are referred to as **lenticels.** These are "breathing pores": they allow for the exchange of gasses between the plant's tissues and the atmosphere. The "dots" on apple skin are also lenticels. Because they penetrate the epidermis, lenticels of fruits can allow for the entrance of decay organisms and act as a point of water loss.

Epidermis

The epidermis is a continuous cell layer that envelops the plant. Except in older stems and roots, in which it may be obliterated, the epidermis sheaths the entire plant. The structure of the epidermis varies, and it may be composed of different kinds of primary tissue. Slightly above the root tips, the epidermal cells form tube-like extensions called root hairs, which function in the absorption of water and inorganic nutrients. Hairs are also found in epidermal cells of the shoot. The velvety feel of rose petals is due to the uneven surface of their upper epidermal cells. The guard cells forming **stomata** are modified epidermal cells (Fig. 3-13).

A significant feature of epidermal cell structure is the **cuticle,** a waxy layer that appears on the exposed surface of the cell. The waxy material, **cutin,** acts as a protective covering that prevents the desiccation of inner tissues (Fig. 3-14). It is particularly noticeable in such fruits as the apple, nectarine, and cherry, where its

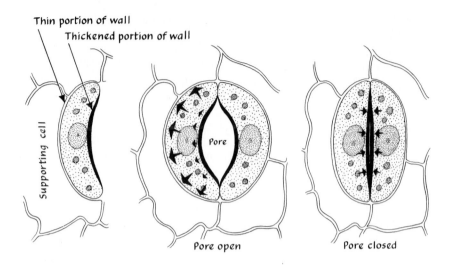

Thin portion of wall

Thickened portion of wall

Supporting cell

Pore

Pore open

Pore closed

FIGURE 3-13. The stomata are epidermal structures composed of two guard cells that form a pore. The portion of the guard cell wall abutting the pore is thicker than the rest of the cell wall. This causes the pore to open when the guard cells become turgid. [Adapted from Bonner and Galston, *Principles of Plant Physiology*, W. H. Freeman and Company, copyright © 1952.]

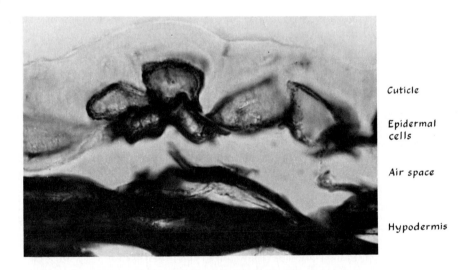

Cuticle

Epidermal cells

Air space

Hypodermis

FIGURE 3-14. The epidermis and cuticle of the 'Stayman Winesap' apple fruit. The skin of an apple is several cells thick and consists of epidermis and hypodermis. The latter is composed of compressed layers of cortical cells. The cause of the separation of the epidermis and hypodermis shown in the photograph is unknown. It produces a white patch on the fruit, sometimes referred to as "scarfskin." [Photograph courtesy D. F. Dayton.]

accumulation results in a bloom that can be polished to a high gloss. The cuticle is responsible for the tendency of water to bead on leaves and fruit, which, because many disease organisms are water borne, may be very important to the plant's ability to resist disease.

Secretory Glands

A number of morphologically unrelated plant structures—some very simple, others quite complex—secrete or excrete complex metabolic products, often in viscuous or liquid form. Many of these substances are of economic importance—for example, resin, rubber, mucilages, and gums.

Secretory functions are carried out by many multicelled, hair-like epidermal appendages called **trichomes** (Fig. 3-15). Moreover, complex secretory structures called glands develop from subepidermal and epidermal tissue in various parts of the plant body. The fragrance of flowers is produced by glands called nectaries. Some specialized glands secrete essential oils (see p. 107). In citrus fruits such glands are found in the peel, as can be verified by squeezing a portion of the rind next to a flame: rupturing the glands releases the volatile and inflammable

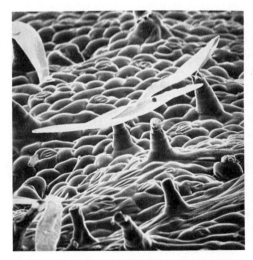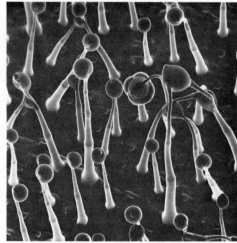

FIGURE 3-15. *Left,* upper surface of chrysanthemum leaf showing T-shaped trichomes (×144). [Stereoscan electron photomicrograph furnished by C. V. Cutting, Long Ashton Research Station, Bristol, England.] *Right,* upper surface of the leaf of *Solanum pennellii,* a relative of the potato, showing globular-tipped trichomes. [From C. D. Claybert, *HortScience* 10:13–15, 1975.

essential oils. Gums, mucilages, and resins are formed in ducts and canals, intercellular spaces between groups of specialized cells. Often these substances are secreted in response to wounding, as in the stem of peach and cherry.

Latex, a milky, viscous substance containing various materials, particularly gums, is found in some angiosperm families. It may be formed in ordinary parenchyma or in a complex series of branching tubes called laticifers. They are not considered distinct tissue but are associated with other tissues, commonly the phloem. Latex is released under pressure when these cells are punctured. Rubber is a constituent of some latex-forming plants. The content varies greatly, but in the rubber tree (*Hevea brasiliensis*) 40 to 50 percent of the latex is rubber. The function of the latex duct network is not clear, but it is felt that it probably serves as an excretory system for the plant and plays some role in wound healing.

MORPHOLOGICAL STRUCTURES

The Root

The root, although visibly inconspicuous, is a major component of the plant both in terms of function and in terms of absolute bulk. It may consist of more than half of the dry weight of the entire plant body. The root is structurally adapted for its major function of absorbing water and nutrients. Owing to its complex branching (which occurs irregularly, rather than at nodes as in stems), and its tip area of root hairs, the root presents a very large surface in intimate contact with the soil. The process of absorption in most plants is carried out mainly in the root hairs, which are constantly renewed by new growth. The growth of the plant is largely limited by the extent of its underground expansion. This vast network of roots anchors the plant and supports the superstructure of food-producing leaves. The older roots may also serve as storage organs for elaborated foodstuffs, as in the sweetpotato and carrot.

The original seedling root, or **primary root,** begins the formation of the root system of the plant by establishing a branching pattern. When the primary root becomes the main root of the plant the network is referred to as a **taproot** system, as in the walnut, carrot, beet, and turnip. In many plants, however, the primary root ceases growth when the plant is still young, and the root system is taken over by new roots that grow adventitiously out of the stem, forming a **fibrous** root system, as is typical of grasses (Fig. 3-16). In addition many taprooted plants (for example, apple) form an upper network of fibrous feeder roots. This permits deep anchorage and a more reliable water supply while providing absorptive capacity in the more fertile upper layers of soil. A fibrous root system may be produced artificially by destroying the taproot. This is accomplished by transplanting or undercutting and is a standard horticultural practice with shrubs and trees.

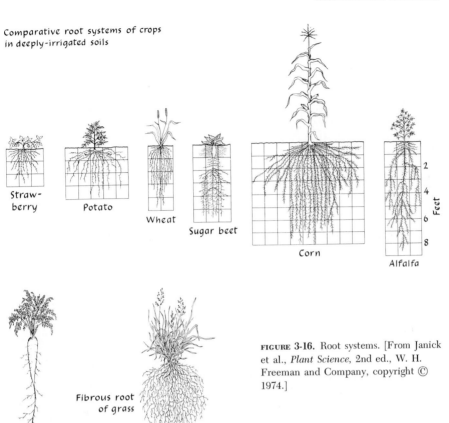

Comparative root systems of crops in deeply-irrigated soils

Straw-berry

Potato

Wheat

Sugar beet

Corn

Alfalfa

Feet

2
4
6
8

Fibrous root of grass

Taproot of carrot

FIGURE 3-16. Root systems. [From Janick et al., *Plant Science*, 2nd ed., W. H. Freeman and Company, copyright © 1974.]

Nurserymen endeavor to build up a fibrous root system concentrated in a "ball" below the plant. This permits even relatively large plants to be successfully transplanted.

In general, plants having a fibrous root system are shallow-rooted in comparison with taprooted plants. Shallow-rooted plants will, of course, be more subject to drought and will show quicker response to variations in fertility treatments.

The morphological structure of the root is shown in Figure 3-17. The arrangement of its vascular system is different from that of the stem. Note the lack of pith and the predominant pericycle and endodermis. A cambium serves to increase the girth of perennial roots, as in the stem.

The major root modifications of horticultural interest are those affecting the storage function. Roots of certain species become swollen and fleshy (Fig. 3-18) with stored food in the form of starches and sugars. Some of these storage roots,

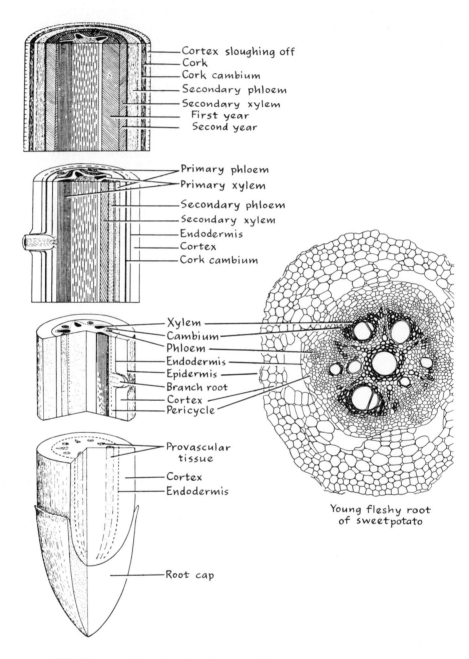

Cortex sloughing off
Cork
Cork cambium
Secondary phloem
Secondary xylem
First year
Second year

Primary phloem
Primary xylem
Secondary phloem
Secondary xylem
Endodermis
Cortex
Cork cambium

Xylem
Cambium
Phloem
Endodermis
Epidermis
Branch root
Cortex
Pericycle

Provascular
tissue
Cortex
Endodermis

Root cap

Young fleshy root
of sweetpotato

FIGURE 3-17. Diagrammatic sections through a root. [Adapted from Weatherwax, *Plant Biology*, Saunders, 1947.]

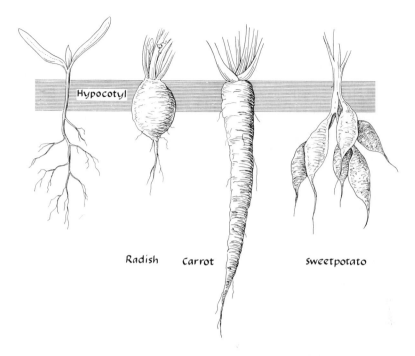

Hypocotyl

Radish   Carrot   sweetpotato

FIGURE 3-18. Root modifications. The swollen roots of the radish and carrot
include the hypocotyl, a transition zone between the rudimentary seedling root
and shoot.

such as the carrot, sweetpotato, and turnip, are edible. In some plants with storage
roots, such as the sweetpotato and the dahlia, the stored food, coupled with the
ability of the roots to form adventitious shoot buds, renders the roots important in
propagation.

The Shoot

The shoot has been described as a "central axis with appendages." The "central
axis"—that is, the stem—supports the food-producing leaves and connects them
with the nutrient-gathering roots. The stem is also a storage organ, and in many
plants its structure is greatly modified for this function. Young green stems also
have a small role in food production because of the chlorophyll they contain.
Plants assume extremely varied forms, ranging from a single upright shoot, as in
the date palm, to the prostrate branched "creepers." It is the structure and growth
pattern of the stem that determine the form of the plant. Basic structural and
anatomical features of herbaceous and woody stems are shown in Figure 3-19.
  The upright growth of plants having one active growing point and a rigid stem

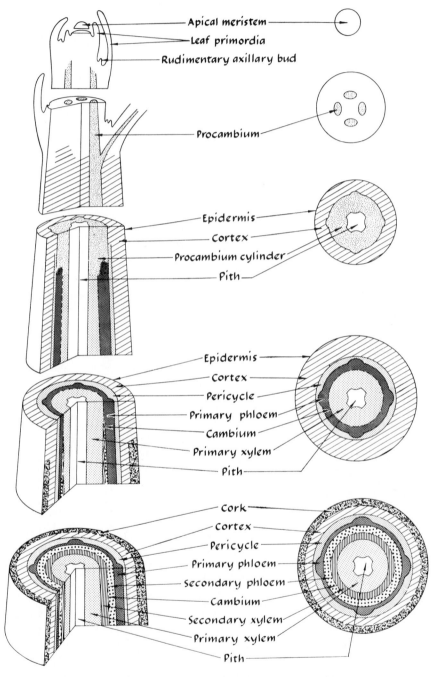

Apical meristem

Leaf primordia

Rudimentary axillary bud

Procambium

Epidermis
Cortex
Procambium cylinder
Pith

Epidermis
Cortex
Pericycle
Primary phloem
Cambium
Primary xylem
Pith

Cork
Cortex
Pericycle
Primary phloem
Secondary phloem
Cambium
Secondary xylem
Primary xylem
Pith

FIGURE 3-19. Diagrammatic sections through a stem. [Adapted from Holman and Robbins, *A Textbook of General Botany*, Wiley, 1939.]

is considered normal, and our descriptive terms are used to differentiate other growth patterns. Typical shrubby or bushy growth is brought about by the absence of a main trunk or **central leader.** Growth is characterized by a number of erect or semierect stems, none of which dominates. The distinguishing feature is form rather than size. Similarly, slender and flexible stems that cannot support themselves in an erect position are known as **vines.** Vines will trail unless mechanical support is used to make them grow upright. They may be either herbaceous (morning-glory, pea) or woody (grape).

Buds

The stem is divided into mature regions and actively growing regions in which growth and differentiation take place. An embryonic stem is called a bud. All buds do not grow actively; many exhibit arrested development or dormancy but are nevertheless potential sources of further growth. Although the buds of some plants may be so embedded in the stem tissues as to be relatively inconspicuous, the buds of others may be fairly elaborate, conspicuous structures. The form, structure, and arrangement of buds prove to be a useful guide in describing woody plants even when the leaves are absent, as in winter. Typical buds are shown in Figure 3-20.

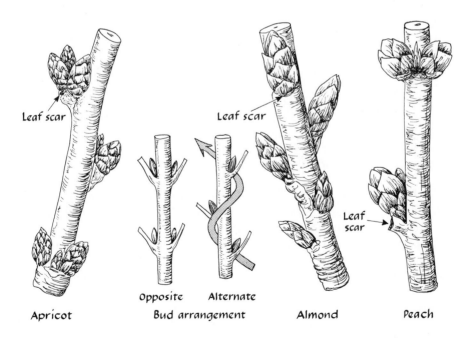

FIGURE 3-20. Dormant fruit and leaf buds of apricot, almond, and peach. The arrangement in each is alternate. [Adapted from Zielinski, *Modern Systematic Pomology*, Wm.C Brown, 1955.]

Growth may originate from a single **terminal** bud or from **lateral** buds that occur in the leaf axis. In addition, buds may be formed in internodal regions of the stems, leaves, or roots, often as a result of injury. These are called **adventitious** buds.

Buds may produce leaves, flowers, or both, and are referred to as **leaf, flower,** or **mixed** buds, respectively. When more than one bud is present at a leaf axis, all but the central or basal bud are called **accessory** buds. The arrangement or topology of buds on a stem is based on the leaf arrangement. It two or more leaves are opposed to each other at the same level, the leaf (and bud) arrangement is said to be **opposite** or **whorled.** When they are at different levels, they are arranged in a spiral and are said to be **alternate** (Fig. 3-20). The spiral pattern of leaf arrangement (**phyllotaxy**) is varied and is expressed as a fraction ($\frac{1}{2}$, $\frac{1}{3}$, $\frac{2}{5}$, $\frac{3}{8}$), where the numerator is the number of turns to get to a leaf directly above another and the denominator is the number of buds passed. Phyllotaxy has taxonomic significance, since it is often the same throughout a genus or even throughout a whole family.

Stem Modifications

The stem may be greatly modified from a basically cylindrical structure (Fig. 3-21). Some of these alterations may appear quite bizarre; yet upon close analysis these modifications can be shown to be basically stem-like in structure; that is, they have nodes, leaves, or similar scale-like structures, and they function in transport or storage. Stem modifications can be divided into above-ground forms (crowns, stolons, spurs) and below-ground forms (bulbs, corms, rhizomes, tubers). Since many stem modifications contain large amounts of stored food, they may be significant in propagation and, as in the tubers of potatoes, important as a source of food.

ABOVE-GROUND STEM MODIFICATIONS. The **crown** of a plant is generally that portion just above and just below ground level.[1] This portion of the plant may be greatly enlarged, as in the bald-cypress (*Taxodium distichum*). Crowns may be thought of as "compressed" stems. The underlying structure of the strawberry crown can be clearly seen by artificially elongating it through treatment with gibberellic acid, as shown in Figure 3-22. Leaves and flowers arise from the crown by buds, as they do in stems. In addition, fleshy buds from crowns may produce a whole new plant, referred to as **crown divisions.** The crown may be modified into a food storage organ, as it is in asparagus.

Short, many-noded, horizontal branches growing out of the crown, bearing fleshy buds or leafy rosettes, are referred to variously as offsets, slips, suckers, and pips. These stem modifications, which can be collectively termed **offshoots,** are important in providing both natural and artificial means of propagation.

Stolons are stems that grow horizontally along the ground. A **runner** is a stolon with long internodes originating at the base or crown of the plant. At some of the

---

[1]But notice that, in the terminology of forestry, *crown* refers to the branched top of a tree.

ABOVE GROUND MODIFICATIONS

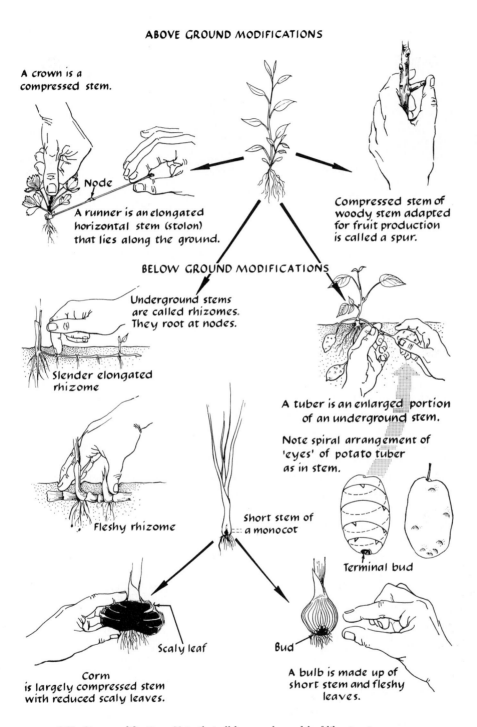

A crown is a compressed stem.

Node

A runner is an elongated horizontal stem (stolon) that lies along the ground.

Compressed stem of woody stem adapted for fruit production is called a spur.

BELOW GROUND MODIFICATIONS

Underground stems are called rhizomes. They root at nodes.

Slender elongated rhizome

A tuber is an enlarged portion of an underground stem.

Note spiral arrangement of 'eyes' of potato tuber as in stem.

Fleshy rhizome

Short stem of a monocot

Terminal bud

Scaly leaf

Bud

Corm is largely compressed stem with reduced scaly leaves.

A bulb is made up of short stem and fleshy leaves.

FIGURE 3-21. Stem modifications. Note that all have nodes and leaf-like structures.

FIGURE 3-22. The crown of a strawberry is a compressed stem. This can be clearly seen by elongating the stem with gibberellic acid (*right*). Note that the runners are formed from the leaf axils. [Photographs courtesy J. Hull, Jr.]

nodes, roots and shoots develop. A well-known example of a plant with runners is the cultivated strawberry.

Spurs are stems of woody plants whose growth is restricted. They are characterized by greatly shortened internodes, and are usually attached to normal branches. In mature fruit trees, such as apple, pear, and quince, flowering is largely confined to spurs. Spurs are not irrevocably static, and may revert to normal stem growth even after many years of fruiting.

BELOW-GROUND STEM MODIFICATIONS. A **bulb** is essentially a compressed modification of the shoot. It consists of a short, flattened, or disk-shaped stem surrounded by fleshy, leaf-like structures called scales, which may enclose shoot or flower buds. Bulbs are found only in some monocotyledonous plants. The scales, filled with stored food, may be continuous and form a series of concentric layers surrounding a growing point, as in the onion or tulip (**tunicate bulbs**), or they may form a more or less random attachment to a small portion of stem, as in the Easter lily (**scaly bulbs**). Bulbs commonly grow under the ground or at ground level, although certain bulb-like structures (**bulbils**) may be formed on aerial stems in association with flower parts, as they are in some kinds of onions.

Corms are short, fleshy, underground stems having few nodes. The corm is almost entirely stem; the few rudimentary leaves are nonfleshy. The gladiolus and crocus are propagated by corms. Like bulbs, corms are found only in some monocotyledonous plants.

Rhizomes are horizontal underground stems. They may be compressed and fleshy, as in *Iris*, or slender with elongated internodes, as in turf grasses (such as Bermuda grass). Normally, roots and shoots develop from the nodal regions. Such weeds as quack grass and Canadian thistle are particularly insidious because they spread so rapidly, owing to their natural propagation by rhizomes.

Tubers are greatly enlarged fleshy portions of underground stems. They are typically noncylindrical. (The word *tuber* is derived from a Latin word meaning "lump.") The edible portion of the potato is a tuber. The "eyes" arranged in a spiral around the tuber are the buds. Each eye consists of a rudimentary leaf scar and a cluster of buds.

The Leaf

Leaves are the photosynthetic organs upon which higher plants depend for the formation of carbon-containing compounds. The leaf is basically a flat appendage of the stem arranged in such a pattern as to present a large surface for the efficient absorption of light energy. The leaf **blade** is usually attached to the stem by a stalk or **petiole.** Leaf-like outgrowths of the petioles, known as **stipules,** are commonly present.

The anatomical structure of the leaf is shown in Figure 3-23. Note that the vascular system in leaves forms a branching network of **veins.** The veining is typically net-like in dicots and parallel in monocots. The leaf blade, although commonly bilaterally symmetrical, is not radially symmetrical, since it has a distinct upper and lower side. Beneath the upper epidermal layer, which is

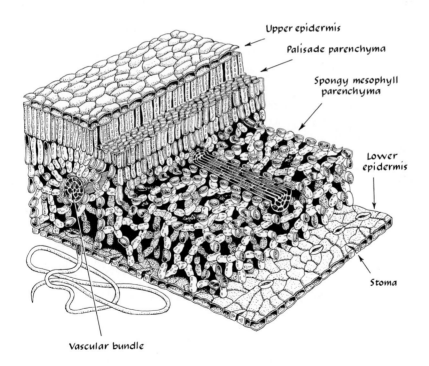

Upper epidermis
Palisade parenchyma
Spongy mesophyll parenchyma
Lower epidermis
Stoma
Vascular bundle

FIGURE 3-23. Structure of an apple leaf.

FIGURE 3-24. The cabbage head consists of large fleshy leaves attached to a compressed stem. [Photograph by J. C. Allen & Son.]

characterized by heavy deposits of cutin, lie series of elongated closely packed "palisade" cells that are particularly rich in chloroplasts. The irregularly arranged cells beneath the palisade cells produce a sponge-like region (**spongy mesophyll**) that provides an area necessary for gaseous exchange in photosynthesis and transpiration. The lower epidermis is interspersed with **stomata**—openings in the leaf that permit the exchange of gasses and water vapor with the environment.

Leaves of plants vary from the flat thin disks described to the stem-like fleshy structures found in the common house plant *Sansevieria*. The tendrils of peas are modifications of the leaf. Leaves are the edible portions of many plants, such as lettuce, spinach, and cabbage (Fig. 3-24), and they are often the chief features of many ornamentals (such as poinsettia), especially when they are rich in red and yellow pigments.

The Flower

The flower shows great variety in structure, composition, and size. The principal flower parts are shown in Figure 3-25.

Sepals (collectively, the **calyx**) enclose the flower in bud. They are usually small, green, leaf-like structures below the petals.

Petals (collectively, the **corolla**) are the conspicuous portion of most flowers. They are often highly colored, though rarely with green pigments, and they may

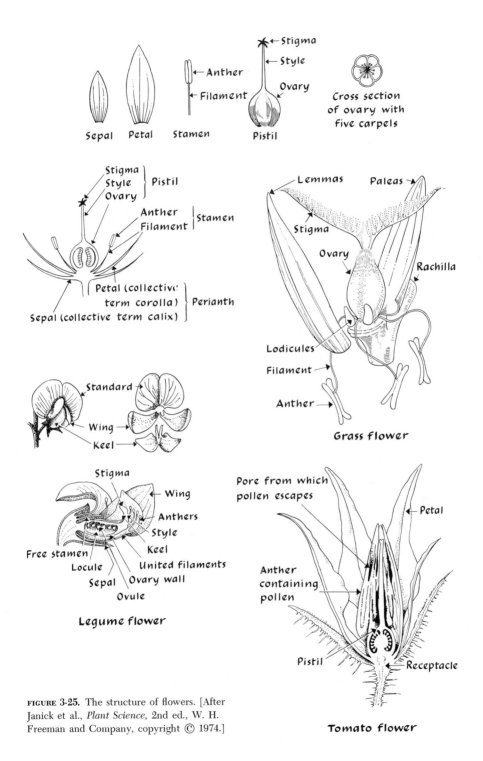

Sepal   Petal   Stamen

Stigma
Style
Anther
Filament
Ovary
Pistil

Cross section
of ovary with
five carpels

Stigma
Style } Pistil
Ovary

Anther
Filament } Stamen

Petal (collective
term corolla) } Perianth
Sepal (collective term calix)

Standard
Wing
Keel

Lemmas   Paleas
Stigma
Ovary
Rachilla
Lodicules
Filament
Anther

Grass flower

Stigma
Wing
Anthers
Style
Keel
Free stamen
United filaments
Locule
Ovary wall
Sepal
Ovule

Legume flower

Pore from which
pollen escapes
Petal
Anther
containing
pollen
Pistil
Receptacle

Tomato flower

FIGURE 3-25. The structure of flowers. [After
Janick et al., *Plant Science*, 2nd ed., W. H.
Freeman and Company, copyright © 1974.]

contain perfume glands as well as nectar glands that produce a viscous sugary substance. The extremely large, showy flowers of many cultivated ornamentals are the result of rigorous selection.

Stamens are often considered the "male" parts of flowers. Each stamen consists of a pollen-bearing **anther** supported by a **filament.** When the pollen is mature, it is discharged through the ruptured anther wall.

The **pistil,** which is often considered the "female" part, consists of an **ovule**-bearing base (or **ovary**) supporting an elongated region (or **style**) whose expanded tip (or surface) is called the **stigma.** The ovule gives rise to the seed. The mature ovary (with or without seeds) becomes the fruit.

The petals and sepals of the flower, as well as the reproductive parts—that is, the stamens and pistils—are essentially modified leaves. The leafy origins of the stamens can be clearly shown in the stamenoids or "extra" petals of the cultivated rose (Fig. 3-26). These flower parts are borne on an enlarged portion of the flower-supporting stem called the **receptacle.**

Flowers composed of sepals, petals, stamens, and pistils are referred to as **complete** (Fig. 3-27). **Incomplete** flowers lack one or more of these parts. For

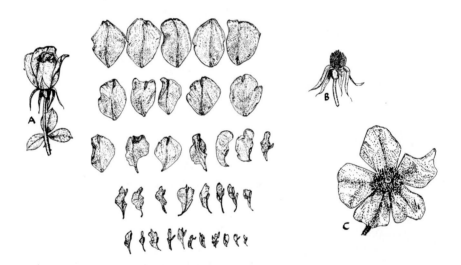

FIGURE 3-26. Roses have five sepals, five petals, numerous stamens or petaloids, and several pistils. A petaloid is a petal-like structure developed from a stamen, forming what is termed semi-double or double flowers. A, double flowering rose with all five petals intact. From this specimen the five petals and 31 petaloids have been removed progressively inward. Each petaloid shows a rudimentary anther. B, the same specimen after removing all the petals and petaloids to expose the numerous stamens. The number of petaloids depends largely on genetic factors. C, a single rose with five petals and no petaloids. [From Honeywell, *Roses* (Extension Circular 427), Purdue University.]

FIGURE 3-27. The flower of the lily has all parts and is therefore perfect and complete. When grown commercially, the anthers are removed because the pollen stains the petals. [Photograph by J. C. Allen & Son.]

example they may lack stamens (**pistillate** or "female" flowers) or pistils (**staminate** or "male" flowers). Those that contain both stamens and pistils (**perfect, bisexual,** or **hermaphroditic** flowers) may lack calyx or corolla.

Similarly, plants are referred to as **staminate, pistillate,** or **perfect** on the basis of the type of flowers they bear. When both staminate and pistillate flowers occur on the same plant, as in corn, the sex type is **monoecious.** Species in which the sexes are separated into staminate and pistillate plants are **dioecious** (date palm, papaya, spinach, asparagus, hemp). Other combinations of flower types also occur. For example, muskmelons have perfect and staminate flowers on the same plant; this sex type is referred to as **andromonoecious.**

There are many ways in which the flowers are arranged on the plant. The term **inflorescence** refers to a flower cluster. Some of the more common types of inflorescence are diagrammed in Figure 3-28.

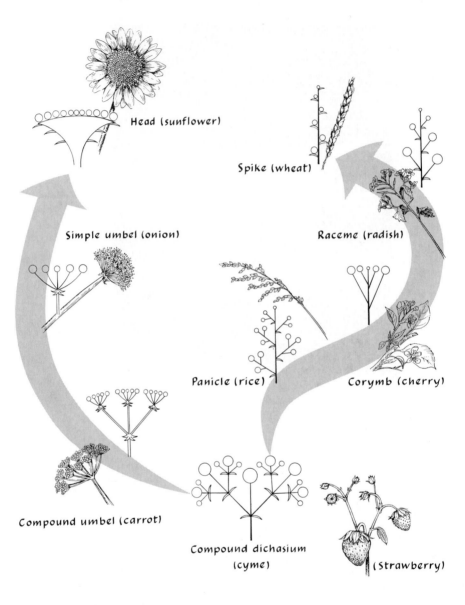

Head (sunflower)

Spike (wheat)

Simple umbel (onion)

Raceme (radish)

Panicle (rice)

Corymb (cherry)

Compound umbel (carrot)

Compound dichasium
(cyme)

(Strawberry)

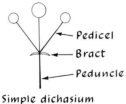

→ Pedicel

→ Bract

→ Peduncle

Simple dichasium

FIGURE 3-28. Common inflorescences.
[After Janick et al., *Plant Science*, 2nd
ed., W. H. Freeman and Company,
copyright © 1974.]

The Fruit

The botanical term "fruit" refers to the mature ovary and other flower parts associated with it. Thus, it may include the receptacle as well as the withered remnants of the petals, sepals, stamens, and stylar portions of the pistil. It also includes any seeds contained in the ovary.

The structure of the fruit is related to the structure of the flower. Fruits are classified, according to the number of ovaries incorporated in the structure, as **simple, aggregate,** or **multiple** fruits. They may also be classified by the nature and structure of the ovary wall; their ability to split apart when ripe (**dehiscence**), as well as the manner in which this occurs; and the way in which the seed is attached to the ovary.

SIMPLE FRUITS. The majority of flowering plants have fruits composed of a single ovary. These are referred to as simple fruits. In the mature fruit (when the enclosed seed is fully developed) the ovary wall may be **fleshy** (composed of large portions of living succulent parenchyma) or **dry** (made up of nonliving sclerenchyma cells with lignified or suberized walls).

The ovary wall, or **pericarp,** is composed of three distinct layers. These are, from outer to inner layer, the **exocarp, mesocarp,** and **endocarp.** When the entire pericarp of simple fruits is fleshy, the fruit is referred to as a **berry** (not the same as the horticultural term for the edible portion of some "small fruits"). The tomato, grape, and pepper are berry fruits. The muskmelon is a berry (specifically, a **pepo**) with a hard rind made up of exocarp and receptacle tissue (Fig. 3-29). Citrus fruits are also berries but of a sort called **hesperidium:** the rind is made up of exocarp and mesocarp, and the edible juicy portion is endocarp.

Simple fleshy fruits having a stony endocarp (such as peach, cherry, plum, and olive) are known as **drupes** (or **stone fruits**). The skin of these fruits is the exocarp;

FIGURE 3-29. The muskmelon is a berry fruit called a pepo. The rind is exocarp; the edible flesh is mesocarp. [Photograph by J. C. Allen & Son.]

the fleshy edible portion is the mesocarp. Simple fleshy fruits in which the inner portion of the pericarp forms a dry paper-like "core" are known as **pomes** (apple, pear, and quince, for example).

The dry, dehiscent, simple fruits include such types as pods (pea), follicles (milk weed), capsules (jimson weed), or siliques (crucifers). The dry simple fruits that do not dehisce when ripe include achenes (sunflower), caryopses (corn), samaras (maple), schizocarps (carrot), and nuts (walnut). These are diagrammed in Figure 3-30.

AGGREGATE FRUITS. Each aggregate fruit is derived from a flower having many pistils on a common receptacle. The individual fruits of the aggregate may be drupes (stony), as in blackberries, or achenes (that is, one-seeded, dry fruits attached to the receptacle at a single point), as in strawberries. In the strawberry, the fleshy edible portion is the receptacle.

MULTIPLE FRUITS. Each multiple fruit is derived from many separate but closely clustered flowers. Familiar examples of multiple fruits are the pineapple, fig, and mulberry. The beet "seed" is really a multiple fruit.

The Seed

A seed is a miniature plant in an arrested state of development. Most seeds contain a built-in food supply (the orchid seed is an exception). Structurally the seed is a matured ovule, although various parts of the ovary may be incorporated in the **seed coat.** The miniature plant, or embryo, develops from the union of gametes, or sex cells. (The details of the fertilization process will be discussed in Chapter 9, Mechanisms of Propagation.) By the time the seed is **mature,** the **embryo** is differentiated into a rudimentary shoot (**plumule**), a root (**radicle**), and one or two specialized seed leaves (**cotyledons**). A transition zone between the rudimentary root and shoot is known as the **hypocotyl.** Diagrams of various seeds are shown in Figure 3-31.

The stored food is present in seeds as carbohydrates, fats, and proteins. Seeds are thus a rich source of food as well as of fats and oils for industrial purposes. This stored food may be derived from a tissue called the **endosperm,** which is formed as a result of the fertilization process. The endosperm may produce a specialized region of the mature seed, as in corn, or it may be absorbed by the developing

FIGURE 3-30 (facing page). Various types of fruits. Simple fruits are made up of a single ovary. Aggregate fruits are made up of many pistils on a common receptacle. In the strawberry, the seed-like structures are achenes—small, dry, indehiscent, one-seeded fruits. Multiple fruits are derived from closely clustered flowers. The individual fruitlets in the pineapple are berry-like. [Adapted from Holman and Robbins, *A Textbook of General Botany*, Wiley, 1939.]

A. SIMPLE FRUITS

DRY
Dehiscent

Pod of pea

Silique of
crucifer

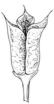

Follicle of
larkspur

Capsule of
Jimson weed

Nondehiscent

Samara of
maple

Caryopsis of
corn

Achene of
sunflower

Schizocarp
of carrot

Nut (acorn)
of oak

FLESHY

Berry of
tomato

Pepo of
squash

Hesperidium
of orange

Stone or drupe
of peach

Pome
of apple

B. AGGREGATE FRUITS

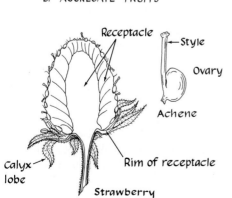

Receptacle

Style

Ovary

Achene

Calyx
lobe

Rim of receptacle

Strawberry

C. MULTIPLE FRUITS

Individual
Berry-like fruitlets

Core

Pineapple

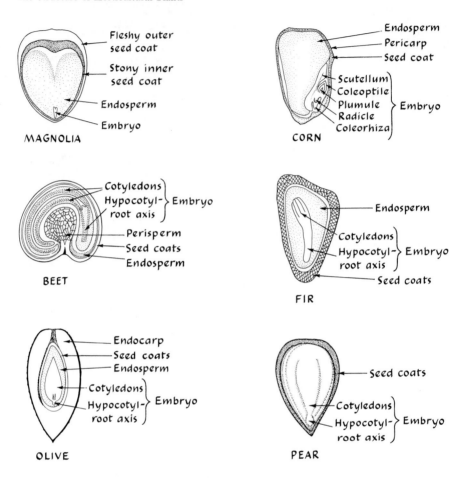

FIGURE 3-31. Structure of seeds and one-seeded fruits. [Adapted from Hartmann and Kester, *Plant Propagation: Principles and Practices*, Prentice-Hall, 1959.]

embryo. In the latter case, the cotyledons serve as food-storage organs (for example, as in beans and walnuts).

Seeds vary greatly in size and shape. Most plants can be identified by their seeds alone. In addition, great variation exists among seeds of a species. Differences include such things as the presence or absence of spines (spinach), color variation (beans), and the chemical composition of stored food (sugary versus starchy corn).

Seed germination refers to the change from the status of arrested development to active growth. The subsequent seeding stage, the interval during which the young foraging plant becomes dependent on its own food manufacturing structures, is diagrammed in Figure 3-32.

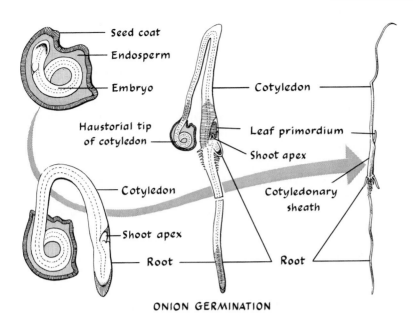

ONION GERMINATION

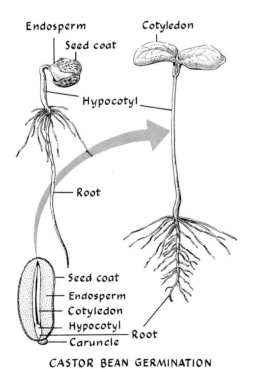

CASTOR BEAN GERMINATION

FIGURE 3-32. Seed germination and seedling morphology in onion (a monocot) and castor bean (a dicot). In some dicotyledonous plants the expansion of the hypocotyl elevates the cotyledon above ground (epigeous germination), whereas in others the hypocotyl fails to expand, and the cotyledons remain below ground (hypogeous germination). [Adapted from Foster and Gifford, *Comparative Morphology of Vascular Plants*, 2nd ed., W. H. Freeman and Company, copyright © 1974.]

Selected References

Bold, H. C., 1967. *Morphology of Plants*, 2nd ed. New York: Harper & Row. (A basic text on the subject.)

Eames, A. J., and L. H. MacDaniels, 1947. *An Introduction to Plant Anatomy*. New York: McGraw-Hill. (An elementary but authoritative text on vascular plant anatomy.)

Esau, K., 1977. *Anatomy of Seed Plants*, 2nd ed. New York: Wiley. (A comprehensive advanced treatise.)

Foster, A. S., and E. M. Gifford, Jr., 1974. *Comparative Morphology of Vascular Plants*, 2nd ed. San Francisco: W. H. Freeman and Company. (A physiogenetic approach to plant morphology.)

Heyward, H. E., 1938. *The Structure of Economic Plants*. New York: Macmillan. (This book is especially noted for its treatment of particular species of plants, including corn, onion, beet, radish, pea, celery, sweetpotato, tomato, squash, and lettuce.)

Weier, T. E., C. R. Stocking, and M. G. Barbou, 1970. *Botany: An Introduction to Plant Biology*, 4th ed. New York: Wiley. (An excellent introductory text.)

4

Plant Growth

When I consider every thing that grows
Holds in perfection but a little moment. . . .

SHAKESPEARE, Sonnet 15

THE CELLULAR BASIS OF PLANT GROWTH

The term "plant growth" refers to an irreversible increase in size. The increase in size and dry weight of an organism reflects an increase in protoplasm. This may occur through increases both in cell size and in the number of cells. The increase in the size of the cell has some limitations imposed on it by the relationship between its volume and its surface area. (The volume of a sphere increases faster than its surface area.) The process of **cell division** provides the basis for growth. Cell division is a biochemically regulated process, however, and is not necessarily directly controlled by any relationship between the volume of a cell and the area of its membrane.

The increase in protoplasm is brought about through a series of events in which water, carbon dioxide, and inorganic salts are transformed into living material. With respect to plant cells, this process involves the production of carbohydrates (**photosynthesis**), the uptake of water and nutrients, and the elaboration of complex proteins and fats from carbon fragments and inorganic compounds (**metabolism**). The required chemical energy is provided by **respiration** at night and photosynthesis by day. These physiological processes are functions of both individual cells and multicellular organisms. They are not unrelated to each other any more than the ignition system is unrelated to the compression stroke of the cylinders in the gasoline engine. Nevertheless, their classification into separate processes is useful.

Photosynthesis

Photosynthesis is the process in which carbon dioxide and water are transformed, in the presence of light, into carbon-containing, energy-rich, organic compounds. This conversion of light energy into chemical energy is the most significant of the life processes. With few exceptions all of the organic matter in living things is ultimately provided through this sequence of biochemical reactions.

Photosynthesis takes place primarily in the presence of two pigments, chlorophyll *a* and chlorophyll *b*, and, as far as we know, it takes place only in the chloroplasts of living cells. Photosynthesis is a complex series of integrated processes that can be stated in abbreviated form by the following chemical reaction:

$$\underset{\text{Water}}{12H_2O} + \underset{\substack{\text{Carbon}\\\text{dioxide}}}{6CO_2} + \underset{\substack{\text{in the presence}\\\text{of chlorophyll}}}{\overset{\text{light energy}}{}} \longrightarrow \underset{\text{Sugar}}{C_6H_{12}O_6} + \underset{\text{Oxygen}}{6O_2} + \underset{\text{Water}}{6H_2O}.$$

The series of photosynthetic reactions can be grouped into a light phase (the reactions that require light) and a dark phase (the reactions that do not require light).

The first step in this series is independent of temperature, and consists in the trapping of light energy, which accomplishes the cleavage of the water molecule into hydrogen and oxygen (**photolysis**). The oxygen is released as gaseous molecular oxygen, and the hydrogen is trapped by a hydrogen acceptor, nicotinamide adenine dinucleotide phosphate (NADP). Thus, the liberation of oxygen in photosynthesis is independent of the synthesis of carbohydrates. This step has been referred to as the **Hill reaction** (the NADP serving as the natural Hill reagent). The trapping of light energy by the conversion of adenosine diphosphate (ADP) to adenosine triphosphate (ATP) occurs in a process known as **photophosphorylation.** The combination of Hill reaction and phosphorylation is known as the **light phase** of photosynthesis.

The conversion of light energy to chemical energy is achieved by the formation of energy carriers such as ATP and $NADPH_2$. ATP has been termed "energy currency" and $NADPH_2$ as "reducing power." ATP is involved in energy transfers in many vital processes of the cell. It is formed from ADP through the addition of a third phosphate group. The captured light energy stored in the third phosphate bond becomes available following the conversion of ATP back to ADP. NADP accepts electrons and hydrogen atoms produced in photolysis and transfers these to other compounds. Indeed the crucial reactions of photosynthesis are the conversion of ADP to ATP and the reduction of NADP to $NADPH_2$.

The **dark phase** of photosynthesis (also called the **Calvin cycle**) is greatly affected by temperature and has been shown to be independent of light. Essentially the hydrogen atoms from water are transferred by the hydrogen-carrying acceptor ($NADPH_2$) to a low-energy organic acid to produce, with the help of the energy of the ATP, a "carbohydrate" of higher energy, from which sugars are formed.

This reduction reaction—that is, the addition of electrons and hydrogen atoms to carbon dioxide—results in the formation of sugar units. The precise pathways by which carbon dioxide is synthesized to sugar are now almost completely known. A substance formed early in this synthesis has been identified as a 3-carbon phosphorus-containing compound, phosphoglyceric acid, two molecules of which eventually give rise to a single 6-carbon sugar and, finally, to starch grains in the chloroplasts, as shown in Figure 4-1.

Photosynthetic efficiency (the rate of net photosynthesis) is equal to the rate of "gross" photosynthesis less any photosynthate that may be lost through respiration:

Net photosynthesis = gross photosynthesis − respiration

Recent studies have indicated that there are two distinct categories of plants in regard to net photosynthetic rate (Table 4-1). One group, the most typical, fixes carbon from ribulose diphosphate into a 3-carbon acid (phosphoglyceric acid) utilizing the enzyme ribulose-diphosphate carboxylase. Thus, this type of photosynthesis is referred to as the C_3 type. The second group of plants, which have a much higher rate of net photosynthesis, include tropical grasses, sugar cane, and species of *Atriplex* and *Amaranthus* (these two genera include many fast-growing weeds). In this group, CO_2 is first fixed into phosphoenolpyruvate (PEP) to yield 4-carbon acids (such as oxaloacetic, malic, and aspartic acid) and involves the enzyme PEP carboxylase; hence it is known as the C_4 type of photosynthesis. C_4 plants carry out both C_3 and C_4 photosynthesis, whereas C_3 plants lack the C_4 pathway.

Although the C_4 pathway requires slightly more energy than the C_3 pathway, this requirement is offset by other features. The most important advantage is the apparent absence of a type of light-dependent respiration (hence the term **photorespiration**) that is linked to the photosynthetic cycle in C_3 plants. Photorespiration lowers the apparent efficiency of CO_2 assimilation in C_3 plants. The apparent absence of photorespiration in C_4 plants occurs because the photosynthetic affinity for CO_2 is so high that photorespiratory CO_2 is recycled. The process is different from normal ("dark") respiration, which is independent of light. (See the discussion under "Respiration," p. 110.) The occurrence of photorespiration was difficult to detect because it is hard to distinguish the CO_2 fixed in photosynthesis from the CO_2 given off by photorespiration.

The difference in photorespiration between the C_3 and C_4 systems can be demonstrated by enclosing plants in a sealed, illuminated container. As photosynthesis increases, CO_2 is taken up by the plant and fixed, and the effect is to lower the amount normally present in the air. For C_3 plants, a steady state of CO_2 concentration in the air (known as the **compensation point**) is reached when there remains about 50 ppm at 77°F (25°C). This amount is as large as it is because photorespiration releases CO_2. With a C_4 plant such as corn, which lacks photo-

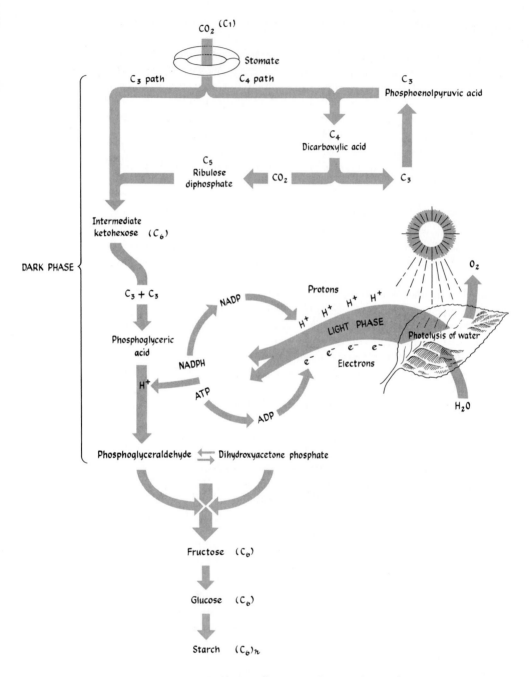

FIGURE 4-1. Photosynthesis. Note the C_3 and C_4 pathways.

TABLE 4-1. Comparative rates of maximum photosynthesis.

Type of plant	Maximum photosynthesis (milligrams CO_2 fixed per square decimeter of leaf surface per hour)
C_3 photosynthesis	
Slow-growing perennials (desert species, orchids)	1–10
Evergreen woody plants	5–15
Deciduous woody plants	15–30
Rapidly growing annuals (wheat, soybean, sugar beet, sunflower)	20–50
C_4 photosynthesis	
Tropical grasses, sugar cane, corn, *Amaranthus, Atriplex*	50–90

SOURCE: Jarvis and Jarvis, 1964, *Plant Physiology* 17:645–666.

respiration, the amount of CO_2 in such a sealed container diminishes to less than 10 ppm. A soybean plant (C_3) enclosed with a corn plant (C_4) will die because the corn exhausts the CO_2.

The rate of photorespiration increases with temperature faster than gross photosynthesis. Thus many C_3 plants are nonproductive at high temperatures (77–95°F, or 25–35°C), whereas C_4 plants such as tropical grasses increase in productivity at these higher temperatures.

Other specific adaptations are associated with the C_4 system. These include enhanced translocation of photosynthetic products, specialized leaf anatomy and chloroplast structure, and greater ratio of chlorophyll *a* to chlorophyll *b*.

Nutrient Absorption and Translocation

Although chemical analysis of plant cells indicates the presence of many different elements, only 16 have been shown to be essential to plant life. The most abundant elements, carbon, hydrogen, and oxygen, are derived largely from carbon dioxide and water. The other 13 (iron, potassium, calcium, magnesium, nitrogen, phosphorus, sulfur, manganese, boron, zinc, copper, molybdenum, and chlorine) are derived ultimately from the soil in the form of inorganic salts (Table 4-2). Plant growth is dependent on the availability of the essential nutrients. Since nutrients and water are ultimately supplied to the cell from the soil, the study of plant nutrition is largely concerned with the biology and chemistry of the soil.

TABLE 4-2. The thirteen essential elements derived
from the soil.

Essential element	Percentage in representative agricultural soils	Amount (lb/acre)
Fe	3.5	70,000
K	1.5	30,000
Ca	0.5	10,000
Mg	0.4	8000
N	0.1	2000
P	0.06	1000
S	0.05	1000
Mn	0.05	1000
Cl	0.01	200
B	0.002	40
Zn	0.001	20
Cu	0.0005	5
Mo	0.0001	2

SOURCE: Bonner and Galston, 1952, *Principles of Plant Physiology*
(W. H. Freeman and Company).

With respect to absorption, the cell can be considered as a mass of protoplasm surrounded by a differentially permeable membrane that permits passage of water and inorganic salts but restrains the passage of most large complex molecules, such as sucrose. Molecules move through a selectively permeable membrane by **diffusion.** The movement of water through such a membrane is referred to as **osmosis,** and involves diffusion as well as bulk flow caused by hydrostatic pressure differences. The osmotic movement of molecules can be demonstrated in nonliving closed systems by immersing a differentially permeable membrane that contains sugar water into a solution of pure water (or one with a lesser amount of sugar). The water moves from the solution of high solvent concentration (pure water) to the solution of low solvent concentration (sugar solution) as is illustrated in Figure 4-2. Living cells, however, are able to accumulate certain ions in a manner unaccounted for by diffusion. The cell appears to act as a metabolic pump. This process, known as active uptake, requires energy, which is supplied by respiration. The ability of molecules to move in and out of plant cells is related to the size of the molecules, their solubility in fats, and their ionic charge; membrane permeability is affected by the ionic concentration of the nutrient medium. Monovalent ions (K^+, Na^+, Cl^-) appear to increase the permeability of membranes, whereas polyvalent cations (Ca^{++} and Mg^{++}) decrease membrane permeability. Furthermore, different ions interact in their effects on membrane permeability.

Translocation may be defined as the movement of inorganic or organic solutes from one part of the plant to another. The transport of water and solutes in and out of single cells and simple multicellular plants is accomplished largely by diffusion. In higher plants, however, this conduction of solutes is carried out largely in distinct tissue systems. Physiological specialization in multicellular plants is made possible because of the rapid, large-scale transport of substances within the plant. This movement is largely a two way stream, in which water and its solutes move up from the roots through the xylem, and synthesized sugars move out of the leaves to other parts of the plant through the phloem (Fig. 4-3). There is, however, some movement of minerals in the phloem, and the xylem of woody stems functions in the upward movement of organic compounds, especially at certain seasons of the year.

The upward movement of water and solutes in the xylem of higher plants is related in part to **transpiration,** the evaporative loss of water vapor from the leaves through their numerous stomata. As water is lost by the cells, a diffusion-pressure deficit draws the water from the xylem elements, which form large numbers of continuous tubes from roots of leaves. Thus, the tension is transmitted through the entire column to the root cells and results in increased water absorption. The rate of transpiration is affected by the degree to which the stomata are opened and by such environmental factors as temperature and vapor pressure of water in the atmosphere, which affect the rate of water evaporation.

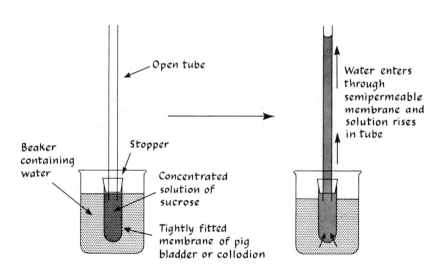

FIGURE 4-2. Diffusion in an artificial osmotic system. [Adapted from Bonner and Galston, *Principles of Plant Physiology*, W. H. Freeman and Company, copyright © 1952.]

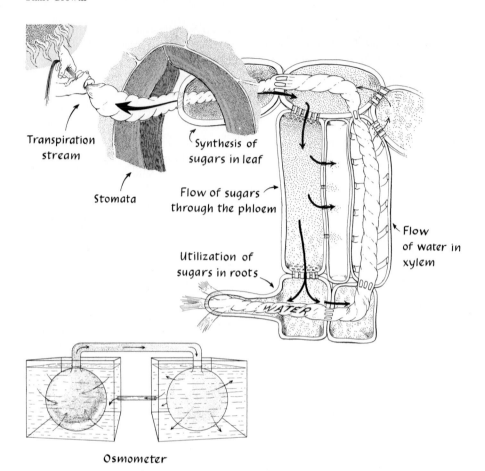

Transpiration
stream

Stomata

Synthesis of
sugars in leaf

Flow of sugars
through the phloem

Utilization of
sugars in roots

Flow
of water in
xylem

WATER

Osmometer

FIGURE 4-3. Diagram of the translocation of water and elaborated sugars in the plant. The upward movement of water through the xylem can be explained on the basis of a tension on the continuous water column in the plant. This tension, produced by the evaporation of water from the leaf (transpiration), is transmitted to the absorbing cells of the root. Sugars synthesized in the leaves move through the sieve tubes of the phloem. Phloem transport is a pressure flow brought about by a high osmotic concentration in the leaf cells and a low concentration in the receiving cells. A model of this system, called the osmometer, is shown at lower left. [Adapted from Bonner and Galston, *Principles of Plant Physiology*, W. H. Freeman and Company, copyright © 1952.]

The opening of the stomata is a mechanical process regulated by the turgidity of the guard cells (see Fig. 3-13).

The movement of sugars occurs principally in the phloem. Phloem transport appears to be accomplished by increased osmotic concentration in the leaf mesophyll cells brought about by the high concentrations of dissolved photosynthates. These sugars then move into the sieve tubes of the phloem by a process that is not clearly understood. The resultant sugar gradient appears to produce a flow, and other substances are swept along with it. The sugars are utilized in the receiving cells for respiration, growth, or storage. There is also evidence of lateral transport between xylem and phloem.

Metabolism

All of the various materials produced in the plant are ultimately derived from the carbon compounds produced by photosynthesis and from the inorganic nutrients and water absorbed from the soil (Fig. 4-4). The synthesis (anabolism) and

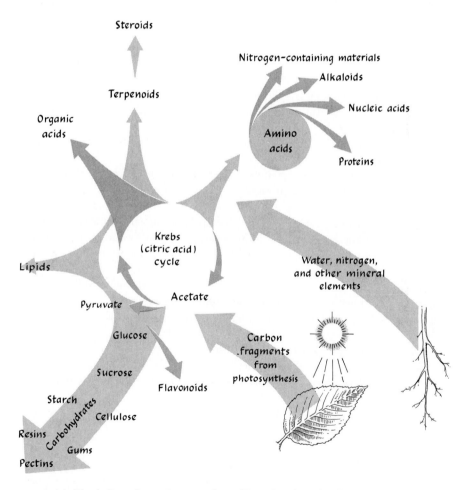

FIGURE 4-4. Metabolic pathways in green plants. [From Janick et al., *Plant Science*, 2nd ed., W. H. Freeman and Company, copyright © 1974.]

degradation (catabolism) of these organic materials is known as **metabolism.** The degradation of sugars and fats and the release of energy in respiration are examples of catabolic metabolism. The step-by-step elucidation of the pathways in plant metabolism comprises some of the most interesting chapters in the history of biochemistry.

Plants are cultivated for the complex molecules they synthesize. Carbohydrates, proteins, and lipids are the compounds of most concern because they are the major constituents of food. The metabolism of carbohydrates has already been dealt with briefly in the various discussions of photosynthesis. Proteins are second in abundance to carbohydrates as plant constituents (Table 4-3). An understanding of the relations between carbohydrate metabolism and nitrogen metabolism is necessary to a thorough comprehension of plant development.

Green plants can survive in an inorganic world. They are autotrophic for nitrogen as well as carbon. Except for the nitrogen-fixing legumes and a few other plants that depend on their bacterial partners, plants are dependent on the nitrate and ammonium ions in the soil solution as a source of nitrogen. Since microorganisms in the soil rapidly oxidize the reduced forms of nitrogen, the most common source of nitrogen taken up by plants is the nitrate ion, NO_3^-. Before combining with organic acids to form amino acids, nitrate is reduced through a series of enzymatic reactions. The reduced nitrogen groups NH_2 and NH_3 combine with the carbon frameworks formed during the oxidation of sugars to form amino acids.[1]

Amino acids join to form proteins through a complex series of reactions regulated by the nucleic acids present in the cell. The primary sites of protein synthesis in plants are the tissues in which new cells are being formed, such as tips of stems and roots, buds, cambium, and developing storage organs. The green leaf is also an important site of protein synthesis, since both the carbon frameworks and the inorganic nitrogen necessary for the formation of amino acids are readily available there. Leaf protein is in a continuous state of turnover. When leaves are placed in the dark, are excised from the plant, or reach the state of senescence, protein breaks down and soluble nitrogen compounds are formed that can be reincorporated into amino acids in other leaves or organs of the plant.

Because the nitrogen metabolism and the carbohydrate metabolism of the plant are closely linked, both processes are markedly affected by gross changes in the environment of the plant, such as heavy nitrogen fertilization, reduced light intensity, or drought. These effects are reflected in altered patterns of growth and development.

A great number of organic compounds, many of great complexity, are formed

---

[1] The ammonia (NH_3) initially combines with the organic acid α-ketoglutaric acid to form the amino acid glutamic acid, an energy-requiring process. Other amino acids are constructed via the transfer of the amino group (NH_2) from this amino acid to other organic acids from the metabolic pool (a process known as transamination).

TABLE 4-3. Nutrient constituents of certain plant parts.

Plant part	Percent dry matter	Nutrients as percentage of dry matter			
		Crude protein°	Fat	Carbohydrate	Mineral
Cabbage leaf	15.8	16.5	2.5	62.0	19.0
Beet root	16.4	9.8	0.6	82.9	6.7
Potato tuber	21.2	10.0	0.5	84.0	5.2
Wheat seed	89.5	14.7	2.1	81.0	2.1
Soybean seed	90.0	42.1	20.0	32.8	5.1

° Includes protein and nitrogen-containing compounds.

in metabolic processes. (Hundreds of different compounds contribute to the flavor of coffee alone!) Some of these metabolites are commonly referred to as essential oils, pigments, vitamins, and so forth. These terms, however, are not specific in the chemical sense, and they often refer to mixtures of materials, many of them unrelated. For example, pigments—substances that preferentially absorb certain wavelengths of visible light—differ structurally and chemically. Not all substances produced in the various metabolic pathways are, so far as is known, essential to the plant. Such substances are characteristically produced only by certain plants or plant groups. Many of them (rubber and menthol, for example) are highly prized and constitute the prime reason for the cultivation of the species from which they are derived.

The following brief sections deal with the various categories of metabolic substances other than carbohydrates, proteins, and lipids.

ORGANIC ACIDS AND ALCOHOLS. Plants manufacture a great variety of organic acids. A number participate in metabolic cycles like those involved in the biological combustion of sugar to carbon dioxide. Organic acids are found in the vascular sap and commonly accumulate in certain plant organs, particularly the fruit (citric acid in lemons, malic acid in apples).

Alcohols occur only in very low concentrations in the uncombined state, and they are generally found combined with organic acids as esters. Fats are esters of alcohol and fatty acids. The characteristic odors and flavors of fruit are due to a combination of the volatile organic acids, esters, and other compounds such as ketones and aldehydes.

AROMATIC COMPOUNDS. Aromatic compounds, in contrast to aliphatic or "straight-chain" carbon compounds, have structures containing at least one benzene ring. Many, but not all, of these compounds have very characteristic odors, hence the name "aromatic." The simple phenolic compounds (those

containing one benzene ring), such as vanillin and methyl salicylate, are responsible for the odors of vanilla and wintergreen.

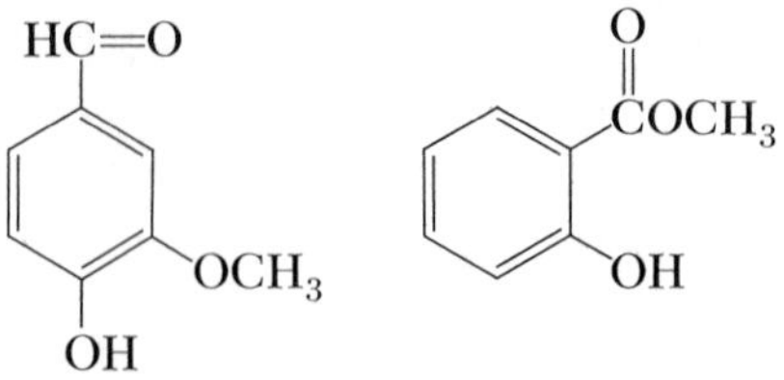

Vanillin   Methyl salicylate

The volatile phenolic compounds (specifically, the phenylpropane derivatives) are
responsible for the characteristic flavors and odors of cinnamon, clover, and
parsley.

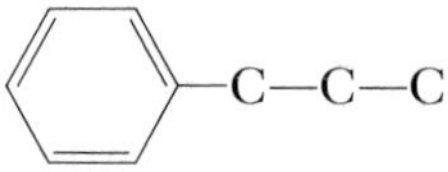

Phenylpropane skeleton

Two commercially important plant constituents, lignins and tannins, are
mixtures of complex aromatic and carbohydrate materials. Lignins (polymers of
phenolic acid) harden the cellulose cell wall into an inelastic and enduring
material resistant to microbial decomposition. Tannins have the property of
precipitating proteins, and they are used to transform animal hides into leather.
Tannins are extracted from a number of plants—the word "tan" originally
referred to the bark of the oak tree. Unripe persimmons are high in tannin, as any
person bold enough to sample one will discover: the extremely bitter, astringent
taste is due to protein precipitation on the tongue.

Flavonoids are a group of aromatic compounds characterized by two substituted benzene rings connected by a 3-carbon chain and an oxygen bridge.

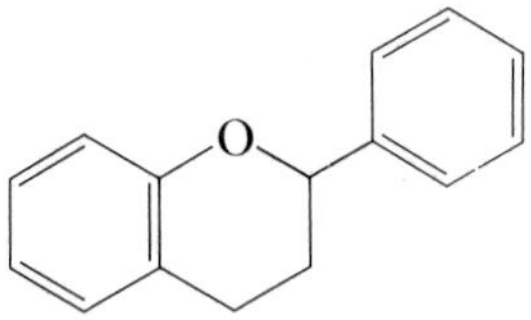

Flavonoid skeleton

The flavonoids are commonly attached to sugars to form glycosides. Among the
flavonoids are many common pigments, including the anthocyanins and flavones.
Anthocyanins are responsible for the reds and blues of many fruits and flowers;
flavones, for the yellow of lemon. The insecticide rotenone is related to the
flavonoids.

TERPENOIDS AND STEROIDS. Such compounds as essential oils, steroids, alkaloids, and various pigments are related in the chemical sense to fused units of the 5-carbon substance isoprene. Two units of isoprene may form a terpene unit, the basic structure of a number of important plant constituents called terpenoids.

Isoprene skeleton

Myrcene,
an open-chain terpene

Menthol,
a cyclic terpene

The essential oils are mixtures of volatile, highly aromatic substances that have distinctive odors and flavors. They are formed in specialized glands, ducts, or cells in various plant parts. Although called "oils," these compounds are not lipids (nor are they necessarily "essential" to the plant—in fact, the term refers to the notion of *scent*, rather than the notion of importance). An example is menthol, which is the major constituent of both peppermint oil and turpentine. The odors, flavors, and other properties of essential oils make them economically important for a variety of uses.

Resins—gummy exudates of many plants—are terpenoids consisting of three to six isoprene units. Resins are often associated with essential oils (and have the ability to harden as the oils evaporate). Resin synthesis occurs near wounded tissue, and the resins serve to retard water loss and entry of microorganisms. The ability of plants to produce resins is lowered if the general vigor of the plant is poor.

Steroids are complex cyclic terpene compounds composed of eight units of isoprene (tetraterpenoids). Although their function in plants is not well under-stood, it is known that many of them have important metabolic effects in animals. Cortisone, sex hormones, and vitamin D are steroids.

The carotenoids are complex tetraterpenes—yellow to red pigments that occur in many different kinds of tissue. Common members of this group include carotene (composed exclusively of carbon and hydrogen) and xanthophyll (com-posed of oxygen in addition to carbon and hydrogen). The most widespread carotenoid is the orange pigment β-carotene, which, when split in the digestive tract of animals, gives rise to two molecules of vitamin A.

Rubber is a high-molecular-weight terpenoid containing 3,000 to 6,000 iso-prene units. Although it is produced in many dicotyledonous plants, only a few

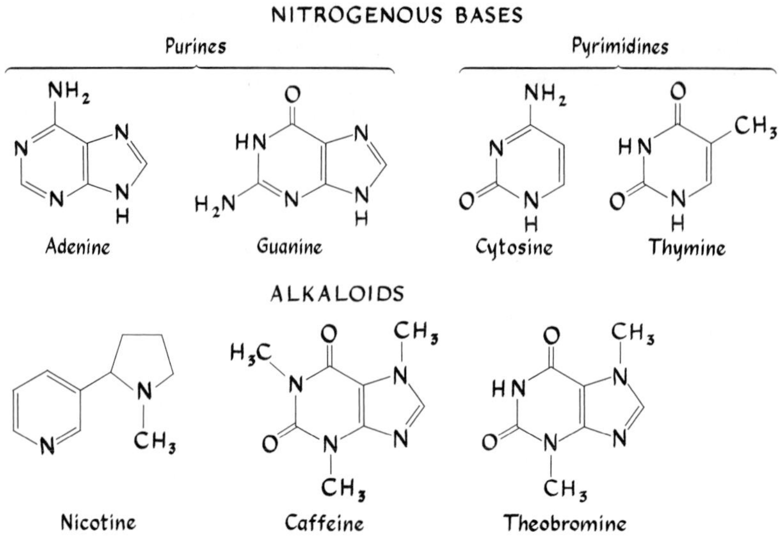

NITROGENOUS BASES

Purines

Pyrimidines

Adenine   Guanine   Cytosine   Thymine

ALKALOIDS

Nicotine   Caffeine   Theobromine

FIGURE 4-5. Cyclic nitrogen compounds.

produce enough for commercial purposes. The most important is the rubber tree, *Hevea brasiliensis*. The rubber is extracted from latex, a milky, sticky liquid exuded by glandular structures associated with the phloem.

NON-PROTEIN NITROGEN COMPOUNDS. Nitrogen is found in a wide variety of compounds in addition to amino acids and proteins. These include the nitrogen bases and a heterogeneous group of compounds called alkaloids. Heterocyclic rings containing both carbon and nitrogen are a common structural form (Fig. 4-5).

The nucleoproteins, the source of the genetic material, are composed of proteins and nucleic acids. Nucleic acids are high-molecular-weight polymers of nucleotides, each formed from three constituents: a sugar (specifically ribose or deoxyribose), phosphoric acid, and a nitrogenous base having the structure of either a purine or pyrimidine ring. In deoxyribonucleic acid, the genetic material found in chromosomes, there are two purines (adenine and guanine) and two pyrimidines (cytosine and thymine). Thus there are four nucleotides, depending on the specific base involved.

The sequence of each of the four possible nucleotides spells out a message that is decipherable by the cell. This genetic code specifies the sequence of amino acids in protein. Enzyme specificity is determined by its sequence of amino acids (as well as by its structural configuration); thus, DNA controls the destiny of the cell. The gene, in essence a functional portion of the DNA molecule, is the genetic information passed from generation to generation through the gametes.

Alkaloids, as their name indicates, are organic bases. (Although the nitrogen bases discussed above fit into the broad definition of alkaloids, they are usually treated separately.) The function of many alkaloids in plants is obscure. Nevertheless, the many effects that these substances have on human physiology render them of extraordinary pharmacological interest. A list of alkaloids derived from plant sources, along with their medicinal properties or other uses, is shown in Table 4-4. It is of interest to note that the most popular beverages—coffee, tea, maté, and cocoa—all contain stimulating alkaloids: the first three contain caffeine; the last, theobromine.

A number of extremely important pigments are complex alkaloids called porphyrins. Chlorophyll is a magnesium-containing porphyrin. The cytochromes are iron-containing porphyrins that function in energy-transfer. Hemoglobin, the oxygen carrier in animal blood, is also a porphyrin.

TABLE 4-4. Origin and uses of some plant alkaloids.

Alkaloid	Common plant source	Medicinal and other uses
Atropine	*Datura stramonium, Atropa belladonna, Duboisia* spp.	Relaxes gastro-intestinal tract, affects parasympathetic nervous system
Caffeine	*Thea sinensis, Coffea arabica, Ilex paraguayensis, Cola acuminata*	Stimulates central nervous system
Cocaine	*Erythroxylon coca*	Surface anesthetic, stimulates central nervous system
Colchicine	*Colchicum autumnale*	Induces chromosome doubling, relieves symptoms of gout
Emetine	*Cephaelis ipecacuanha*	Emetic effect, an amoebicide
Ephedrine	*Ephedra gerardiana*	Stimulates central nervous system, relieves nasal congestion
Hydrastine	*Hydrastis canadensis*	Antihemorrhagic effect
Morphine	*Papaver somniferum*	Analgesic effect
Nicotine	*Nicotiana tabacum*	Insecticide
Pelletierine	*Punica granatum*	Vermifuge (tapeworm)
Pilocarpine	*Pilocarpus microphyllus*	Diaphoretic (sweat inducing) effect
Quinine	*Cinchona* spp.	Antimalarial medicine, a cardiac depressant
Reserpine	*Rauwolfia serpentina*	Tranquilizer, sedative effect
Strychnine	*Strychnos nux-vomica*	Poison, stimulates central nervous system
Tubocurarine	*Chondodendron tomentosum*	Relaxes muscles
Theobromine	*Theobroma cacao*	Diuretic effect, a stimulant
Yohimbine	*Pausinystalia yohimba*	Aphrodisiac effect

Respiration

Energy is required to run the machinery of the cell. The energy incorporated in the chemical bonds of the sugars formed from photosynthesis cannot, of course, be harnessed by the cell from high-temperature combustion, but must be provided at low and constant temperatures in delicately controlled reactions. Respiration, the process of obtaining energy from organic material, is accomplished with great efficiency in the cell. It is, in a superficial sense, the reverse process of photosynthesis:

$$C_6H_{12}O_6 + 6O_2 \longrightarrow 6H_2O + 6CO_2 + \text{energy}.$$

The captured energy of light is released from the low-temperature oxidation (removal of hydrogen) of sugars. Although a small part is lost as heat, the useful energy is channeled into chemical work, initially as high-energy phosphates, and later in the synthesis of organic materials required in growth and development (Fig. 4-6).

The biologic combustion of sugar is accomplished through an extremely

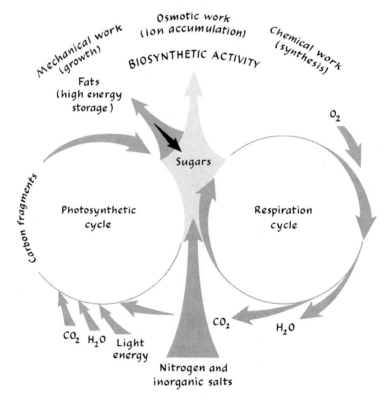

FIGURE 4-6. The respiration cycle in green plants.

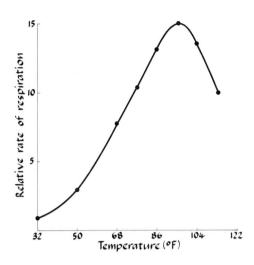

FIGURE 4-7. The relationship between temperature and the respiration rate of germinating pea seedlings. [Adapted from Bonner and Galston, *Principles of Plant Physiology*, W. H. Freeman and Company, copyright © 1952; after data of Fernandes.]

complicated series of reactions involving many specific enzyme systems and energy carriers. The steps from 6-carbon sugar to carbon dioxide involve the transformation of phosphorylated sugars to 3-carbon pyruvic acid ($CH_3COCOOH$). This step, known as **glycolysis,** is common to many organisms. Organisms that do not require oxygen (**anaerobes**) transform pyruvic acid to alcohol or lactic acid. Organisms that do require oxygen (**aerobes**) transform pyruvic acid to water and carbon dioxide. This involves the participation of a number of plant acids in the cyclic series of steps known as the citric acid cycle, or (in honor of its discoverer) **Krebs cycle.**

The rate of respiration depends on many factors. It is highest in rapidly growing tissues and lowest in dormant tissues. The rate of respiration is greatly influenced by temperature: as Figure 4-7 shows, it approximately doubles for each $18°F$ ($10°C$) rise over a range of $40–97°F$ ($4–36°C$). Other factors influence the respiration rate, including the availability of oxygen and carbohydrates and the age and condition of cells and tissues. Respiration is a common feature of all living material, and consequently the detection of evolved carbon dioxide is often utilized as a test for life.

PLANT DEVELOPMENT

The plant is more than the sum of its physiological processes. The orderly cycle of development that the whole plant undergoes involves complex patterns of change in cells, tissues, and organs. This cycle begins with seed germination and progresses through juvenility, maturity, and flowering. Upon fruiting the essential cycle of plant development is completed. In perennials the plant is ready to recycle after a period of quiescence. In annuals and biennials fruiting is a signal to the organism to enter the final phases of plant growth—senescence and death.

Differentiation

Explaining differentiation is one of the great problems of biology. There are two levels of complexity to the problem. The first concerns development of individual cells. Unicellular organisms are capable of undergoing complex transformations with no apparent internal stimulus other than their own genetic makeup. The differentiation of individual cells must involve some systematic turning on and off of the multiplicity of complex instructions potentially receivable from the genetic material. Differentiation in multicellular organisms comes about as a result of differential growth within and among cells. This is accomplished in an orderly and systematic way, with the mitotic process in cell division insuring genetic continuity of all cells. The differentiation of genetically identical cells in multicellular organisms is indeed a mystifying process. The way in which this is accomplished appears to depend on the interaction between the cell's genetically controlled processes and their external environment. In a multicellular organism the environment of one part of a cell may be quite different from that of another part. Investigations of tissue and organ differentiation have shown that many of these differences involve the interaction of substances produced by different parts of the organs. For example, to culture tomato roots artificially, not only must root meristem be present, but also thiamine and pyridoxine (familiar as vitamins in the B complex), which are normally provided by the leaves.

In higher organisms particular cells take over the control of differentiation through the action of "chemical messengers." Naturally occurring organic substances that affect growth and development in very low concentrations and whose action may be involved in sites far removed from their origin are known as **hormones** (Fig. 4-8). A great many such substances occur in plants. Minute amounts (as low as one part per billion) exert measurable physiological effects. The term **growth regulators** has been coined to include all naturally occurring and synthetically copied, or created, substances that affect growth and development. They may be either inhibitors or promoters of growth; and a single such substance may be variously an inhibitor *and* a promoter, depending upon its concentration.

Growth Regulators

From continuing research in the chemical control systems affecting differentiation in plants, a number of important groups of hormone-like substances have been defined: **auxins, gibberellins, cytokinins, abscisic acid,** and **ethylene.** These substances are so grouped on the basis of structural and physiological similarities, but their functions overlap and interact. There are also undoubtedly other groups. The basis for the chemical control of many distinct growth patterns, such as flowering, remains unknown. In addition, many natural occurring compounds known to inhibit growth and developmental processes fit no well-defined group.

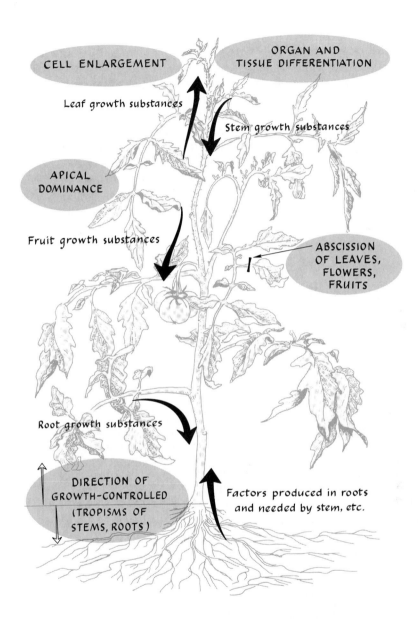

CELL ENLARGEMENT

ORGAN AND
TISSUE DIFFERENTIATION

Leaf growth substances

Stem growth substances

APICAL
DOMINANCE

Fruit growth substances

ABSCISSION
OF LEAVES,
FLOWERS,
FRUITS

Root growth substances

DIRECTION OF
GROWTH-CONTROLLED
(TROPISMS OF
STEMS, ROOTS)

Factors produced in roots
and needed by stem, etc.

FIGURE 4-8. Plant growth and development is directed by organic substances produced in various parts of the plant and translocated to others. One group of such substances, called auxins, are produced in the plant extremities. Auxins are associated with various growth functions, as indicated by the shaded portions of the diagram. [Adapted from Bonner and Galston, *Principles of Plant Physiology*, W. H. Freeman and Company, copyright © 1952; and Leopold, *Auxins and Plant Growth*, University of California Press, 1955.]

The chemical nature of some of these regulators will be discussed here, and later, in Chapter 7, we will discuss how application of them can be used to direct the growth of plants.

AUXINS. The auxins, the first class of growth regulators to be discovered, have received considerable attention from plant physiologists. Auxins are growth-promoting plant hormones. Cell elongation, the simplest example of anatomical differentiation, is directly affected by auxin concentration. Their mode of action appears to involve alterations in the plasticity of the cell wall. This fundamental property of auxins has been used in assaying their activity. The basic bioassay for auxin activity consists of measuring the elongation of dark-grown oat coleoptile sections. Similar tests assay the rate of curvature of longitudinally halved stems in response to auxin application, as in the split-pea test (Fig. 4-9). The most common natural auxin is indoleacetic acid (IAA).

CH_2COOH

Indoleacetic acid (IAA)

FIGURE 4-9. The split-pea test measures the biological activity of auxins. Sections of stems are split with a razor blade and placed in the solution to be tested. The amount of inward curvature is proportional to the auxin concentration. The petri dish on the left contains auxin; the dish on the right is the control and shows no activity. [Photograph courtesy Purdue University.]

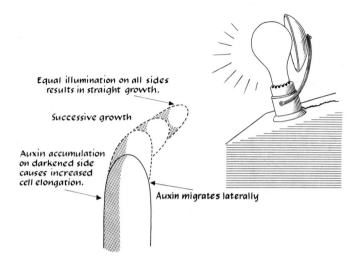

Equal illumination on all sides
results in straight growth.

Successive growth

Auxin accumulation
on darkened side
causes increased
cell elongation.

Auxin migrates laterally

FIGURE 4-10. Phototropism results from the redistribution and inhibition of auxin in the growing point by light. The subsequent accumulation of auxin on the darkened portion elongates cells on that side and bends the seedling toward the light.

Some very fundamental growth responses have been shown to be controlled by auxins. Phototropism, the bending toward light of the stem apex (growing point) can be explained as the result of differential cell elongation caused by the accumulation of auxin on the darkened side of the meristem (Fig. 4-10). The cells on that side elongate and cause the stem to bend toward the light. Similarly, auxins greatly affect growth patterns in the plant. For example, **apical dominance**—the inhibition by the growing point of the growth of axillary buds below it—appears to be a function of auxin distribution. Auxins are produced in greatest abundance in a vigorously growing stem apex, and high concentrations emanating therefrom have been shown to inhibit bud break. Removal of the auxin-producing stem tip increases lateral bud break and subsequent branching, usually directly below the cut. Thus the form of plants can be changed by the manipulation of apical dominance through pruning.

Auxins are at present the best understood of the many substances affecting plant development. They are formed in the stem and possibly in the root apices, whence they move to the rest of the plant. Their distribution, however, is not uniform. The resultant concentration of auxins in various parts of the plant has been correlated with inhibition and stimulation of growth (Fig. 4-11) as well as with differentiation of organs and tissues. Such processes as cell enlargement, leaf and organ abscission, apical dominance, and fruit set and growth have been shown to be influenced by auxins. Auxins have been associated with flowering and sex expression in some plants. Research on auxins has had a deep impact on agricul-

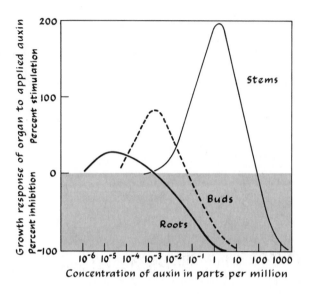

FIGURE 4-11. The effects of auxin concentration on the growth of roots, buds, and stems. [Adapted from Machlis and Torrey, *Plants in Action*, W. H. Freeman and Company, copyright © 1956; after Thimann.]

ture. They are used as herbicides, to promote rooting and fruit set, to control fruit drop, and to thin fruits with a minimum of labor.

The role of auxins has been associated with so many diverse growth systems in plants that it has at times been tempting to suggest that they must behave as "master" hormones affecting growth and differentiation. However, the importance ascribed to any particular developmental material may be influenced by the intensity of study that it receives. Growth and development appear to depend on interactions of many factors. The conclusion is now inescapable that auxins, though important substances, constitute only one class of many significant materials involved in differentiation and growth.

GIBBERELLINS. The gibberellins are a group of at least 45 closely related, naturally occurring terpenoid compounds. They were discovered through studies of excessive growth of rice occurring in response to a fungal disease. The gibberellins affect cell enlargement and cell division in subapical meristems. Their most startling effect is the stimulation of growth in many compact plant types. Minute applications transform bush beans to pole beans or dwarf maize to normal maize (Fig. 4-12). This effect is utilized as a biological assay. In addition to the dwarf-reversing response, gibberellins have a wide array of effects in many developmental processes, particularly those controlled by temperature and light (photoperiod),

FIGURE 4-12. Giberellin treatment stimulates normal growth in dwarf plants. In beans, 20 μg changes the bush to a vine habit; in corn, 60 μg causes a change from dwarf to normal growth.

Untreated   Gibberellin-   Untreated   Gibberellin-
                 treated   treated

Bush bean   Dwarf corn

including seed and plant dormancy, germination, and seed-stalk and fruit development. In brewing, barley is treated with gibberellic acid in the malting stages to increase the enzyme content of the malt. The use of gibberellins has had impres-

$$\text{Gibberellic acid (GA}_3\text{)}$$

sive agricultural applications, particularly in grape production. In Japan, seedlessness is induced in grapes, and in California the 'Thompson Seedless' cultivar ('Sultanina') shows a remarkable increase in berry size after gibberellin treatment.

CYTOKININS. The term cytokinins (or simply kinins) has been applied to a group of chemical substances that have a decisive influence in the stimulation of cell division. Many cytokinins are purines. Kinetin, the original cell-division stimulant isolated from yeast, has been identified as 6-furfurylaminopurine. The synthetic cytokinin benzyladenine, or BA (6-benzylaminopurine) stimulates shoot initiation in woody plants, including conifers.

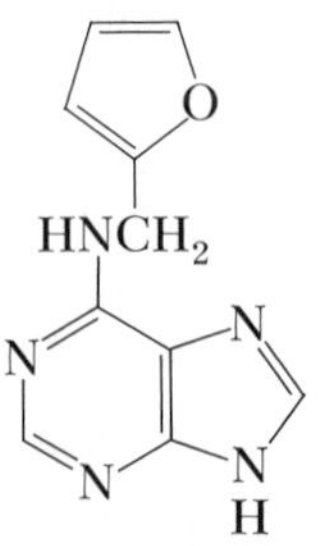

6-Furfurylaminopurine (Kinetin)

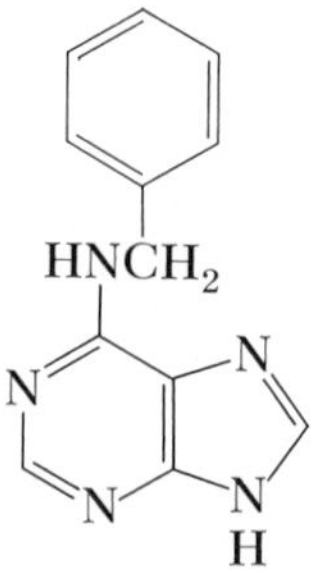

6-Benzylaminopurine (BA)

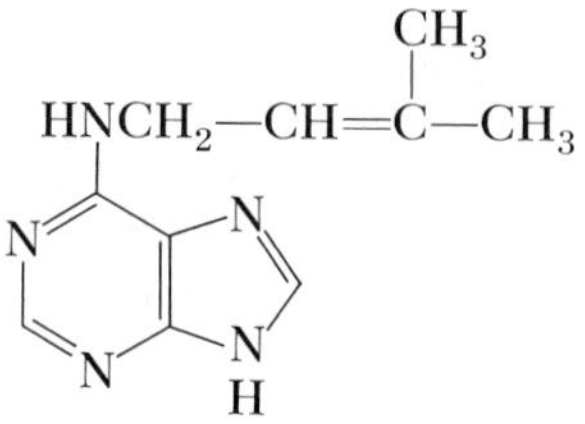

6-(γ,γ-Dimethylallylamino)purine (2iP)

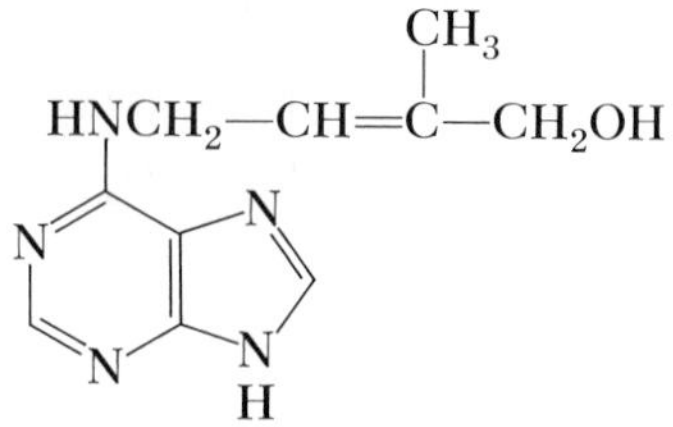

Zeatin

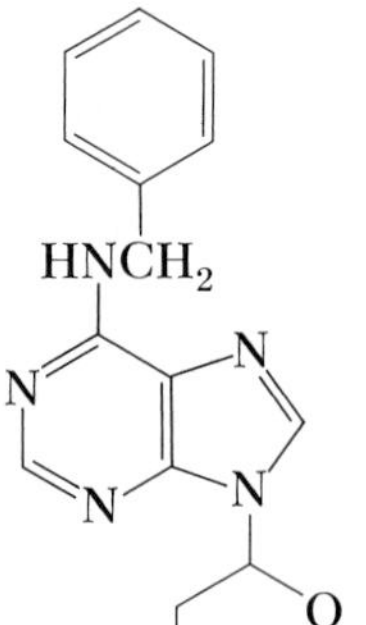

6-(Benzylamino)-9-(2-tetrahydropyranyl)-9*H*-purine (PBA)

Many cytokinins have been detected in research involving tissue culture. Vascular strands laid on top of cultured pith tissue stimulate differentiation in otherwise nondividing cells, an effect also induced by adenine and kinetin. Cytokinins and auxins interact to affect differentiation (Fig. 4-13): high auxin and low cytokinin gives rise to root development; low auxin and high cytokinin gives rise to bud development; equal amounts result in undifferentiated growth. Cyto-

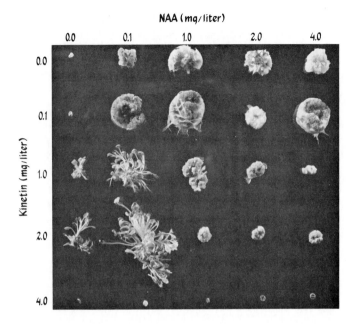

NAA (mg/liter)

Kinetin (mg/liter)

FIGURE 4-13. Effect of various levels of a cytokinin (kinetin) and an auxin (naphthaleneacetic acid, or NAA) on organogenesis in tissue culture. Leaf discs of salpiglossis were cultured for 40 days.

kinins are found in abundance in fruit and seeds (for example, corn endosperm, coconut milk), and are probably important in promoting growth and differentiation of the embryo. Tumor cells are also rich in cytokinins.

Cytokinins affect such diverse physiological mechanisms as leaf growth, light response, and aging. Recently, cytokinins and other plant hormones have been found to be active in regulating protein synthesis, possibly by turning gene transcription on and off or acting through transfer RNA to control translation of the gene product.

ETHYLENE. Although it has been known for many years that ethylene has a number of striking effects on plant growth and development, only recently has it been considered to be a regulatory hormone. Thus, it has been a standard practice to achieve ripening in banana through the introduction of ethylene, and ethylene (as well as such closely related compounds as acetylene and calcium carbide) has

$$H_2C=CH_2$$

Ethylene

been used to obtain uniform flowering in pineapple. Ethylene has enormous effects on dark-grown seedlings, causing swelling and disorientation; such effects were used as a method of assay before the days of gas chromatography. Ethylene also influences cell division; thus, tomatoes grown in high concentrations of ethylene show extensive rooting up and down the stem. It is wise not to store budwood or hardwood cuttings in places where apples are stored: apples produce ethylene as they ripen, and ethylene will cause splitting and bark peeling. In the past decade it has been clearly shown that the natural or "endogenous" ethylene produced by plants influences (1) the natural course of development in etiolated seedlings, (2) abscission, (3) floral initiation (in some plants), and (4) fruit ripening. Ethephon, or (2-chloroethyl)phosphonic acid, a material that is transformed into ethylene in the plant, has a wide array of interesting effects, including the induction of uniform ripening in pineapple and tomato, sex conversion, induction of pistillate flowers in androecious (staminate) plants of cucumbers, and the stimulation of latex flow in rubber trees.

ABSCISIC ACID. Abscisic acid (ABA), referred to at one time as "dormin" and "abscisin II," is a natural inhibitory compound that affects bud and seed dormancy and leaf abscission. Similar synthetic substances have been reported. It has a wide range of effects: in addition to promoting dormancy in buds and seeds and accelerating leaf abscission, it also promotes flowering in some short-day plants. The dormancy response may be through an effect of RNA and protein synthesis. Some effects of ABA seem to be reversed by gibberellins.

Abscisic acid

INHIBITORS. Natural as well as synthetic inhibitory substances are often placed together as a diverse class of growth regulatory materials (Fig. 4-14). Natural inhibitors help control such processes as seed germination, shoot growth, and dormancy. Several synthetic inhibitors have found important agricultural applications. Maleic hydrazide has been effective in preventing the sprouting of onions and potatoes. A number of materials that inhibit the natural formation of gibberellins by the plant act to dwarf plants. CBBP and CCC show promise for use in reducing the height of many ornamental flowering plants without unduly interfering with flowering time or flower size. A gibberellin suppressor, SADH (also called daminozide), in addition to dwarfing, affects fruit maturity, hardiness and many other plant processes.

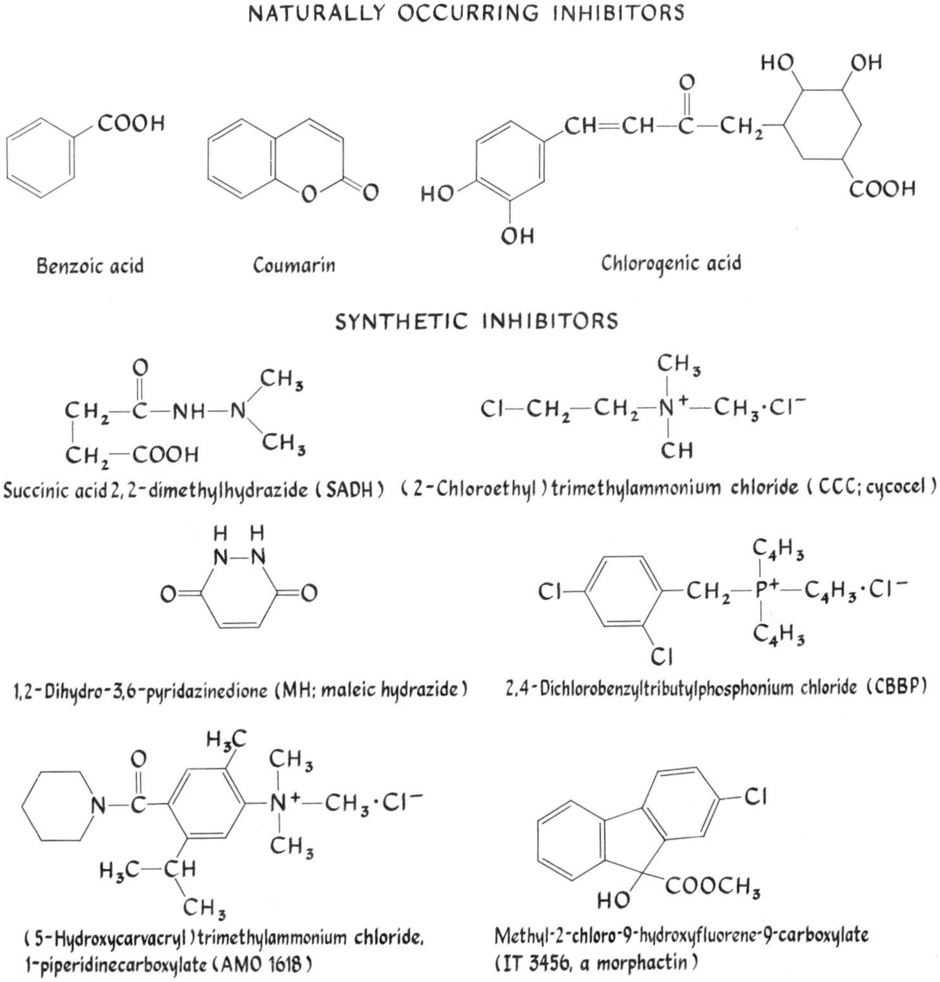
NATURALLY OCCURRING INHIBITORS
Benzoic acid
Coumarin
Chlorogenic acid
SYNTHETIC INHIBITORS
Succinic acid 2, 2-dimethylhydrazide (SADH)
(2-Chloroethyl) trimethylammonium chloride (CCC; cycocel)
1,2-Dihydro-3,6-pyridazinedione (MH; maleic hydrazide)
2,4-Dichlorobenzyltributylphosphonium chloride (CBBP)
(5-Hydroxycarvacryl) trimethylammonium chloride,
1-piperidinecarboxylate (AMO 1618)
Methyl-2-chloro-9-hydroxyfluorene-9-carboxylate
(IT 3456, a morphactin)

FIGURE 4-14. Structural formulas of some naturally occurring and synthetic inhibitors. [After Weaver, *Plant Growth Substances in Agriculture,* W. H. Freeman and Company, copyright © 1972.]

ENVIRONMENTAL FACTORS IN PLANT GROWTH

The primary environmental factors in plant growth are **soil,** which provides nutrients and moisture in addition to mechanical support; **radiant energy** in the form of heat and light; and **air,** which provides both carbon dioxide and oxygen. Soil and radiant energy vary greatly over the surface of the earth. Although the composition of air over the earth is fairly uniform above the ground, the amount of air in the soil varies greatly.

Certain areas in the temperate and tropical regions of the earth are capable of supporting luxuriant plant growth. In these favored locations the plant becomes adjusted to its environment and becomes an integral part of it. When these plants are cultivated the delicate balance of nature is often disturbed. Any factor of the plant's environment that becomes less than optimum will limit its growth.

The Soil

The plant and the **soil**—the reservoir of nutrients and moisture—are in intimate association. The soil is far more than an inorganic mass of debris—it is a biological system in a state of dynamic equilibrium. The genesis of soil from the earth's crust begins with a disintegration process whereby the parental rock becomes finely subdivided. Leaching and the subsequent action of leached materials on the original mineral substances form entirely new substances. It is the biological action of plant and microorganism, however, that transforms the subdivided minerals into the complex material known as soil.

Soil genesis is a continuing process. It can be seen from a vertical slice through shallow soil, where bedrock is just slightly beneath the soil surface. The three rather distinct gradations from bedrock to "topsoil" are referred to as horizons (Fig. 4-15). The morphology of these horizons makes it possible to classify soil into types, in order that its structure and potential fertility may be predicted.

The Physical Properties of Soil

ORGANIC MATTER. A fertile soil is literally alive. Although insects and earthworms are the most obvious of the living things that inhabit the soil, the bacteria, fungi, and other microorganisms constitute the great bulk of soil organisms (Table 4-5). The organic matter of the soil is derived not only from decomposed plant and animal tissue but from the microorganisms themselves.

The decomposition of plant and animal material is accomplished by enzymatic digestion carried out by soil microorganisms. The decomposition of simple carbohydrates (starches and sugars) is a fairly rapid process and results in the release of carbon dioxide in the soil. Water-soluble proteins are decomposed readily to amino acids and then to ammonium compounds. Ammonium compounds, under the action of certain "nitrifying" bacteria, are transformed to nitrates, in which form they are again available to plants. The decomposition of organic materials, however, is not complete. Certain substances, such as lignins, waxes, fats, and some proteinaceous materials resist decomposition, but, through complex biochemical processes, form a dark, noncrystalline, colloidal substance called **humus.** Humus has absorptive properties for nutrients and moisture that are even higher than those of clay. Yet, unlike clay, it has extremely low plasticity and cohesive properties. Thus, small amounts of humus greatly affect the structural and nutritive properties of soil.

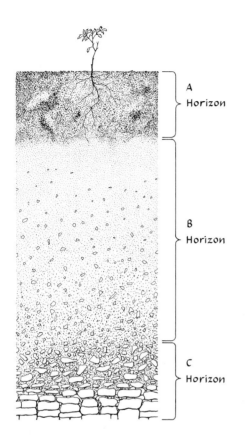

A
Horizon

B
Horizon

C
Horizon

FIGURE 4-15. A soil profile. [From Bonner and Galston, *Principles of Plant Physiology*, W. H. Freeman and Company, copyright © 1952; after Lyon and Buckman.]

TABLE 4-5. The weight of soil organisms per acre-foot of fertile agricultural land.

Organism	Weight (lb/acre-ft)	
	Low fertility	High fertility
Bacteria	500	1000
Fungi	1500	2000
Actinomycetes	800	1500
Protozoa	200	400
Algae	200	300
Nematodes	25	50
Other worms and insects	800	1000
Total weights	4025	6250

SOURCE: Data from Allen, 1957, *Experiments in Soil Bacteriology* (Burgess).

SOIL TYPES. There are two basic types of soil—mineral and organic. **Mineral soils** are composed of inorganic substances and varying amounts of decaying organic matter (from a trace to 20%). **Organic soils** (for example, muck and peat) are formed from partly decayed plant materials under marshy or swampy conditions. When such soils contain more than 65% organic matter, they are referred to as peat; those containing 20–65% organic matter are called muck. Organic soils are dark brown to nearly black in color. These soils cannot be cultivated unless they are drained and their fertility problems are corrected. Properly managed organic soils are highly productive (Fig. 4-16). These soils are porous and well aerated, and they have a high water-absorption capacity.

The mineral substances of the soil consist of particles of different size; in decreasing order, these are stone, gravel, sand, silt, and clay. The proportion of these substances determines the soil **texture** (Fig. 4-17). Such names as *clay loam* and *silty clay* are textural classifications of soils.

Although the physical properties of the coarser materials do not differ greatly from those of the rocks from which they are derived, materials composed of particles of submicroscopic size, known as **clays,** show distinct physical properties. The clays, the most chemically and physically active portion of the soil, are of colloidal size and are crystalline in structure. They are formed from the parent

FIGURE 4-16. Properly managed muck soils are among our most productive agricultural lands. Almost 1800 bushels of hand-crated onions per acre is an impressive sight seldom seen in the age of automatic harvesters. [Photograph by J. C. Allen & Son.]

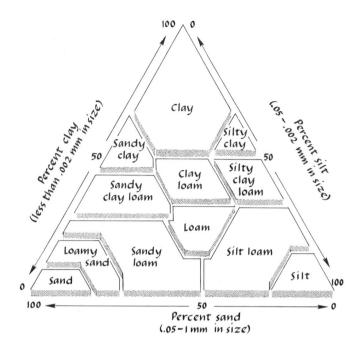

FIGURE 4-17. Soil texture triangle. To find the textural name of a soil, locate the points corresponding to the percentage of clay and silt on each side of the triangle. The silt line is projected inward parallel to the clay side; the clay line is projected inward parallel to the sand side. The lines will meet in the class name of the soil. [Adapted, with permission of the publisher, from Lyon, Buckman, and Brady, *The Nature and Properties of Soils*, 5th ed., copyright 1952 by the Macmillan Company.]

minerals by a crystallization process and are not merely finely subdivided rock. The clay particles are made up of "flakes," or sheet-like units, held together by O—H linkages or by ions between the flakes. Their significant structural characteristic is their tremendous surface area relative to their volume. The clay particle carries a negative charge. Thus, clay particles are electrically active, attracting positively charged ions (H^+, Na^{++}, K^+, Ca^{++}, Mg^{++}, and others). The adsorbed water on clays acts both as a lubricant and as a binding force. When wet, clay platelets behave like a stack of wet poker chips. This, to a large degree, explains the plasticity of clay. Wet clay soils low in organic matter and low in weakly hydrated cations, such as calcium, become **sticky** or **puddled.**

SOIL STRUCTURE. The "structure" of soil refers to the arrangement of the soil particles into aggregates. The factors determining good structure are the size and arrangement into granules of the soil particles. Soil granules are masses of mineral particles of various sizes interspersed with organic material or some cementing compound. **Granulation** is largely brought about by the aggregation of soil particles by exudates of microorganisms. In addition, such environmental factors as freezing and thawing or wetting and drying help to break up larger aggregates into granular size. The granulation of soils is particularly important to proper plant growth because of the effect on interstitial spaces, known as **pores.**

The pore space of soil is occupied by water and air in varying proportions, the soil acting as a huge sponge. The total pore space of soil, which is usually about 50% of the total volume, is not as important as the characteristic size of the pore spaces. Clay soils have more total pore space than sandy soils, yet their small size allows only slow movement of air and water. When the small pores of clay soils become filled with water, the lack of aeration so essential to root growth becomes limiting. When filled with water, the larger pore spaces of soils will soon drain out by gravity, whereas the smaller pore spaces will hold the water by capillary action. This **capillary water** is of the utmost importance to the plant: it is the soil solution upon which the plant depends for water and nutrients.

The main factors contributed by good soil structure are proper aeration and water-holding capacity. The crumbly nature of good agricultural soils depends directly upon soil texture and the percentage of organic matter. Under field conditions, only organic matter is amenable to variation. In potting soils, both the texture and the percentage of organic matter may be modified.

The Chemical Properties of Soil

CATION EXCHANGE. In plant nutrition, the most significant feature of the colloidal particles, clay and humus, is their capacity for attracting cations and the subsequent exchange of one ion for another—a process known as **cation exchange.** Thus, nutrients that otherwise would be lost by leaching are held in reserve by the clay particles. When exchanged, these ions become available to the plant.

The process of base exchange is not a random process. The cations differ in their replacement process such that, if present in equal amounts,

$$H^+ \text{ replaces } Ca^{++} \text{ replaces } Mg^{++} \text{ replaces } K^+ \text{ replaces } Na^+.$$

The addition of large amounts of one cation may cause it to replace another by "sheer force of number" (**mass action**). This is largely what occurs with the addition of inorganic fertilizer.

Hydrogen ions are made continually available by the dissociation of carbonic acid formed from the dissolved CO_2 released by living roots in respiration and from the biological decay of carbohydrates.

$$CO_2 + H_2O \rightleftharpoons H_2CO_3 \rightleftharpoons H^+ + HCO_3^-.$$

The steady release of H^+ tends to promote the exchange of cations, making them available for plant growth. The cations are replenished by the decomposition of rocks, the degradation of organic materials, and the application of fertilizers. The

TABLE 4-6. Cation-exchange capacity ranges for various soil types.

Soil type	Cation-exchange capacity (milliequivalents/100g)
Sands	2–4
Sandy loams	2–17
Loams	7–16
Silt loams	9–30
Clay and clay loams	4–60
Organic soils	50–300

SOURCE: Adapted from Lyon, Buckman, and Brady, 1952, *Nature and Properties of Soils* (Macmillan).

cations in a productive soil exhibit an equilibrium between the soil particle, the soil solution, and the plant.

The **cation-exchange capacity** of a soil is expressed in milliequivalents (meq) per 100 g and it is equivalent to the milligrams of H^+ that will combine with 100 g of dry soil. The exchange capacity of soil differs with the percentage of humus and with the percentage and composition of clay. The clays differ markedly in their ability to exchange cations. Montmorillonite clays have an exchange capacity of about 100 meq/100 g, whereas the kaolinite clays have a low exchange capacity of about 10 meq/100 g. In contrast, the exchange capacity of humus ranges from 150–300 meq/100 g. The ranges of exchange capacity for various soils are presented in Table 4-6.

SOIL REACTION. **Soil reaction** refers to the acidity or alkalinity of the soil. It is expressed in terms of pH, the logarithm of the reciprocal of the hydrogen ion concentration, and is usually expressed in units from 0 to 14:

$$pH = \log \frac{1}{[H^+]}.$$

Table 4-7 gives the concentration in moles of H^+ and OH^- for pH values of 0 to 14. Note that the molar concentration of H^+ × the molar concentration of OH^- equals a constant of 10^{-14}. The pH of the soil is regulated by the extent of the colloid fraction charged with hydrogen ions. For example, a clay particle charged with abundant H^+ acts as a weak acid and imparts an acid reaction or low pH. Similarly, a clay particle charged with mineral cations imparts an alkaline reaction, or high pH (Fig. 4-18).

The proper soil pH (6–7) is vitally important in plant growth. Abnormally high soil pH (above 9) or low pH (below 4) are, in themselves, toxic to plant roots.

TABLE 4-7. The concentration of H^+ and OH^- with varying pH.

pH	Soil reaction			H^+ Concentration (moles/liter)°	OH^- Concentration (moles/liter)†	Reaction of common substances
0				10^{-0}	10^{-14}	
1				10^{-1}	10^{-13}	
2				10^{-2}	10^{-12}	
3		very strong		10^{-3}	10^{-11}	lemon juice
4	Acidity	strong		10^{-4}	10^{-10}	orange juice
5		moderate		10^{-5}	10^{-9}	
6		slight	soil range	10^{-6}	10^{-8}	milk
7		neutral		10^{-7}	10^{-7}	pure water
8		slight		10^{-8}	10^{-6}	sea water
9	Alkalinity	moderate		10^{-9}	10^{-5}	soap solution
10		strong		10^{-10}	10^{-4}	
11		very strong		10^{-11}	10^{-3}	
12				10^{-12}	10^{-2}	
13				10^{-13}	10^{-1}	
14				10^{-14}	10^{-0}	

° 1 mole of H = 1 g   † 1 mole of OH = 17 g

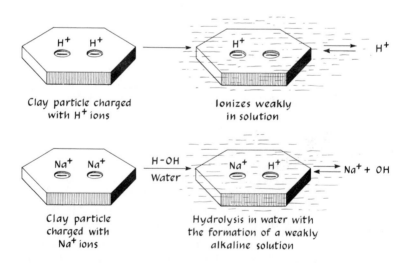

Clay particle charged with H^+ ions

Ionizes weakly in solution

Clay particle charged with Na^+ ions

Hydrolysis in water with the formation of a weakly alkaline solution

FIGURE 4-18. The soil reaction depends on whether the clay particles are charged with hydrogen ions or mineral cations. [From Bonner and Galston, *Principles of Plant Physiology*, W. H. Freeman and Company, copyright © 1952.]

Within this range the pH determines the behavior of certain nutrients, precipitating them or making them unavailable (Fig. 4-19). For example, the chlorosis found in some plants grown in soil with a high pH is a result of iron deficiency caused by the precipitation of iron compounds. Soil organisms, especially bacteria, are also affected by pH. Vigorous nitrification and nitrogen fixation require a pH above 5.5.

Soil Fertility

The fertility of the soil is only indirectly related to the chemical composition of the primary inorganic minerals. The most important factor in soil fertility is the level of the forms of the nutrients available to the plant. Such levels are related to many factors, among which are the solubility of the nutrients, the soil pH, the cation-exchange capacity of the soil, the soil texture, and the amount of organic matter present.

Nitrogen is the nutrient most limiting to plant growth. The main available forms of nitrogen in the soil are nitrate (NO_3^-) and ammonium (NH_4^+) ions. The nitrite ion (NO_2^-) can be utilized by the plant, but it tends to be unstable and toxic in high amounts. The transformation of nitrogen-containing compounds to available forms is referred to as the **nitrogen cycle.** This circuitous route of nitrogen from element to protein and back is largely biological, as shown in Figure 4-20.

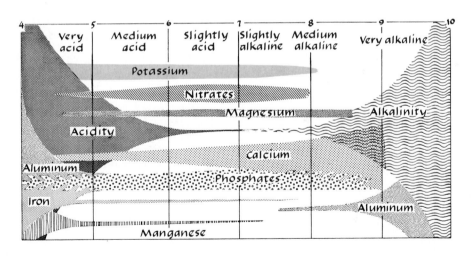

FIGURE 4-19. The relation between soil reaction and the availability of plant nutrients to crops. The thickness of the various bands indicates the relative availability of the nutrients they represent. [Courtesy of Virginia Polytechnic Institute and State University.]

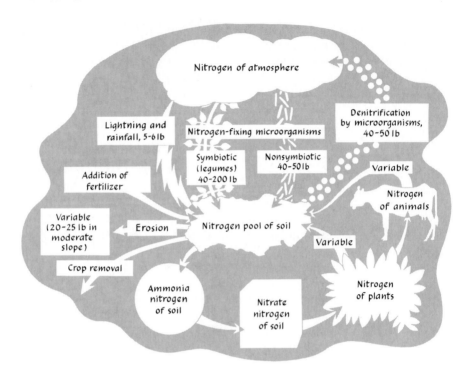

FIGURE 4-20. The nitrogen cycle. Nitrogen removal by crops must be compensated for by nitrogen addition. [Adapted from Bonner and Galston, *Principles of Plant Physiology*, W. H. Freeman and Company, copyright © 1952.]

Nitrogen added		Nitrogen removed	
Method	lb/acre	Method	lb/acre
Plant and animal residues	Variable	Crop harvested	Variable
Nitrogen fixation		Leaching	
Symbiotic	40–200	Crop rotation	5–10
Nonsymbiotic	40–50	Bare soil	60–70
Lightning and rainfall	5–6	Erosion (moderate slope)	20–25
		Denitrification	40–50

The transformation of atmospheric nitrogen into forms available to plants, **nitrogen fixation,** is accomplished by certain species of bacteria (Table 4-8). The most efficient of these bacteria are symbiotic; that is, they convert atmospheric nitrogen to combined forms only in association with the roots of legumes. The breakdown of the complex proteins of organic material into amino acids is also accomplished largely by bacterial action. But the nitrogen from this process is only available after the death and disintegration of the bacteria involved in this

TABLE 4-8. Nitrogen-fixing bacteria.

Type	Genus	Requirements
Symbiotic	*Rhizobium*	Carbohydrates, inhibited by nitrates and ammonium
Nonsymbiotic		
Anaerobic	*Clostridium*	Carbohydrates, inhibited by nitrates
Aerobic	*Azotobacter*	Calcium, traces of molybdenum

decaying process. Soil microorganisms have the first call on nutrients. This is especially true for material with a carbon to nitrogen ratio by weight greater than 10:1. The breakdown of amino acids to forms of nitrogen available to plants takes place by transformations referred to as **ammonification** and **nitrification.**

Ammonification

$$\text{Amino acids from degradation of protein} \xrightarrow[\text{Many bacteria}]{} NH_4^+ \text{ (ammonium ion)}$$

Nitrification

$$NH_4^+ \xrightarrow[\substack{\text{Nitrosococcus} \\ \text{Nitrosomonas}}]{} NO_2^- \text{ (nitrite ion)} \xrightarrow[\text{Nitrobacter}]{} NO_3^- \text{ (nitrate ion)}$$

The bacteria involved in nitrification are autotrophic and aerobic; that is, they do not require organic nutrition, but they do require oxygen. Thus, they are greatly affected by soil aeration, temperature, and moisture.

The removal of nitrogen from the soil is partly biological. In addition to its removal by plants (which is permanent when a crop is harvested), certain bacteria convert nitrates back to atmospheric nitrogen. This process of denitrification is an anaerobic process. Thus, loss of proper aeration results in the loss of available nitrogen. Furthermore, nitrates are readily soluble in water and, if they are not utilized by microorganisms or higher plants, they may be lost by leaching. In summary, the level of available nitrogen is dependent upon the content of organic matter and the microbiological activity of the soil. Consequently the amount of nitrogen available is related to cropping practice. The available soil nitrogen is, of course, greatly affected by the application of fertilizer. Quickly available forms of inorganic nitrogen probably account for most of the nitrogen in intensively cropped soils today.

Phosphorus, unlike nitrogen, is relatively stable in the soil (Fig. 4-21). Phosphorus is "tied up," or fixed, in compounds containing calcium, magnesium, iron, or aluminum. The availability of phosphorus to the plant is low and is primarily related to pH. At very low pH (2–5), applied phosphorus is precipitated out of the

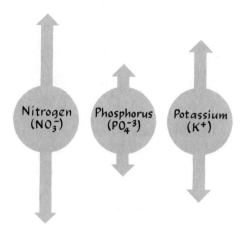

FIGURE 4-21. The relative mobility of nitrogen, phosphorus, and potassium in the soil. The high movement of nitrogen is due to the complete solubility of nitrates in the soil solution. The movement of phosphorus is regulated by the low solubility of the phosphorus compounds that are formed in the soil. Although the potassium compounds in the soil are soluble, the movement of potassium is controlled by its exchange properties with the colloidal fraction.

soil solution as complex aluminum and iron compounds. At high pH (7–10) phosphorus becomes fixed in complex calcium compounds. At pH 5–7 it is in the form of mono- or dicalcium phosphates and is most available for plant use. The concentration of phosphorus in the soil solution is low at best. In fertile agricultural soils only 0.5–1 part per million of phosphorus is in solution, as compared with 25 parts per million of nitrogen. However, because the mobility of phosphorus in the soil is low, leaching is slight.

Potassium is available largely as exchangeable ions on the soil colloid. Although potassium is plentiful in mineral soils, the low solubility of the primary potassium-rich minerals results in low availability from this source. There is, however, a continual renewal by the primary mineral of exchangeable ions. Potassium tends to be low in organic soil. The leaching of potassium varies greatly, depending upon the type of clay and the amount of organic matter in the soil.

Calcium is seldom deficient as a nutrient. However, its many effects on soil microbial activity, pH, and the subsequent absorption of other ions make it a common soil amendment. It is present in the soil both in water-soluble form as an exchangeable cation and in combination with organic compounds.

Magnesium, like calcium, is absorbed as an ion. It occurs in the soil solution in soluble form and as an exchangeable cation. Like calcium, it is sometimes deficient in acid, sandy soils in humid regions.

Sulfur is not present in large amounts in the soil. It is continually leached, but there appears to be a continuous turnover in the soil. It is added by rainfall near industrial regions, where rain absorbs sulfur dioxide from the air. The chief source, however, lies in organic material; thus, deficiencies occur in soils that are either low in organic matter or distant from industrial areas. Actual sulfur deficiencies in horticultural crops are rare under present practices. Sulfur is commonly added to soils through the application of such compounds as superphosphate.

Manganese is available in the soil in ionic form. However, in alkaline soils high in organic matter and under aerobic conditions, manganese is oxidized from the manganous to the manganic form ($MnO \longrightarrow MnO_2$; that is, from Mn^{++} to Mn^{++++}), which renders it unavailable. On the other hand, soil acidity, low content of organic matter, and anaerobic conditions may result in manganese toxicity.

Boron, zinc, copper, molybdenum, iron, and **chlorine** are definitely trace elements, since they are required by the plant only in minute amounts. In areas under intensive production, deficiencies of these nutrients are neither common nor extensive.

Moisture

Water is a constituent of all cells, the amount varying with the tissue in question. It may be as low as 3% in shelled peanut seed, about 40% in dormant wood, and up to 95% in such succulent fruits as the watermelon. Water is the solvent system of the cell and it provides a medium for transfer within the plant. It maintains the turgor necessary for the intricacies of transpiration and plant growth. In addition, water itself is required as a nutrient for the production of new compounds. One third of the weight of carbohydrates and proteins is derived from chemically combined water.

The water in a plant is in a continual state of flux. A net loss of water causes growth to stop, and a continued water deficiency causes irreversible alterations of the plant that result in death. This may occur quite rapidly under hot, dry conditions in plants that are not structurally adapted to prevent water loss.

The high percentage of water in plants and its capacity as a nutrient carrier and solvent do not alone explain the high rate of water utilization by plants. The water requirements of plants, expressed as the number of units of water absorbed per unit of dry matter produced, varies from about 50 in conifers to 2500 in leafy vegetables! Most crop plants range from 300 to 1000. While growing, the plant continuously absorbs water from the soil and gives it off in transpiration. This loss of water is a byproduct of carbon fixation. Carbon dioxide, which provides the carbon necessary for growth, enters the plant through the water films surrounding the spongy mesophyll of the leaf. As this film evaporates, it is replenished from the tissues of the plant, which in turn draw water from the vascular system.

The transpirational loss of water by the plant can be considered as an exchange of water for carbon, and in this sense transpiration is necessary for plant growth. Rapidly growing plants thus require great amounts of water, greatly in excess of the amount found in the plant itself. The rate of water loss depends largely on the temperature, the relative humidity, and the amount of air movement. Radiation from the sun provides the energy required to change the state of the water from film to vapor. This "boiling off" of water is responsible for the dissipation of a large part of the total energy received by the plant from the sun.

Soil Moisture

The amount of soil moisture that is of benefit to plants has definite limits. Too much water may be as troublesome as not enough. Excess water is not toxic; rather, it is the lack of aeration in waterlogged soils that causes damage. Plants can be grown satisfactorily in water solutions when aeration is provided (Fig. 4-22).

The amount of water in a soil may be expressed in a number of ways. The expression of soil moisture in inches of water per foot of soil is useful for some purposes (1 acre-inch is equivalent to approximately 27,000 gallons). Expressing soil moisture in terms of the **field capacity** of a soil takes into account the physical condition of the soil, and thus has greater agricultural significance. The field capacity of a soil is the maximum amount of moisture that is retained after the surface water is drained and after the water that passes out of the soil by gravity (free water) is removed.

The water content of a soil can also be expressed in terms of the availability of water to the plants that grow in it. The moisture content at which irreversible wilting occurs is known as the **permanent wilting point.** The percentage of water present depends upon the soil, but it is relatively independent of the test plant.

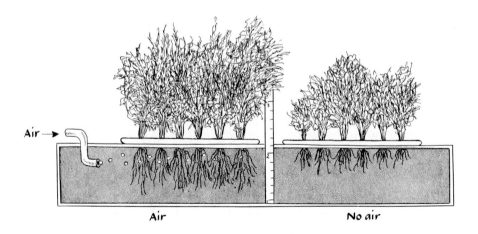

Air ➝

Air   No air

FIGURE 4-22. Effect of aeration on asparagus plants grown in a nutrient solution containing all the essential elements. [Adapted from Hoagland and Arnon, California Agricultural Experiment Station Circular 347, 1950.]

The moisture left in the soil but unavailable to the plant is known as **hygroscopic water** and **chemically combined water.** The hygroscopic water is held tenaciously by the soil in "atomically" thin films. The amount of hygroscopic water varies with the surface area of the soil particles, and is therefore highest in clay and organic soils.

The total amount of soil moisture present is not as important as its availability. The **available moisture** is the difference between the permanent wilting point and the field capacity. This water is often referred to as **capillary water.** It is retained in the smaller soil pores, where the capillary forces prevent water drainage, and as films around the soil particles (Fig. 4-23). Soils differ in their ability to hold moisture, and this ability depends upon their texture (Table 4-9). Although sandy soils afford better drainage and aeration, they have a lower water-holding capacity than clay soils. The total amount of capillary water can be increased in a sandy soil by increasing its content of organic matter. The total amount available to a crop will depend on many factors, among which are the type and depth of soil, the depth of rooting of the crop, the rate of water loss by evaporation and transpiration, the temperature, and the rate at which supplemental water is added. In addition, the amount of available water itself is a factor: the smaller the amount of water in a soil, the greater is the tenacity with which the water is held.

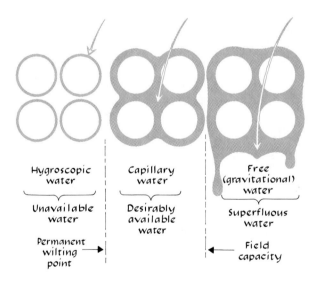

FIGURE 4-23. The classes of soil moisture. All the capillary water is not equally available to plants. As the capillary water is depleted, the tension by which this water is held in the soil increases from 1 atmosphere of pressure at field capacity to about 15 atmospheres at the permanent wilting point. The amount of capillary water present increases with the fineness of the soil pore space. [Adapted from Bonner and Galston, *Principles of Plant Physiology,* W. H. Freeman and Company, copyright © 1952.]

TABLE 4-9. General range of available-moisture holding capacities for normal soil conditions.

Soil texture	Available moisture	
	Range (in/ft)	Average (in/ft)
Very coarse-textured sands	0.4–0.7	0.5
Coarse-textured sands, fine sands, and loamy sands	0.7–1.0	0.8
Moderately coarse-textured sandy loams and fine sandy loams	1.0–1.5	1.2
Medium-textured very fine sandy loams, loam, sandy clay loams, and silt loams	1.5–2.3	1.9
Moderately fine-textured clay loams and silty clay loams	1.7–2.5	2.1
Fine-textured sandy clays, silty clays, and clay	1.6–2.5	2.0

SOURCE: Shockley, 1956, in *Sprinkler Irrigation Manual* (Wright Rain, Ringwood, England).

This tenacity is measured in atmospheres of pressure required to drive off the water. At field capacity, water is held with a force of about 15 atmospheres.

Because the rate of water extraction of soil is a function of root concentration, it decreases with the depth of the root zone. About 40% of the total water is extracted from the upper quarter of the root zone, 30% from the second quarter, 20% from the third quarter, and 10% from the bottom quarter. Under maximum transpiration, sufficient water for maximum growth cannot be obtained when the upper quarter of the root zone is depleted.

Water Movement

The movement of water through a soil is related to the amount of water present. Water applied to a soil moves through the soil only as fast as field capacity is attained (Fig. 4-24). The rate of water movement depends to a great extent on soil texture. Because the pore size is smaller in heavy clay soils, water moves much slower through these than through loamy or sandy soils.

The upward movement of water through a soil is caused by capillary action. Since this is a surface-tension phenomenon, the height the water will reach is inversely related to the diameter of the openings in the soil. Thus, the finer the soil spaces, the greater the distance of capillary movement. The upward rise of capillary water from the water table (the depth at which all the soil is at field capacity) is a factor in the replenishment of water lost to the plant and evaporated from the soil. This evaporative loss of water is restricted to the upper portion of the soil, since it takes more and more pressure to pull the water, depending on the height of the water column. During a period of extended drought, it becomes easy to recognize the shallow-rooted plants.

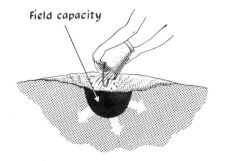

Field capacity

FIGURE 4-24. The rate of movement of water added to the soil is related to the speed at which field capacity is attained.

Radiant Energy

The sun is the primary source of energy available to the earth and its atmosphere. This energy is transferred across 93 million miles of space in the form of radiation. Solar radiation reaches the earth in the form of electromagnetic waves travelling at the speed of 186,000 miles/sec. This energy is described in terms of its wavelength, just as sound is described in terms of its pitch. The radiant energy occurring as visible light is but a small fraction of the frequency range of the electromagnetic spectrum (Fig. 4-25).

The amount and quality of solar energy that any portion of the earth's surface receives is dependent upon its duration and intensity. The seasonal difference in duration of the intercepted radiation is a consequence of the variation in day length and cloud cover. The intensity of the intercepted radiation is related to the angle at which the solar rays penetrate the earth's atmosphere. The water vapor in the atmosphere (and, to a smaller extent, the air and dust) diffuses, reflects, and absorbs this radiant energy. The damaging ultraviolet radiation at the short-wavelength end is absorbed by the ozone layer. Because the earth is spherical, the rays of the sun falling on the poles are oblique as compared to those falling on the equator. These rays are spread over a larger surface of the earth and pass through a thicker layer of atmosphere. When the sun is directly over the equator its rays must penetrate an air mass at the poles equivalent to 45 air masses at the equator.

Absorption of Solar Radiation

The quantity and quality of radiation depend on the temperature of the radiating body. The higher the temperature, the greater the rate of radiation and the richer the proportion of short-wavelength (high-frequency) radiation. Thus, the high-temperature solar radiation consists mostly of short-wave radiation in the visible or near-visible portions of the spectrum. This short-wave solar energy is absorbed at the earth's surface, where it is transformed into heat. The earth then becomes a radiating body at a lower temperature (average, 57°F, or 14°C). The earth's radiation is in the form of long waves (low-frequency radiation).

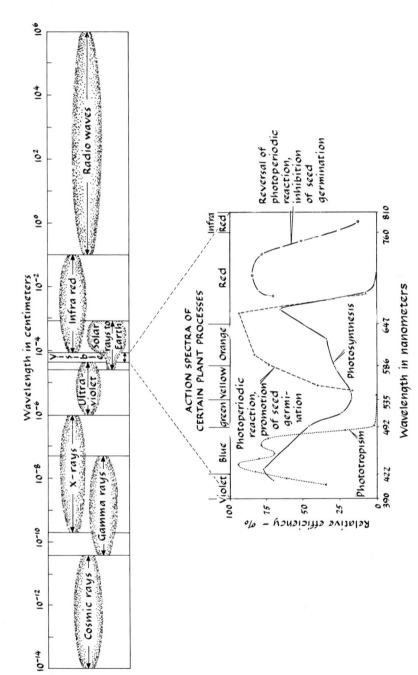

FIGURE 4-25. The electromagnetic spectrum and action spectra of certain plant processes. [Adapted from Machlis and Torrey, *Plants in Action*, W. H. Freeman and Company, copyright © 1959.]

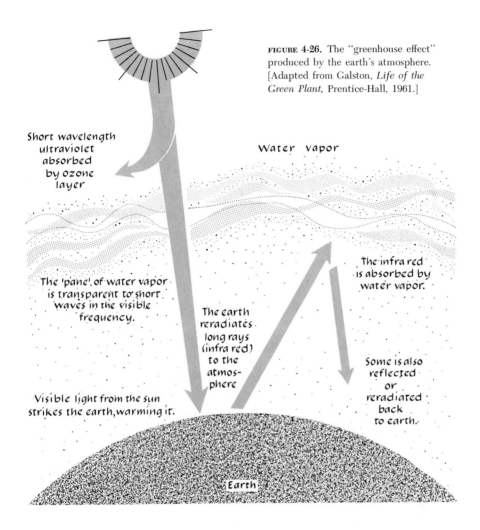

FIGURE 4-26. The "greenhouse effect" produced by the earth's atmosphere. [Adapted from Galston, *Life of the Green Plant*, Prentice-Hall, 1961.]

Short wavelength ultraviolet absorbed by ozone layer

Water vapor

The 'pane' of water vapor is transparent to short waves in the visible frequency.

The earth reradiates long rays (infra red) to the atmosphere

The infra red is absorbed by water vapor.

Visible light from the sun strikes the earth, warming it.

Some is also reflected or reradiated back to earth.

Earth

Water vapor, the most significant of the atmosphere's absorbing gases, absorbs only about 14% of the incoming short-wave radiation, but it absorbs 85% of the earth's long-wave radiation. This tends to keep surface temperatures much higher than they otherwise would be. The atmosphere thus acts as a pane of glass, transparent to the sun's short waves, but opaque to the earth's long waves; hence, the name **greenhouse effect** for this phenomenon (Fig. 4-26). The earth thus receives most of its heat only indirectly from the sun.

Heat Transfer

The transfer of heat energy is accomplished by **radiation, conduction, convection,** and **reflection. Radiation** is an organized flow of energy through space. It does not

TABLE 4-10. Heat conductivity of various substances.

Substance	Value (cal/cm-sec-°C)
Silver	1.0
Iron	0.1
Water	0.0013
Dry soil	0.0003
Sawdust	0.0001
Air	0.00005

travel in the form of heat, for heat involves molecular motion, but in the form of electromagnetic waves. When radiation is absorbed on a surface it usually produces a rise in temperature. In this case, radiation is transformed into heat energy. In **conduction,** the energy flows through the conducting medium from the warmer point to the cooler one. The transfer of heat through the soil takes place by conduction. The ability of a substance to conduct heat (**conductivity**) varies with the material, as shown in Table 4-10. The movement of heat by **convection,** or circulation of warmed air or water, is related to its change in density as a result of heat. Air near a stone radiating heat warms and becomes less dense than the cooler air farther away and is pushed upwards. Similarly, cool water sinks. Heat, as well as light, may be **reflected** from a surface. A sheet of polished metal will reflect both heat and light and will reflect both in the same way. The persistence of snow in mild weather is a result of its high reflective property.

The Plant in Relation to Temperature

The minimum and maximum temperatures to support plant growth generally lie between 40 and 97°F (4.5 and 36°C). The temperature at which optimum growth occurs varies with the plant and differs with the stage of development (Fig. 4-27). In addition, different parts of the same plant will withstand varying minimum temperatures. Roots of cold-acclimated plants are more sensitive to low temperature than stems, and flower buds are more tender than leaf buds.

A number of growth processes show a quantitative relationship to temperature. Among these are respiration, part of the photosynthetic reaction, and various maturation and ripening phenomena. In addition, such plant processes as dormancy, flowering, and fruit set are temperature-critical. The optimum temperature for plant growth depends, then, on the species and cultivar, and on the particular physiological stage of the growth process. Plants grown under uniform, constant temperature do not grow or produce fruit as rapidly as do plants grown under alternating night and day temperatures. Most plants require a lower night than day temperature. Some plants require a period of cold temperature to

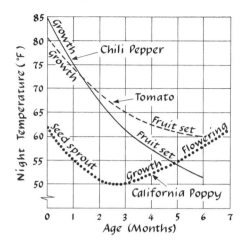

complete their annual or life cycle. This is discussed more fully in the following chapter.

There are various types of injury associated with extremes in temperature. Temperatures close to or below the freezing point of water may cause permanent damage resulting in death. This is dramatically seen after the first fall freeze, when most herbaceous plants are killed. Cold injury brought about by freezing is thought to be due to the formation of ice crystals, which cause mechanical injury to the cell. Another possibility is that such crystal formation causes desiccation of the cell. The withdrawal of water from the cells results in protein precipitation. Cold injury during the winter is also associated with tissue desiccation brought about by decreased water absorption by the roots. This type of **winter kill** is common injury to evergreens where transpiration during the winter is a contributing factor. Evergreens may be protected from winter kill by covering them with plastic, which cuts down on transpirational losses.

Some plants are found to be sensitive to temperatures slightly above freezing. This **chilling injury** is noted in peanut, velvet bean, sweetpotato, and many of the cucurbits.

Plants that are resistant to cold injury (**hardy** plants) as compared to those susceptible to cold injury (**tender** plants) appear to have an increased proportion of unfreezable water (**bound water**) in the cells, an accumulation of soluble carbohydrates, and a lower water content. The free water associated with succulent tissue freezes at 32°F (0°C). The **osmotically held water,** caused by an increase in sugar-like substances, has a lowered freezing point depending on the concentration and acts as an "antifreeze." The bound, or colloidally held, water freezes at a still lower temperature. The winter injury that occurs in grapes after a season of unusually heavy production (often due to inadequate pruning) is associated with a low sugar content of the tissues, which renders them susceptible to cold injury. Thus differences in cold resistance of particular plants may be induced by factors that tend to increase sugar accumulation. This will be discussed under "Hardening" in Chapter 7 (p. 265). The variation in cold resistance among plant species is probably related to their ability to bind water in nonfreezable forms. The greater the proportion of bound water, the hardier the plant.

The "heaving" of soil caused by alternate freezing and thawing injures the plant by the mechanical ripping of the root system. This may be overcome by procedures such as mulching that tend to prevent premature thawing. A substantial portion of freezing injury is associated with unseasonably high temperatures in the winter. In temperate regions, unseasonably high temperatures in late winter often initiate growth prematurely. This renders the plant extremely susceptible to subsequent cold weather. This is often noticed on the southern side of trees (in the Northern Hemisphere), where insolation is greatest. Similarly, the early blooming of fruit trees brought about by unseasonably warm weather is feared because of the increased danger of frost injury to flower buds.

High-temperature injury is often related to desiccation. The "burning up" of plants during unusually hot weather is usually a result of excessive water loss in transpiration that is not balanced by water uptake. This is very noticeable when unusually warm, dry, windy weather occurs after transplanting. Soil surface temperatures under these conditions may be high enough to interfere with root growth. Young transplants often "burn off" at the soil line. Extremely high air temperatures (115–130°F, or 46–54°C) may be lethal to the plant as a result of the coagulation of protein. The cessation of growth under hot weather is a reflection of an altered metabolic balance. When the respiration rate increases more rapidly than the rate of photosynthesis, there will be a resultant depletion of food reserves. As in all biologic processes, the critical temperatures vary with the material.

The Plant in Relation to Light

Plants grown in the absence of light but provided with a source of food from storage organs (for example, seed, tuber, or bulb) are yellow and have greatly elongated spindly stems (Fig. 4-28). The same plants, when provided with light, develop the green color associated with the development of chlorophyll and the initiation of photosynthesis, and assume the normal stem structure. The morphological expression of light deficiency is called **etiolation,** and is related to the effects of light on auxin distribution and synthesis. The dependence of chlorophyll development on light is utilized in the production of **blanched,** or white, asparagus and celery. In Europe there is a preference for white asparagus and celery. Asparagus is dug instead of cut to obtain blanched spears. Blanched celery is produced by mounding the base of the growing plant with some opaque material such as soil or paper. In the United States, self blanching types are grown that are naturally somewhat lighter and produce a thicker stalk, which shades the inner portion of the plant.

Some anthocyanin pigments also require light for their development. The 'Sinkuro' cultivar of eggplant only develops purple pigment in the presence of

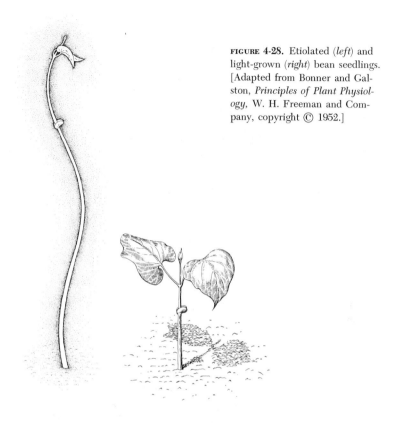

FIGURE 4-28. Etiolated (*left*) and light-grown (*right*) bean seedlings. [Adapted from Bonner and Galston, *Principles of Plant Physiology*, W. H. Freeman and Company, copyright © 1952.]

light, and the fruit is white under the calyx. Similarly, apples produced on the inside of the tree do not develop as intense a pigmentation as do those on the outside. Apple fruits can be imprinted with slogans or designs by means of tapes that have transparent letters or lines on an opaque background (Fig. 4-29).

Light influences a great many other plant responses. These include germination, tuber and bulb formation, flowering, and sex expression. This effect of light on plant development is often related to the length of the light and dark periods (**photoperiod**). These aspects of light will be discussed in the following chapter.

Quality and Quantity of Light

The radiant energy required by plants is confined almost entirely to the visible spectrum. Growth is optimum when the entire range of the visible spectrum (that is, white light, or sunlight) is provided. Light energy, described in terms of particles called **photons** (quanta), is inversely proportional to the wavelength. Thus, visible light of different wavelengths, which we see as different colors, provides different energy requirements. The light reactions of the plant (photo-

FIGURE 4-29. Light influences the formation of some anthocyanin pigments. These labeled apples were produced by affixing the tapes shown below to the fruits on the tree before the natural formation of pigment. [Photographs courtesy Purdue University.]

synthesis, phototropism, photoperiodism) are based on photochemical reactions carried on by specific pigment systems that respond to various wavelengths (Fig. 4-25). For example, those portions of the visible spectrum that result in phototropism are the violet, blue, and green regions. The red portion, most effective in photosynthesis, is ineffective in phototropism. The pigments that absorb wavelengths effective in phototropism are yellow, perhaps carotenoids or flavonoids. It is intriguing to consider that these pigments manufactured by plants provide animals with compounds involved in *their* photoreceptive reactions—vision.

Light quantity or intensity refers to the concentration of light waves. It can be expressed in terms of electrical energy (watts) per unit area or in terms of luminosity (foot-candles). Since these units are used to express intensity, they are not completely satisfactory when considering plant irradiation. The foot-candle, a measurement of luminosity, is the intensity of radiation based on the sensitivity of the human eye. Thus, the same energy at a wavelength of 555 nm will have a higher foot-candle rating than light with a wavelength of 650 nm because the eye is less sensitive to this part of the spectrum. The use of radiation intensity in terms of power units, such as watts per unit area, also does not take into account the spectral composition. This must be kept in mind when interpreting intensity requirements. The range in light intensity over the earth is enormous. Light intensity at full sun is a billion times brighter than starlight. The intensity of various light conditions expressed as foot-candles is shown in Table 4-11. The different light reactions of the plant vary in their requirements with respect to both the intensity required to initiate the reaction and the effect of the intensity on the rate of the reaction.

TABLE 4-11. Intensity values in foot-candles for various light conditions and plant photoreactions.

Light condition or photoreaction	Foot-candles
Starlight	0.0001
Moonlight	0.02
Photoperiodic induction (cocklebur)	0.3
Indoors near window	100
Overcast weather	1000
Maximum photosynthesis (individual leaf)	1200
Direct sunlight	10,000

The intensity and quality of light reaching the plant vary with the season, the latitude, and the weather conditions affecting the water vapor in the atmosphere. Thus, during the winter, light often becomes a limiting factor in greenhouses, although heat may be provided artificially. The northern areas, which are in almost continuous light for a part of the year, provide abundant photosynthesis where temperature is not limiting. The enormous size of the potatoes and cabbages produced in Alaska is due to the abundant light energy available in this region during the summer.

Light and Photosynthesis

The rate of photosynthesis is related to the availability of the raw materials, water and carbon dioxide, and to the energy available in the form of light and heat. These simple requirements are abundantly provided in the temperate and tropical areas of the earth and sea.

The photosynthetic rate is proportional to the intensity of light up to about 1200 foot-candles.[2] Chlorophyll thus is able to use efficiently only a portion of the incident light energy on a sunny day, which may be more than 10,000 foot-candles. However, due to shading effects, a maximum amount of light intensity is required to provide all of the leaves in a plant with optimum amounts of energy. The rate of photosynthesis is sharply curtailed during low light intensity with cloudy weather. Not all plants, however, respond to high light intensity. Some require as little as one-tenth of full sunlight. These differences in light intensity

---

[2] The lux (*abbr.* lx) is the metric counterpart of the foot-candle (1 lx = 10.8 ft-c). Terminology for light has undergone transformation since the adoption of SI units (Systemè International). "Light intensity" (which is measured in lumens, *abbr.* lm) is defined as the radiant flux *emitted* from a light source, while "illuminance" (which is measured in luxes, or lm/m^2) is defined as the radiant flux *intercepted* per unit area. When reporting illuminance, the measured range of wavelength should be reported (400–700 nm is the photosynthesis range).

requirements enable the classification of plants as sun plants or shade plants.

Only about 1% of the light received by the leaf during sunny days is utilized in photosynthesis. The remainder is reflected, reradiated, transformed into heat, or utilized for transpiration (Fig. 4-30).

The photosynthetic reaction is specific in its light-quality requirements. Chlorophyll absorbs the red and blue portions of the spectrum, permitting the green light to go through. Thus, chlorophyll appears green. The absorption qualities of the chlorophylls of higher plants show a higher absorption of the red light than the blue.

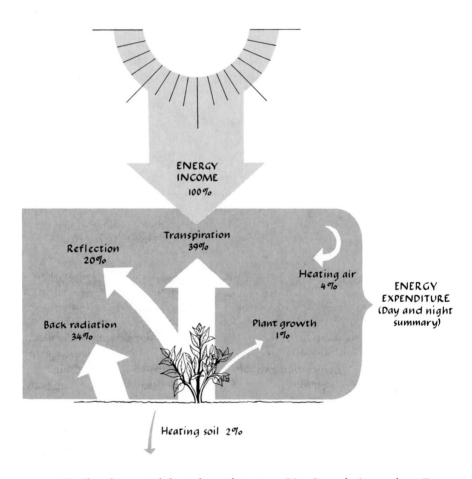

FIGURE 4-30. The solar energy balance during the summer (May–September) in southeast England. The energy expenditure is expressed as the percentage of the total that is sufficient to evaporate 36 inches of water. [Adapted from Penman, in *Sprinkler Irrigation Manual*, Wright Rain, Ringwood, England, 1956.]

Air Composition

Air is not usually considered a limiting factor in plant growth. There is relatively little variation in the composition of the atmosphere, and normal movement of the air tends to keep its constituent gases in equilibrium. But under optimum growing conditions and high light intensities, low concentrations of carbon dioxide may limit photosynthesis. However, with wind, the tremendous volume of air (of which 0.03% is CO_2) normally provides carbon dioxide sufficient for the available light. Furthermore, there is a limit to the amount of carbon dioxide a plant will tolerate. Carbon dioxide is being added to the air in carnation greenhouses in Colorado, where the natural light intensity is very high. Similarly, the addition of carbon dioxide would be necessary for optimum production of algae, which has been suggested as a means of providing food for an overcrowded world.

Air pollution is dangerous to plants just as it is to people, and the quality of the air is becoming an important factor in plant growth. This is especially true for horticultural operations that are close to urban and industrial areas. About 125 million tons of air pollutants are released annually in the United States. These pollutants include carbon monoxide, 52%; oxides of sulfur, 18%; hydrocarbons, 12%; particulate matter, 10%; and oxides of nitrogen, 6%. Air pollution can be traced to its principal sources: 60% of it to transportation vehicles, largely the automobile; 19% to industry; 12% to power generating plants; and 9% to space heating and refuse burning. Particulate materials can damage plants, but the gaseous contaminants are more injurious. The most damaging to plants are ozone and peroxyacetyl nitrate (the main phototoxicants in photochemical smog) and sulfur dioxide (released by combustion of sulfur-containing fuels and the smelting of certain ores). Other materials causing injury to crops include fluorides (produced by processes in the aluminum, glass, ceramic, and phosphate industries), ethylene, nitrogen dioxide, chlorine, and hydrogen chloride.

The availability of oxygen in the soil is often a critical factor in plant growth. Poorly aerated soils have a low oxygen and high carbon dioxide content. This reduces respiration in roots and limits root growth, resulting in a reduction of water and nutrient uptake by limiting the absorption surface and impeding the active absorption process. High carbon dioxide levels also have a toxic effect on roots. The death of large trees when their roots are covered over with extra soil is a dramatic example of the effects of an altered ratio of oxygen to carbon dioxide in the soil.

Aquatic plants, or plants adapted to marshy or boggy conditions, may be structurally altered so that the oxygen requirement of the roots is satisfied by oxygen absorbed through the leaves. Other water-loving plants apparently have adapted in some manner to low oxygen concentrations. Oxygen and carbon dioxide levels have a great effect on fruit and plant storage. This topic will be discussed in Chapters 5 and 11.

Selected References

Bonner, J., and J. E. Varner (editors), 1976. *Plant Biochemistry*. New York: Academic Press. (An advanced treatise.)

Brady, N. C., 1974. *The Nature and Properties of Soils,* 8th ed. New York: Macmillan. (A basic and widely used text on agricultural soils.)

Rabinowitch, E., and Govindjee, 1969. *Photosynthesis*. New York: Wiley. (A broad introduction to the physiology of photosynthesis and the enzymatic processes associated with it.)

Salisbury, F. B., and C. Ross, 1969. *Plant Physiology*. Belmont, Calif.: Wadsworth. (An excellent modern text in plant physiology.)

5

Plant Development

by CHARLES E. HESS,
University of California, Davis

EMILIA: Of all flow'rs
            Methinks a rose is best.
WOMAN: Why, gentle madam?
EMILIA: It is the embleme of a maid;
            For when the west wind courts her gently,
            How modely she blows and paints the sun
            With her chaste blushes! When the
            north winds near her,
            Rude and impatient, then, like chastity
            She locks her beauties in her bud again,
            And leaves him to base briers.

SHAKESPEARE, *Two Noble Kinsmen* [II. 2]

After a consideration of the plant in terms of physiological processes, it is now possible to consider the organism as an integrated mechanism capable of an irrevocable increase in size and complexity. Starting with the germination of a seed, the developmental history of a plant will be traced through juvenility, maturity, flowering, and fruiting. At fruiting the essential cycle of plant growth is completed. In perennials, the plant is ready to recycle after a period of quiescence. In annuals, fruiting is a signal to the organism to enter the final phases of plant growth—senescence and death (Fig. 5-1).

VEGETATIVE PHYSIOLOGY

Germination

Germination includes all the sequential steps from the time the seed imbibes water until the seedling is self-sustaining (Fig. 5-2). In the simplest terms, germi-

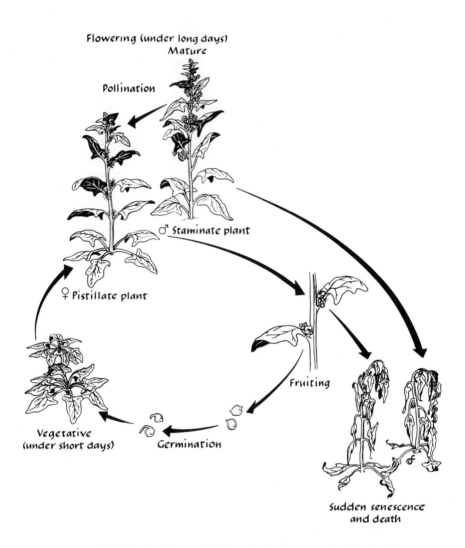

FIGURE 5-1A. The developmental history of a herbaceous annual (spinach): The seed of spinach quickly germinates under proper environmental conditions for growth. If the seedling is grown under short-day conditions, it forms a distinct vegetative stage (rosette). Under long-day conditions, the stem elongates to form a seedstalk and initiates flowers. Since spinach is normally dioecious, staminate and pistillate flowers form on separate plants. Soon after the staminate plants flower, and soon after the pistillate plants fruit, they undergo rapid senescence and die.

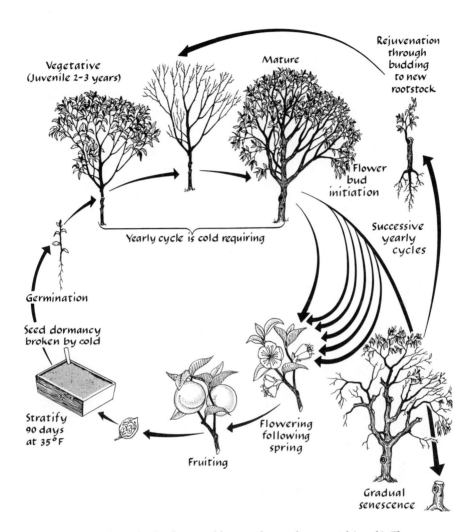

FIGURE 5-1B. The developmental history of a woody perennial (peach): The peach seed germinates only after a period of cold treatment. The juvenile stage lasts about 2–3 years. The yearly cycles must be interrupted by a period of cold. Usually flower buds are initiated in the third or fourth year and open in the following spring. The plant may live for 50–60 years, and it undergoes a very gradual deterioration. However, the plant may be rejuvenated from buds close to the base of the tree. If vegetative growth is continually budded onto new rootstocks, the "original" plant may remain, in effect, immortal.

151

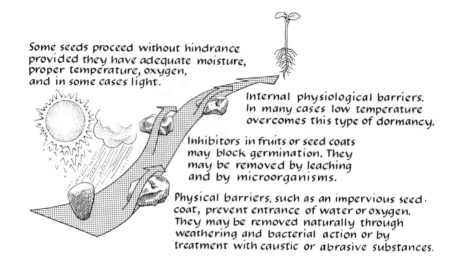

Some seeds proceed without hindrance provided they have adequate moisture, proper temperature, oxygen, and in some cases light.

Internal physiological barriers. In many cases low temperature overcomes this type of dormancy.

Inhibitors in fruits or seed coats may block germination. They may be removed by leaching and by microorganisms.

Physical barriers, such as an impervious seed coat, prevent entrance of water or oxygen. They may be removed naturally through weathering and bacterial action or by treatment with caustic or abrasive substances.

FIGURE 5-2. Germination is the route from seed to seedling. Barriers to germination (dormancy) may prevent its occurrence even though environmental conditions favorable to growth are present.

nation involves the enzymatic conversion of complex reserve substances to simple soluble substances that are readily translocated to the embryonic plant. Here some substances are oxidized through respiration to release energy, and others are utilized in synthesis. The seed must be provided with ample supplies of oxygen (to satisfy the respiratory requirements) and water (to provide a medium for enzymatic activity and synthesis). The temperature of the environment must be such that the biochemical processes of degradation and synthesis can operate. Although light is not usually required, in some plants (for example, in certain cultivars of lettuce) it can either trigger or inhibit the germination process.

Seed Dormancy

Seed that is viable and yet fails to germinate in the presence of favorable environmental conditions is said to be **dormant.** The cause of dormancy may be physical or physiological.

Physical dormancy may be caused by hard, impervious seed coats, which provide a barrier to the entrance of water and, in some plants, oxygen. The legume family provides the greatest number of examples of this type of dormancy. Some hard seeds resist water imbibition by means of a "one-way valve" in the hilum, the scar formed from the attachment of the seed to the pod. Although the seed may lose moisture in a dry environment, a moist environment effectively closes the valve. Under natural conditions these seeds do not germinate until soil microorganisms or weathering have sufficiently weakened the seed coat to permit the entrance of water. Thus, setting the stage for germination may require a number of years rather than only a small part of one season. Seeds having an impervious coat may be artificially worn or weakened in order that germination can take place uniformly and without delay (see Chapter 9).

Physiological dormancy may be due to a number of mechanisms. These commonly involve growth-regulating systems of inhibitors or promotors. Inhibiting substances that block the germination process may be present in the flesh of the fruit, in the seed coat, or even in the endosperm of the seed. Germination of seed within a tomato is a rare occurrence, yet as the flesh of the tomato is removed and the seeds are rinsed, germination takes place without delay. This inhibition of germination is due not to the low pH or high osmotic values of the tomato flesh but to a specific chemical entity. Although the inhibitor in tomato fruit has not been isolated and characterized, there is evidence that it belongs to a group of compounds known as the unsaturated lactones. A naturally occurring member of this group is coumarin, a substance responsible for the aroma of freshly cut hay.

Coumarin

The sesquiterpenoid compounds also contain inhibitors. The best known is abscisic acid (ABA), which is described in the section on growth regulators in Chapter 4. Dormancy in apple seeds and in buds of sycamore maple (*Acer pseudoplatanus*) has been correlated with the presence of ABA. The mode of action of such substances as coumarin consists of blocking or inactivating enzymes essential to the germination process. For example, it has been shown that the activity of α- and β-amylases is blocked in the presence of a germination inhibitor. The amylases are essential for the hydrolysis of starch; that is, the conversion of complex insoluble carbohydrates to simple soluble forms. The inhibitory effects of abscisic acid are attributed to its inactivating effect on substances that promote germination, such as gibberellic acid. As with dormancy imposed by the impervious seed coat, inhibitors delay and extend the period during which germination takes place. Leaching by rain and degradation by microorganisms are examples of how germination inhibitors are removed from the fruit and seed coats.

Physiological dormancy may also be due to such internal factors as an immature embryo. In some seeds, at the time the fruit is ripe, the embryo consists of only a few undifferentiated cells. The seed of the American holly (*Ilex opaca*) is an example. Therefore, an **afterripening** process must occur during which the embryo differentiates at the expense of the endosperm. In other seeds, particularly in woody species of the temperate zone, the causes of internal dormancy are more complex. In addition to the inhibitors that block germination, a germination stimulator is also required, which is produced as a prerequisite for, or concurrently with, germination. Both the removal of the blocking action of the inhibitor and the production of the stimulatory substance occur during a condition known as cold stratification. The prerequisites for cold stratification are the presence of moisture and a temperature above freezing but below 50°F (10°C). A temperature of approximately 41°F (5°C) appears to be optimum. The fate of the inhibitors during the cold stratification period is not completely clear, but the production of a germination stimulator has been demonstrated.

The morphological site of internal dormancy is primarily in the plumule. In some seeds, however, it is found in the radicle; in others, both the plumule and the radicle are dormant. Under such conditions it is often necessary to expose the seed to alternating cold and warm temperatures. For example, with *Viburnum* it is necessary to provide a warm temperature (70–80°F, or 21–27°C) for the radicle to develop. Exposure to low-temperature stratification is then necessary. As a result, the inhibitor content decreases, and growth-promoting substances increase. Germination then takes place when the seeds are re-exposed to warmth.

More complex are the germination requirements of the tree peony (*Paeonia suffruticosa*). The seed must first be exposed to low-temperature stratification to break the dormancy of the radicle. Then a warm temperature is required for the radicle to develop. As soon as it emerges, the seed must be returned to low-temperature stratification to break the dormancy of the plumule. After the second stratification period, the seed germinates upon return to warm temperatures.

The fact that the germination of a seed is blocked by one form of dormancy does not exclude the possibility that another form may also be present. Such a condition is called **double dormancy** and is characteristic of several members of the legume family. A well-known example is the redbud (*Cercis canadensis*). Both physical and physiological blocks to germination are present, the former being a seed coat that is impervious to water and the latter being an internal dormancy that can be broken by exposure to cold. The sequence of events to which seeds with double dormancy must be exposed is very precise. The physical barrier to germination must be removed before any attempt is made to remove the physiological barrier. If a seed coat is impervious, water essential for biochemical reactions cannot enter, and cold treatments to remove inhibitors or to promote the synthesis of a stimulator will not be effective. In nature, seeds with double dormancy often require two years for germination. The first year is required for soil microorganisms and weathering to remove the physical barrier by rendering the seed coat permeable to water and oxygen. The seed is not yet ready to germinate, however, because the internal dormancy has not been broken. During the second winter internal dormancy is broken, and germination occurs the following spring.

Dormancy of seeds is a biological mechanism that provides protection against premature germination when environmental conditions may not be favorable for seedling growth. Thus, in nature, seeds that have a cold-requiring internal dormancy need an exposure to the low temperatures of winter. Seeds from woody plants of the temperate zone will not germinate during the late fall to face the unfavorable growing conditions of winter, but are delayed by internal dormancy until the following spring.

The germination of seeds of most of our common vegetable crops are often not blocked by either physical or physiological forms of dormancy. In contrast, such plants as the woody ornamentals and many weed species possess, almost without exception, one or more of the major types of dormancy. Thus, it would appear that dormancy mechanisms may be eliminated through an intensive breeding program.

Juvenility

A seed is considered germinated when it has produced a plant that, under proper environmental conditions, is potentially capable of continuous and uninterrupted growth. From the time this stage is reached until the first flower primordium is initiated, the plant is considered to be in the vegetative phase of growth. If during the vegetative phase the plant cannot be made to flower, regardless of the environmental conditions imposed, it is said to be **juvenile.** Juvenility and maturity, however, are relative terms. In many species these growth phases blend into each other. The end of the juvenile period is indicated when the plant responds to flower-inducing stimuli.

The juvenile phase is characterized by the most rapid rate of growth in the plant's lifetime and, in some plants, by distinct morphological and physiological features. The juvenile phase varies in length from one to two months for annuals to a period of a few years for most fruit trees, but some plants, such as bamboos, require scores of years to reach maturity. Among the morphological features that are associated with juvenility, and which are lost or altered at maturity, are the presence of thorns (*Citrus*), the presence of leaf lobes (ivy), the lack of leaf lobes (*Philodendron*), and the angle of the branches with respect to the main axis of the plant (spruce). The geotropic response varies with developmental stage in some plants. In its juvenile phase of growth, English ivy (*Hedera helix*) is a trailing vine, climbing only with support; in the mature stage of development, the plant grows upright (Fig. 5-3). The ability of juvenile plants to initiate adventitious roots readily is a common feature of many species. This ability decreases or is lost entirely in mature forms.

One physiological concept of juvenility postulates that the apical meristem, although constantly laying down new cells, actually goes through an aging process. This assumes that no correlative growth substances are involved. A more likely hypothesis is that the cause of juvenility is the presence of substances emanating from the seed or from the juvenile root system. An explanation for the transition from the juvenile to the mature state is that the "juvenile factor"

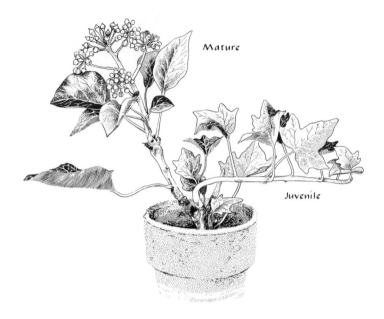

FIGURE 5-3. Morphological differences between juvenile and mature phases of growth in English ivy (*Hedera helix*). In the juvenile state, the leaves are lobed and growth is horizontal. When the plant becomes mature, the leaves are entire and shoots grow upright and bear flowers.

gradually becomes exhausted as the plant grows or is rendered ineffective as the distance from the apex to the root system increases. In apple, a certain height (about 6 feet) must be achieved before seedlings flower. The most effective way to reduce the time to first flower is to grow the seedlings in such a way that they attain a large size as rapidly as possible. However, reducing the time to the appearance of first flower may not necessarily mean that the juvenile phase has been reduced.

Some support for the notion of a disappearing juvenile factor is found in grafting experiments involving mature shoots and a juvenile stock of *Hedera helix*. If the mature branch is grafted on a juvenile stock, juvenile shoots develop at first on the mature branch. After a few weeks' growth the juvenility gradually disappears, and the shoot again becomes mature (Fig. 5-4). This particular experiment lends support to the "exhaustion" concept, which assumes that the new shoots on the mature scion first utilize the juvenile factors in the stock. As growth proceeds the juvenile factors become exhausted, and the mature phase is then resumed.

Additional support for the idea of hormonal regulation of juvenility is found in the fact that the application of gibberellic acid to the mature phase of *Hedera helix* will cause rejuvenation. In apples, growth retardants such as SADH (succinic

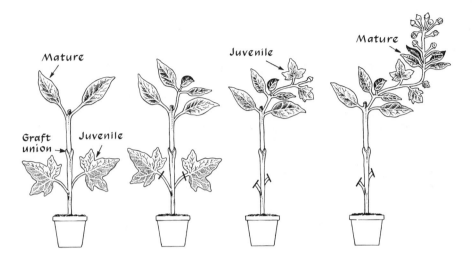

FIGURE 5-4. Induction of juvenility in English ivy, and reversal to the mature state, by grafting.

acid 2,2-dimethylhydrazide) have been used to reduce the juvenile phase and induce flowering one or two years earlier than normal. The mode of action of the growth retardants may be the blocking of gibberellin synthesis.

Environmental factors also can influence the duration of the juvenile phase. Growing birch (*Betula verrucosa*) in uninterrupted periods of long daylight, or crab apples (*Malus hupehensis*) in a CO_2-enriched atmosphere, accelerated vegetative growth and thereby shortened the time required to reach the mature phase of development. Ultimately, however, the duration of the juvenile phase is under genetic control. In a breeding program, therefore, it is possible to reduce the juvenile phase by using parents that transmit precocious flowering and by selecting for this character.

Bud Dormancy

Vegetative growth is not a continuous process but is associated with periods of arrested development. One type of arrested growth is brought about by unfavorable environmental conditions. For example, the growth of bluegrass, a common turf species, ceases under moisture stress, and the above-ground parts may die under continued drought. When conditions are more favorable, growth resumes from underground rhizomes. Many plants similarly survive periods of temperature extremes by undergoing a period of quiescence.

Bud dormancy, as is true of seed dormancy, implies that growth is temporarily suspended even though all the external conditions normally required for growth are provided. For example, woody plants of the temperate climates develop vegetative buds at each node throughout the growing season. The lack of growth of these buds after formation is initially an expression of apical dominance controlled by auxin distribution (see Chapter 7). At this time, these buds can be induced to grow by pruning, which removes the growth inhibiting apical meristem. With the onset of autumn, however, the buds develop a true dormant condition and will not grow even if the plant is moved to a warm greenhouse. The degree of dormancy varies not only between species but also between buds on the same plant. The flower buds of many trees such as peach, cherry, and apple have a lower chilling requirement than do the vegetative buds, and therefore flower before the leaves emerge. In *Forsythia* the cold requirement is so minimal that it may be seen to flower in the fall after a brief period of cold weather.

In woody plants of the temperate climates, the onset of dormancy is conditioned by short day length and low temperature. Bud dormancy in these plants is probably internally regulated by the formation of growth-inhibiting substances. It is broken naturally by cold temperatures. This cold reaction is localized, for if an isolated stem of a dormant plant is cold-treated while the remainder of the plant is kept at a warm temperature, only the dormancy of the treated stem buds are broken.

The survival value of dormancy is clear. A physiological mechanism that prevents growth is a biological check on the perversity of weather. If woody plants lacked internally imposed dormancy they might initiate growth under favorable periods in the late fall only to have their succulent new growth succumb to succeeding severe weather. After there has been enough cold weather to break dormancy, growth is limited only by favorable temperature.

The period of cold treatment required to break dormancy not only varies with the species but is sensitive to selection within species. For example, peaches have been selected for cold requirements that vary from 350 to 1200 hours below 45°F (7°C). Low-chill-requiring cultivars are selected for southern areas where the periods of cold weather may be brief. If cultivars having high chill requirements are grown too far south, they will leaf out poorly or not at all in the spring (Fig. 5-5). It is the cold-requiring dormancy brought on by short day length that prevents the production of temperate fruit crops in subtropical regions.

Maturity

When a plant becomes potentially capable of reproduction it is said to be **mature.** The maturity of a plant is unquestionably attested to by the development of flowers. In many plants, however, physiological and morphological changes characteristic of maturity take place before the macroscopic expression of flowering becomes apparent. In English ivy, for example, leaf shape is greatly modified as the mature state is reached. Leaves lose the lobed shape characteristic

FIGURE 5-5. The effects of insufficient chill in the peach. *Left*, a peach seedling photographed at the South Coast Field Station in California on May 31, 1961, when only 300 hours below 45°F (7°C) had accumulated. The tree, which has a relatively long chilling requirement, shows severe injury as a result of insufficient chill. The tree has flowered and fruited because flower buds have a lower chill requirement than leaf buds. *Right*, the same seedling on June 6 at Yucaipa, where the minimum temperatures are significantly lower as a result of the elevation (2500–2900 feet), has leafed out normally. [Photographs courtesy J. W. Lesley, University of California, Division of Agricultural Sciences.]

of the juvenile state and become entire as maturity is attained. The last leaf formed prior to the flower bud is almost reduced to a bract.

When a plant reaches maturity, it is capable of flowering, but it will not necessarily do so. The environment to which the plant is exposed at the time of maturity determines whether the plant will exhibit the ultimate expression of the mature state—the flowering response.

REPRODUCTIVE PHYSIOLOGY

Flowering

Flowering is a term that refers to a wide spectrum of physiological and morphological events. The first event, the most critical and perhaps the least understood, is the transformation of the vegetative stem primordia into floral primordia. At

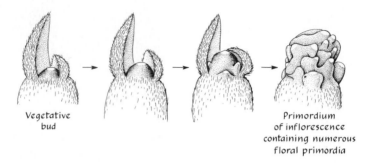

Vegetative
bud

Primordium
of inflorescence
containing numerous
floral primordia

FIGURE 5-6. The transition of a vegetative bud to a floral primodium in the cocklebur. [From Bonner and Galston, *Principles of Plant Physiology*, W. H. Freeman and Company, copyright © 1952; after Bonner and Thurlow.]

this time subtle biochemical changes take place that dramatically alter the pattern of differentiation from leaf, bud, and stem tissue to the tissues that make up the reproductive organs—pistil and stamens—and the accessory flower parts—petals and sepals. Among the events that follow initiation are development of the individual floral parts, floral maturation, and anthesis.

Once a meristem has been biochemically signaled to change from the vegetative to the reproductive state, microscopic changes in its configuration become apparent (Figs. 5-6 and 5-7). Growth of the central portion is reduced or inhibited, and the meristem becomes flattened in contrast to the conical shape characteristic of the vegetative condition. Next, small protuberances develop in a spiral or whorl arrangement around the meristem. Although this phase of reproductive differentiation is quite similar to vegetative differentiation, a basic difference does exist in that there is no elongation of the axis between the successive floral primordia as there is between the leaf primordia. In most plants, once the transformation from the vegetative to the reproductive state has been made, the process is irreversible and the floral parts will continue development until **anthesis**—the point at which the flower is fully open—even though the environmental conditions that existed during initiation are changed. By the time anthesis takes place, meiosis has already occurred, and pollen and embryo-sac development are complete. At this stage the plant is prepared for the next major step in its development—fruiting.

Carbohydrate–Nitrogen Relationship

Floral initiation has been studied primarily through manipulations of elements of the plant's environment, particularly nutrition, light, and temperature. The rapid advances in the knowledge of the purely nutritional aspects of plant physiology in the latter part of the nineteenth century created an atmosphere in which it was believed that perhaps all aspects of plant growth could be explained or regulated by an alteration or adjustment of a plant's nutrition. This approach culminated,

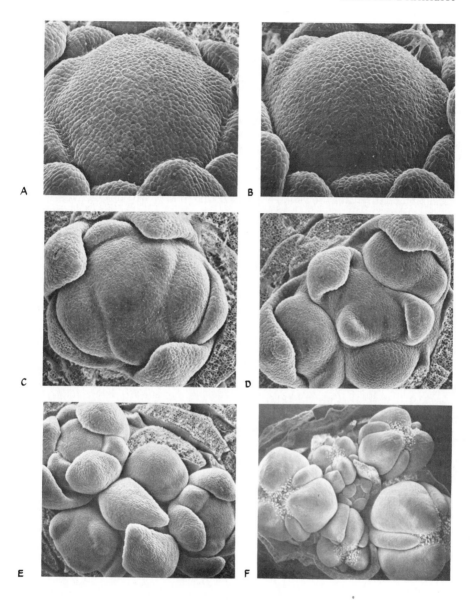

FIGURE 5-7. Morphological changes associated with the transition from vegetative to reproductive buds in Easter lily: A, vegetative meristem surrounded by leaf primordia in late October. B, the prefloral meristem becomes dome-like in early January. C, early reproductive stage in mid-January showing floral primordium regions. Each primordium differentiaties into a bract and a floral bud. D, reproductive meristem showing five floral buds, late January. E, F, reproductive meristem showing continued development of floral buds. [From De Hertogh, Rasmussen, and Blakely, J. Amer. Soc. Hort. Sci., 101:463–467, 1976.]

with respect to flower initiation, with the concepts of E. J. Kraus and H. R. Kraybill, who proposed, in 1918, that the initiation of flowering was regulated by the **carbohydrate–nitrogen relationship** of the plant. When tomato plants were grown under conditions favoring photosynthesis, and at the same time were supplied with an abundance of nitrogen fertilizers, vegetative growth was lush and flowering was reduced. But when the nitrogen supply was reduced while photosynthesis was maintained at a high level, vegetative growth was reduced and flowering was abundant. With the combination of low nitrogen and low photosynthesis, both vegetative growth and flowering were reduced. This concept of nutritional control of flowering was readily accepted, and it stimulated investigations on the effect of the carbon–nitrogen relationship in other plants. The results of such studies, however, indicated that plants will flower over an extremely wide range of carbon–nitrogen ratios. In view of our present appreciation of the tremendous physiological effects of minute quantities of growth-regulating substances, it is not difficult to understand why the gross ratios of total carbon compounds to total nitrogen compounds does not provide a consistent indication of the physiological condition of the plant. This is not to say that the nutritional status of a plant lacks importance in regard to flower initiation. The nutritional status can directly influence the degree or quantity of flowering, but can only indirectly affect the qualifying event of initiation. As an example, it is possible, through the continued removal of new growth, to cause a tomato plant to initiate a flower when only the cotyledons are present (Fig. 5-8). It can be demonstrated

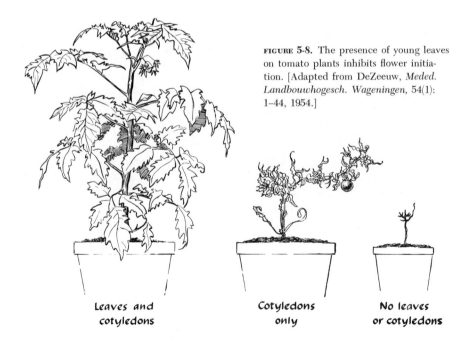

FIGURE 5-8. The presence of young leaves on tomato plants inhibits flower initiation. [Adapted from DeZeeuw, *Meded. Landbouwhogesch. Wageningen,* 54(1): 1–44, 1954.]

Leaves and
cotyledons

Cotyledons
only

No leaves
or cotyledons

that the presence of new growth has an inhibitory effect upon floral initiation. From these results it is tempting to suggest that the reason nitrogen reduces flowering in the tomato is only incidentally associated with the carbohydrate–nitrogen relationship. Perhaps the stimulation of new growth by nitrogen inhibits flower initiation. In support of this concept is the observation that almost any means by which growth can be reduced, such as bending a branch from an upright to a downward position, results in increased floral initiation.

The concept that flower initiation is triggered by minute chemical changes was suggested by J. von Sachs in 1865. However, because of the popularity of the nutritional concepts of growth control, Sachs's theories received little support. In 1920, two years after the Kraus and Kraybill hypothesis was published, a discovery was made that caused a revolution in the concepts of flower initiation and provided direct support to the postulations of Sachs.

Photoperiodic Effect

The discovery of photoperiodism by W. W. Garner and H. A. Allard, scientists working for the United States Department of Agriculture, was made in conjunction with a breeding experiment. A new cultivar of tobacco was developed that flowered only in the greenhouse during the fall. Since most cultivars of tobacco flower in the summer, Garner and Allard attempted to provide the new variety with environmental conditions that would cause it to initiate flowers in the summer in order to make additional crosses. But no matter what method was used in an attempt to reduce vegetative growth and initiate flowering—altering the nutrition of the plant, allowing it to become pot-bound, withholding water—all attempts were failures. An attempt was then made to vary an environmental factor that had not been previously considered—namely, day length. The result was the discovery that, by artificially shortening the daily exposure to light during the summer, the new variety of tobacco could be made to flower as profusely as it did in the fall. Garner and Allard's research stimulated a great amount of investigation in the field now called **photoperiodism**—the growth response of a plant to the length of day, or, more precisely, to the length of the light and dark periods. It was soon found that a great number of plants responded to variations in day length. Some plants responded exactly as did the new cultivar of tobacco, but others responded in exactly the opposite way; that is, they flowered only when the days were long or were artificially lengthened. As the results of these many investigations accumulated, it became apparent that a majority of plants fell into one of three categories: **short-day, long-day,** and **day-neutral** plants. Short-day plants initiate flowers only when the day length is below about 12 hours. These include many of the spring- and fall-flowering plants, such as chrysanthemum, salvia, cosmos, and poinsettia. Long-day plants initiate flowers only in day lengths exceeding 12 hours. They include almost all of the summer-flowering plants of the temperate zones, such as beet, radish, lettuce, spinach, and potato. Day-neutral plants apparently can initiate flowers during days of any length. They include the dandelion, buckwheat, and many tropical plants that either flower on a year-round basis or, if they do not, can be shown to be affected by other environmental conditions. The tomato is a typical example of a day-neutral plant, but it has been demonstrated that, with proper control of other environmental factors and under certain temperatures, the tomato will initiate a greater number of flowers under short day length. As the study of plant response to day length continues other categories have been added. For example, there are nonobligate long- or short-day plants. These are plants that will flower regardless of the day length but that will flower earlier or more profusely when the day is either long or short. A petunia will flower either during long or short days, but it will flower better during long days. It is therefore classified as a nonobligate long-day plant. In still another category are the plants that flower only after an alternation of day lengths and are known as "long-day, short-day plants." Such plants require first an exposure to long days and then to a period of short days.

The discovery of photoperiodism is particularly significant in that it clearly demonstrates the hormonal control of flower initiation. The mature or newly expanded leaf is the perceiver of changes in day length. In some plants the leaves need only to be exposed to one light–dark cycle of the proper day length to cause flower initiation. In the majority of plants, however, several to many cycles are required. Once the leaves have received the photoperiodic message, it has been postulated, they produce a substance, or a precursor of a substance, called **florigen.** Unfortunately, however, florigen remains only a name, for it has proven to be one of the most elusive of all plant-growth substances. Its transport from the leaves to the growing point can be demonstrated up and down stems, across graft unions, and from one plant to another, but the substance has not been isolated. Present thinking is that florigen *per se* may not exist, but that it may be represented by a particular balance of already known growth factors.

A major advance that may lead to an understanding of the flowering process is the isolation of the pigment system in the leaf called **phytochrome,** which specifically receives the photoperiodic message. Phytochrome, a blue-green pigment present in small amounts in all plants, was discovered by H. A. Borthwick and S. B. Hendricks, scientists with the United States Department of Agriculture, and first reported in 1959. Three important characteristics of photoperiodism aided in the discovery of phytochrome. First, interruption of the dark period with a small amount of light prevents flowering in short-day plants and permits flowering of long-day plants (Fig. 5-9). Interruption of the light period with brief intervals of darkness has no effect on flowering. The night-interruption reaction is extremely critical in some plants. It can be shown that a very weak light (0.3 foot-candles) is enough to interfere with the dark period in cocklebur. Second, red light is the most effective portion of the light spectrum for producing the night-interruption effect. Third, light from the far-red portion of the spectrum can completely reverse the effects of the red light. Furthermore, the effect of the far-red exposure is reversed again by exposure to red light. The direction of the photoconversion of phytochrome is determined by the light quality (that is, red or far-red) of the final exposure.

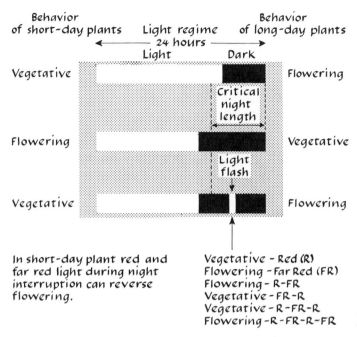

FIGURE 5-9. Photoperiodism is the control of flowering by the length of the light and dark periods to which the plant is exposed. The length of the dark period rather than the length of the light period is the critical factor. Short-day plants flower under periods of long night. Long-day plants flower under periods of short night. The flowering effect can be reversed by substituting red and far red light during night interruption. [Adapted from Galston, *The Life of the Green Plant*, Prentice-Hall, 1961.]

The reaction is completely reversible over a considerable period of time, as long as the time lapse between exposures does not exceed a specific amount. Under short-day conditions an interruption of the dark period with red light causes a molecular shift in phytochrome that brings about flowering in long-day plants and prevents flowering in short-day plants. The reason for this differential response is not known. The action of far-red light can reverse the red effect. In addition, this shift is also influenced by the length of the dark period and by temperature. This may explain the light–temperature relationship in photoperiodism and may explain why the light break is most effective at the middle of the dark period. The characterization of phytochrome is still under way. It is known to be a protein with chromophoric (pigment) groups. The basic molecular unit of phytochrome is estimated to have a molecular weight of 120,000 or 240,000 with three chromophores per molecule. The chromophore is a pigment very similar to the chromophore of c-phycocyanin, an algal chromoprotein.

Phytochrome apears to have many effects upon growth and development. It has been shown to be responsible for the promotion and inhibition of germination in seeds of some plants. In lettuce, seed germination can be inhibited or promoted by alternating red and far-red light. Although all seeds contain phytochrome, many plants do not require this "light signal" to germinate.

Although the discovery of phytochrome has made a very significant contribution toward explaining the mechanism of flower initiation, a vast amount of experimentation must be done before the substances specifically involved in the transition of the apical meristems from the vegetative to the flowering condition can be isolated and their mode of action determined.

The emphasis placed upon photoperiodism should not be interpreted as being an indication that it has exclusive control over flowering. Temperature, for example, has both an indirect and a direct effect upon flower initiation. It can influence flowering indirectly by modifying the plant response to a given photoperiod. Thus, a poinsettia will initiate and develop flowers in 65 days when grown under short-day conditions at 70°F (21°C). When grown under the same conditions at 60°F (15.5°C), however, 85 days are required before initiation occurs. A more striking example is that of the strawberry. At temperatures above 67°F (19.5°C) the June-bearing strawberry behaves as a short-day plant and will not initiate flowers in day lengths longer than 12 hours. At temperatures below 67°F (19.5°C) its response is that of a day-neutral plant, and the plant will initiate flowers even in continuous illumination.

Vernalization

The direct effect of temperature upon flowering was recognized more than one hundred years ago by the American J. H. Klippart. He described a process whereby it was possible to "convert" winter wheat into spring wheat. (Winter wheat is planted in the fall and produces a crop the following year; spring wheat produces a crop in the same year it is planted.) The process consisted essentially of reproducing the process that occurs in the field. Klippart partially germinated the winter wheat and then prevented further development by maintaining the seed at a temperature near freezing. The treated seed was planted in the spring, and it produced a crop in the same year. Therefore, the normal biennial habit of the winter wheat had been eliminated, or, more correctly, the cold requirement was satisfied. Klippart's research did not receive much attention until his work was repeated by the controversial Russian agronomist T. Lysenko.[1] The phenomenon was characterized as being the effect of temperature, during one or more of the developmental phases of plant growth, upon future flowering behavior, and was named **vernalization** by Lysenko. Since Lysenko's work in 1928, a considerable amount of research has been devoted to mechanisms of the cold response. A vernalized plant—that is, a plant which has been given a cold treatment—can be

---

[1] In the late 1940s Lysenko was placed in charge of Soviet genetics research. He became notorious for his autocratic methods and his controversial (now discredited) views about genetics and plant breeding.

grafted onto a nonvernalized plant, and both will flower. The implication is that a substance has passed from the vernalized plant, across the graft union, to the nonvernalized plant. A similar experiment has been conducted with annual and biennial forms of the same plant, such as henbane. The annual form requires no cold treatment to flower, and, when grafted onto the biennial form, causes the latter to flower earlier than it would normally. In wheat, this difference between annual (spring) and biennial (winter) forms of wheat has been shown to be controlled by a single gene, which might indicate that a single compound is involved. Although the flower-inducing substance produced during vernalization has not been isolated, a clue to its identity is found in the fact that the cold requirement of some plants can be partially or completely replaced with gibberellic acid.

Further evidence that temperature can have a profound effect upon flowering is provided in the phenomenon of **devernalization.** Plants exposed to a cold period of the proper temperature (41 °F, or 5 °C) for a period of time normally sufficient to induce flowering (usually at least six weeks) can be reverted back to their original nonflowering condition by exposure to high temperature. Onion growers take advantage of devernalization on a commercial scale. Onion sets are stored during the winter at temperatures near freezing to retard spoilage. The onion sets are vernalized at this temperature, and, if they were planted directly from cold storage in the spring, the set would quickly flower and no bulb would be formed. Therefore, the onion sets are exposed to temperatures above 80 °F (27 °C) for 2–3 weeks before planting. The sets, now devernalized, will form bulbs instead of flowers.

Although photoperiodism and vernalization appear to have many similarities, and are definitely interrelated, the stimuli produced in the two responses to environment do not seem to be identical. Evidence for this supposition is taken from experiments in which it is possible to separate the effect of vernalization from those of photoperiodism. For example, most biennial plants when supplied the required exposure to low temperature still will not flower unless they are given the proper photoperiod. Similarly, where gibberellins replace the cold requirement, flower initiation will not occur unless the day length is correct.

In addition to these temperature effects, it has been shown that the alternation of warm and cool temperatures also influences flowering. This is the phenomenon of **thermal periodicity.** The classic example is the tomato, which will initiate more flowers when grown at a cycle of about 80 °F (27 °C) during the day and 63–68 °F (17–20 °C) during the night than when grown at higher or lower night temperatures. This phenomena has been utilized in greenhouse tomato culture.

Moisture Effects

Moisture is another environmental factor that may influence flower initiation. For example, if rhododendrons are subjected to a period of rainy weather during the fall, which is when they normally initiate flower buds, most of the buds will be vegetative. Many other woody plants respond similarly. It can be regularly observed that flowering in the spring is much more abundant after a dry summer and fall than after a wet summer and fall. Those who emphasize the nutritional role in flower initiation interpret this effect of moisture as being due to a favorable carbohydrate–nitrogen relationship. In rainy weather, the production of carbohydrates is reduced to a level that is too low to provide the balance necessary for flowering. As pointed out before, however, this interpretation appears to be an oversimplification of a rather complex phenomenon.

Time of Flower Initiation

The time at which the initiation of flowers takes place is of particular importance to the horticulturalist. Many perennial plants, both woody and herbaceous, initiate flower primordia from several to many months before flowering (Fig. 5-10). In the apple, for example, flower buds are initiated from June to August, depending upon the cultivar. Flowering occurs during the following spring. In June-bearing strawberries the flowers are initiated in August and September, when the days become short. Attempts to obtain two crops from plants that normally produce one have been made in Holland. There, the growers subject the plants to a short photoperiod immediately after the first crop is harvested rather than waiting for it to occur naturally in late summer or fall.

Regulation of Flowering

Evidence that flowering is regulated by an endogenous chemical stimulus led many investigators to attempt to induce flowering by external applications of growth regulators. In some cases the results have provided substantial practical applications. For example, auxins (see p. 114) have been used extensively to

FIGURE 5-10. The crocus flowers in the early spring from buds initiated the previous summer. [Photograph by J. C. Allen & Son.]

induce flowering in pineapple, although they are now being replaced by the ethylene-generating compound known as ethephon. Apparently, the auxins caused flowering indirectly by stimulating the formation of ethylene by the pineapple tissues.

Gibberellins also can cause flowering in some long-day plants or in some plants that require a cold treatment prior to flowering. Examples are cabbage, beet, carrot, and endive. However, gibberellins have not been used extensively for this purpose on a practical basis.

Sex Expression

The environment, besides having a profound influence upon flower initiation, can also influence the subsequent differentiation of the flower. This phenomenon is best seen in sex expression. In the normal sequence of the growth of cucurbits, for example, the first flowers produced are male (Fig. 5-11). As growth continues there is an alternation of male and female flowers; eventually only female flowers are initiated. If the plants are grown under long-day conditions and cool temperatures there is an increase in the ratio of female to male flowers. The effect of nutrition, photoperiod, and temperature has been demonstrated to affect sex expression in many plants.

Sex expression can also be modified by the application of growth regulators. Auxins and ethephon induce femaleness, whereas gibberellins induce maleness. Some of the effects of specific growth regulators on various plants are discussed in Chapter 7.

Fruit Development

Fruit development can be conveniently divided into four phases: (1) initiation of the fruit tissues, (2) prepollination development, (3) postpollination growth, and (4) ripening, maturation, and senescence. The origin of the fruit is found in the initiation of the floral primordia, which usually develops concomitantly with the flower. The increase in fruit size in the prepollination phase of development is primarily the result of cell division. After pollination, cell enlargement is responsible for the major portion of size increase. However, in some large-fruited plants (for example, watermelon and squash) cell division continues for some time after pollination, the final size being a consequence of increases both in cell number and in cell size. The stimuli and nutrients for prepollination growth are supplied primarily by the main body of the plant. In plants having perfect flowers, the stamen primordia are often differentiated before the ovary primordium, and they have been shown to be a source of growth stimulus. Surgical removal of the immature stamens in the flower bud adversely affects the growth of the ovary if the operation is performed at an early stage of development. Extraction of the unripe anthers reveals the presence of large amounts of auxin.

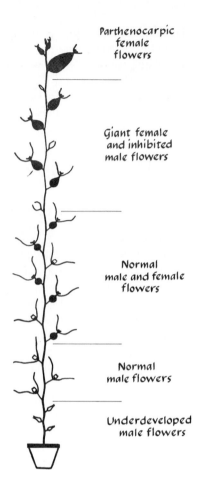

Parthenocarpic
female
flowers

Giant female
and inhibited
male flowers

Normal
male and female
flowers

Normal
male flowers

Underdeveloped
male flowers

FIGURE 5-11. The normal sequential development of male and female flowers on the acorn squash. [Adapted from Nitsch, Kurtz, Liverman, and Went, *Amer. J. Bot.*, 39:32–43, 1952.]

Pollination

One of the most critical points in the growth and development of a fruit is pollination. Pollination has at least two separate and independent functions. The first is the initiation of the physiological processes which culminate in "fruit set" or, more precisely, in inhibition of fruit or flower abscission. The second function of pollination is to provide the male gamete for fertilization. That these two functions are indeed separate can be demonstrated by the use of dead pollen. In the orchid the use of dead pollen results in fruit set and some growth, but fertilization is not possible. A more precise demonstration of the multiphase function of pollen is the use of a synthetic auxin (Fig. 5-12). Here again fruit set is obtained. The fact that water extracts of pollen are also effective in inducing fruit set has led to the postulation that the pollen contains an auxin. But the minute amount of auxin present in the pollen that lands on a stigma usually cannot

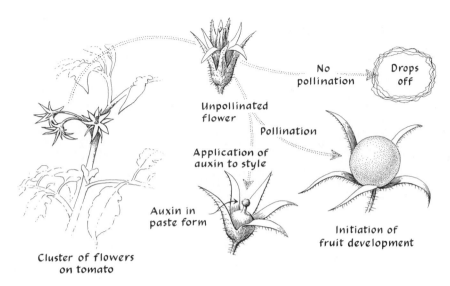

No
pollination

Drops
off

Unpollinated
flower

Pollination

Application of
auxin to style

Auxin in
paste form

Initiation of
fruit development

Cluster of flowers
on tomato

FIGURE 5-12. In addition to providing the male gametes, pollination prevents abscission of the flower. This role of the pollen may be replaced by the application of auxins. [From Bonner and Galston, *Principles of Plant Physiology*, W. H. Freeman and Company, copyright © 1952; after Avery and Johnson.]

account for the auxin response obtained. Instead it seems that the pollen contributes either an enzyme that converts auxin precursors present in the stigma to auxin or provides a synergist that renders effective the auxin already present.

Even if pollination takes place and fruit set is obtained, fertilization is not absolutely assured. Sometimes pollen does not germinate, or if it does, the pollen tube may burst in the style. The germination of pollen is dependent upon the presence of a medium of the proper osmotic concentration (as shown by the effect of various sugar concentrations upon germination) and is stimulated by the presence of certain inorganic substances, such as manganese sulfate, calcium, and boron. In the lily, it has been shown that the highest concentration of boron occurs in the style and stigma.

In addition to the presence of organic and inorganic substances present in and on the stigma, which stimulate germination, there is also evidence for the existence of substances that chemically attract the growth of the pollen tube. If, for example, a slice of lily stigma is placed on an agar medium containing sucrose and boron, and is surrounded by a ring of pollen placed at a distance from the stigma, all the pollen tubes will grow in the direction of the stigma.

If the growth of the pollen tube is very slow, the style, or even the entire flower, may be shed. This may be artificially prevented, however, by the application of auxins such as α-naphthalene acetamide. This technique is particularly valuable to plant breeders, who through the use of auxins are now able to obtain seed from strains of petunia, cabbage, and marigold that previously had been considered

FIGURE 5-13. The effect of pollination on fruit set and the influence of the resultant seed on fruit growth make pollination a crucial phase in the production of many crops grown for their fruits. For example, in apple and pear, where genetic incompatibility prevents self-pollination within the same clone, more than one pollen-fertile clone must be present to assure pollination. Since pollen is transferred by insects, bees are often placed in the orchard for this purpose. [Photographs by J. C. Allen & Son.]

self-sterile. When fruit set and growth are obtained either by pollination or by the application of an auxin, but fertilization does not take place, the fruit is said to be "parthenocarpic." Although no seeds are present, seedlike structures may develop, as in the case in some seedless cultivars of holly, orange, and grape.

The effect of pollination on fruit set and the influence of the resultant seed on fruit growth make pollination a crucial phase in the production of many fruit crops. For example, in apple and pear, in which genetic incompatibility prevents self-pollination within the same clone, more than one pollen-fertile cultivar must be present to assure pollination. Because pollen is transferred by insects, bees are often kept for this purpose (Fig. 5-13).

Although pollination has the two major functions of fruit set and fertilization, there is still another interesting but uncommon effect called **metaxenia,** the direct influence of pollen on the maternal tissues of the fruit. In the date palm, a dioecious plant, the pollen, in addition to causing fruit set and pollination, affects both the size of the fruit and the date of ripening. This influence of the pollen variety can be demonstrated by using one strain of pollen to fertilize a number of female inflorescences of different date cultivars. This effect is in contrast to the direct effect of pollen on embryonic or endosperm tissue, which is referred to as **xenia.** The different colored kernels of Indian corn are a good example of xenia.

Postfertilization Development

After fertilization, the plant enters into a phase of physiological activity that is second in intensity only to germination. The developing fruit no longer depends primarily upon the parent plant for a source of growth stimuli. Instead, these stimuli are provided by the developing seed within the fruit. The role of the seed can be empirically demonstrated by observing that misshapen fruits result from uneven distribution of seeds. In many fruits, a direct correlation exists between weight (or length) and seed number. This effect of the seed on fruit development is mediated through chemical substances. For example, extracts of immature seeds can stimulate growth of unpollinated tomatoes. Furthermore, it is possible to correlate various physiological events in the development of a fruit with the presence of growth substances. It has been demonstrated that the auxin levels reach a low at the time of flower drop, and particularly during the natural abscission of partially developed fruit that occurs when an especially heavy crop is obtained in fruit trees, known in horticultural terminology as the "June drop." The relationship between natural auxin level and fruit development is shown in Figure 5-14. The growth of the strawberry receptacle also has been prevented by removal of the achenes, which are one-seeded fruits. If one or two achenes were left intact, the receptacle would grow only in the area directly under the achene (Fig. 5-15). The achenes were extracted and found to contain high levels of auxin, whereas the receptacle tissue was comparatively low in auxin content. Finally, the addition of an auxin in the form of indoleacetic acid (IAA) dramatically stimulated the growth of an acheneless receptacle.

Other growth regulators are involved in fruit development. Grapes have two peaks of fruit growth; the first period has been correlated with increases in auxin content and the second with the biosynthesis of gibberellins. A sequence of growth regulators has also been suggested, with cytokinins playing a key role during cell division, followed by auxins and gibberellins during subsequent periods of growth. Practical use of the role of gibberellins in fruit development is seen in the table-grape industry. Application of a spray containing 20 ppm of gibberellic acid doubles the size of 'Thompson Seedless' grapes.

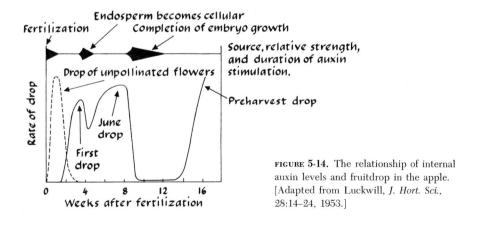

FIGURE 5-14. The relationship of internal auxin levels and fruitdrop in the apple. [Adapted from Luckwill, *J. Hort. Sci.*, 28:14–24, 1953.]

Growth of receptacle
induced by one
achene

Growth of receptacle
induced by three
achenes

Normal growth of
receptacle induced
by many achenes

FIGURE 5-15. Developing seeds are a source of stimulus for fruit growth. In the strawberry, when all the achenes except one are removed, only the receptacle tissue directly under the achene will enlarge. If additional achenes remain, additional areas of the receptacle will develop. The stimulus from the seeds can be replaced in part by the application of auxins. [From Bonner and Galston, *Principles of Plant Physiology*, W. H. Freeman and Company, copyright © 1952; after Nitsch.]

Nutrition and Fruit Growth

Although the control center of fruit growth is located in the seed, the raw materials for fruit development are supplied by other parts of the plant. Thus, the nutrition and moisture available to the plant directly affect fruit size. It has been calculated that at least 40 leaves on a mature apple tree are required to support the growth of one apple. If the forty-to-one ratio is substantially reduced by an abnormally high fruit set, the quality and size of the individual fruits is greatly reduced. Therefore, it is now a common orchard practice to reduce the number of fruit artificially. This process, known as **fruit thinning,** will be discussed in Chapters 7 and 8.

Ripening

A final, dramatic physiological event marks the end of maturation and the beginning of senescence of the fruit. This is the **climacteric.** It is characterized by a marked and sudden rise in the respiration of a fruit prior to senescence, and it takes place apparently without the influence of external agents. The respiration rate then returns to a level equal to or below the level that existed prior to the climacteric. The climacteric is associated not only with the quantitative burst in carbon dioxide production but also with qualitative changes related to ripening, such as pigment changes. The transition from green to yellow in certain cultivars of apple, pear, and banana takes place during or immediately following the climacteric. The peak of acceptability, or "edible ripeness," of pears coincides with the peak of the climacteric. In apple, banana, and avocado, maximum acceptability is reached immediately after the climacteric. Finally, a marked increase in the susceptibility of fruits to fungal invasion follows the climacteric.

The occurrence and causes of the climacteric are currently under study. Almost all fruits studied exhibit a characteristic rise in respiration after the harvest, with the exception of most citrus fruits. (The lemon, when held in an atmosphere containing at least 33% carbon dioxide, will exhibit a climacteric.) The degree and duration of the rise varies considerably between species, from a short intense peak for the avocado to a longer, less definite peak for the apple. Explanations for the climacteric rise in respiration are varied. One hypothesis holds that, prior to the climacteric rise, the acidity of the cytoplasm decreases to a critical level, which in turn increases the permeability of the cell membranes. Then, fructose, which previously had been accumulated in the vacuole, can move back into the cytoplasm and provide a substrate for increased respiration.

A second hypothesis involves adenosine diphosphate (ADP) and adenosine triphosphate (ATP), compounds that are instrumental in energy transfer in the cytoplasm. During respiration, a part of the energy released is preserved in the conversion of ADP to ATP; that is, the addition of a phosphate to ADP in the presence of energy results in the formation of a high-energy bond. This bond can later be broken to release energy for use in synthesis. The rate of respiration may be limited by the amount of ADP available for accepting a high-energy phosphate bond. It is therefore postulated that, during fruit maturation, when there is rapid cell enlargement and a high demand for protein synthesis, there is a shortage of ADP. But as the fruit matures, ADP becomes available and the respiration rate increases. Evidence in support of this hypothesis is that the addition of ADP to tissues of immature fruit causes an increase in respiration, but as maturation progresses the response decreases until it is completely lost during the climacteric.

As with any physiological event, temperature has a profound effect upon maturation and the climacteric. For example, a comparison of respiration has been made between apples held at $36°F$ ($2°C$) and apples held at $73°F$ ($23°C$). The maximum respiratory activity is 5–6 times as high at $73°F$ as at $36°F$, and it takes 25 times as long to reach the climacteric at the lower temperature. However, the total amount of CO_2 liberated during the time between harvest and the end of storage life was approximately the same for both temperatures, equivalent to 16–20% of the reserve carbohydrates initially present in the fruit. In pears a pattern was established similar to that for apples at the two temperature extremes, but the rates of CO_2 evolution were much higher and the storage life much shorter.

It is becoming clear that fruit ripening, like other phases of plant development, is a DNA-controlled process. There is a rise in RNA in fruits at climacteric followed by increases in enzyme proteins such as the synthetases, hydrolases, and oxidases. Therefore, the climacteric is associated with both degradative and synthetic reactions. Ethylene and perhaps other volatiles appear to play key roles in the ripening process, and ethylene has been referred to as a fruit ripening hormone. It is true that ethylene can accelerate fruit ripening; treatments to remove ethylene from within tissues of preclimacteric fruit can delay ripening. How ethylene exerts its effect on the initiation of ripening is not yet clear.

SENESCENCE

Senescence refers to the erosive processes that accompany aging prior to death. This process is one of the most baffling and least understood of the developmental processes.

Senescence in plants may be **partial** or **complete** (Fig. 5-16). **Partial senescence** is the deterioration and death of plant organs such as leaves, stems, fruits, and flowers. Examples are the death and abscission of cotyledons in bean plants, the death of two-year-old raspberry canes, or the death of the entire shoot of tulip in the early summer. **Complete senescence** is the aging and death of the entire plant except for the seeds. The termination of the life cycle of true annuals and biennials is often sudden and dramatic. After fruiting, whole fields of such crops as spinach and corn die in a synchronized pattern during the early or middle part of the growing season. In contrast, the senescence of such perennial plants as the apple or peach appears as a gradual erosion of growth and viability. In addition, perennials can be rejuvenated. Mature apple trees may be revitalized by severe pruning and fertilization or by encouraging the growth of adventitious buds at the base of the tree. The senescence of annual plants is relatively irreversible, and is associated with flowering and fruiting. Spinach plants kept vegetative under short days will not become senescent. After having been induced to flower, however, they will surely die.

The older concepts explaining the senescence of annual plants are associated with a depletion hypothesis. It is suggested that during flowering and fruiting

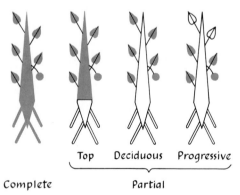

Top   Deciduous   Progressive

Complete   Partial

Senescence

FIGURE **5-16.** Patterns of plant senescence. The dead portions of the plant are indicated by shading. [After Leopold, *Plant Growth and Development,* McGraw-Hill, 1964.]

FIGURE 5-17. The separation of bolting and flowering effects on senescence. Treatments from left to right: (1) short day; (2) short day + gibberellic acid; (3) long day; (4) long day + gibberellic acid. Note that the phenomenon of bolting does not lead to senescence unless it is associated with flowering induced by long days. Plants shown are spinach. [From Janick and Leopold, *Nature*, 192:887–888, 1961.]

essential metabolites are drained from the main plant and accumulate in the fruit and seed. By the time the fruit is mature, the plant is depleted to the point that further growth is impossible, and death rapidly follows. It can be shown that the removal of fruits can significantly postpone senescence and death. It has also been observed that the greater the number of flowers or fruits on an individual plant the more rapid is senescence and death.

However, the depletion concept of senescence is an oversimplification of a complex process. Unpollinated pistillate plants and staminate plants of a dioecious plant such as spinach also become senescent. In these plants senescence can be postponed by the removal of developing flowers, but death cannot be averted. Clearly, there is no depletion of the plant to account for this effect. It can further be shown that the seed-stalk formation that accompanies flowering is distinct from senescence, since spinach plants induced to bolt through the use of gibberellic acid, but kept vegetative, do not become senescent (Fig. 5-17). Apparently, the biological signal that causes an annual plant to become senescent is associated only with the flowering process and is merely accentuated through fruiting. The relationship between this signal and flowering is not at all clear.

Selected References

Audus, L. J., 1972. *Plant Growth Substances*, 3rd ed. New York: Barnes & Noble. (An advanced treatise on the chemistry and physiology of growth regulating materials.)

Galston, A. W., and P. J. Davies, 1970. *Control Mechanisms in Plant Development*. Englewood Cliffs, N.J.: Prentice-Hall. (A brief review of plant developmental processes written for the beginning student.)

Leopold, A. C., and P. E. Kriedemann, 1975. *Plant Growth and Development*, 2nd ed. New York: McGraw-Hill. (An excellent text on plant physiology.)

Mayer, A. N., and A. Poljakoff-Mayber, 1974. *The Germination of Seeds*, 2nd ed. New York: Pergamon Press. (An up-to-date work on the physiology, ecology, and environmental biology of seeds.)

Salisbury, F. B., 1963. *The Flowering Process*. New York: Pergamon Press. (A biological framework for understanding conversion from the vegetative to the reproductive state in higher plants.)

Smith, H., 1975. *Phytochrome and Photomorphogenesis: An Introduction to the Photocontrol of Plant Development*. New York: McGraw-Hill. (Photobiology of plants.)

Thimann, K. V., 1977. *Hormone Action in the Whole Life of Plants*. Amherst, Mass.: University of Massachusetts Press. (An advanced text on phytohormones and plant growth and development.)

Torrey, J. G., 1967. *Development in Flowering Plants*. New York: Macmillan. (The ontogeny of plant development.)

Vince–Prue, D., 1975. *Photoperiodism in Plants*. London: McGraw-Hill. (An advanced treatise.)

Wareing, P. F., and I. D. J. Phillips, 1970. *The Control of Growth and Differentiation in Plants*. New York: Pergamon Press. (An introduction to the processes underlying and controlling development in plants.)

Wightman, F., and G. Setterfield (editors), 1968. *Biochemistry and Physiology of Plant Growth Substances*. Ottawa: Runge Press. (Proceedings of the 6th International Conference on Plant Growth Substances held at Carleton University, Ottawa, July 24-29, 1967.)

Wilkins, M. B., 1969. *Physiology of Plant Growth and Development*. New York: McGraw-Hill. (A text on advanced plant physiology and plant biochemistry containing contributions from eighteen specialists in selected areas of whole-plant physiology.)

Zimmerman, R. H. (editor), 1976. *Symposium on Juvenility in Woody Perennials* (Acta Horticulturae 56). International Society for Horticultural Science. (Recent advances in juvenility research.)

II

THE TECHNOLOGY
OF HORTICULTURE

Climatron, St. Louis, Missouri. [*Photograph by Piaget, Courtesy Frits Went.*]

6

Controlling the
Plant Environment

Shall I compare thee to a summer's day?
Thou art more lovely and more temperate:
Rough winds do shake the darling buds of May,
And summer's lease hath all too short a date;
Sometime too hot the eye of heaven shines,
And often is his gold complexion dimm'd,
By chance or nature's changing course untrimm'd . . . .

SHAKESPEARE, Sonnet 18

As human beings increase their dominion over the earth they must not only satisfy their own narrow limits of survival but must also control the environment of their plants. This may require providing moisture where none is expected to occur naturally, or building elaborate structures that literally house plants for their entire productive life. The environmental factors most amenable to modification are soil, light, temperature, and moisture. This chapter is a discussion of environmental control in relation to plant growth. The control of environment in relation to storage is discussed in Chapter 11.

Although in this chapter the modification of the environmental factors affecting plant growth are discussed more or less individually, their interrelations must be kept in mind. For example, we have seen that temperature and light interact in their effects on plant growth and development. Thus, temperature must be considered not only in relation to day–night fluctuations but with respect to day length. The control of temperature in a greenhouse by shading affects light availability. Similarly, soil fertility and water availability are intimately related. The alteration of one factor very often affects another. The successful culture of the plant depends on the proper synchronization of these environmental modifications.

SOIL MANAGEMENT

The soil provides support for the plant and is the storehouse of plant nutrients, water, and oxygen for root growth. The ability of the soil to support plant growth is often referred to as its **productive capacity.** The soil's productive capacity must be considered in terms of fertility and physical condition. It is not enough that the nutrients necessary for growth be contained in the soil; they must be released in a form readily available to the plant. Furthermore, the soil must be conserved; it is a renewable but not easily replaceable natural resource. Soil management is concerned with the sustained use of land and with economic crop production.

Maintenance of Soil Fertility

Soil fertility, the nutrient-supplying capacity of the soil, depends upon the amount and availability of plant nutrients (see Chapter 4). The recognition of soil fertility as a factor in crop response is recorded in ancient Greek writings, as is the supplemental use of manure on soils to improve plant growth. The use of cover crops, the mixing of soils, and the addition of lime and salts to increase the productivity of soils are mentioned in Roman agricultural treatises. Nevertheless, it was not until well into the nineteenth century that the role of inorganic nutrients in soil fertility was understood. The study of soil fertility today is intimately involved with microbiology, chemistry, and physics.

The maintenance of soil fertility is concerned with adjusting the current supply of available nutrients to optimum levels for economic crop production. The inherent fertility of the soil is related to the factors that contributed to its formation—namely, the parent minerals from which the inorganic part of soil was formed, the topography, the climate, the natural vegetation, and time. The fertility of "virgin" soil—soil that has not been disturbed by cultivation—reaches an equilibrium such that the nutrients released into the soil equal those that are lost from it. The inherent fertility of virgin soils, which varies greatly, may become depleted when the soil is brought under cultivation (Fig. 6-1). This is

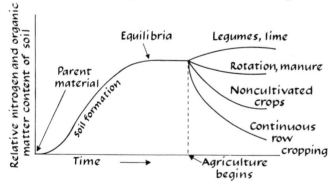

FIGURE 6-1. Influence of human activity on the fertility and organic matter of the soil. [Adapted from Jenny, *Factors in Soil Formation*, McGraw-Hill, 1941.]

TABLE 6-1. Approximate amounts of nutrients removed from soil by fruit and vegetable crops.

Crop	Yield per acre	Nutrient (in elemental form) removed (lb/acre)					
		N	P	K	Ca	S	Mg
Apple (fruit)	600 bu	34	3	32	3	12	5
Bean (seed)	40 bu	86	7	28	5	5	4
Cabbage (heads)	20 tons	120	14	133	16	32	5
Onion (bulbs)	1000 bu	154	26	121	18	48	12
Peach (fruit)	500 bu	30	3	73	3	4	5
Potato (tubers)	600 bu	126	19	150	7	14	14
Spinach (tops)	1000 bu	88	10	35	17	8	10
Sweetpotato (roots)	400 bu	53	7	93	7	9	13
Tomato (fruit)	20 tons	80	9	117	16	5	9
Turnip (roots)	500 bu	64	13	72	14	20	5

SOURCE: Data from American Plant Food Council.

brought about by crop removal (Table 6-1) and mineralization of organic matter. In addition, the removal of a permanent plant cover, row cropping, and the loss of soil structure by cultivation hasten erosion by water and wind. The nutrients lost may be replaced by supplemental additives in the form of fertilizers and manures. Nitrogen also may be added through the process of nitrogen fixation by bacteria on the roots of a leguminous crop. Chemical fertilizers have been the most economical source of nitrogen.

Fertilization as a Horticultural Practice

Fertilization refers to the addition of nutrients to the plant. The primary objective of crop fertilization is to achieve an optimum plant response. This may not necessarily be the greatest response; in commercial crop production it is that point at which the value of the increased response is equal to the cost of the additive (Fig. 6-2). Fertilization beyond this level must be considered a wasteful practice. Not only is an excess subject to loss by leaching and volatilization, but it may actually be toxic to crops. Overfertilization of greenhouse soils has become a serious problem.

Materials that supply nutrient elements to plants are known as **fertilizers.** Those that supply nitrogen, phosphorus, and potassium, the major plant nutrients, are called **complete fertilizers.**

The **grade** or **analysis** of these fertilizers is a three-part number that shows the percentages of nitrogen (expressed as elemental N), phosphorus (expressed as P_2O_5), and potassium (expressed as K_2O), *in that order*. The reason phosphorus and

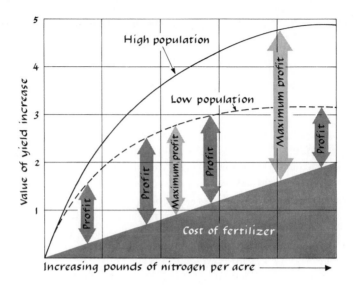

FIGURE 6-2. The yield response diminishes with increasing quantities of fertilizer. The optimum rate is obtained when the value of the yield response is just equal to the last increment of fertilizer added. At this point the profit is maximized. But the response to fertilizer may depend on other factors. As is shown above, increasing the plant population changed the response curve and made a higher rate of nitrogen profitable. [Adapted from *Soil* (USDA Yearbook, 1957).]

potassium are not expressed in their elemental form is historical. A movement is currently in progress to have this changed, especially in scientific writing. To change oxide analyses to elements and *vice versa*, multiply by the following conversion factors.

Element	Oxide to element	Element to oxide
Phosphorus	0.43	2.33
Potassium	0.83	1.20

The oxide analysis is still used commercially. Thus a 10–10–10 fertilizer is 10% elemental nitrogen by weight, but only 4.3% elemental phosphorus and 8.3% elemental potassium. Similarly, an 80-lb bag of a 5–15–30 fertilizer will contain 4.0 lb of N, the amount of phosphorus in 12 lb of P_2O_5 (5.2 lb of P) and the amount of potassium in 24 lb of K_2O (19.9 lb of K).

The fertilizer **ratio** is simply the analysis expressed in terms of the lowest common denominator. A 6–10–4 analysis has a 3:5:2 ratio, and a 10–10–10 analysis has a 1:1:1 ratio. Fertilizers are referred to as **low-analysis** fertilizers when the amount of available nutrients is below 30% by weight and as **high-**

analysis fertilizers when the amount of available nutrients is 30% or more. Because of the greater weight and the extra handling required, low-analysis fertilizers tend to be more expensive per nutrient unit than high-analysis fertilizers. High-analysis fertilizers, however, require special accuracy and care in their application. Although by law the analysis of nitrogen, phosphorus, and potassium must be stated on the fertilizer bag, many fertilizers contain, or are restricted to, other plant nutrients.

Fertilizers may be classified as **natural organics** or **chemicals.** Natural organics (for example, manure, blood, fish scraps, and cottonseed meal) are compounds derived from living organisms. Chemical fertilizers, such as ammonium nitrate and superphosphate, are synthesized from inorganic minerals. A number of forms of nitrogen-containing compounds are now synthesized, including urea and cyanamid. Urea can be synthesized directly from the air, which requires only a source of electrical power. Although urea and cyanamid are organic compounds in the chemical sense, they are not necessarily derived from living systems.

At present, most nitrogen fertilizer is synthesized by the Haber process, in which atmospheric nitrogen is reacted with hydrogen to form ammonia. The ammonia can be used directly as a fertilizer, or it can be used as a raw material for the manufacture of urea, nitrates, or other nitrogenous compounds. The hydrogen required by the Haber process is generally extracted from natural gas, and the cost of that fuel makes up most of the cost of the manufactured nitrogen fertilizer. The synthesis of a ton of anhydrous ammonia requires 30,000 cubic feet of natural gas.

The nutrients in natural organic fertilizers and some synthetic organics undergo gradual chemical transformations to available forms. "Urea-forms" are combinations of urea and formaldehyde—the greater the proportion of formaldehyde, the lower the solubility of urea and hence the slower the "release." Other types of slow-release materials are composed according to specifications for particle size, chemical composition, and solubility. A special slow-release system has been developed by encapsulating fertilizer pellets with a plastic coat. The coat acts as a membrane to release the plant nutrients gradually. These materials are comparatively expensive per pound of nutrient, but they have special uses in horticulture, such as for container-grown nursery stock, because they allow application costs to be reduced.

Modern fertilizers can be compounded to satisfy the different needs of the user. Thus they may not only be made up of different nutrients but may be a mixture of organic and inorganic forms. In this way some or a portion of the applied nutrients can be made available immediately whereas the rest can be released slowly, commensurate with the needs of the crop. Decisions about whether a slow-releasing fertilizer or a number of applications of a fast-releasing fertilizer should be used are usually based on economic considerations.

The physical characteristics of fertilizer materials vary greatly. Although many fertilizers are solids, they may be applied in dissolved form as a liquid, as in irrigation water. Nitrogen may be applied in the ammonia form as a gas. In addition to soil application, nutrients may be applied directly through the foliage.

Nitrogen can be efficiently applied through the leaves by spraying them with urea. The application to the foliage of such trace elements as manganese and boron has also proved practical.

LEVEL OF FERTILIZATION. The kind and level of fertilization is based on crop need in relation to current fertility levels and the alternative sources of plant nutrients. The prediction of plant-nutrient needs has been one of the main goals of plant-fertility studies. The techniques that are now used consist of correlating plant responses with chemical tests on the soil or with tests made on the plant tissues themselves. However, the total nutrient content of the soil does not give a true picture of nutrient availability. The available nutrients are related to the ex-changeable cations, the soil reaction or pH, and the organic cycles. Some biological assays of the soil have been made by utilizing the responses of particularly sensitive plants or microorganisms. The relationship between these tests, soil type, and climate have been correlated for many crop plants. Quick tests have been developed, although these are often not too accurate. In many plants severe deficiencies of certain nutrients produce characteristic responses in the plant called **deficiency symptoms,** which often can be used to diagnose the trouble (Table 6-2). The good plantsman will not permit nutrient shortages to become so severe that deficiency symptoms appear.

TABLE 6-2. A key to plant nutrient-deficiency symptoms based on vegetable crop responses.

Symptoms	Deficiency
A. Symptoms on leaves, stems, or petioles.	B
Flowering or fruiting affected.	M
Storage organs affected.	N
Variable plant growth throughout the field. Some plants appear normal, some show severe marginal leaf necrosis, while others are stunted. Determine soil pH.	ACID or ALKALINE SOIL COMPLEX
B. Youngest leaves affected first.	C
Entire plant affected or oldest leaves affected first.	I
C. Chlorosis appears on youngest leaves.	D
Chlorosis is not a dominant symptom. Growing points eventually die and storage organs are affected.	H
D. Leaves uniformly light green, followed by yellowing and poor, spindly growth. Most common in areas with acidic, highly leached, sandy soils low in organic matter.	SULFUR
Uniform chlorosis does not occur.	E
E. Leaves wilt, become chlorotic, then necrotic. Onion bulbs are undersize and outer scales are thin and lightly colored. May occur on acidic soils, on soils high in organic matter, or on alkaline soils.	COPPER
Wilting and necrosis are not dominant symptoms.	F

Symptoms	Deficiency

F. Distinct yellow or white areas appear between veins, and veins eventually become chlorotic. Symptoms rare on mature leaves. Necrosis usually absent. Most common on calcareous soils ("lime induced chlorosis").   **IRON**

Yellow/white areas are not so distinct, and veins remain green.   **G**

G. Chlorosis is less marked near veins. Some mottling occurs in interveinal areas. Chlorotic areas eventually become brown, transparent, or necrotic. Symptoms may appear later on older leaves. In peas and beans, the radical and central tissue of cotyledons of ungerminated seeds become brown ("marsh spot"). Most common on soils with pH over 6.8.   **MANGANESE**

Leaves may be abnormally small and necrotic. Internodes are shortened. Beans, sweet corn ("white bud" of maize), and lima beans most affected; potatoes, tomatoes and onions somewhat affected; uncommon with peas, asparagus and carrots. Reduced availability in acidic, highly leached, sandy soils, in alkaline soils, and in organic soils.   **ZINC**

H. Brittle tissues. Young, expanding leaves may be necrotic or distorted followed by death of growing points. Internodes may be short, especially at shoot terminals. Stems may be rough, cracked, or split along the vascular bundles (hollow stem of crucifers, cracked stem of celery). Most likely on highly leached, acidic soils and on organic soils with free lime.   **BORON**

Brittle tissues not a dominant symptom. Growing points usually damaged or dead ("dieback"). Margins of leaves developing from the growing point are first to turn brown or necrotic, expanding corn leaf margins are gelatinous and necrotic, expanding cruciferous seedling leaves are cupped and have necrotic margins; old leaves remain green. Common on acidic, highly leached, sandy soils. May result from excess Na, K, or Mg from irrigation waters, fertilizers or dolomitic limestone. (Celery blackheart, brown heart of escarole, lettuce tipburn, internal tipburn of cabbage, internal browning of Brussels sprouts, hypocotyl necrosis of snap beans.   **CALCIUM**

I. Plant exhibits chlorosis.   **J**

Chlorosis is not a dominant symptom.   **L**

J. Interveinal or marginal chlorosis   **K**

General chlorosis. Chlorosis progresses from light green to yellow. Entire plant becomes yellow under prolonged stress. Growth is immediately restricted and plants soon become spindly and drop older leaves. Most common on highly leached soils or with high organic matter soils at low temperatures. Soil applications of N show dramatic improvements.   **NITROGEN**

K. Marginal chlorosis or chlorotic blotches which later merge. Leaves show yellow chlorotic interveinal tissue on some species, reddish purple progressing to necrosis on others. Younger leaves affected with continued stress. Chlorotic areas may become necrotic, brittle,

TABLE 6-2. A key to plant nutrient-deficiency symptoms based on vegetable crop responses (*continued*).

Symptoms	Deficiency
and curl upward. Symptoms usually occur late in growing season. Most common on acidic, highly leached, sandy soils or on soils with high K or high Ca.	MAGNESIUM
Interveinal chlorosis, with early symptoms resembling N deficiency (Mo is required for nitrate reduction); older leaves chlorotic or blotched with veins remaining pale green. Leaf margins become necrotic and may roll or curl. Symptoms appear on younger leaves as deficiency progresses. In brassicas, leaf margins become necrotic and disintegrate, leaving behind a thin strip of leaf ("whiptail," especially of cauliflower). Common on acidic soils or highly leached alkaline soils.	MOLYBDENUM
L. Leaf margins tanned, scorched, or have necrotic spots (may be small black dots which later coalesce). Margins become brown and cup downward. Growth is restricted and dieback may occur. Mild symptoms appear first on recently matured leaves, then become pronounced on older leaves, and finally on young leaves. Symptoms may be more common late in the growing season due to translocation of K to developing storage organs. Most common on highly leached, acidic soils and on organic soils due to fixation.	POTASSIUM
Leaves appear dull, dark green, blue-green, or red-purple, especially on the underside, and especially at the midrib and veins. Petioles may also exhibit purpling. Restriction in growth may be noticed. Availability reduced in acidic and alkaline soils, and in cold, dry, or organic soils.	PHOSPHORUS
Terminal leaflets wilt with slight water stress. Wilted areas later become bronzed, and finally necrotic. Very infrequently observed.	CHLORINE
M. Fruit appears rough, cracked or spotted. Flowering is greatly reduced. Tomato fruits show open locule, internal browning, blotchy ripening or stem-end russeting. Occurs on acidic soils, on organic soils with free lime, and on highly leached soils.	BORON
Cracking and roughness are not dominant symptoms. Fruits exhibit water-soaked lesions at blossom end, later becoming sunken, dark or leathery (blossom-end rot of tomato, pepper, and watermelon). Common on acidic, highly leached soils.	CALCIUM
N. Internal or external necrotic or water soaked areas of irregular shape (hollow stem of crucifers, internal browning of turnip and rutabaga, canker or blackheart of beet, water core of turnip). May occur on acidic soils, on alkaline soils with free lime, or on highly leached soils.	BORON
Cavities develop in the root phloem, followed by collapse of the epidermis, causing pitted lesions. (Cavity spot of carrots or parsnips.) Common on acidic, highly leached soils.	CALCIUM

SOURCE: English and Maynard, 1978, *HortScience* 13:28–29.

The relationship between fertility level and plant performance varies with the species and the nutrient. For example, 100–150 lb of nitrogen applied prior to planting will promote optimum production of tomatoes on mineral soils, whereas this level of nitrogen will reduce muskmelon yields as a result of decreased production of perfect flowers. However, the plant response to fertility level can be discussed in general. At one end of the scale are **deficiency levels,** at which plants show definite symptoms of deficiency. At somewhat higher levels, although they may not show obvious deficiency symptoms, crop plants may show reduced yield. This has been termed **"hidden hunger."** At levels above which no response to fertilizers may be demonstrated, the plant may continue to show an increasing level of nutrient absorption. This is termed **"luxury consumption."** At abnormally high levels, growth is reduced and death may even occur. Maximum production presumably occurs in that state of soil fertility in which a slight luxury consumption exists.

The level of crop response to fertilization is related in part to the productive capacity of the soil. Crops on soils of low productive capacity show a maximum response at a lower level of fertility than on soils of high productive capacity. Productive capacity is based on long-term nutrient availability and soil condition. Owing to the nature of forces in the soil that establish an equilibrium between the soil and the soil solution, optimum fertility cannot be achieved in one quick step. When larger amounts of fertilizer are placed on soils of a low productive capacity, much of it is wasted. These excess nutrients may be leached, may be tied up in forms unavailable to the plant, or may be poorly distributed throughout the soil with respect to the needs of the plant. However, continued applications of fertilizer at the level of optimum plant response tend to increase the productive capacity of the soil, ultimately raising its yield potential.

PLACEMENT AND TIMING. One of the important factors in the use of fertilizer is proper placement and timing. This is related to the efficiency of the plant's utilization of the fertilizer, the prevention of injury, and convenience and economy of application. To be effective, fertilizer must applied where and when the plant needs it. Single yearly applications may not be sufficient for some nutrients, such as nitrogen, and they may not be necessary for others. Large amounts of highly soluble fertilizers should not be applied to growing plants, especially when young, because of salt injury (Fig. 6-3). In perennials, or in long-season annuals, it may be more efficient to control carefully the availability of nutrients throughout the season, and for this reason repeated applications are made. This is especially important with nitrogen fertilization, because excess amounts of nitrogen fertilizers are often irretrievably lost to the plant through leaching or other means.

There are various methods of fertilizer placement. A fertilizer may be applied prior to planting by scattering it uniformly over the surface of the ground (**broadcast application**). It may be dropped behind the plow at the bottom of the furrow (**plow-sole placement**), or placed in a band under the seed or to one or both sides (**band placement**) during planting (Fig. 6-4). Another mode of application

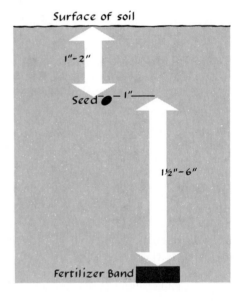

FIGURE 6-3. An excess of soluble fertilizers applied to growing plants such as turf grasses may burn the foliage or may kill the entire plant. [Photograph courtesy W. H. Daniel.]

FIGURE 6-4. The proper placement of fertilizer band for many vegetable crops. Large amounts of fertilizer banded too close to the seed may inhibit growth because of salt injury. Fertilizer placed too far away may be insufficient to stimulate the early seedling growth that is so important to optimum plant performance. [Adapted from *Soil* (USDA Yearbook, 1957).]

consists of applying fertilizer directly over the crop after emergence (**top dressing**) or beside the row (**side dressing**). Side-dressed applications of fertilizer are often made along with a cultivation, and are thus mixed into the soil. Fertilizer may also be applied along with mechanical transplanting, either added as a band under the plant or dissolved in the supplemental transplanting water (**starter solution**).

Nutrients may be applied in solution to the leaves (**foliar application**). Absorption of nutrients applied in this fashion is very rapid, and the plant's response may be evident within a day or two; but applications of this kind must be more frequent than applications to the soil because there is little residual effect.

The timing of fertilizer applications depends upon the nutrient and the crop as well as on the soil type and the climate. Nitrogen fertilizers must be supplied as close to the plant as possible to be of any use. The nitrate forms of nitrogen are water soluble and move rapidly within the soil. Although the ammoniacal forms of nitrogen are held on the soil colloids, they become "mobile" as soon as they are converted to the nitrate form. Thus, nitrogen is often in short supply, especially in sandy soils. Soils high in applied organic matter may be temporarily short of nitrogen as the result of a buildup of microorganisms that feed on the organic matter. These microorganisms pre-empt available soil nitrogen for their own use. In addition, the nitrogen concentration in the soil solution may be relatively low in the spring, especially in cold, wet, poorly drained soils, owing to a lack of nitrification by aerobic bacteria and to excess denitrification by anaerobic bacteria. Consequently, many crops show a good response to spring applications of nitrogen. It is unwise to apply all of the nitrogen at planting time because of possible injury, so it is often also applied as a side dressing during the growing season.

In such perennial fruit crops as apple and peach, excess nitrogen is associated with poor fruit color and soft fruit, as well as with undesirable vegetative growth that occurs late in the season and leaves the plant vulnerable to winter injury. Consequently, nitrogen is usually applied only once, early in the spring, in order that any excess will have been used up by summer. In small-fruit crops that ripen early in the summer, such as the strawberry, nitrogen is not applied to bearing patches in the spring because of the undesirable effect of fruit softening.

In contrast to nitrogen, phosphorus moves very little in the soil. Consequently, the total quantity needed during the season can be applied at one time. Because of the high phosphorus requirement of seedlings it is important that adequate levels be made available close to the seed or transplant. The use of starter solution, transplanting water supplied with liberal phosphorus (about 1500 ppm), is recommended for many transplants. The phosphorus is supplied by soluble phosphate salts such as monoammonium phosphate, diammonium phosphate, and monopotassium phosphate. A popular starter solution uses three pounds of a mixture of diammonium phosphate and monopotassium phosphate (10–52–17 analysis) in 50 gallons of water. When an extensive root system is established, the plant requirements are satisfied by lower levels of phosphorus. Phosphorus applications are often banded under the seed to achieve the same effect (Fig. 6-5). Perennial plants usually do not respond to phosphorus application because the root systems are extensive and active throughout most of the growing season.

Because of the low mobility of phosphorus and the low efficiency of phosphorus uptake by plants (less than 25%), it has been found profitable to build up soil phosphorus levels prior to planting horticultural crops. Once phosphorus levels are brought up, supplemental additions of phosphorus need not be frequent. In turf, for example, there is practically no net loss of phosphorus.

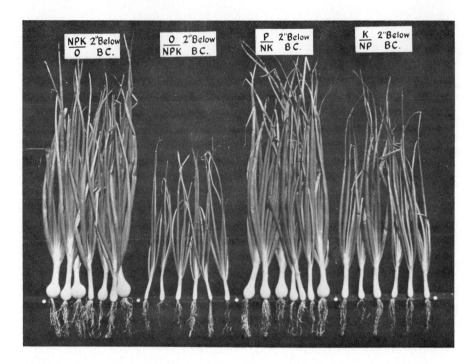

FIGURE 6-5. Response of onions to fertilizer placement. In all cases equal amounts of fertilizer were added, either banded 2 inches below the seed or broadcast over the surface of the soil. The greatest response was obtained when the fertilizer was placed below the seed, where it was available to the young plant. The treatment series indicate that phosphorus is the critical nutrient with respect to placement. [Photograph courtesy J. F. Davis.]

Potassium salts are intermediate in mobility to phosphorus and nitrogen. Owing to their solubility, potassium salts cannot be placed close to seeds or plants in any great amounts. Since potassium is not as critical for seedling growth as nitrogen or phosphorus, broadcast applications are usually made before planting.

Regulation of Soil Reaction

Soil reaction, so important to nutrient availability and root growth, is an important concern of soil management. Although plants vary in their response to pH (Fig. 6-6), most horticultural crops do well with a pH between 6.0 and 6.5. There is a group of "acid-loving" plants that require conditions of low pH (4.5–6.0).

FIGURE 6-6 (facing page). Suitable pH ranges from various crops and ornamental plants. [Adapted, with permission of the publisher, from Lyon, Buckman, and Brady, *The Nature and Properties of Soils*, 5th ed., copyright 1952 by the Macmillan Company.]

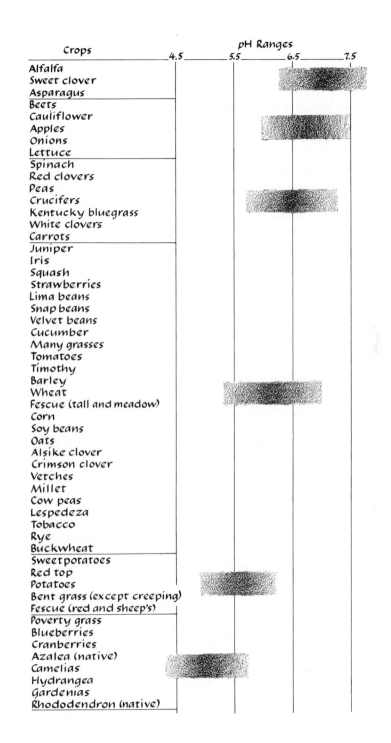

Crops	pH Ranges

Alfalfa
Sweet clover
Asparagus
Beets
Cauliflower
Apples
Onions
Lettuce
Spinach
Red clovers
Peas
Crucifers
Kentucky bluegrass
White clovers
Carrots
Juniper
Iris
Squash
Strawberries
Lima beans
Snap beans
Velvet beans
Cucumber
Many grasses
Tomatoes
Timothy
Barley
Wheat
Fescue (tall and meadow)
Corn
Soy beans
Oats
Alsike clover
Crimson clover
Vetches
Millet
Cow peas
Lespedeza
Tobacco
Rye
Buckwheat
Sweet potatoes
Red top
Potatoes
Bent grass (except creeping)
Fescue (red and sheep's)
Poverty grass
Blueberries
Cranberries
Azalea (native)
Camelias
Hydrangea
Gardenias
Rhododendron (native)

4.5 5.5 6.5 7.5

These include gardenias and camellias, in addition to most members of the family Ericaceae (which includes the azaleas, rhododendrons, cranberries, and blueberries). A low soil reaction may be used to control soil diseases in crops that prove less sensitive to low pH than the corresponding disease-producing organism. Thus, potatoes may be grown at pH 5.2 to control scab, a disease caused by a fungus that is not adaptive to acid soils. Potatoes, however, grow equally well at a higher pH in disease-free soil.

The natural reaction of soils is due to the interaction of climate with the parent materials of the soil. In general, acid soils are common where the precipitation is high enough to leach appreciable amounts of exchangeable bases from the surface layers. Thus, in the humid climates, the areas of most intense horticulture, soil acidity is often a problem. Soil alkalinity occurs generally in the more arid regions, where there is a comparatively high degree of cation accumulation. Alkaline soils may result in a high salt accumulation (salinity). This can be a problem in the irrigated soils of the American Southwest.

Soils become acid because the basic cations on the soil colloid are replaced by hydrogen ions. This process can be reversed, and the soil pH increased, by adding basic cations—for example, calcium, magnesium, sodium, and potassium. Calcium is the most economic cation for increasing soil pH. In addition, calcium has other beneficial effects. It is an essential element in plant nutrition; it is thought to promote good soil structure by promoting granulation; and it encourages certain soil microorganisms, especially the nitrifying and nitrogen-fixing bacteria. The addition of calcium or calcium and magnesium compounds to reduce the acidity of the soil is known as **liming.** Although the term *lime* refers correctly to CaO, it is used in the agricultural sense to include oxides, hydroxides, carbonates, and silicates of calcium or calcium and magnesium. Soil liming has resulted in significant increases in plant growth. The amount of liming required depends upon the degree of pH change desired, the base exchange capacity of the soil, the amount of precipitation, and the liming material and its physical form in relation to particle size.

Soil may be made more acid by placing hydrogen ions on the soil colloid. This is done by adding substances that tend to produce strong acids in the soil. Some nitrogen fertilizers will increase soil acidity, but elemental sulfur is by far the most effective substance to use for this purpose. In warm, moist, well-aerated soils, bacterial action converts sulfur to sulfuric acid.

Maintenance of Soil Condition

Tilth

Maintaining the physical condition of the soil (**tilth**) is a significant part of soil management. Soil structure and texture have a direct effect on the water-holding capacity and aeration of the soil. This influences root growth as well as the soil

FIGURE 6-7. A puddled soil (*left*) compared with a well-granulated soil (*above*). [Adapted, with permission of the publisher, from Lyon, Buckman, and Brady, *The Nature and Properties of Soils*, 5th ed., copyright 1952 by the Macmillan Company.]

microorganisms that play an important part in making available the nutrients in organic matter. Crusting and puddling of the soil are indications of poor tilth (Fig. 6-7).

The physical condition of the soil is largely conditioned by the amount of organic matter it contains. Organic matter may be maintained (and in some cases increased) in field soils by altering crop rotation and by adding supplemental organic matter in the form of manure. In potting soils organic matter is often added in the form of peat or ground bark. The use of chemical soil conditioners such as Krilium have been suggested, although this has not proven practical. The problem of maintaining soil condition is complicated because most cultivation practices, contrary to first impression, do not aid in improving soil structure. Even the gains made by distributing plant material through the soil may be more than offset by compaction, by the loss of organic matter as a result of oxidation, and by erosion. Clay soils in particular must be handled carefully to maintain soil structure. If they are cultivated when too wet the soil becomes puddled. The use of heavy equipment (such as mechanical harvesters) on wet clay soils seriously impairs soil structure and often leads to compaction. This is less of a problem on sandy or peat soils, whose structure is generally good under most conditions.

Increasing Organic Matter

Organic matter affects both the fertility and the physical condition of the soil. Organic matter acts as a storehouse of nitrogen and other nutrients, and it greatly influences the exchange capacity of the soil. It improves the physical condition of the soil by increasing its water-holding capacity, so important to sandy soils, and by increasing aeration, which is especially necessary in clay soils.

FIGURE 6-8. Good soil structure under an alfalfa sod. Note how the roots have penetrated the soil. [Photograph by J. C. Allen & Son.]

The accumulation of organic matter reaches an equilibrium in undisturbed soil. Because the organic material of the soil is largely under the control of biologic and climatic forces, it is important to differentiate between temporary and long-term increases, in organic matter. Unfortunately, it is easy to reduce the organic matter of a soil, and it is relatively difficult to build it up. The major loss of organic matter comes about from increased oxidation as a result of cultivation and from crop removal.

Perhaps the best way to effect a long-term increase in organic matter is through the extended use of a legume or grass sod (Fig. 6-8). The organic matter produced as a result of root disintegration is protected from excessive oxidation by the constant cover and lack of cultivation. Permanent orchard plantings are kept in good tilth by the use of sod. Mixtures of shallow- and deep-rooted grasses, such as 'Alta' fescue (shallow-rooted) and 'Ladino' clover (deep-rooted) are often used to obtain a resilient floor under heavy equipment as well as to prevent compaction and to improve drainage. Sods, of course, have certain disadvantages. They are not generally used in peach culture because of frost injury (see Chapter 12) and because the grasses tend to compete with the trees for nitrogen. In annual and perennial plantings organic matter can be built up by rotating row crops with legume sods. A standard soil management for potatoes in the northern United States involves a three-year rotation of potatoes, grain, and clover. The grain is used to control weeds and to act as a "nurse crop" for the establishment of the clover.

The plowing down of a growing crop as a "green manure" temporarily increases the organic content of the soil, but this practice cannot be expected to produce a long-term increase in organic matter. The buildup of high populations of microorganisms may actually reduce the net organic-matter content of the soil through an unexplained breakdown of the normally more resistant fractions of the organic materials in the soil. The rapid breakdown of a green manure does release tied up nutrients; but such materials as millet or sudan grass, which have a high carbon content in relation to their nitrogen or protein content, may create a temporary nitrogen shortage because of the utilization of the available nitrogen by microorganisms. Nitrogen must then be added to compensate for this shortage. Sweet clover and other legumes that have a high nitrogen content in proportion to carbon are destroyed more rapidly by microorganisms, and they release a steady supply of nitrogen as they are broken down.

The use of grasses or legumes planted in middle or late summer provides the advantages of a sod for part of the year. These **cover crops** are probably of most value in preventing winter erosion. They are usually plowed under in the spring as a green manure. This should be done early to prevent the loss of excessive moisture from the soil.

The use of manures to supplement soil organic matter is at present an expensive procedure. Manure is probably more valuable for its nutrients than for its contribution of organic matter. Similarly, the use of mulches indirectly adds to organic matter, though only in the upper surface of the soil. Their greatest contribution, however, lies in reducing the amount of cultivation, which in effect limits the oxidation of organic matter.

Regulation of Nutrient Cycles in the Soil

Soil fertility may be maintained by intervention in the nutrient cycles that connect plants, soil, animals, and microorganisms. There is the potential for a large increase in the biological fixation of nitrogen. Alfalfa, a legume, has been grown with annual yields of 16 tons per acre without the application of nitrogen fertilizers. Alfalfa is about 3% nitrogen; thus, in a single acre of alfalfa, the bacteria associated with the plant roots must fix at least 1000 pounds of nitrogen per year, about five times the amount generally accepted as typical. Nitrogen fixation in the soil might be further improved by artificially selecting the most efficient strains of symbiotic bacteria, and by encouraging the widespread adoption of improved legumes, particularly in the tropics. It was recently discovered that bacteria capable of nitrogen fixation live in partially symbiotic associations with certain tropical grasses, including maize. The genetic manipulation of bacteria offers hope that the capacity to fix nitrogen may eventually be conferred to all crops! A recent development that shows promise is the discovery of an inexpensive substance (nitropyrin) that retards the bacterial conversion of ammonia to nitrite in the soil. Since ammonia is a cation and is retained by soil colloids, whereas nitrite and nitrate are readily leached away, nitropyrin could retard the loss of nitrogen from the soil. Finally, the absorptive capacity of some plant roots is increased by an intimate association, called a **mycorrhiza,** between the roots and a fungus. The encouragement of mycorrhizal associations might benefit certain crop plants, particularly in soils of generally low fertility or where specific nutrients may be fairly scarce.

Soil Conservation

Agriculture is conducted in a finite and decreasing land area throughout the world's temperate climates. Its basis is a layer of topsoil that averages only seven inches over the earth's surface. This delicate mantle of topsoil cannot be exploited indefinitely; it must be conserved and renewed.

Although the reduction of the productive capacity of soils through the loss of fertility and structure is considerable, the most serious problem is erosion. Nutrients can be added by fertilization, but the loss of topsoil cannot be so easily or quickly remedied. The loss of soil due to wind and water is a national problem. Erosion clogs rivers with silt, complicates droughts and floods, and compounds poverty. Yet soil conservation need not be practiced for purely altruistic reasons. Soil conservation yields immediate rewards in terms of plant growth, and it must be considered the basis of sound soil management.

Erosion of the soil is a natural process influenced by climate, topography, and the nature of the soil itself. Where permanent and undisturbed plant cover exist, erosion is more or less gradual and in equilibrium with soil-formation processes. Erosion is accelerated in the absense of plant cover. Areas that, because of climate or topography, are unable to support a permanent plant cover undergo a "geologic" erosion, such as is found in the Grand Canyon. The accelerated soil erosion brought about by cultivation or overgrazing comes about principally through the action of water in humid climates (Fig. 6-9) and through the action of wind in arid climates.

The maintenance of vegetative cover is basic to soil management. Vegetative cover retards erosion by breaking and cushioning the beating force of the rain (Fig. 6-10), by increasing the absorptive capacity of the soil, and by holding the soil against both water and wind. The soil cover increases the infiltration of water through the soil by preventing the clogging of the soil pores by fine surface particles. The techniques used for increasing soil cover include sod culture (as in orchards), crop rotation, cover cropping, and mulching.

FIGURE 6-9. Mountain runoff has severely eroded this vineyard in Santa Fe Springs, California. [Photograph courtesy USDA.]

Water erodes the soil by literally carrying it away. The carrying power of moving water increases with its speed and volume. The volume of water depends upon the amount of rainfall and the rate at which it is absorbed by the soil. The speed with which this water moves is directly related to the slope of the land and the amount of cover. Any technique that either increases absorption or reduces the speed of the runoff will help to prevent soil erosion.

The absorptive capacity of the soil may be increased by deep plowing, by increasing the amount of organic matter in the soil, or by increasing drainage. Thus, the burning or removal of organic matter is a poor conservation practice. Where natural drainage is poor, the installation of drainage tiles beneath the surface may be necessary to remove water and introduce air.

Controlling erosion by reducing the speed of runoff may be accomplished in a number of ways. Most basic is **contour tillage,** in which plowing, cultivation, and the direction of the "row" follow the contour of the land rather than its slope. This slows runoff and therefore increases the ability of the tilled soil to absorb water (Fig. 6-11). The use of **intertillage,** or **strip cropping,** which alternates strips

FIGURE 6-10. The beating force of a raindrop striking wet soil. Soil particles and globules of mud are hurled in all directions. [Photograph courtesy U.S. Soil Conservation Service.]

FIGURE 6-11. Contour planting of a field of pineapple in Hawaii. [Photograph courtesy Dole Corp.]

199

of sod and row crops planted along the contour, helps to slow runoff by interposing barriers with high absorptive capacity, and this alternation also provides the benefits of rotation. On steeper slopes, where greater amounts of surface water must be accommodated, the use of **waterways** (permanently sodded areas), facilitates water removal and minimizes erosion.

Where contour cultivation and strip cropping are not sufficient to check erosion, **terraces** constructed on the contour must be used. Terracing, an ancient practice, consists of cutting up a slope into a number of level areas. Terraces appear as steps on the hillside. Although the boundaries of ancient terraces were made of stone, modern terraces are made by building low, rounded ridges of earth across the sloping hillside (Fig. 6-12). Terracing slows down the speed of surface runoff, and, although it is designed primarily to prevent erosion, it also facilitates the storing of available water. Thus, terracing is an important practice in areas of low rainfall.

Some of the erosion caused by wind, especially in open prairies or plains, may be checked by planting **windbreaks**—one or more rows of trees or shrubs planted at right angles to the prevailing winds (Fig. 6-13). The effectiveness of a windbreak is local and is related to the thickness and the height of the trees. The maintenance of a permanent plant cover in conjunction with windbreaks effectively reduces wind erosion. On organic soils, small grain rows are used as temporary windbreaks to protect vegetable seedlings.

FIGURE 6-12. A terraced peach planting on a fine sandy loam. The terrace ridges have been left in the rye cover in order to protect the land against wind erosion. [Photograph courtesy U.S. Soil Conservation Service.]

FIGURE 6-13. Tree plantings act as windbreaks to reduce wind velocity and thus protect crops as well as homes and livestock from cold and hot winds. Shelterbelts consisting of 5- to 10-row strips of trees help to control wind erosion, contribute substantially to the saving of moisture, and protect crops from damage by high winds. [Photograph courtesy USDA.]

WATER MANAGEMENT

Water management in agriculture is principally concerned with the regulation and use of water to effect plant growth. In addition to the conservation of water, it involves controlling the effects of water excesses and water shortages. Because of the close association between water and soil, the control of moisture must be an integral part of soil management.

Irrigation

And a river went out of Eden to water the garden; and
from thence it was parted, and became into four heads.
GENESIS 2:10

The great ancient civilizations of Egypt, Babylon, China, and Peru were dependent upon irrigation for the abundant agriculture necessary to support their large populations. They developed complex systems of water distribution that involved ditches, canals, and waterways. Mechanical devices such as the dragon wheel and Archimedes' screw were used to move water uphill. Although irrigation has long been a vital part of the agriculture of the semiarid, subtropical climates, it has not been widely practiced in the humid temperate climates until recently. At present, supplemental irrigation is increasing as a component technology of horticulture. This has been brought about by improvements in irrigation technology, such as increased pump efficiency, light-weight tubing, and sprinkler systems.

Types of Irrigation

There are four general methods of soil irrigation: surface irrigation, sprinkler irrigation, subirrigation, and trickle irrigation. In **surface irrigation,** the water is conveyed directly over the field, and the soil acts as the reservoir for moisture. In

sprinkler irrigation, water is conveyed through pipes and is distributed under pressure as simulated rain. In **subirrigation,** the water flows underground as a controlled water table over an impervious substratum and provides moisture to the crop by upward capillary movement. In **trickle irrigation,** water is applied slowly to individual plants through small tubes.

SURFACE IRRIGATION. The application of water by surface irrigation is utilized in arid and semiarid regions where the topography is level (Fig. 6-14). The water is conveyed to the fields in open ditches at a slow, nonerosive velocity. Where water is especially scarce, pipelines may be used, since they eliminate losses due to seepage and evaporation. The distribution of water is accomplished by various control structures or by siphons. The flow of water must be carefully controlled, for water is usually fairly expensive in dry regions, and excessive leaching of water-soluble nutrients and erosion of soil may occur with too rapid a flow. Drainage canals must be provided to remove waste water and eliminate ponding. The advantages of surface irrigation over sprinkler irrigation are that it requires less power and less water is lost to evaporation. But since the system depends on gravity flow, it is inefficient in distribution because more water is supplied to those areas closest to the source. Another serious objection to surface irrigation is the deleterious effect that it has on soil structure. Heavy soils become puddled under the heavy load of water, which results in a loss of soil aeration and in subsequent baking and cracking when the soil dries out.

The distribution of water over the field in surface irrigation may be accomplished either by flooding the entire field in a continuous sheet (**flood irrigation**) or by restricting the water to some type of furrow (**furrow irrigation**). The field must be almost level, or low spots will get too much water and high spots will get none.

FIGURE 6-14. Furrow irrigation in a lettuce field in the Salinas Valley, California. [Photograph by Ansel Adams, courtesy Wells Fargo Bank, San Francisco.]

Flood irrigation is common in horticultural crops that are tolerant of excessive water, as are cranberries. Furrow irrigation is the most widely used method of applying water to row crops. Although furrow irrigation is fairly efficient in water utilization, it requires high labor costs. Constant supervision is required to prevent furrow streams from uniting and forming large channels. On rolling land, and with closely growing crops, water may be applied to small furrows that guide rather than carry the water. This is referred to as **corrugating irrigation. Soakers** (perforated hoses that allow water to seep into the soil at a low, uniform rate) provide a form of surface irrigation that can be employed in greenhouses and in home and gardens.

SPRINKLER IRRIGATION. Sprinkler irrigation, although not new, has come into widespread use since 1945, owing to the introduction of light aluminum pipe, quick couplers, and improved nozzles. Until then, sprinkler systems consisted of permanent "overhead" installations that were confined to use in small market gardens. Portable, light-weight pipe has put sprinkler irrigation into a prominent position in agricultural technology. It has proved practical as a means of providing supplemental water in the so-called "humid" climates. The advantages of sprinkler irrigation lie in the even and controlled application rate. Although in arid climates evaporation may be higher with sprinkler irrigation than with surface irrigation, the controlled application rate results in a more efficient use of water. The slower rate of application reduces runoff, erosion, and compaction of the soil. In addition, sprinkler irrigation can be used on land that is too steep to permit the proper use of other methods, and makes full use of land by eliminating the need to devote some space to a permanent distribution system. The limitations of sprinkler irrigation lie in the high initial cost, although operational labor costs are reduced. The power requirements are high, for water pressures of 15–100 pounds per square inch (psi) must be maintained (Fig. 6-15). The chief operational

FIGURE 6-15. Three large diesel engines of this type provide power for sprinkler irrigation on a large Midwestern sod farm. [Photograph courtesy International Harvester Corp.]

FIGURE 6-16. Sprinkler irrigation is practical as a result of portable, light-weight, aluminum pipe. The sprinkler pattern must be overlapped by about 40% in order to achieve uniform application of water. [Photograph courtesy USDA.]

difficulty is high wind, which disturbs the sprinkler patterns and results in uneven water distribution.

Sprinkler-irrigation equipment operates with nozzles or perforated pipe. Rotating sprinkler heads, the most widely used type, apply water to circular areas at rates from 0.2 inch/hour to more than 1.0 inch/hour (Fig. 6-16). In order to obtain proper coverage, each wetted circle should overlap adjacent circles from one-fourth to one-half.

SUBSURFACE IRRIGATION. Subsurface irrigation consists of creating and maintaining an artificial water table. In order for such a system to function properly, the ground must be level, and subsurface soil must be permeable enough to permit the rapid movement of water laterally and vertically. A bottom barrier, such as an impervious layer of soil or rock, must be present to prevent the loss of water through deep percolation. A distribution system of ditches and laterals permits the artificial water table to be raised or lowered by pumping water into or out of the system. There are, however, relatively few places in the United States where these specialized conditions exist. A good example is the San Luis Valley of Colorado, an important potato-producing area. Subirrigation is also fairly common in organic soils.

The level at which the water table is maintained in organic soils can be very important. For most plants, the most favorable plant growth is achieved with a water table 24–36 inches from the surface of the soil, but such crops as mint and celery require a water table about 18 inches from the surface. The water table also affects organic soil subsidence, which is the "loss" of soil due to drying, settling, compaction, slow oxidation, and wind erosion. The rate of subsidence increases as the water table drops. One of the problems associated with a high water table, however, is caused by excessive, unexpected rainfall. Unless the water table can be quickly lowered, uncontrolled flooding may result and cause extensive crop damage.

TRICKLE IRRIGATION. Trickle irrigation (also called drip, high-frequency, or daily-flow irrigation) employs a ramifying system of water-conducting plastic tubes that deliver water, by means of "emitters" or leaks, to individual plants. Water is applied frequently, but very slowly and in small amounts so that little is lost by evaporation or to portions of the field not occupied by plant roots. Water is generally applied at a rate of 1 to 2 gallons per hour under low pressure (15 psi or less). Since moisture stress tends to equalize throughout the plant, as little as 25% of the root system need be wetted.

Trickle irrigation is an extremely versatile method. It has proved useful in the floriculture industry to automate the watering of plants in individual pots (Fig. 6-17). In the greenhouse, each terminal tube of the system ends in a valve that permits individual control and is weighted to keep it in place in the pot. The entire system can be automated for watering and fertilizing with a timing or sensing device (such as a scale to weight the watered pot).

Trickle irrigation has proved to be of greatest value to orchard systems, although it is also used extensively for strawberries and row vegetable crops. The system is flexible because fast and simple coupling and uncoupling are accommodated by either piercing or plugging of the plastic tubes that carry the water. Various types of emitters precisely regulate the flow and reduce the pressure from the "leader" line. A popular type of emitter is a thin tube or pathway through a nozzle with an inside diameter of about 0.9 mm. The length of the tube regulates the pressure—the longer the tube, the slower the rate of delivery. In this way, the delivery of water can be equalized to plants located at different distances from the water source.

Drip irrigation is attractive economically in many horticultural situations. The principal benefit is that plants can be supplied with very nearly the precise amount of water they require. In arid regions, trickle irrigation may have dramatic consequences because evaporation from traditional irrigation systems causes salts to accumulate on the soil surface. Additional water supplied in excess of the amounts required for plant needs will leach these potentially damaging salts from the soil zone close to the roots. Providing this water by trickle irrigation is far more efficient than either wetting the entire field or supplying water through furrows. The waste of water between rows and through deep percolation are eliminated, which eliminates fertilizer contamination of deep underground water supplies.

When saline water is used for irrigation, salt buildup can be controlled by continuous leaching. Salts are pushed out to the periphery of the root profile by an advancing front of water emitted from the emitters. The roots are able to take up water freely from the middle of the wet zone, where soil-moisture tension is low and the salt level remains nearly the same as it is in the irrigation water.

The delivery system of trickle irrigation is very light compared to other systems, and can easily be moved by one person. Fertilization is efficiently combined with irrigation by the addition of soluble fertilizers to the system. Trickle irrigation has been used on slopes as high as 50% without erosion problems because delivery is in a closed system under pressure, only small amounts of water are applied, and there is no runoff.

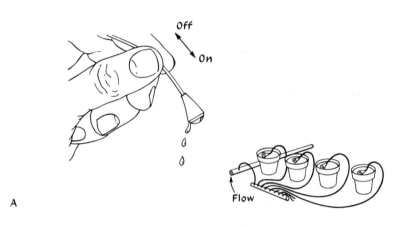

FIGURE 6-17. Trickle irrigation: A, the Chapin System of trickle irrigation for greenhouse watering uses weighted valves (*left*) to deliver water to individual pots (*right*). B, trickle irrigation installation for tomatoes in a plastic greenhouse. C, trickle irrigation systems used in the field. Delivery systems may be modified for different crops. In an orchard densely planted to small trees, pipes can be laid parallel to the rows of trees and fitted with numerous emitters. With large trees, the pipes can be looped around each tree and fitted with several emitters per loop. For row crops, the pattern may be one pipe per row or one pipe in every other row. D, nozzle-like emitters reduce water pressure by causing the water to move along an elongated helical path. The emission holes are made larger when pressures are low so that clogging is reduced. [Courtesy Merle Jensen.]

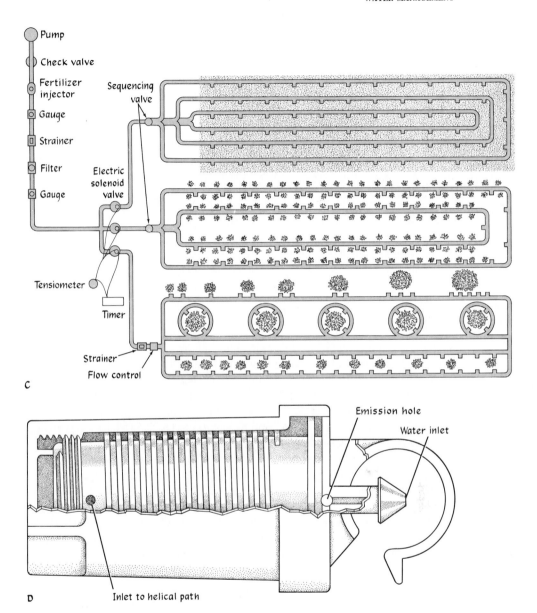

The emitters are preset during construction to emit from 1 to 2 gallons of water per hour. The hook at the right attaches to the delivery pipe at a right angle; water enters the conical tip from the center of the pipe. [Parts *C* and *D* after Shoji, "Drip Irrigation," copyright © 1977 by Scientific American, Inc., all rights reserved.]

Because most of the soil surface stays dry, especially between plants, the crop field is essentially free of weeds. Weeds can be a problem in wet areas near emitters, but they can be controlled by selective herbicides.

Some of the problems of drip irrigation include the high costs of installation and difficulties in keeping equipment operational. Bacterial slime may build up in the plastic tubes, and small soil particles may block the water outlets because the low water pressure used in the system is frequently insufficient to unclog them. In some areas, ants seeking water cut the tubes open, and rodents have been known to use the plastic tubing to sharpen their teeth. Additives in water to decrease bacterial growth, and in the plastic tubing to repel insects and mammals, should solve these problems.

Trickle irrigation, first developed in Israel, has gained wide acceptance world-wide, especially in arid areas with high labor and water costs. It is now the fastest growing method of irrigation, with extensive acreage in the United States, (particularly California), Australia, Israel, South Africa, and Mexico. In the United States, trickle irrigation increased from barely 100 acres in 1960 to 134,000 acres in 1975.

Determining Irrigation Requirements

Irrigation, though capable often of yielding enormous benefits, can also be a wasteful and harmful practice if applied incorrectly. Determining when to water and how much water to supply is one of the main problems in irrigation. It requires the establishment of an accounting system of sorts that will indicate whether or not the available water will meet crop needs in spite of losses due to evaporation, transpiration, runoff, and percolation. The net deficiency not compensated by natural precipitation may be made up by irrigation. Timing can be extremely critical. In snap beans, for example, moisture stress during flowering and pod formation seriously depresses the yield by promoting flower abscission and ovule abortion.

There are two approaches to the determination of irrigation requirements. One is the measurement of soil moisture, from which the moisture available to the plant is determined. The other is calculation of water availability from meteorological data.

Measuring soil moisture accurately is essential for determining irrigation requirements. Although the experienced person can evaluate soil moisture from the "feel" of the soil (Table 6-3), this rough test varies with the soil and the person making the test, and there are now various objective methods available. Soil moisture may be calculated from the weight of soil taken before and after oven drying. Although exact, this gravimetric procedure is slow and laborious. More rapid methods are based on the ability of a sorption block made of gypsum or other porous material to absorb water in proportion to the amount present in the soil. The percentage of the soil moisture in the plaster block may be determined

TABLE 6-3. Feel chart for the determination of moisture in medium- to fine-textured soils. With sandy soil, the balls are more friable and fragile throughout the whole range.

Degree of moisture	Feel	Percent of field capacity
Dry	Powder dry	0
Low (critical)	Crumbly, will not form a ball	Less than 25
Fair (usual time to irrigate)	Forms a ball, but will crumble upon being tossed several times	25–50
Good	Forms a ball that will remain intact after being tossed five times, will stick slightly with pressure	50–75
Excellent	Forms a durable ball and is pliable; sticks readily; a sizable chunk will stick to the thumb after soil is squeezed firmly	75–100
Too wet	With firm pressure, some water can be squeezed from the ball	In excess of field capacity

SOURCE: Strong, 1956, in *Sprinkler Irrigation Manual* (Wright Rain, Ringwood, England).

by weight or by direct measurement of the electrical conductance or resistance between electrodes inserted in the block. Another device, called a **tensiometer,** gives an indication of water availability. It consists of a porous cup filled with water and attached to a vacuum gauge or mercury manometer that measures the tension at which the water is held to the soil. In addition, other sophisticated procedures are available that are based on the thermal properties or on the neutron-scattering potential of the soil. At present, the gravimetric method must be considered the most accurate. The tensiometer and other rapid moisture-measuring devices have not proved altogether successful, owing in part to the random variation of soil moisture, the difficulty of achieving intimate contact of a sorption block with the soil, and the problem of determining the best place in the root zone for making the measurement.

Meteorological and climatological data offer a powerful tool for estimating the amount of available water. The procedure involves the calculation of **consumptive use**—the water lost by evaporation and transpiration—which is probably the best index of irrigation requirements. Consumptive use varies with a great number of factors: temperature, hours of sunshine, humidity, wind movement, amount of plant cover, the stage of plant development, and available moisture.

High rates of water consumption are associated with a high percentage of plant cover and with hot, dry, windy conditions. Because optimum plant health is associated with an adequate uninterrupted supply of water, the peak requirements must be considered. Crops have the highest water requirements in the

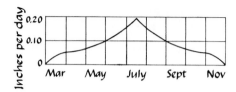

FIGURE 6-18. Rate of water use by peaches in a cool, dry climate. [Adapted from Strong, in *Sprinkler Irrigation Manual*, Wright Rain, Ringwood, England, 1956.]

fruiting or seed-forming periods (Fig. 6-18). An approximate relationship between peak moisture use by horticultural crops and climate is given in Table 6-4. Plants differ in their water requirements largely in relation to their ground covering ability and their depth of rooting (Table 6-5). The relationship between water requirements and depth of rooting might be somewhat less than expected because deep-rooted plants obtain most of their moisture from the upper part of the root zone.

The amount of available moisture provided by rainfall may be very much less than the total rainfall. Owing to evaporation from the soil and the slow rate of infiltration, showers of less than $\frac{1}{4}$ inch during hot summer days may contribute very little to available soil moisture. On the other hand, a high proportion of water from heavy precipitation may be lost by runoff. The effectiveness of precipitation, therefore, depends upon the intensity of rainfall, as well as upon the amount, in relation to temperature and the absorbing capacity of the soil.

It has been possible to determine a satisfactory consumptive-use index for a particular area by using monthly averages of mean temperature and hours of sunshine. Empirically derived constants are available for adjusting these values for

TABLE 6-4. Average peak moisture use (in inches of water per day) for commonly irrigated horticultural crops.

Crop	Cool climate		Moderate climate		Hot climate	
	Humid	Dry	Humid	Dry	Humid	Dry
Potato	0.10	0.16	0.12	0.20	0.14	0.24
Tomato	0.14	0.17	0.17	0.22	0.23	0.27
Bean	0.12	0.16	0.16	0.20	0.20	0.25
Vegetables	0.12	0.15	0.15	0.19	0.20	0.23
Deciduous orchard	0.15	0.20	0.20	0.25	0.25	0.30
Deciduous orchard with cover	0.20	0.25	0.25	0.30	0.30	0.35
Citrus orchard	0.10	0.15	0.13	0.19	0.18	0.23

SOURCE: Strong, 1956, in *Sprinkler Irrigation Manual* (Wright Rain, Ringwood, England).

TABLE 6-5. Normal root-zone depths of mature irrigated crops grown in a deep, permeable, well-drained soil.

Crop	Root depth (ft)	Crop	Root depth (ft)
Alfalfa	5–10	Grass pasture	3–4
Artichoke	4	Ladino clover	2
Asparagus	6–10	Lettuce	$\frac{1}{2}$
Bean	3–4	Onions	1
Beet (Sugar)	4–6	Parsnip	3
Beet (Table)	2–3	Pea	3–4
Broccoli	2	Potato	3–4
Cabbage	2	Pumpkin	6
Cantaloupe	4–6	Radish	1
Carrot	2–3	Spinach	2
Cauliflower	2	Squash	3
Citrus	4–6	Sweetpotato	4–6
Corn (Sweet)	3	Tomato	6–10
Corn (Field)	4–5	Turnip	3
Cotton	4–6	Strawberry	3–4
Deciduous orchard	6–8	Walnut	12+
Grain	4	Watermelon	6

SOURCE: Shockley, 1956, in *Sprinkler Irrigation Manual* (Wright Rain, Ringwood, England).

different crops. Any water deficit can be calculated by subtracting the consumptive-use requirements from the available water present. Potential evapotranspiration (the combination of evaporation and transpiration when the surface is completely covered with vegetation and there is an abundance of moisture) can be calculated by using the Bellani black-plate atmometer, a relatively simple instrument used to measure evaporation (Fig. 6-19), and by estimating the percentage of ground cover.

Because not all of the irrigation water applied is available to the crop, the amount applied must be based on **irrigation efficiency,** the percentage of applied irrigation water that actually becomes available for consumptive use. Water should be applied to bring the soil up to field capacity at a depth commensurate with the bulk of the feeder root system. The rate must be consistent with the absorptive properties of the soil. Irrigation is best applied when the water tension in the zone of rapid water removal goes above four atmospheres or when 60% of the available water in the root zone is depleted. The amount of water that can be

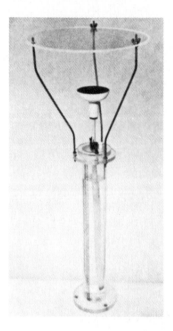

FIGURE 6-19. The Bellani black-plate atmometer measures evaporation. Potential
evapotranspiration can be calculated by
taking readings with this instrument and
estimating total plant cover. [Photograph
courtesy W. H. Gabelman.]

efficiently utilized is primarily related to the level of soil fertility. The maximum
benefits of irrigation are dependent upon the existence of a readily available
nitrogen supply.

Drainage

Drainage is the removal of excess gravitational water from the soil. Under
conditions of good natural drainage, surplus surface and soil water is rapidly
removed to streams and rivers. The poor natural drainage of some areas is a result
of several factors. Such areas may have a high natural water table caused by an
impervious layer that prevents downward percolation, resulting in **waterlogged
soils.** Others may simply be low-lying in relation to surrounding drainage. Some
areas are subject to flooding brought about by the overflow of streams and rivers.
Flooding can be averted either by building protective levees or by controlling the
rate of water movement. This can be accomplished, as part of a program of
upstream watershed management, by controlling excess runoff or by constructing
dams and reservoirs to restrict the flow in times of excess water movement.

The facilitation of natural drainage is both a land-reclamation practice and a
cultural practice. The permanent drainage of wetlands has been a significant
factor in the expansion of agriculture in the eastern United States, where some of
the most productive cropland was formerly "worthless" marsh and swamp. Not all

wetlands are suitable for drainage, but they still remain valuable for wildlife, forest, and recreational use. As a cultural practice, drainage consists of removing the excess water that interferes with plant development and with the performance of such operations as tillage and harvesting. It is necessary when the natural removal of water by runoff, percolation, and evapotranspiration is too slow. Drainage extends the potential growing season by permitting earlier tillage in lands that are otherwise too wet in the spring.

Excess water can be removed by surface or subsurface drainage. Surface drainage refers to the removal of surface water by developing the slope of the land. Subsurface drainage is accomplished by the construction of open ditches and tile fields to intercept ground water and carry it off. The water enters the tiling through the joints, and drainage is achieved by the effect of gravity on the water tiles. Drainage design, which requires determining the depth, size, and number of drains to be installed, is an application of the physics of ground-water movement.

Water Conservation

Water conservation is of national concern. For a nation to prosper, an abundant source of high-quality water must be available for agricultural and industrial use as well as for human consumption and sanitation. The misuse of water resources leads to alternate flood and drought, problems that affect all of us.

Water conservation implies the proper stewardship of our water resources as a whole. It may involve large programs to control flooding, to develop hydroelectric power, and to facilitate navigation. These are projects that require national effort. Water conservation also involves the control of water resources on a smaller scale. It must therefore be a part of the water management of every individual enterprise.

Soil-management practices developed for the efficient utilization of water involve the control of soil erosion and the conservation of soil moisture through the control of runoff and the implementation of methods to increase the water-absorbing capacity of the soil. It may include such practices as mulching or the close mowing of sod in orchards, both of which are designed to reduce the removal of water from the soil. Because horticultural crops are great users of water, the extreme practices of dryland farming, such as fallowing to conserve moisture without resorting to irrigation, are usually not practical. In fallowing, the ground is left unplanted for a whole year and is cultivated only to eliminate weeds, in an effort to build up soil moisture. For horticultural crops to be grown where water is insufficient, irrigation is essential. But irrigation depends on large sources of water; a source of water supply must be developed if organized irrigation facilities are not available and if lakes or streams with sufficient flow do not adjoin the property. Irrigation wells offer one possibility. These are large-volume wells capable of supplying great quantities of water required. Storage ponds or reservoirs are becoming increasingly important as sources of irrigation

FIGURE 6-20. Artificially created storage ponds are an important source of irrigation water. [Photograph courtesy U.S. Soil Conservation Service.]

water. They are usually made by constructing an earthen dam across a gulley or an intermittent or spring-fed stream (Fig. 6-20).

The problem of water rights has social as well as economic implications, which are reflected in our laws. The **Riparian law,** the common law involving water rights with respect to rivers and streams, has established a legal framework for disputes concerning water diversion and distribution. In this common law, property rights do not involve complete water rights except for personal use. Neither the landowner nor anyone else owns the water or may divert it from its normal flow. However, because of differences in water availability from one region to another, the common law has been modified throughout the United States. The right to use water from streams for irrigation is variable and depends on state law. Similarly, the right to pump underground water for irrigation differs widely from state to state. The legal codes must be clearly understood in situations concerning irrigation and drainage procedure.

TEMPERATURE CONTROL

Plant growth shows a marked response to small changes in temperature, and if extremes in temperature persist for even short periods, they will lead to irreversible changes of state, resulting in the death of the plant or parts of it. Methods for the control of temperature in the culture of horticultural crops vary greatly. For the great majority of crops there is no active control but rather an adaptation through selection of location, site, and choice of plant. This is discussed further in Chapter 12. For some field-grown horticultural plants an active attempt may be made to modify and ameliorate extremes in temperature through cultural practices such as mulching and various techniques of frost control. The regulation of temperature in greenhouse culture can be complete, including artificial heating and cooling.

Cultural Practices

Mulching

Mulches are insulating substances spread over the surface of the soil (Fig. 6-21). Although one of their chief purposes is the regulation of soil temperature, they serve many other functions. Mulches conserve soil moisture because they reduce evaporation by lowering the soil temperature and by increasing the absorptive capacity of the upper layer of the soil. Erosion is reduced as a result of decreased surface runoff and the shielding effect of the mulch to driving rain. Mulch is commonly applied for this reason to newly planted lawns and seed beds, especially on sloping areas. Mulches may control weeds and eliminate the need for cultivation by smothering weed growth and cutting off light from the soil surface. They offer protection to flowers and fruit from mud-splattering rain. This is especially important in such low-growing crops as strawberries. In addition, mulches may be a source of organic matter and nutrients for the soil. Mulching is often desirable for its own sake, since its pleasing appearance provides an attractive background for flowers and other plant materials.

Mulches may be applied during the period of active growth (**summer mulch**) or be restricted to late fall to provide cold weather protection (**winter mulch**). Although the benefit of the summer mulch is attributed to a number of factors (for example, moisture conservation and weed control), the principal benefit of winter protective mulch is its influence on the temperature of the soil.

The temperature-stabilizing effect of summer mulches is due to insulation, heat absorption, and shading. The surface of bare, dark-colored soils on a sunny

FIGURE 6-21. Corncobs make an inexpensive mulch for apple orchards in the Corn Belt states. The mulch is spread around the drip line of the tree.

FIGURE 6-22. Strawberries grown on a plastic mulch in California. [Photograph courtesy Victor Voth.]

midsummer day may be higher by 30°F (almost 17°C) than the air temperature. The reduction in soil temperature attained as a result of mulching appears to increase nutrient availability. It also improves root growth and, ultimately, the performance of many plants.

The practice of using plastic sheeting as a summer mulch has shown a tremendous increase throughout the world in the production of fresh market vegetables. This system is a standard practice for strawberry production in California (Fig. 6-22). Plastic mulch has a great influence on soil temperature, soil moisture, and weeds. Clear plastic increases soil temperatures, encouraging early production; but it also stimulates weed growth unless it is used in conjunction with chemical weed-control measures. Opaque black plastic shades the soil and controls weeds.

The use of foams to insulate plants temporarily to control frost damage is a recent innovation that incorporates the principles of a mulch (Fig. 6-23). The foam is a combination of surfactant, stabilizer, and protein material (such as gelatin). Application is made the day before frost is expected; the foam dissipates a few hours after sunrise.

The temperature-regulating effect of a winter protective mulch is two-fold. One effect is to temper extremely low winter temperatures. This is achieved through the insulation effect provided by the mulch, which conserves ground heat. The other effect is to stabilize and buffer soil temperature and prevent recurring freezing and thawing, which rips and injures plant roots through soil heaving. During winter warm spells in cold climates, a mulch tends to keep the

FIGURE 6-23. Foam applied to young vegetable seedlings for frost control. [Photograph courtesy USDA.]

ground frozen by providing insulation and shading. Thus, not only does a mulch "warm" the plant under extreme winter weather, but it also keeps it cold during unseasonable warm spells. By keeping them under a winter mulch, spring-flowering plants may be delayed from early blooming to avoid the damaging effect of spring frost.

The application of winter mulch is usually made after a light freeze so as not to delay dormancy. Tender plants such as roses may be protected by mounding the crown with soil. After the mound is frozen it is covered with an insulating organic mulch. Winter mulching is a standard practice in strawberry culture. After the plants have become dormant in the fall but before heavy injurious freezes, the entire planting is usually covered with 2 to 3 inches of straw (Fig. 6-24). The plants are uncovered in spring, when growth can no longer be prevented. The excess straw is then moved to the middle of the rows. If frost is expected during flowering, re-covering the plant provides a measure of protection.

Most mulching materials consist of plant refuse or by-products: leaves, straw, sawdust, corn cobs, peat, tobacco stems, pine needles, wood chips, or paper. The main virtues of summer mulches are relatively independent of the material. Inorganic substances, such as rockwool or gravel, are also effective. A good mulch must be economical, available, and easy to handle. It must also be stable so that it will not easily wash or blow away. Mulches used around the home must be unobjectionable in odor and appearance.

Some of the problems associated with mulching materials arise from their tendency to act as sources and harborers of plant pests—weeds, disease-producing microorganisms, and rodents. Because of the disease problem, the refuse of the plant being protected—such as its own fallen leaves—should not be used as mulch. Straw that has been improperly handled may contain weed or grain seed, which may contribute to the weed population the following spring. Fresh straw should be prespread and moistened during warm weather to induce germination of any seeds it contains before it is used as a mulch. The use of mulch in orchards

FIGURE 6-24. Mechanized mulching of strawberries. [Photograph courtesy Friday Tractor Co., Hartford, Mich.]

must be accompanied by vigorous measures for rodent control, lest rodent populations build up to damaging levels in the favorable environment that a deep mulch affords.

Many mulching materials are highly inflammable and present a fire hazard. Straw mulch in particular should not be placed too close to buildings. Unless partially decomposed, fresh leaves make unsatisfactory mulch because they tend to pack closely, and may smother plants. Although organic mulches decompose and will eventually contribute plant nutrients, the high carbon content of many of these materials may contribute to nitrogen deficiency. This is especially true if the mulch is later plowed under. This can be avoided by applying extra nitrogen.

Frost Control

A number of techniques can be used to avoid the destructive consequences of spring frost. They may, for example, be avoided by late planting. Although frost conditions may be predicted on a probability basis, it is not always practical to plant at what is calculated to be the last frost-free date. Even this date is only a statistic. The spring culture of seedlings in protective structures circumvents the dangers of frost. Frost control for perennial plantings, however, must depend on more substantial procedures.

The judicious choice of location and site remain the main bulwark against frost. Cultural practices used in the control of frost involve techniques that either encourage the conservation of heat or add heat directly to the immediate environment of the plant. The conservation of heat is brought about by any method that will increase daytime absorption of heat by the soil or prevent its loss at night. This can be accomplished by using hot caps (Fig. 6-25), by cultivating, or by fogging. The addition of heat can be accomplished in a number of ways—for

FIGURE 6-25. Hot caps protect early tomatoes in California's San Luis Rey district. The hot cap is made of a translucent paper and acts as a miniature greenhouse. [Photograph courtesy USDA.]

example, by using heaters, flooding, spray irrigation, or artificial air movement. These methods will be discussed more fully along with the meteorological aspects of frost in Chapter 12, "Horticultural Geography."

Plant-Growing Structures

Cold Frames

An inexpensive form of temperature control for seedlings and transplants during the early spring can be achieved with the **cold frame.** A cold frame is an enclosed ground bed, usually sunken, with a removable sash. Heat is provided through the trapping of solar energy. Temperatures inside the cold frame increase relative to the air during the day when the sash is in place due to the "greenhouse effect" discussed in Chapter 4. Heat is stored in the soil during the night, and plants can be protected even though outside air temperatures dip below freezing. With especially low temperatures, insulating material such as straw is sometimes placed over the sash. Temperatures are maintained during the day by raising or removing the sash. Cold frames are commonly used for starting early transplants from seed or as a means of hardening off greenhouse-grown transplants.

Hotbeds

Hotbeds are essentially cold frames provided with a supplemental source of heat. Additional heat may be provided by fermentation, hot water, steam, or electricity. Fermentation heat is provided from decaying organic matter, most commonly strawy manure, placed under the plants. Hot air, steam, or hot water systems are arranged to heat the soil by conduction. Electrical heating also provides ground heat through the use of a soil-heating cable. Thermostatically regulated electrical heating provides precise temperature control. Such systems can be easily installed; the operating cost, of course, depends upon local electrical rates.

Greenhouses

Greenhouses (in England they are referred to as glass houses) are usually elaborate, permanent structures equipped not only to regulate temperatures but to provide increased environmental control of plant growth (Fig. 6-26). Because of the great amount of control that must be achieved in greenhouses, this type of culture becomes an extremely specialized operation.

In ordinary greenhouses, temperature is regulated through a heating and

FIGURE 6-26. Experimental greenhouse used by the National Aeronautics and Space Administration (NASA) to investigate growth of plants in space stations.

FIGURE 6-27. Fan-and-pad installation for cooling greenhouses. Cooling pads are at right, fans are in roof at left. [Photograph courtesy Acme Engineering and Manufacturing Corp., Muskogee, Okla.]

ventilation system similar to that of the hotbed. In cold climates, a central coal or oil furnace supplies the heat. In Europe, portable "steam plants" are available for this purpose. Peripheral steam heating is the most commonly used distribution system, although heating pipes may be placed under benches in large greenhouses. Ventilation is provided at the sides and top of the structure. Automatic controls are available for both heating and ventilation.

Temperature control during cold weather is a matter of adjusting heating and ventilation to take maximum advantage of solar heat. In warmer weather, however, it becomes increasingly difficult to maintain reasonable temperatures for plant growth with an ordinary ventilation system. Some greenhouse cooling is achieved by shading the glass with a whitewash spray. The whitewash is made in such a way that it will weather off naturally by fall. Fan-and-pad cooling provides an economical system for lowering summer greenhouse temperatures. In this method, cooling is achieved by the evaporation of circulated water through a pad of excelsior or some other coarsely porous material with a high ratio of surface area to volume (Fig. 6-27). Fans opposite the cooling pads draw the cooled air across the greenhouse. The efficiency of the system increases as the humidity goes down. Even in the hot, humid midwestern United States, temperatures can be kept at least on a par with the outdoor shade.

The use of refrigeration equipment is not economical for commercial greenhouse cooling, although it is used to obtain uniform temperatures for experimental

FIGURE 6-28. Plastic greenhouses are convenient, inexpensive structures. The polyethylene plastic is removed in the spring, when temperatures get too high, and is replaced in the fall. The frame can then be covered with shade cloth and the structure converted to a shade house. *Top,* a scissor-type, truss-rafter plastic greenhouse in the process of being covered. *Middle,* a gothic-rafter plastic greenhouse. The insulated pipe carries steam from the greenhouse range. *Bottom,* a gothic-type plastic greenhouse constructed from aluminum pipe used to overwinter strawberries. The strawberries are grown in plastic tunnels with plastic mulch. [Photographs courtesy P. H. Massey, Jr., and Merle H. Jensen.]

FIGURE 6-29. Rigid plastic (fiberglass) covered greenhouses. [Photographs courtesy University of California.]

conditions. Refrigeration equipment is widely used in greenhouses for storage purposes.

Plastic films have proved to be a convenient and inexpensive substitute for glass and have found a ready market in construction of cold frames, hotbeds, small sash houses, and greenhouses (Fig. 6-28). However, the high initial construction cost of glass greenhouses still compares favorably with the cost of plastic greenhouses on a long-term depreciation basis. The light-absorbing qualities of plastic are similar to those of glass. At present, a number of different types of plastic coverings are available. These vary from polyethylene films to the more rigid plastics (Fig. 6-29). Since some polyethylene films disintegrate under the influence of ultraviolet light during the summer, they must be replaced each fall. This provides a unique advantage in that, with the plastic removed, the summer cooling problems are eliminated entirely. Ultraviolet-resistant polyethylene is now available. Shade cloth may be substituted for plastic on the frame during the summer. This means of temperature control is, of course, not possible with the more permanent plastic coverings.

Air-supported plastic greenhouses are being developed to produce crops in inhospitable desert climates using an integrated power, water, and food system (Fig. 6-30).

A

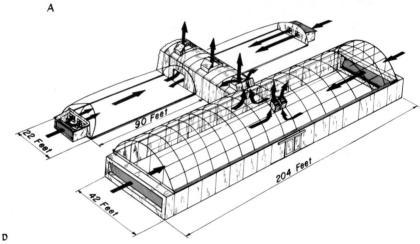

D

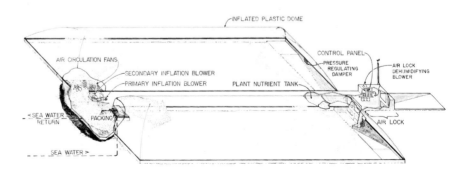

FIGURE 6-30. Controlled environment in an integrated system providing power, water, and food for desert coast areas: a concept pioneered by the University of Arizona, the University of Sonora, and the Rockefeller Foundation. Fresh water is produced from sea water by desalting facilities that harness heat from small engine-driven generators. Heating and cooling is achieved with sea water, which is always about 76–78°F (24–26°C). Humidity is very high and conserves moisture; diseases have not been a problem since spores are removed by seawater spray. Yield per acre is often higher than field production. A, an aerial view of greenhouses built by the University of Arizona's Environmental Research Laboratory and the Arabian principality of Abu Dhabi on the Persian Gulf. Half of the greenhouses are air inflated (*center*) and the remainder are supported structures. B, close-up of the supported greenhouses. C, close-up of the air-inflated greenhouses. D, diagrams of greenhouse system. Water is supplied by a desalting plant. Arrows indicate the air-flow pattern. E, F, vegetable production in the greenhouses. [Courtesy Carl O. Hodges.]

Cloches and Plastic Tunnels

The use of a portable, tent-like glass sash (**cloche**) over individual plants has long been used in European market gardens to facilitate early vegetable production. The use of this technique declined because of the tremendous labor inputs required. However the principle has been revived on a large scale, for winter vegetable production, with the introduction of **plastic tunnels** (Figs. 6-31 and 6-32). Such tunnels are made from sheets of polyethylene laid down mechanically, usually over wire hoops. Soil may be used to seal the sides of the plastic. The increase in temperature within the tunnel encourages early production. A number of systems ameliorate built-up heat as temperatures increase with the approach of summer. In one system the plastic may be temporarily slipped off the hoops during periods of warm weather. In others, the plastic is first perforated for ventilation and then increasingly slit at intervals as temperatures rise. Plastic

FIGURE 6-31. Production of melons under plastic tunnels in Israel. *Top left,* early muskmelon production. Polyethylene is slit as temperatures increase. *Top right,* broad tunnels held down by elastic bands for watermelons. *Bottom left,* covering wire hoops with polyethylene to create narrow tunnels. *Bottom right,* combination of plastic mulch and plastic tunnels. [Photographs courtesy J. Rudich.]

FIGURE 6-32. Use of plastics in California: *Top*, plastic tunnels. *Bottom*, plastic row covers to protect staked tomatoes from rain and wind from November to January in San Diego County. [Courtesy B. J. Hall.]

tunnel and plastic mulch may be combined; in some cases the plastic tunnel is converted to a plastic mulch.

Shade Houses

Shade houses may either be large walk-in structures (Fig. 6-33) or low covered frames. Although shading is commonly used to reduce temperature, it is also used to protect such shade-loving plants as chrysanthemum, hydrangea, azalea, and various foliage plants from leaf damage caused by high light intensity. This is accomplished through the use of such materials as lath or screening. In addition, various types of "shading cloth" are available that can be used to cut down light intensity by different amounts.

FIGURE 6-33. Shade house in Florida used for research on ornamentals. The plants shown are leatherleaf fern (*Polystichum adiantiforme*). [Photograph courtesy Charles A. Conover.]

Propagation beds are often located in shade houses to reduce excessive transpiration. Owing to the inadequacy of their root systems, excessive heat is especially injurious to newly rooted cuttings and transplants. Less watering is required under shade, since transpiration and soil evaporation are reduced.

LIGHT MODIFICATION

The control of light has become a significant part of the technology of horticulture. The manifold effects of light must be considered in terms of the quality, intensity, and duration of the light and their relation to the many physiological processes of the plant.

Satisfying Photosynthetic Requirements

Plant growth depends on the fixation of carbon during photosynthesis. Although most plants grow best in the high light intensities of full sun (5000–10,000 foot-candles), a single leaf is light-saturated at about 1200 foot-candles. The higher intensities are needed, however, to provide sufficient light energy to compensate for leaf shading. Growth is much reduced at lower light intensities. Most plants cannot grow below 100–200 foot-candles, the level of light in an average room. The **compensation point** is that light intensity at which plants will maintain themselves but will not grow. Foliage plants grown for decor are selected for their ability to maintain themselves at this level. For optimum appearance they must usually be replaced within the year unless more light is provided for growth. During the winter the light intensity available above plants in a greenhouse is often between 300 and 1000 foot-candles. As a result of this low light intensity and the short day length, plant growth is often severely limited.

Because of the high energy requirements of photosynthesis and the present cost of power, it is not economically feasible in most situations to use supplemental light to increase photosynthesis. The use of supplemental illumination for this purpose is practical in the greenhouse only where large numbers of "valuable" plants, such as seedlings, are grown in a small area. Supplemental light is used widely to increase growth in experimental studies or for indoor decorative plantings when cost is not a limiting factor.

In outdoor cultivation the efficiency of light utilization may be increased by such cultural practices as spacing, training, and pruning. These techniques are discussed in the following chapter. Rows running in an east–west direction will utilize light more efficiently than will rows running in a north–south direction, in which plants shade each other. In most situations, however, the direction of the row is usually governed by the prevailing slope of the land or by convenience.

Control of Day Length

The control of day length by utilizing either supplemental illumination or shading has become a standard practice in florist-crop production. The artificial lengthening of the day, or interruption of the dark period, makes it possible to promote flowering in long-day plants or to prevent or delay flowering of short-day plants. Similarly, under natural long days, shading with black opaque cloth prevents flowering of long-day plants and promotes flowering in short-day plants. In the culture of chrysanthemums, the most important florist crop in the United States, it is standard practice to control flowering by manipulating the photoperiod. The extension of the photoperiod by illumination is economical on a commercial basis because of the low light intensity required for the process. In this way plants can be induced to flower "out of season." The commercial control of flowering makes it possible to produce a continuous supply of many florist crops. It allows flower production to be synchronized more closely with market demand, which, in the United States, is governed by the season and by proximity to particular holidays. The alteration of photoperiod is a valuable tool for the breeder, who may wish to cross plants that do not normally flower simultaneously.

An increase in photoperiod is achieved by extending the day length to about 17–18 hours. The same effect can be achieved, however, by interrupting the middle of the dark period for about 3 hours. Thus, in terms of power, alteration of the dark period is more efficient than extending the day length. This effect can be made even more efficient by the use of brief light flashes (4 sec/min).

The reduction of photoperiod is achieved by screening the plants with black cloth. Because of the low intensity of light required to stimulate the plants, care must be taken to darken them completely. The common arrangement for indoor and outdoor culture alike makes use of black curtains that can be moved along fixed tracks.

Light Sources

Artificial light sources differ greatly in their spectral distribution (Figs. 6-34 and 6-35). Tungsten lamps, which emit light from a filament heated to extremely high temperatures (about 4670°F, or roughly 2850°C), produce a continuous spectrum from blue to infrared. The radiation within the visible spectrum lies mainly in the red and far red, although the greater part of the overall emission is in the invisible infrared. Fluorescent lamps emit light from both low-pressure mercury vapor and fluorescent powder. Their emission spectrum contains both the continuous spectrum from the fluorescent material and the line spectrum of the mercury vapor.

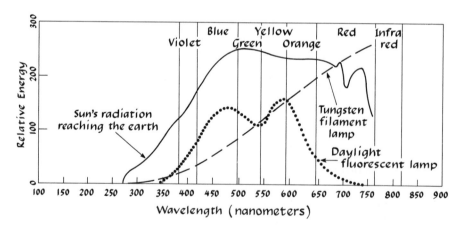

FIGURE 6-34. Spectral emission of tungsten filament lamps and daylight fluorescent lamps compared with the spectrum of the sunlight reaching the earth. The weak line spectra of the mercury discharge from the fluorescent lamps is not shown. The chart offers no quantitative comparison with respect to the energy output of the sources. [Based on data from General Electric Co.]

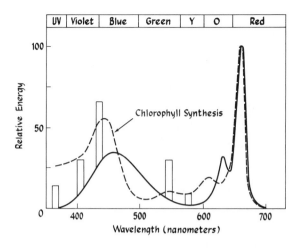

FIGURE 6-35. Spectral energy distribution for fluorescent lamps (Gro-Lux) designed to enhance vegetative plant growth compared with the spectral requirement for chlorophyll synthesis. [Courtesy Sylvania Electric Products.]

Light from ordinary fluorescent lamps is low in red and deficient in far red. This is why fluorescent bulbs are cool. It should be emphasized that different types of lamps and light sources will vary as to their spectral distribution. For example, special fluorescent lights are available that will produce light richer in red.

Because of the energy lost in heat (the infrared radiation), tungsten lights are rather inefficient, since only about 5% of the energy input is transformed into the light range required by plants (400–700 nm), as compared to more than 15% for fluorescent lights. Consequently, fluorescent lights are more efficient in providing the energy required for the acceleration of photosynthesis. Fluorescent lights have not been widely used in greenhouses, however, owing to the relatively high cost of fixtures and installation.

Tungsten lamps have proven very efficient for controlling flowering and for promoting vegetative growth in many woody plants, because the amount of red light so important in the photoperiodic effect is large. Fluorescent light that is poor in red is preferable for supplemental illumination of those plants that respond strongly to far-red light with respect to elongation and become etiolated and spindly. Tungsten light is used where elongation is desirable, as in the culture of asters or hyacinths.

In order to achieve satisfactory growth in completely artificial light, the best results are obtained with a combination of light sources. Tungsten and fluorescent lights complement each other to produce a spectrum closer to that of sunlight

FIGURE 6-36. *Top*, a movable high intensity lighting system designed for winter production of lilies in the Netherlands. *Bottom*, close-up of high-intensity lamp used to increase photosynthesis and plant growth in greenhouses. [Photographs courtesy T. C. Weiler and P. A. Hammer.]

than does either separately. In experimental growth chambers, where high light intensities are desirable, combinations of tungsten, fluorescent, and mercury lamps may be used. Special care must be taken, however, to dispose of the infrared radiation (heat).

Selected References

Bickford, E. D., and S. Dunn, 1972. *Lighting for Plant Growth*. Kent, Ohio: Kent State University Press. (A handbook of light and plant growth, including many applied uses.)

Black, C. A., 1968. *Soil-Plant Relationships*, 2nd ed. New York: Wiley. (An in-depth discussion of soil fertility.)

Downs, R. J., and H. Hellmers, 1975. *Environment and the Experimental Control of Plant Growth*. New York: Academic Press. (A discussion of controlled environments, such as air-conditioned greenhouse, plant-growth chambers, and phytotrons, especially for research.)

Evans, L. T. (editor), 1963. *Environmental Control of Plant Growth*. New York: Academic Press. (A collection of papers dealing with the effects of many environmental factors on plant growth.)

Hagan, R. M., H. R. Haise, and T. W. Edminster (editors), 1967. *Irrigation of Agricultural Lands*. Madison, Wisc.: American Society of Agronomy. (A broad compilation of irrigation practices.)

Langhans, R. W. (editor), 1978. *A Growth Chamber Manual: Environmental Control for Plants*. Ithaca, N.Y.: Cornell University Press. (All about plant growth chambers.)

Russell, E. W., 1973. *Soil Conditions and Plant Growth*, 10th ed. New York: Longmans. (The most famous work on agricultural soils. The first seven editions were written by Sir John Russell.)

Tisdale, S. L., and W. L. Nelson, 1975. *Soil Fertility and Fertilizers*, 3rd ed. New York: Macmillan. (This college text covers the fundamental concepts of soil fertility and fertilizer manufacture.)

U.S. Department of Agriculture, 1955. *Water* (USDA Yearbook, 1955). (An excellent discussion of water in relation to agriculture.)

U.S. Department of Agriculture, 1957. *Soil* (USDA Yearbook, 1957). (A broad, nontechnical treatment.)

Went, F. W., 1957. *The Experimental Control of Plant Growth*. Waltham, Mass.: Chronica Botanica. (Classic studies of controlled plant growth carried out at the Earhart Laboratory. Particularly valuable for data on climatic responses of individual crop plants.)

7

Directing Plant Growth

Go bind thou up young dangling apricocks,
Which like unruly children make their sire
Stoop with oppression of their prodigal weight;
Give some supportance to the bending twigs.
Go thou, and like an executioner
Cut off the heads of too fast growing sprays,
That look too lofty in our commonwealth:
All must be even in our government.

SHAKESPEARE, Richard II [III. 4]

The growth of plants can be modified to suit human desires, and these modifications may be achieved by direct manipulation of the plant itself, as distinct from manipulation of the plant's environment. The direct control of growth by pruning and grafting are among the oldest of horticultural practices. Recently, however, chemical substances (**growth regulators**) that affect growth and development have found increasing application in horticulture.

The direct modification of growth is effectively limited by our knowledge of plant development. The more complete our knowledge and the more refined our techniques, the more sophisticated is our control. It becomes possible to affect not only the amount of growth, but the form and pattern of growth, as well as differentiation in such physiological processes as flowering and rooting.

PHYSICAL CONTROL

Growth may be controlled by purely physical methods. Physical techniques that control the shape, size, and direction of plant growth are known as **training**. Training is in effect the orientation of the plant in space. This may involve merely providing a support on which plants may naturally grow or, in addition, it may include the bending, twisting, or fastening of the plant to the supporting structure

A

B   C

FIGURE 7-1. Training orients a plant in space and is an integral part of the culture of many plants. A, tomatoes are twisted around twine to maximize growing space in greenhouse production. (Tomatoes grown for early market in the field may be trained to wooden stakes.) B, cucumbers are trained to a trellis in greenhouse production. C, a young pear orchard in England grown for pear cider (perry) with two types of training: the tall trees tied to the stake are permanent trees; the bushlike trees between them are temporary. See also Figures 17-18 through 17-21. [Photographs courtesy Merle H. Jensen, P. B. Lombard, and R. R. Williams.]

(Fig. 7-1). Training often is combined with the judicious removal of plant parts, or **pruning.** Pruning may also be performed for other purposes—for example, to adjust fruit load, the subsequent change in form being only incidental.

The object of altering the spatial form or size of a plant is to improve its appearance or usefulness. Certain woody shrubs can be trained and pruned in a great variety of shapes, limited only by the skill of the person wielding the shears. Plant sculpture, known as **topiary,** is considered beautiful by some people and ugly by others, but it illustrates, in any case, the plasticity of the growing plant (Fig. 7-2). The usefulness of a particular spatial arrangement may result from the

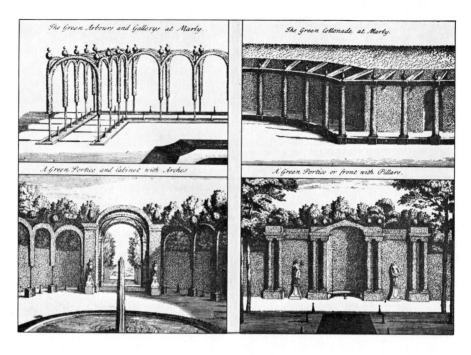

FIGURE 7-2. Examples of topiary art from *The Theory and Practice of Gardening* by Alexander Le Blond, translated by John James in 1728. Topiary refers to the art of training plants to resemble unnatural, ornamental shapes. This type of "bush sculpture" was very popular in the seventeenth and eighteenth centuries for ornamental plantings, but is no longer in fashion. [From Wright, *The Story of Gardening*, Garden City Publishing Co., 1938.]

increased efficiency of light utilization or from the facilitation of cultural operations, such as harvesting or disease control. Furthermore, training and pruning may enhance the productiveness of plants and the quality of plant products.

Training and Pruning as Horticultural Practices

Training and pruning are well known but by no means universal practices. Herbaceous annuals or biennials are usually grown without any attempt to alter their growth patterns. The lack of training is not so much a matter of satisfaction in their performance as it is of practicality. Since there are usually many such plants in relation to the space they occupy, it is not practical to handle each one individually. Perennials, and especially woody plants, are often trained to some degree. Each individual is relatively valuable, since there are few plants per unit area, and since they are grown for extensive periods (the productive life of an apple tree may be forty years). As plant size continually increases, the control of growth through pruning becomes a necessity. The framework of a woody tree in relation to pruning is shown in Figure 7-3.

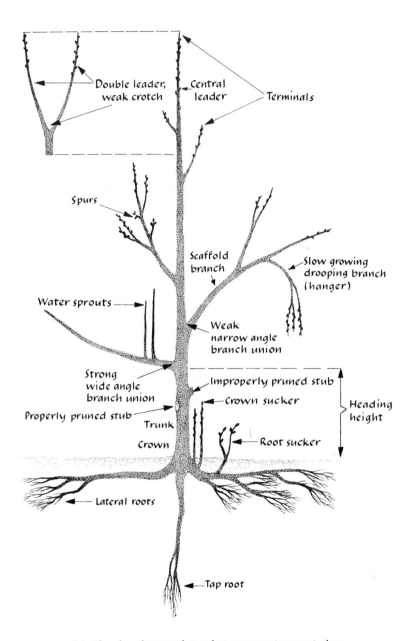

FIGURE 7-3. The plant framework in relation to pruning terminology.
[Adapted, with permission of the publisher, from Christopher, *The Pruning Manual*, copyright 1954 by the Macmillan Company.]

Physiological Responses to Training and Pruning

The orientation of the plant in space has a marked physiological effect on growth and fruiting. Fruit trees planted at an angle of about 45° to the ground have been shown to become dwarfed and to flower earlier. The training of branches in a horizontal position encourages the same effect. This decrease in growth rate and increase of flowering occurs naturally when the weight of a heavy crop load bends a limb down. Thus, fruiting acts as a triggering device to keep the plant reproductive. A clear explanation of this phenomenon has not been made. It has been suggested, however, that the effect is due to a disturbance of the normal auxin movement, which in turn affects phloem transport. This assumes that gravity affects the pattern of auxin distribution along the stem. The effect of the disruption of phloem transport on growth and fruiting is discussed later in this chapter.

The response of the plant to pruning is a result of the altered relationship of the remaining plant parts and the disturbed pattern of auxin production. The effect differs to some extent, depending on whether the plant is dormant or growing when pruned.

Altered Relationship of Plant Parts

An explosion of vegetative growth normally occurs after extensive shoot pruning. This is because severe shoot pruning radically alters the balance between root and shoot. The flush of growth following pruning is caused by the diversion of water, nutrients, and stored food from an undisturbed root system into a reduced bud area (Fig. 7-4). Although there is also some reduction in the amount of stored food (along with some reduction in photosynthetic area), this is negligible because reserve food in the form of sugars and other carbohydrates is stored mainly in the roots and older portions of the shoot, especially during dormancy.

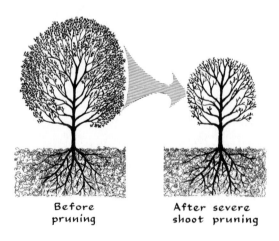

FIGURE 7-4. Shoot pruning alters the balance of root and shoot and results in increased growth of the remaining parts.

Before
pruning

After severe
shoot pruning

The increased growth that occurs after extensive pruning might indicate that the technique has a rejuvenating effect. But the additional growth does not compensate for the removed portion of the plant; the plant pruned of vegetative growth never quite makes up the loss. Thus, pruning is in reality a dwarfing process, although some plant parts may be selectively increased.

Pruning and Flowering

In general, severely shoot-pruned plants, especially if they are young, tend to remain vegetative. Conversely, root pruning encourages flowering. This can be explained in a number of ways. It has been interpreted by an extension of the carbohydrate–nitrogen "theory" of flowering (see Chapter 5). This assumes that the severely shoot-pruned plant draws on its carbohydrate reserve in the promotion of growth. The resulting low carbohydrate–nitrogen balance encourages vegetative growth. Root-pruned plants reduce nitrogen accumulation, but, by slowing down vegetative growth, they conserve carbohydrates. The carbohydrate surplus supposedly promotes flowering. Another equally valid explanation may be that actively growing leaves produce substances that inhibit flowering. Thus, a rapid increase of vegetative growth would be antagonistic to flowering. The encouragement of flowering by root pruning is a direct result of slowing down vegetative growth. The precise relationship, however, is unclear; the explanation awaits a more precise elucidation of the flowering process.

Auxin Imbalance

APICAL DOMINANCE. The role of the apical meristem in inhibiting the growth of dormant buds (bud break) behind it is known as **apical dominance.** This dominance of the apical meristem differs from species to species. Thus, an actively growing bamboo is basically an unbranched stem, whereas a sprawling shrub such as the Pfitzer juniper grows as a many-branched structure. Both the degree of branching and the subordination of lateral growth to the main growing stem, the **central leader,** appear to be functions of apical dominance.

The branching of shoots (and roots) has been shown to be influenced by auxins, which are produced in greatest abundance in a vigorously growing apex. A high auxin concentration moving down from the stem tip has been shown to inhibit lateral bud break. (Rapidly growing unbranched shoots called watersprouts have very high auxin levels and represent an extreme example of apical dominance.) Removal of the stem tip results in an increased amount of lateral bud break and subsequent branching, usually directly below the cut (Fig. 7-5). This is explained by the destruction of the auxin-producing meristem, although it must be admitted that the precise mechanism has not been established. Thus, pruning that merely removes the tip of the stem (heading back) can create new form changes by the destruction of apical dominance (Fig. 7-6). Similarly, pruning that merely removes laterals but leaves the stem tip undisturbed (**thinning out**) not only

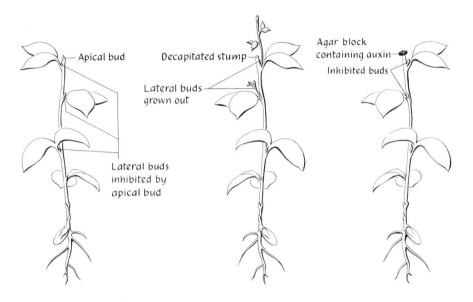

FIGURE 7-5. Apical dominance refers to the effect of the apical bud in inhibiting bud break below. Removal of the apical bud encourages lateral breaks. The growth of lateral buds can be inhibited by auxin application to the cut portion of the stem. [Adapted from Bonner and Galston, *Principles of Plant Physiology*, W. H. Freeman and Company, copyright © 1952.]

FIGURE 7-6. These coleus plants are of the same age, and were grown under the same conditions with the exception of pruning. Removal of the apical meristem of the plant on the right has stimulated the growth of lateral shoots. [Photograph courtesy E. R. Honeywell.]

FIGURE 7-7. Wide-angled scaffold branches are encouraged by proper pruning in the peach. *Left*, the central leader was summer pruned (pinched) above a bud during the first growing season to channel growth into selected scaffold branches and to encourage wide-angled attachment. *Right*, during winter pruning, the central leader was cut to a stub. The stub will be removed in a few years. [Photographs courtesy of F. H. Emerson.]

eliminates branching but, by increasing the vigor of stem tip and presumably its auxin content, also limits future lateral bud break.

BRANCH ANGLE. There is also evidence that the angle of branching is controlled by auxins. Branches produced below an actively growing, auxin-producing apex form a wider angle with the main stem than do branches formed where the growing point has been removed. A heavily pruned young fruit tree tends to produce narrow-angled branches. In peach pruning, wide-angled scaffold branches are encouraged by permitting growth of a bud from the stub (Fig. 7-7).

Pruning Techniques

Heading Back and Thinning Out

The two basic pruning cuts, heading back and thinning out, are illustrated in Figure 7-8. **Heading back** consists of cutting back the terminal portion of a branch to a bud, whereas **thinning out** is the complete removal of a branch to a lateral or main trunk. Because the heading back of a stem destroys apical dominance, it is usually followed by the stimulation of several lateral bud breaks, depending on the species and the distance from the tip to the cut. To encourage spreading growth, the branch is usually cut back to an outward-pointing bud. Heading back tends to produce a bushy, compact plant. The shearing of a hedge is an extreme example of this type of pruning. Heading back actively growing plants is referred to as **pinching.**

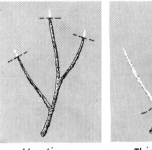

Heading   Thinning

FIGURE 7-8. Heading encourages lateral growth. Thinning results in a more open structure.

Thinning, in contrast to heading, encourages longer growth of the remaining terminal branches. The net result of thinning is a reduction of laterals. Thinning of weak growth tends to "open up" the tree. It usually results in producing a larger rather than a bushier plant. The rejuvenation of older trees by reducing and thereby stimulating the remaining growing points is accomplished by thinning. The thinning out of growing wood is referred to as **deshooting.**

Timing of Pruning

> Prune when the knife is sharp.

The proverbial saying might have been true long ago, but today the time to prune is influenced by a number of other factors, including convenience, the peculiarities of the species, and the effect desired. Although it is best to keep some plants more or less continually pruned, this is seldom practical. Fruit trees are usually **dormant pruned.** Not only is this most convenient in the cycle of orcharding, but the framework of the plant can be more easily seen with the foliage off. Where winter temperatures are low, the pruning operation is usually delayed until the severest weather is past in order to reduce winter injury to fresh cuts. The pruning operation is best not carried on into the growing season because of the additional loss of translocated foods. **Summer pruning** of new growth, however, makes it possible to avoid structural faults before too much growth is wasted. This is especially important when the tree is young. With proper heading, little photosynthetic area need be lost. However, extensive dormant pruning has been shown to be less devitalizing than summer pruning. In addition, pruning wounds made in the early spring heal better than those made at other times of the year. Extensive pruning should be avoided in the late summer since this may initiate abundant, succulent vegetative growth, which may render the plant subject to winter injury. However, diseased growth or dead wood is best pruned away at once, regardless of the season. This wood, besides being unattractive or dangerous, may become a harboring place for disease-producing pests.

It is only sensible to synchronize the time of pruning such that it does not interfere with the principal functions of the plant. Thus, ornamental flowering shrubs that bloom from buds laid down the previous year should be pruned after they bloom. Similarly, it would be absurd to prune large limbs of fruit trees supporting a maturing crop.

Objectives of Pruning

Pruning to Control Size

Probably the most obvious effect that pruning has on perennial plants is to control their size. Since perennials grow continually, an optimum size can be maintained only by the selective removal of plant parts. Thus, the lawn is cut, the hedge is clipped, and shrubs and fruit trees are pruned in an effort to keep them within bounds (Fig. 7-9).

The size of a plant may be controlled for esthetic or utilitarian reasons. In fruit production, where crops are hand picked, large tree size makes harvesting (as well as effective spraying) extremely difficult. Certain pruning techniques can be used to reduce the height of large trees over a period of years without excessive injury.

The compensating effect of pruning on growth may increase the size of particular plant parts. This diversion of growth may be utilized to achieve an actual increase in height or spread, even though total growth is reduced. The selective removal of buds, flowers, or fruits to increase the size of the remaining parts must be considered a specialized part of pruning. This, discussed further in

FIGURE 7-9. The fairway is kept in bounds by constant mowing. [Photograph courtesy International Harvester Co.]

Chapter 8, is also known as **thinning** (but should not be confused with the pruning technique **thinning out**). The removal of buds to increase the size of the flowers produced by the remaining buds is known as **disbudding.**

Pruning to Control Form

The art of training and pruning to control plant form has received much attention in horticultural writings. In this context, form refers not only to the gross shape of the plant but also to its structural makeup, which involves the number, orientation, relative size, and angle of branches. The natural form characteristic of different species may be greatly modified with pruning. Plants may be trained to grow upright or to spread, and branching may either be increased or decreased.

Woody plants, especially those bearing heavy loads of fruit, must be considered as structural units because they may be torn apart in a high wind. Structural strength in fruit trees is obtained by pruning to eliminate narrow-angled branches and to achieve a well-spaced arrangement of wide-angled scaffold branches. Narrow-angled branches are weak and tend to break under pressure because of the lack of continuous cambium and the inclusion of squeezed-off bark in the crotch (Fig. 7-10). For maximum strength, only one branch should develop at any point on the main stem, and the branches should be well distributed around the tree (Fig. 7-11). Owing to the increase in diameter of branches, it is necessary to select branches carefully when young.

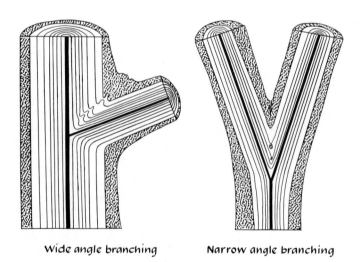

Wide angle branching   Narrow angle branching

FIGURE 7-10. Narrow branch angles are weak because of the enclosure of bark and the formation of wood parenchyma in the crotch. [Adapted from Eames and MacDaniels, *An Introduction to Plant Anatomy*, McGraw-Hill, 1947.]

FIGURE 7-11. For maximum strength scaffold branches of apple should be well spaced and evenly oriented around the tree. *Left*, in the unpruned tree the branches to be removed have been marked with paint. *Right*, the same tree after pruning. [Photographs courtesy Purdue University.]

The control of plant form may be utilized to achieve increased quality through better light distribution. The center of an unpruned apple tree is almost impervious to light and, as a result, produces few fruit, those that are produced being poor in color and quality. Opening up the tree is also important for disease control in that it permits good spray distribution and facilitates rapid drying.

Mechanical harvesting of fruit crops requires specialized training and pruning to adapt the plant to the machine. For plants harvested by shaking, this involves the development of high trunks (to allow for a single "grab" by the machine when the tree is young) and only two or three main scaffolds for use when the tree is older. Mechanical harvesting of grapes by the use of a cutterbar requires special trellising and training of the vines.

Pruning for Plant Performance

ESTABLISHMENT OF TRANSPLANTS. The transplanting of large plants from natural growing sites is usually very difficult. Root pruning or repeated transplanting when the plant is young encourages a fibrous root system and allows the plant to be moved safely when large.

Proper root and shoot pruning greatly aids in reducing transplanting shock and promotes successful plant establishment. This is especially true in bare-rooted transplants. Light root pruning stimulates root initiation; shoot pruning conserves moisture by reducing the transpiration surface in relation to the root area.

PRODUCTIVITY AND QUALITY. Pruning is often a necessary step in the control of productivity. Where vigorous bud wood is desired, as in scion orchards, heavy pruning stimulates vegetative growth. On the other hand, where flower or fruit production is the desired aim, selective pruning that eliminates weak, nonproductive wood will aid in channeling the plant's energy into flowering and fruiting.

In addition, fruit and flower quality is greatly affected by the vigor of the wood that bears it as well as its location in the tree. Shoot growth (suckers) on the understock of grafted plants must be continually removed to eliminate nonproductive growth. Similarly, forest trees may be pruned of unnecessary lower branches to produce knot-free lumber.

Training Systems

Training systems are carried on to control form throughout the life of the plant. Consequently, special attention must be given in the formative years. The objective is to obtain some predetermined shape in an attempt to achieve greater productivity, quality, ease of culture, or beauty.

Branch Orientation and Leader Training

The main factors that determine form are the location of the points on the main stem from which branches form and the subsequent orientation of the branches (Figs. 7-12 and 7-13). The branches may be oriented around the stem to produce a "natural" shaped tree, or they may be oriented in a single plane to provide a flat shape known as an **espalier** (from the French word for shoulder). There are many variations on these two general shapes that differ principally in the height of the stem before the first branch (heading height), the angle of the branches from the main stem, and the distribution and relative length of the branches.

In the **central-leader system** of training, the trunk is encouraged to form a central axis with branches distributed laterally up and down and around the stem. The central axis, or leader, is the dominant feature of the tree's framework, and the main direction of growth is upward. In the **open-center** or **vase system** of training, the main stem is terminated and growth is forced through a number of branches originating rather close to the upper end of the trunk. Special pruning is required to prevent a lateral from becoming dominant—that is, from forming a new central leader. Although the open-center tree is a lower tree than the central-leader tree, it has inherent mechanical weaknesses, due to its narrow crotches and close branching. The **modified-leader** system is somewhat intermediate between these two types.

Originally, an espalier was a railing or trellis along which plants (usually fruit trees or vines) were trained to grow flat. Plants trained in this manner than came to be known also as espaliers. An espalier restricted to one shoot, or two shoots growing in opposite or parallel directions, is called a **cordon.** Because of the extensive pruning labor required, tree fruits have not been commercially grown as espaliers in the United States. Grapes, however, are commonly grown as espaliers, as in the widely used Kniffin system of training. Properly executed espaliers are extremely attractive as ornamentals. They are created with a combination of pruning and actual bending of the shoots while they are still succulent (Fig. 7-14).

Heading height
High   Low

Leader training

Central

Modified

Open center

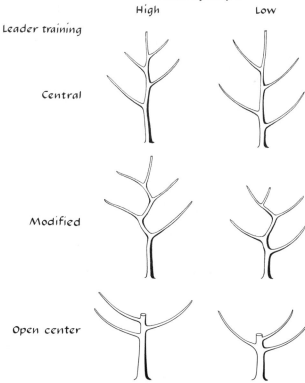

FIGURE 7-12. Training and tree form.

Tree shape   Top view   Side view

Natural

Flat
(espalier)

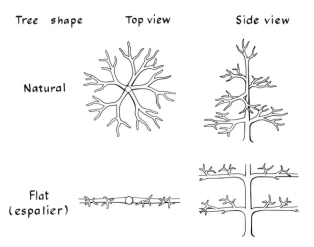

FIGURE 7-13. Branch orientation and tree form. Branches may be distributed spirally around the stem or in a single plane.

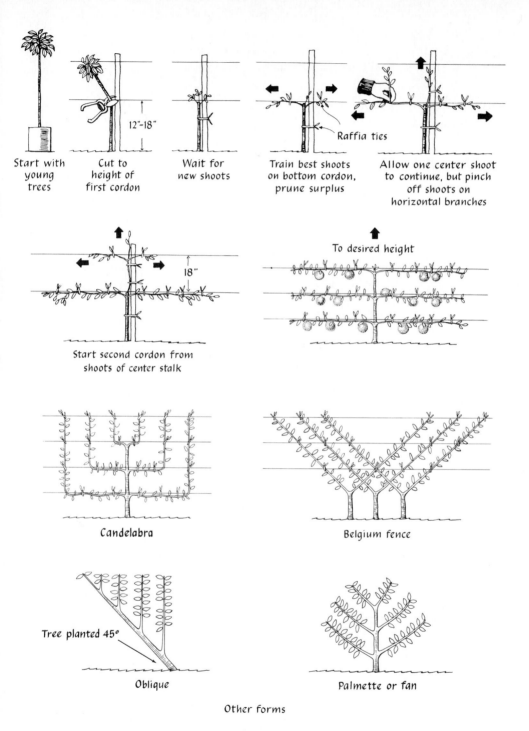

Start with young trees

Cut to height of first cordon

12"–18"

Wait for new shoots

Train best shoots on bottom cordon, prune surplus

Raffia ties

Allow one center shoot to continue, but pinch off shoots on horizontal branches

Start second cordon from shoots of center stalk

18"

To desired height

Candelabra

Belgium fence

Tree planted 45°

Oblique

Palmette or fan

Other forms

FIGURE 7-14. The creation of an espalier. [Adapted from Hudson, *Sunset Pruning Handbook*, Lane Book Co., 1952.]

FIGURE 7-15. The peach is commonly pruned to an open center and is usually composed of two or three widely spaced main scaffold limbs. *Top left*, a well-formed three-scaffold crotch on a three-year-old tree. *Top right*, a six-year-old tree on which six scaffold branches were allowed to develop. All but two should have been eliminated at the start of the second year's growth. Note the weakness that has developed between the scaffold branches. *Bottom left*, a "two-story" tree showing severe winter injury as a result of delayed maturity. The center leader should have been removed at the start of the second season's growth. *Bottom right*, this tree was growing next to the one illustrated at the left. The two-branch wide-angle crotch shows no evidence of winter injury. [Photographs courtesy Purdue University, from Extension Circular 426, 1956.]

Such shoots will retain the imposed shape when lignification sets in. Well-planned espaliers can be strong enough to be self-supporting eventually, although frames are required in the early years.

The particular system by which a plant is trained depends to a large extent on the species. Peaches and apricots can be pruned to an open center because of the broad angle of attachment of the branches, which produces strong crotches. In addition, the central leader in peach trees is subject to winter injury. The reason for this is not clear, but it appears to be related to the failure of the central leader to harden off (Fig. 7-15). Because the narrow-angled branching of apples and pears

FIGURE 7-16. An eight-year-old apple tree trained to a modified leader. *Above left,* before corrective pruning. *Above right,* after corrective pruning. The removal of the main portion of the central leader has opened up the center of the tree. Many of the pruning cuts were made to correct structural weaknesses. *Facing page left to right,* narrow-angled, forked branches were eliminated.

makes the open-center tree unfeasible, they are usually trained to either the central-leader or the modified-leader system (Fig. 7-16). Cherries and plums are trained to either the open-center or modified-leader system. Citrus and other evergreen fruits may be pruned lightly to establish a stronger framework, but usually little subsequent pruning is performed except to eliminate dead wood after a freeze. Mechanical hedgers are now widely used.

Tree Geometry and Planting Systems

The three-dimensional shape of the tree is ultimately based on the architecture of the framework. Various tree shapes are diagrammed in Figure 7-17. The optimum shape differs, depending on the efficiency of light interception, ease of harvest, structural strength, and pruning and training costs.

Trees may be either free-standing or supported. Free-standing trees must by necessity be based on a strong framework of structurally sound main limbs called **scaffolds.** To overcome weakness caused by a shallow root system or that inherent in espalier training, trees may be supported with the aid of individual stakes (often

Closely spaced branches, especially those growing toward the center of the tree, were removed. Intertwining branches were corrected. (Note that limb-rub injury has girdled the upright branch.) Watersprouts were removed. (The watersprout growing within the crotch would have resulted, if neglected, in extensive injury.) [Courtesy Purdue University.]

necessary when a tree is young) or on trellises constructed of narrow strips of wood or of posts and wire. In one type of trellis, two to four wires are strung, one above the other, across the row of supports; in another, T-shaped standards, each consisting of a post and a cross-piece, support several wires in a horizontal plane.

Planting systems involve a combination of individual plant training and patterns of plant placement in the field (Fig. 7-18). In perennial fruit crops various planting systems are used to achieve maximum efficiency in the production of fruit. The variables in the systems are tree size (often modified by a combination of pruning and rootstock), tree density, and management strategies based on mechanization of harvest, pruning and training costs, irrigation, and cultivation. The fruit industry has gone through an evolutionary period as a result of changes in rootstocks, cultivars, land values, and marketing practices. The trend has been to high-density plantings of smaller trees to increase early production and maintain high-quality fruit (Fig. 7-19). Planting systems based on intensive training, the use of tree supports, close spacing, and dwarfing rootstocks, which have long been practiced in Europe, have recently been modified to North American conditions, especially for the production of apples.

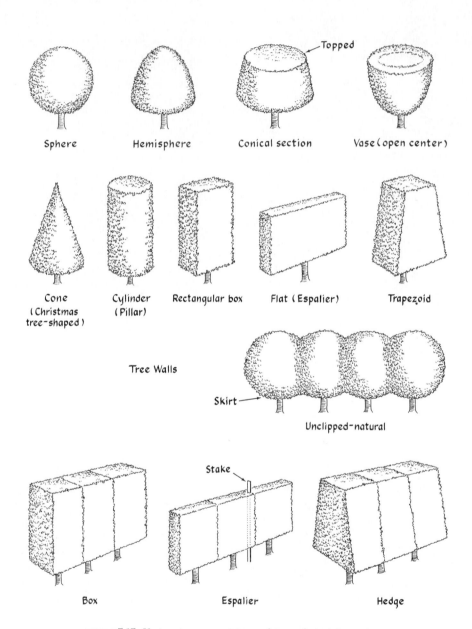

FIGURE 7-17. Various tree geometries used in apple training systems.

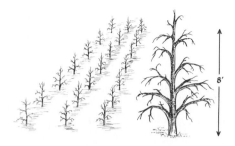

"Free standing" tree trained to a central leader and maintained in a "Christmas tree" form.

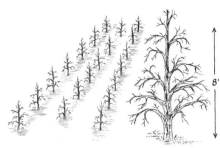

"Off-set" system using free standing tree forming double or triple rows.

"Spindle Bush," or slender spindle bush, each tree staked for support and branch training.

"Bed" planting system with drive rows every 5th or 6th row.

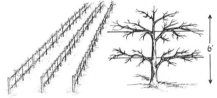

"Two Wire Trellis," each tree trained in espalier form into slender rows 10, 12, or 15 feet apart.

"Palmette" tree form trained on 3-wire trellis.

"Oblique" form trained on 4-wire trellis, each branch slanted at a 45° angle on a central leader (The Oblique form can also be made by slanting the tree trunk at 45° and training vertical branches.)

"Pillar" training system which calls for renewal of 3 and 4 year fruiting branches in alternate years to produce younger bearing branches.

FIGURE 7-18. Various tree planting systems used in apple. [Courtesy R. F. Carlson.]

A

B

C

D

FIGURE 7-19. Evolution of high-density tree planting systems. *A*, an old pear orchard (for pear cider or perry) planted before 1760 in Gloucestershire, England. Note the high heading, to accommodate grazing, and very wide spacing, about 50 feet apart (17 trees per acre). *B*, a more closely spaced apple orchard in Ontario, Canada (about 1950), showing low-headed, modified-leader training. Spacing is approximately 40 × 30 feet (72 trees per acre). [Photograph courtesy Ontario Ministry of Agriculture and Food.] *C*, hedgerow peach in West Lafayette, Indiana, planted 10 × 14 feet apart (383 trees per acre). *D*, a high-density planting of nectarines, trained on the free spindle system, in Italy in the 1970s (538 trees per acre). *E*, an experimental hedgerow training system in Corvallis, Oregon, showing the shape of the tree walls after pruning. Spacing is 15 × 4 feet (726 trees per acre). [Photographs courtesy P. B. Lombard, R. R. Williams, E. W. Franklin, F. H. Emerson, R. A. Hayden, F. Loreti, and M. N. Westwood.]

E

Renewal Pruning

In order to promote superior performance in perennial plants grown for flowering or fruiting, pruning must stimulate the most reproductive growth; it must continually renew growth to produce wood of the optimum reproductive age. Depending upon the species, this may be wood of the current season, or it may be wood one or more years old.

The factors to be considered before pruning are (1) the time at which the buds are differentiated in relation to blooming and (2) the age of the wood that produces the most abundant and highest-quality buds. Flower buds may be initiated in the year of flowering, as in summer- or fall-blooming plants (for example, the rose and chrysanthemum), or on the previous year's growth, as in spring-flowering plants (for example, apple, lilac, peach, and brambles). Buds that differentiate the year previous to flowering may be produced on that previous year's new growth or on older wood. Many species form buds from spurs on older wood. Spurs may bear more or less irregularly for as long as twenty years. In apple, however, they are most productive for two to five years.

Plants that flower on current growth, as does the rose, are often severely dormant-pruned to encourage vigorous reproductive growth. If unpruned, the abundance of buds produces inferior individual blooms. The degree of pruning is related to the vigor of the plant; the more vigorous the plant, the greater the number of buds that are retained. During the growing season, overvigorous canes that tend to remain vegetative are headed back. Roses grown for mass effect, such as climbers, are less severely pruned, although thinning of older growth is required to stimulate vigorous new canes.

The cutting of blooms in the summer serves to invigorate the remainder of the plant. Senescent flowers should be removed to prevent fruit development, since fruits drain nutrients from the plant.

Brambles (blackberry, raspberry, and related fruits) produce fruit on year-old canes. Although the roots are perennial, the canes are biennial and either weaken severely or die right after fruiting. Thus, pruning has a number of functions. Immediately after bearing, the fruiting canes are removed to encourage new shoot growth. In red raspberry (*Rubus idaeus*) new canes arise as root suckers, hence the old canes may be completely removed. Because black raspberries (*R. occidentalis*) do not produce suckers, their fruiting canes are removed above the crown. The dormant year-old canes are thinned if necessary and headed back to remove the weaker buds and to encourage branching. Black and purple raspberries are further pinched, when summer growth is two or three feet tall, to increase lateral branching.

Grapes bear fruit on the current growth from buds laid down the previous season. The greatest production is achieved from the fourth to eighth bud, and the quantity and quality of production is based on the vigor of the plant in relation to the number of remaining buds. If too many buds are left in relation to vigor, the grapes will be small and of poor quality. If too few buds are left, yield will be

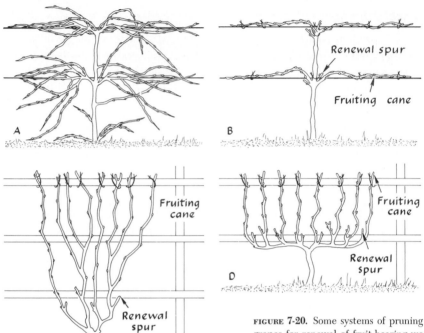

FIGURE 7-20. Some systems of pruning grapes for renewal of fruit-bearing wood. The severity of pruning is related to the vigor of the previous year's growth. *A*, a mature dormant vine before pruning. *B*, the same vine after pruning according to the four-cane Kniffen system. *C*, the fan system. *D*, the horizontal-arm spur system.

reduced. Therefore, renewal pruning is an important practice in controlling yield, size, and quality in grapes. A pruning formula has been devised for the 'Concord' grape on the basis of plant vigor as determined by the weight of the prunings. Thirty buds are left for the first pound and ten buds for each additional pound of wood removed. This formula of "30 + 10" results in a "moderately" pruned vine, which gives optimum production in Ohio. In addition to fruiting canes, stubs of one or two buds, called **renewal canes,** are retained. These provide growth from which fruiting canes may be selected the following year. Thus, from a single trunk, growth is renewed each year (Fig. 7-20). Other training methods differ merely with respect to the form of the plant; the renewal principle is essentially the same.

An improved system of training grapes, called the **Geneva Double Curtain,** employs a divided canopy (Fig. 7-21). In essence, the system converts a wide canopy achieved by the traditional Kniffin training system into two narrow "curtains" (the leaf and shoot system formed along a single vertical plane). This is

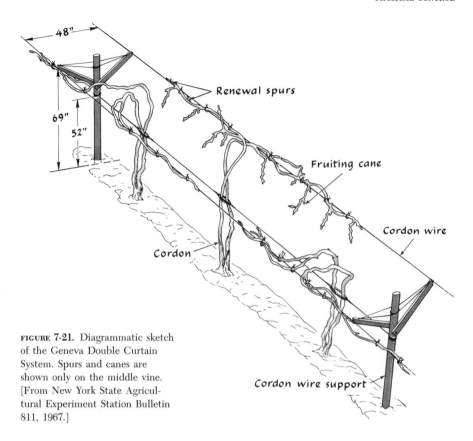

48"

69"

52"

Renewal spurs

Fruiting cane

Cordon wire

Cordon

Cordon wire support

FIGURE 7-21. Diagrammatic sketch of the Geneva Double Curtain System. Spurs and canes are shown only on the middle vine. [From New York State Agricultural Experiment Station Bulletin 811, 1967.]

achieved by hand positioning or combing horizontal shoots so that all trend vertically downward. In a wide canopy where leaves shade each other as well as fruit, grape maturation is retarded and vines are less productive. Converting a wide canopy into narrow "curtains" exposes the leaves on the basal portion of the shoot to higher light intensities and thus to higher temperatures, which leads to more rapid maturation of the fruit and to an increase in soluble solids (sugar). In comparison to the Kniffin system, yield increases of more than 50% and increases in soluble solids of more than 1% have been obtained with the Geneva Double Curtain. A further advantage is that the positioning of the shoots away from the posts of the trellis permits mechanical harvesting. This system appears to be a useful one for all grapes, and it is especially valuable in areas of high humidity and low light intensity.

The pruning of fruit trees is done in stages. When the tree is young, pruning is principally a training operation to control form and to produce a structurally sound framework. Because of the adverse effects on early bearing, pruning must

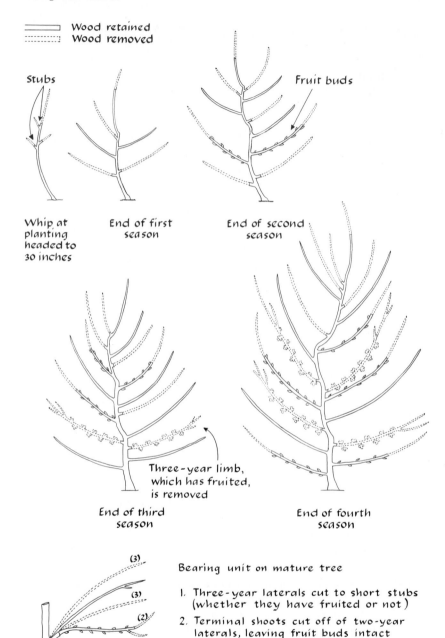

Wood retained
Wood removed

Stubs

Fruit buds

Whip at
planting
headed to
30 inches

End of first
season

End of second
season

Three-year limb,
which has fruited,
is removed

End of third
season

End of fourth
season

Bearing unit on mature tree

1. Three-year laterals cut to short stubs
(whether they have fruited or not)

2. Terminal shoots cut off of two-year
laterals, leaving fruit buds intact

3. One-year laterals are thinned. (approx.
20-25 laterals are left per tree)

FIGURE 7-22. Pruning pillar apple trees. [Adapted from Weiss and Fisher,
Canadian Department of Agriculture, 1960.]

be limited at this stage. After four to six years, when the major scaffold limbs are established, renewal pruning insures a continuing bearing surface of two- to four-year-old wood, from which the bulk of the crop develops.

The pillar system of training the apple—a system developed in England—is a good illustration of renewal pruning (Fig. 7-22). This technique can be used with a tree structure consisting of a single leader 10–12 feet high. The number of bearing units maintained depends upon the vigor of the tree. Each unit consists of a two-year fruiting limb, a one-year-old shoot, and the current growth. Dormant pruning consists of (1) removing the spent fruiting limb (now three years old), (2) heading back the two-year terminals of the limb, but leaving the fruit buds intact in anticipation of fruiting, and (3) thinning out all but 20–25 year-old shoots. This pruning pattern is repeated each year; thus the productive two-year-old fruiting wood is continually renewed. High quality and annual production are achieved by controlling the bearing area. Close spacing of the trees (6 × 12 feet or 605 trees per acre) results in high production per acre. By controlling size, mechanical production practices may be facilitated and harvesting is greatly simplified. The pruning operation, although extensive, is routine; no difficult decisions need to be made. Whether the pillar system of renewal pruning will prove to be practical in the United States remains to be seen. The renewal systems presently used in the United States for pruning apples differ from the pillar system in regard to the age of wood removed and the pattern of the framework.

BIOLOGICAL CONTROL

Graft Combination

One way of controlling a plant biologically is through grafting, which is practiced to modify growth as well as for propagation. The interaction of two or more plants in a graft combination may affect both growth and productivity. Moreover, improved disease resistance and hardiness can be achieved by the creation of a plant composed of more than one genetic component.

The practice of grafting as a means of growth control is used most extensively with fruit trees. Graft combinations of herbaceous plant material have not been fully explored, for unless the plant itself is relatively valuable, grafting is not an economical horticultural practice. In Japan, however, watermelon is grafted onto the gourd *Lagenaria* to control Verticillium wilt, and eggplant is grafted onto *Solanum integrefolium* to increase productivity.

Fruit trees are normally composed of **scion** of a particular cultivar grafted onto a **rootstock,** although more complex combinations are possible (Fig. 7-23). The rootstock may either be grown from seed (**seedling rootstock**) or it may be asexually propagated (**clonal rootstock**). Some rootstocks, even though produced

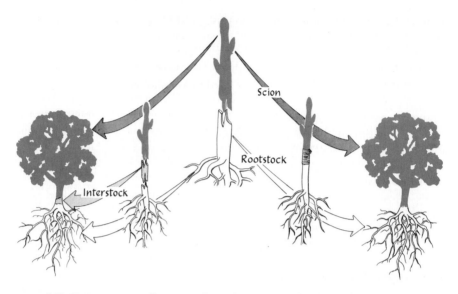

FIGURE 7-23. Fruit trees normally consist of two distinct parts, the rootstock and the scion. More complex trees may be made up of three or four components. [Adapted from Bonner and Galston, *Principles of Plant Physiology*, W. H. Freeman and Company, copyright © 1952.]

from seed, are in effect clonal because of apomixis, as in citrus. Seedling rootstocks are often derived from a particular clonal cultivar. Thus, the "western" pear seedlings are usually derived exclusively from the 'Bartlett' cultivar. Some of these terms will be encountered again in the discussion of grafting in Chapter 9, but there the emphasis will be on the techniques of grafting; here we will discuss the effects of grafting.

Restriction of Growth

The use of specific rootstocks to restrict the growth of the scion variety is an ancient practice. The degree of **dwarfing** achieved varies with the species and the rootstock (Table 7-1). The physiological explanation of the dwarfing effect has not been fully established. Evidence exists that it may be related to a number of causes, among which are the restriction of upward translocation of inorganic nutrients through the rootstock, the restriction of downward phloem transport, or some physiological disturbance caused by graft incompatibility.

There are abundant sources of dwarfing rootstocks for the apple (Fig. 7-24). These rootstocks are all from various species of apple, and many of them derive from old European clones grown under the names 'French Paradise' and 'Doucin.' These older clones were collected at the East Malling Research Station (in East Malling, Kent, England), where they were sorted out, standardized, and coded—EM 1, EM 2, and so on. Crosses of EM 9 and others with 'Northern Spy'

TABLE 7-1. Some dwarfing rootstocks.

Fruit crop	Rootstock	Effect
Apple	*Malus* clones East Malling series Malling Merton series	Complete range of dwarfing
	Malus sikkimensis "seedlings"°	Slight dwarfing
Pear	'Angers' quince clones	True dwarfing
Sweet cherry	'Stockton' Morello cherry	Height reduced to half as compared to Mazzard stocks
Peach	*Prunus domestica* clones	Slight dwarfing
	Prunus institia clones	Slight dwarfing
Plum	*Prunus besseyi* clones	Slight dwarfing
Orange	Palestinian sweet lime seedling°	Slight dwarfing
	Sour orange seedling°	Slight dwarfing

SOURCE: Adapted from Brase and Way, 1959, *New York Agricultural Experiment Station Bulletin* 783.
° Apomictic

apple have produced a number of rootstocks that show varying degrees of dwarfing and are resistant to woolly aphids. These are standardized as the Malling Merton series (MM 101, MM 102, and so on). EM 8 and EM 9 are extremely dwarfing rootstocks. These rootstocks produce the dwarfing effect even if interposed between a nondwarfing rootstock and a scion variety (Fig. 7-25). Thus, it is possible to avoid the shallowrooted characteristics of EM 9, which produces poorly anchored trees that are liable to tip over in wet ground after a strong wind, by using it as an **interstock** rather than a rootstock.

Pears are dwarfed by certain clones of quince, which belongs to a closely related genus (*Cydonia*). To avoid the graft incompatibilities of certain varieties of pear and quince, an interstock mutually compatible to both scion and stock is

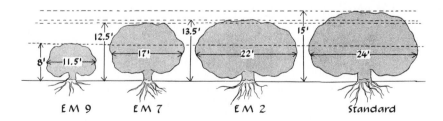

FIGURE 7-24. Comparative sizes of 17-year-old McIntosh apple on EM 9, EM 7, EM 2, and standard seedling rootstocks. [Adapted from Ontario Department of Agriculture, Circular 334, 1958.]

FIGURE 7-25. A comparison of fruiting and size of seven-year-old 'Delicious' apple on seedling rootstock (*left*) and on 'Clark' dwarf apple (*right*). The 'Clark' dwarf is obtained by using a stem piece of 'Clark' (a selection of 'French Paradise') as an interstock between the seedling rootstock and the scion cultivar.

used as a "bridge." The pear cultivar 'Hardy' is often used as such a bridge. When budding is employed rather than grafting, a shield bud of 'Hardy' inserted under the scion bud serves the same function. Dwarfing rootstocks also exist for stone fruits.

Stimulation of Growth

Rootstocks may be used to compensate for poor root growth. The experimentally produced pear-apple hybrids had to be grafted on pear or apple rootstocks to survive. The upright cultivars of *Juniperus virginiana* have poor root systems, which makes them difficult to propagate. By grafting them onto the sturdy root system of *Juniperus chinensis* 'Hetzii,' a superior plant is created. In apple, the use of 'Virginia Crab' as a rootstock or body stock to increase the vigor of the scion was once common, but this practice is no longer recommended because of the sensitivity of 'Virginia Crab' to a virus disease known as "stem pitting" (Fig. 7-26).

Flowering and Productivity

The induction of flowering in the "nonflowering" Jersey group of the sweetpotato by grafting to several species of related genera of the Convolvulaceae (morning

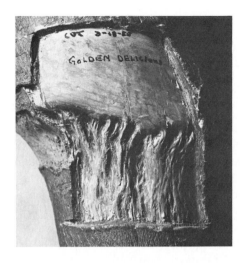

FIGURE 7-26. Symptoms of stem-pitting virus on a Virginia crab-apple bodystock. The scion cultivar, 'Golden Delicious,' does not show the symptoms, but it may carry the virus. [Photograph courtesy R. B. Tukey.]

glory family) is a striking use of grafting to effect differentiation (Fig. 7-27). Apparently, a specific flowering substance produced by the morning glory is transferred to the sweetpotato. This substance is produced in the leaves of the morning glory species in response to photoperiod and temperature effects. This technique allows the use of hybridization as a breeding method for the sweet-potato. It is of course unnecessary in the commerical production of sweetpotatoes,

FIGURE 7-27. Sweetpotato flowering after being grafted to a species of morning glory (*Ipomaea nil*). [Photograph courtesy S. Lam.]

since they are propagated from adventitious shoots that grow from the roots.

The age at which fruit trees will begin to bear can be affected by the rootstock. The severely dwarfing rootstocks of apple and pear also encourage early bearing. Dwarfing rootstocks have been used to induce fruiting in pears, which are notoriously late bearing, without a permanent dwarfing effect. Pears grafted or budded to quince are planted with the union 6–8 inches below the ground. Early fruiting is stimulated by the quince rootstock. Scion rooting, however, eventually overcomes the dwarfing influence of the quince rootstock.

Although the yield of dwarfed trees is smaller per tree than that of standard trees, their reduced size allows for closer spacing. This factor, coupled with the tendency for early bearing, often results in greater per-acre yields with dwarf apples than has been achieved with standard trees on the standard spacing. Some dwarfing rootstocks actually produce more efficient fruit trees than do seedling rootstocks. In general, dwarfing rootstocks seem to have no effect on fruit size.

Some characteristics of the scion fruit may be affected by the rootstock. For example, the rough lemon rootstock lowers the sugar content of scion cultivars of orange more than other rootstocks, and rootstocks may affect such characteristics as blooming date and maturity.

Phloem Disruption

The induction of early fruiting and the control of growth by techniques that disrupt the phloem, such as girdling, scoring, or ringing, are also ancient horticultural practices. Inverting a ring of bark accomplishes the same effect (Fig. 7-28). These practices, performed in early July (in the temperate regions of the Northern Hemisphere), are used to initiate flower-bud formation in two- or three-year-old clonally propagated apple trees in order that they will flower and bear fruit the following year. Phloem disruption does not overcome seedling juvenility, but, like the use of dwarfing rootstocks, it is practiced to increase flowering on four-year-old apple seedlings. The injured phloem retards the downward movement of the synthesized organic materials. The induction of flower bud initiation is apparently caused by the accumulation of some substance above the injured phloem, for a single branch can be induced to flower on an otherwise barren tree. The effect is temporary, owing to the regeneration of phloem in the cuts, and to the seam in the inverted ring. Although the inverted ring effectively blocks the downward movement through the phloem, according to Karl Sax, a proponent of the technique, this one-way movement (polarity) may eventually be reversed.

The fact that phloem disruption also produces a dwarfing effect in addition to the induction of early flowering suggests that some types of rootstock dwarfing may be related to the interference of downward phloem transport with the subsequent accumulation above the graft union of organic substances. This, however, cannot explain the dwarfing effects of all rootstocks.

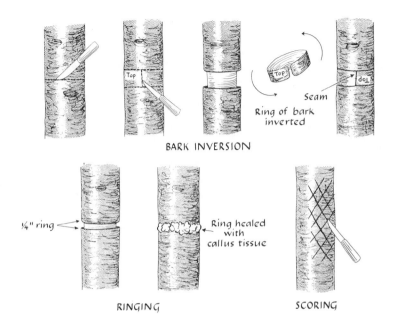

Top

Seam
Ring of bark
inverted

Top

do1

BARK INVERSION

¼" ring

Ring healed
with
callus tissue

RINGING

SCORING

FIGURE 7-28. Dwarfing and the induction of early fruiting may be accomplished by phloem disruption.

Hardening

Hardening, in the broad sense, refers to processes that increase the ability of a plant to survive the impact of unfavorable environmental stress. In its more restricted meaning, hardening refers to processes that enable plants to withstand cold injury, just as the term *hardiness* usually refers specifically to cold-hardiness.

Cold-hardiness is a variable characteristic that differs greatly with species and with seasonal change. For example, many plants that survive extreme winter freezing may be severely injured by spring frost. The natural change in hardiness in woody plants is related to temperature, water availability, and day length. The onset of cooler weather and shorter days in the fall brings about dormancy and a physiological toughening in some plants.

Cold resistance of the whole plant develops in a number of steps. As the end of the growing season approaches, the rate of growth slows down and carbohydrates begin to accumulate. Reserve foods accumulate during this period partly because they are not being utilized in growth and respiration. They are translocated to other plant parts, particularly roots, where they remain over the winter as insoluble fats and starches, and are used at least to some extent in respiration. Also during this period, water in the individual cells moves from the protoplasm to the vacuoles. These changes in water relations and in the availability of certain carbohydrates promote stability of the protoplasmic proteins, which have a higher proportion of bound water during winter than during the summer. There seems to be a surface effect on protoplasmic colloids, so that they do not clump or precipitate so readily. During the onset of winter the concentrations of colloidal and dissolved materials become much greater than during the summer, which may contribute to cold resistance. Experimentally, stem sections of mulberry (*Morus alba*), a woody plant, have survived temperatures of −320°F (−195°C) for 160 days.

There are several ways to condition plants for cold weather. One of the simplest techniques is to withhold water and fertilizer, especially nitrogen, toward the end of the growing season. Nitrogen stimulates vegetative growth, and, if applied late in the season, may cause plants to continue to grow until late in the season rather than **harden off** prior to winter. Succulent green shoots, stimulated by the addition of nitrogen, are not hardy, but are quickly killed as temperatures approach freezing. On the other hand, additions of phosphorus and potassium seem to increase frost-hardiness, but even these must not be applied too late in the growing season.

Promoting plant vigor during the growing season increases the supply of carbohydrates. Plants with high levels of reserve carbohydrates are better able to withstand cold weather than those in poor condition. Respiration continues to some extent throughout the winter, and there must be a supply of carbohydrates to maintain life processes during this period. Some grape cultivars overbear, and if the excess fruit is not removed before it matures, the plants are subject to winter injury.

As a general rule, resistance to cold injury increases with age until senescence. When exotic ornamental shrubs and other plants are introduced into a region, seeds are not planted directly in the field. Rather, they are planted in nurseries and carefully sheltered and protected for several years to get them past the juvenile stage. After they have ceased to be tender young plants and are hardened to some extent, they survive much better than if they are planted directly in the field. Some species of ornamentals must be six to eight years old before they will survive field planting.

The hardening of transplants may be achieved by any treatment that materially checks new growth. This may be accomplished by gradually exposing plants to cold, by withholding moisture, or by a combination of these two treatments. In general, a period of about ten days is sufficient to harden plants. This treatment is designed to produce a stocky, toughened plant in contrast to a soft, tender, "leggy" plant. Hardened plants are often darker green in color, and hardened crucifers have a greater amount of waxy covering on the leaves. The induced cold resistance brought about by the hardening treatment in such cool-season crops as cabbage or celery may be considerable. Unhardened cabbage plants show injury at 28°F (about 2°C), whereas hardened plants can withstand temperatures as low as 22°F (5.5°C). In such warm-season crops as tomato, the degree of cold-hardiness imposed is slight, but hardening inures the plant to transplanting shock and hastens plant establishment under adverse conditions. The reduced growth rate enables the plant to withstand desiccation until the root system becomes established. Tender plants, which respire rapidly, have little chance of survival if warm, windy, dry conditions follow transplanting. The cessation of growth under the hardening treatment, however, may severely interfere with subsequent performance; thus care must be taken to avoid overhardening. Under ideal transplanting conditions, hardening may not be necessary.

CHEMICAL CONTROL

The control of plant growth and differentiation through the use of chemical substances is a modern development in horticulture, although examples can be found of the early use of various substances (such as salt, wine, urine, and preparations made from germinating seeds) to this end. In general only a few isolated examples of these ancient practices have been shown to have a real physiological basis. The advances in this area have not come about through empirical methods but are instead largely a by-product of investigations into growth and development. This field was given great impetus by the impact of auxin studies on horticultural technology. A number of substances are now known that have a relatively broad spectrum of effects (for example, indoleacetic acid and the gibberellins). Others merely mediate or block specific metabolic pathways. Such substances as ethyl methane sulfonate (a mutagen) and colchicine (an alkaloid from the autumn crocus, *Colchicum autumnale*, that causes chromosome doubling) have been used to achieve permanent genetic change (see Chapter 10). It can be assumed that, as knowledge expands in this area, an increasing number of growth-regulating substances will be found.

At present, the substances used to kill plants account for the greater part of chemical control. This is discussed further in Chapter 8. The following section will review chemical substances that affect physiological processes important in horticultural practice.

Rooting

The rooting of cuttings has been shown to be influenced by auxins, although auxins are by no means the only substances involved. In a cutting, the natural auxin produced in young leaves and buds moves naturally down the stem and accumulates at the cut base along with sugars and other food materials. The natural formation of roots is apparently triggered by the accumulation of an optimum auxin level in relation to these substances. In a wide variety of plants, rooting is markedly increased by the addition of synthetic auxins. Although a wide variety of such compounds has been used, the greatest degree of success has been achieved with indolebutyric acid (Fig. 7-29):

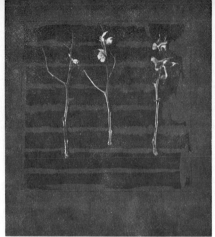

Control group
no treatment

10mg added
potassium salt of
indolebutyric acid

FIGURE 7-29. The potassium salt of indolebutyric acid markedly increases the rooting of *Chaenomeles* (flowering quince). [Photographs courtesy J. S. Wells.]

$$CH_2CH_2CH_2COOH$$

Indolebutyric acid

Other auxins have a very narrow range of effective concentrations. Concentrations below the critical level are ineffective in root initiation, whereas those above the critical level not only inhibit root growth and bud development but may cause gross morphological damage. Indoleacetic acid is ineffective, probably because it is readily destroyed by the plant.

Cuttings from many plants that are naturally difficult to root, such as apple, do not respond to auxin application. An interesting facet of this problem is that the transition of cuttings of some plants from "easy-to-root" to "difficult-to-root" is associated with the plant's change from juvenility to maturity. This may be due to the formation at maturity of inhibitors that block rooting. Thus, difficult-to-root dormant grape cuttings become easy to root when leached with water to remove the inhibitor. Other plants become difficult to root not because inhibitors are present but because they become deficient in, or lack, certain required substances. If these substances are applied in combination with an auxin, rooting can be promoted. This problem will be discussed further in Chapter 9.

Bolting

Bolting (or seed-stalk formation) in biennial plants is induced by cold. Thus, carrots and onions do not flower unless growth is interrupted by a cold-induced dormancy. This cold requirement may be replaced under certain conditions by treatment with gibberellins, a group of substances whose primary morphological effects are associated with stem elongation. Gibberellins apparently replace or substitute for the natural compounds that are produced (or that accumulate) during the cold period and are responsible for bolting. Gibberellins will also induce bolting in those plants in which the process does not require cold but that are photoperiod sensitive, such as spinach. The bolting stimulus provided by gibberellins is independent of the flower-inducing stimulus. Thus, flowering in some chrysanthemums is dependent not only upon a short photoperiod but upon elongation induced by cold treatment. Gibberellins can replace the cold requirement, but the plants remain vegetative if exposed to a long photoperiod. This use of gibberellins in promoting bolting has far-reaching effects in breeding, where time is often a limiting factor.

Gibberellins have the useful effect of facilitating normal seed-stalk formation. For example, the heads of some lettuce varieties are so tight that the seed stalk cannot push through and may break or rot unless the head is cut. The tight head is a desirable trait in crop production, but it interferes with seed production. Treatment with gibberellins can be used to encourage normal seed-stalk formation in these lettuce types.

Modification of Sex Expression

One of the most dramatic changes achieved with growth regulators has been the alteration of sex expression. Modification of sex has great potential for plant breeding. In cucumber, ethephon delays staminate flowering and thus transforms monoecious lines into all-pistillate (all-female) lines, thereby increasing yields and facilitating hybrid seed production. Similar effects have been reported with muskmelon and squash (but not watermelon). Gibberellins increase maleness in cucumbers; staminate flowers have been induced on genetically, all-pistillate (gynoecious) lines.

Although grapes have hermaphroditic flowers, certain parts of the flowers may be nonfunctional. Some clones have only functionally staminate flowers (the pistil is undeveloped). It has been possible to induce female organs (and viable seeds by selfing) in these "staminate" clones with cytokinins as well as with ethephon. Ethephon is also effective in inducing seed formation in some normally seedless clones, some of which are viable. This has promise for the breeding of seedless grapes.

Flower Induction

The biochemical basis of flowering still remains unknown. That the triggering mechanism is hormonal, however, is indicated by the translocation of the photoperiodic stimulation from leaf to bud and across graft unions. It has been demonstrated that the suspected flowering hormone ("florigen") is not an auxin, although an auxin is involved in the process. Thus, auxins applied to many plants after the initiation of flowering may effectively promote flowering.

Flower induction in pineapple has been achieved with ethylene as well as with auxins. Recent evidence suggests that auxins act by affecting the natural ethylene-generating mechanism of the pineapple fruit; nevertheless, the use of auxins to induce flowering in pineapple has been a standard culturing procedure. Different auxin derivatives have been used, such as sodium naphthalene acetate in Hawaii, and 2,4-D (2,4-dichlorophenoxyacetic acid) in the Caribbean area.

Biennial bearing, a serious problem with apples, can be controlled quite effectively with SADH (succinic acid 2,2-dimethylhydrazide), a growth retardant that suppresses gibberellin formation. This effect is achieved by the reduction of shoot growth that promotes flower formation in the following season.

Fruit Set

Promotion of Fruit Set

The practice of chemically inducing fruit set has followed from studies of the relation of natural auxins to fruiting (see Chapter 5). The use of auxin derivatives to set fruit in the absence of pollination (parthenocarpy) has had some commercial utilization in winter production of tomatoes in the the greenhouse, where fruit set is often poor, and in fig and grape production. The auxin substances generally used for tomatoes are p-chlorophenoxyacetic acid and β-naphthoxyacetic acid. This practice is limited, however, since it causes fruit abnormalities, such as puffiness and premature softening. The use of auxins to set fruit in 'Calimyrna' fig eliminates the need for male trees and the practice of caprification. However, the most effective auxin for fruit set (p-chlorophenoxyacetic acid) produces a seedless fruit that has not proved to be acceptable, and the auxin that permits the development of the seed coat (benzothiazole-2-oxyacetic acid) is not as effective for fruit set. The use of auxin sprays to promote fruit set in some grape cultivars has eliminated the need for girdling. This has become a widely adopted practice in California for use with the 'Black Corinth' and 'Thompson Seedless' cultivars. The induction of parthenocarpy in grapes with gibberellins has made possible the commercial production of seedless grapes of the 'Delaware' cultivar in Japan. Such clones receive two gibberellin treatments, the first to induce seedlessness, and the second to increase fruit size.

Flower and Fruit Thinning

The removal of flowers and fruits to reduce crop loads has been referred to as thinning. The relationship of thinning to fruit quality and size is discussed further in Chapter 8. The reduction in crop load by **chemical thinning** has become a standard practice in a large part of the fruit industry. It is one of the best examples of the chemical control of growth. The physiological action of these chemicals consists in preventing the completion of fertilization or to induce embryo abortion, both of which result in natural abscission. Chemical thinning may be performed prior to fertilization (**flower thinning**) or after fertilization (**fruit thinning**).

The materials effective in thinning may be referred to either of two groups, depending on their mode of action. Flower-thinning compounds are composed of caustic and toxic substances (for example, phenols, cresols, and dinitro-compounds) that kill off the blossoms or render them sterile. The principal effect of an interesting substance called Mendok (sodium dichloroisobutyrate) is to induce pollen sterility. This male gametocide has been used experimentally in tomatoes (a self-pollinated crop) to induce sterility in the early clusters and thereby to concentrate fruit set. The concentrated ripening of fruits would be desirable for "once-over" mechanical harvesting. The fruit-thinning materials are auxin derivatives, and they bring about thinning largely through embryo abortion. It is interesting to note that auxins, which set fruit in some species, are used to remove fruit in others.

All auxins are not necessarily effective in thinning; in fact, only naphthaleneacetic acid and its derivatives are effective in fruit thinning. This auxin is widely used in apples and is effective in peaches, pears, olives, and grapes, although N-1-naphthylphthalamic acid has been more widely used in the stone fruits. Chemical thinning with auxins is also employed to prevent fruiting in trees used as ornamentals, where only flowering is desired and fruit is considered a nuisance.

The way in which auxin derivatives cause embryo abortion is not clear. For example, the principal absorption of auxins is not through the fruit but through the foliage. The degree of thinning with auxins is greatly affected by the concentration used, the timing of the application in relation to fruit development, and the species and cultivar, as well as by such environmental factors as temperature and humidity.

The production of 'Thompson Seedless' grapes for table use has been revolutionized through the use of gibberellins, which loosen fruit clusters, greatly increase fruit size, and thin berries. In California, all table-grape acreage of 'Thompson Seedless' is treated with gibberellins (Fig. 7-30). On seeded grapes, the berry-thinning effect reduces cracking and preharvest diseases. The auxin p-chlorophenoxyacetic acid has also been used to increase the size of pineapple fruits.

Ripening

It has long been known that ethylene and acetylene stimulate fruit ripening. Ethylene gas applied in ripening rooms is a standard practice to accelerate banana ripening. The discovery that an ethylene-generating material, ethephon, stimu-

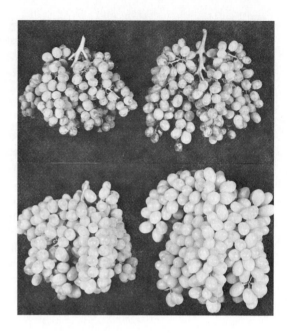

FIGURE 7-30. Effect of gibberellic acid (GA) on 'Thompson Seedless' grapes: *top left*, control; *top right*, 5 ppm GA; *bottom left*, 20 ppm GA; *bottom right*, 50 ppm GA. [Photograph courtesy R. J. Weaver.]

lates fruit ripening has made the induction of the ripening process possible in the field. The use of ethephon in pineapple encourages uniform ripening of entire fields and has tremendous implications because it makes complete mechanical harvesting a possibility. A similar effect is also achieved with tomato. Ethephon appears to have tremendous commercial applications in horticulture in the control of ripening as well as other effects. It is now used to stimulate latex flow in rubber trees.

Preharvest Fruit Drop

The effect of auxins in inhibiting abscission has found an important horticultural application in the control of preharvest fruit drop. The natural auxin, which prevents abscission and is produced by the seed, apparently decreases with fruit maturity. A number of synthetic substances are now used to delay fruit drop, among which are naphthaleneacetic acid, 2,4,5-trichlorophenoxyacetic acid, and 2-(2,4,5-trichlorophenoxy)propionic acid. Chemical control of preharvest drop is widely used in the fruit industry for apple, apricot, pear, prune, almond, and citrus fruits, and especially in cultivars of these fruits that are prone to drop prematurely. It is an effective means of preventing fruit drop that is ordinarily accentuated after frost. The use of preharvest drop control to increase red color development after maturity cannot be recommended because overmature fruit does not store well.

Red color development can be increased by SADH, an antigibberellin, which also delays preharvest drop in apples and appears to have other interesting beneficial effects, including prolonging of storage and shelf life and reduction of water core and storage scald.

Dormancy

The modification of seed and plant dormancy promises to be an important area for chemical control, because the extension of dormancy in woody plants to avoid damage by spring frost would provide great economic benefits. With the discovery of abscisic acid, the control of dormancy has received increased attention.

In the past much effort was expended in prolonging dormancy in horticultural products during storage. For example, serious losses of potatoes and onions result from the sprouting that occurs with the breaking of dormancy. High concentrations of auxins have been successful in prolonging dormancy, and a number of other substances that prolong dormancy without killing the tuber or bulb are being isolated. Many of these substances cannot be used on foods; however, the

H H
N—N
O=⟨   ⟩=O

Maleic hydrazide

use of the growth inhibitor maleic hydrazide has been effective in inhibiting sprouting in onions and potatoes and can be applied to the growing plant.

Although a complete biochemical basis for the chilling requirement has yet to be defined, several materials have been used to induce or terminate dormancy. Gibberellic acid has been the most effective material to substitute for low temperature in satisfying the chilling requirement, and it has been used for breaking dormancy in potatoes. It is a commercial practice in Florida to produce an early crop using freshly harvested potatoes for "seed." Gibberellins appear to overcome abscisic acid, the natural dormancy promoter.

Growth Promoters and Inhibitors

Chemical materials that promote or inhibit plant growth have promising uses in horticulture. For example, kinetin (6-furfurylaminopurine) acts as a promoter of cell division and appears to have value in promoting callus formation. Similarly, the growth-stimulating effects of gibberellins are used to increase the size of celery. Substances that inhibit or retard growth are equally desirable. The effect of maleic hydrazide in slowing down the growth of turf to reduce the frequency of cutting has not proven practical, but this work has stimulated a search for other

273

FIGURE 7-31. The growth-retardant properties of CBBP. [Adapted from Agricultural Research Service 22–65, USDA, 1961.]

Untreated

Soil treated
with CBBP
(Phosphon)

compounds that might perform better. Several compounds have been found that dwarf plants effectively by retarding stem growth. Examples are CBBP, or phosphon (2,4-dichlorobenzyltributylphosphonium chloride) and CCC ((2-chloroethyl)trimethylammonium chloride). Such substances show promise for use in reducing the height of many ornamental flowering plants without unduly interfering with flowering time or flower size (Fig. 7-31).

SADH, a growth retardant that acts as an antigibberellin, has a wide array of effects already discussed. Its growth-retardant properties may be of use in tomato culture to improve transplant durability and to concentrate fruit maturity by decreasing early yield.

Ancymidol (α-cyclopropyl-α-(4-methoxyphenyl)-5-pyrimidinemethanol) is an extremely powerful growth retardant. Its most important commercial use is to control the height of Easter lilies, but it is extensively used in other florist crops because of its effectiveness over a wide spectrum of plant species.

Pinching, Disbudding, and Sprout Control

Pinching (removal of the apical meristem) to facilitate branching is a standard practice with many florist crops (for example, carnation and chrysanthemum). It is possible to accomplish the same end through the application of such materials as methyl esters of C_8 to C_{12} fatty acids (for example, methyl decanoate) and C_8 to C_{12} alcohols. These "chemical pinching agents" destroy the growing point and facilitate bud break below.

FIGURE 7-32. Control of sprouting in lemon by sprays of 1% NAA (naphthaleneacetic acid) plus 5% EHPP (ethylhydrogen 1-propylphosphonate): *left*, control; *right*, treated. [From R. L. Phillips and D. P. H. Tucker, *HortScience* 9:199–200, 1974.]

Disbudding (removal of axillary buds below a terminal) is used to increase the size of the terminal flower and to alter plant form in such florist crops as chrysanthemums, carnations, and roses. There is an active search now going on for chemical disbudding agents; some materials show promise, but results so far have not been consistent enough for commercial use.

The control of root suckers, watersprouts, and excessive shoot development after pruning is a recurring problem in many fruit trees. The use of high (1%) concentrations of NAA (naphthaleneacetic acid) in paints or applied as a spray (with dormant oil and wetting agents to increase absorption) has proven effective for inhibiting these unwanted shoots (Fig. 7-32).

Abscission

Substances that influence abscission of leaves and fruits have various uses in horticulture, such as the defoliation of nursery stock to improve storage, and defoliation to encourage fruiting and to facilitate mechanical harvest. Various materials defoliate plants by burning off their leaves. These include chlorates, cyanimids, borates, urea, and potassium iodide. The abscission of fruits as well as leaves is encouraged by ethylene and thus also by ethylene-generating materials and materials that increase the natural ethylene level, such as auxins and abscisic acid. At present, ethylene-generating materials have the most promise for loosening fruits. These include ethephon for cherries, Alsol (2-chloroethyl-tris-(2-methoxyethoxy)silan) for olives, and Release (5-chloro-3-methyl-4-nitro-1*H*-pyrazole) for citrus.

Storage Disorders

A number of physiological disorders are associated with postharvest storage of fruits. Scald, a disorder of the peel, is a common hazard in the storage of apples. Materials applied before storage to decrease this disorder (such as diphenylamine and ethoxyquin) are widely used.

Selected References

Annual Review of Plant Physiology. Palo Alto, Calif.: Annual Reviews, Inc. (Recent advances in plant physiology are brought up to date in timely, technical reviews. Published each year since 1950.)

Christopher, E. P., 1954. *The Pruning Manual.* New York: Macmillan. (A nontechnical, thorough treatment.)

Galston, A. W., and P. J. Davies, 1970. *Control Mechanisms in Plant Development.* Englewood Cliffs, N.J.: Prentice-Hall. (A brief treatment of development control through phytochrome systems and various growth-regulating substances.)

Tukey, H. B., 1964. *Dwarfed Fruit Trees.* New York: Macmillan. (A review of dwarfing rootstocks. Reprinted 1978 by Cornell University Press, Ithaca, N.Y.)

Weaver, R. J., 1972. *Plant Growth Substances in Agriculture.* San Francisco: W. H. Freeman and Company. (A text on the applied uses of plant-growth substances.)

8

Biological Competition

Why should we in the compass of a pale
Keep law and form and due proportion,
Showing as in a model our firm estate,
When our sea-walled garden, the whole land,
Is full of weeds, her fairest flowers chok'd up,
Her fruit-trees all unprun'd, her hedges ruin'd,
Her knots disordered, and her wholesome herbs
Swarming with caterpillars?

SHAKESPEARE, *Richard II* [III. 4]

The earth is covered with a variety of life forms in competition for food, light, and space. To a great extent, we, the ascendant species, now direct this competition to our own advantage. The efficiency of this control is commonly thought of in terms of civilization or culture. The standard of living of a people is directly related to their ability to compete with other organisms.

The two great professions that deal directly with our control over our biological competitors are medicine and agriculture. Human diseases, inborn errors of metabolism not included, in large part are due to competitions by microorganisms for ascendancy in the human body. Agriculture, on the other hand, is concerned with human efforts to control and exploit plant and animal life for food and fiber. The term horticulture, the intensive cultivation of "garden" crops, implies man's interference with the natural competition and interaction among living things. The forms of biological competition that concern us in plant cultivation are plant pests (pathogens, predators, and weeds) and the crop plants themselves.

Plant pests take a heavy toll from world agriculture. In the United States the yearly loss in farm crops has been estimated at more than 7 billion dollars. Moreover, the annual cost of pesticides and related equipment amounts to 430 million dollars for insects, 230 million dollars for diseases, and 2.5 billion dollars for weeds! Some species of pests could wipe out the crops in an entire section of the country, so awesome is their destructive power. Others that may not be as spectacular in the short run are, in the long run, equally destructive. These pests continually "peck away" at our abundance, but their damage tends to go unnoticed. To a particular grower or home owner a particular pest may mean the difference between feast or famine, sun or shade. Eventually, however, all of us share the cost of this waste.

COMPETITION BETWEEN CROP AND PATHOGEN

Disease

The word disease means literally "lack of ease." Broadly defined, a disease is any injurious abnormality. A distinction between these abnormalities—physiological or anatomical—is made on the basis of cause. Diseases caused by some biological agent are referred to as **pathogenic** diseases. **Nonpathogenic** diseases may include the adverse effects of abnormal physiological disorders; environmental factors, including extremes of heat or cold, soil fertility, water availability, and pollution (Fig. 8-1); graft incompatibilities; spray injury; or disorders due to unknown causes.

The use of the word *disease* to refer to insect injury, much less to nonpathogenic disorders of plants, may be objectionable to some. Nevertheless, considera-

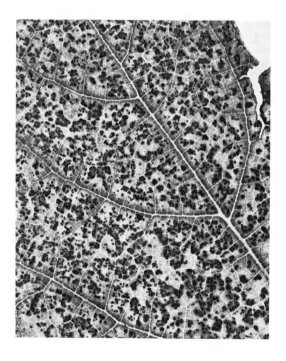

FIGURE 8-1. Ozone (O_3) injury to grape leaves results in lesions (called stipple) that are distributed interveinally over the entire leaf surface. Excessive exposure to ozone can reduce photosynthesis and cause leaf injury and death. Ozone is most abundant in urban environments where there are large quantities of the air pollutants that promote its formation. Although 1 and 2 parts per hundred million (pphm) is a normal level, occasionally ozone has reached levels as high as 100 pphm in the Los Angeles basin of California. [Courtesy W. J. Kender.]

tion of the term *mental disease* will indicate that the word is certainly not a narrow one. The confusion arises from the common use of the term in a restrictive sense to refer to the injurious effect caused by a pathogenic microorganism in intimate association with the host plant for extended periods of time. This use of the term would apply to the detrimental effects that viruses, bacteria, and fungi have on plants. The detrimental effects of insects, mice, or birds would then be classified as injury rather than disease, the term *predator* rather than *pathogen* being applied to these pests. The distinction between a predator and a pathogen, however, is not clear-cut. Many plant-attacking nematodes as well as insects (such as the peach-tree borer) are in intimate contact with the plant for extended periods of time and could well be considered pathogens even in the sense of the restrictive definition. In the ensuing discussion, no special distinction will be made between pathogen and predator.

Pathogenic plant diseases may be discussed either in terms of the agent (pathogen) causing the disease or in terms of the plant's response. The specific responses of the plant are known as **symptoms,** which, together with evidence of the pathogen, **signs,** permit the diagnosis of the disease—that is, the association of the disease with its cause. Care must be taken not to confuse the cause of the disease, the pathogen, with the disease itself.

Symptoms

Plants respond to the irritation produced by an external biological agent in a limited number of ways. The most extreme response is death, or **necrosis.** The entire plant may die, or the necrosis may be limited to specific organs, such as leaves, branches, flowers, or fruits. It may even be restricted to very small areas, resulting in spots or holes. Decline may be gradual and incomplete. For example, chloroplast breakdown, which produces yellowing or mottling, does not necessarily result in the immediate death of the plant. Another basic plant response is a reduced growth rate, which may affect the entire plant or certain of its parts, resulting in stunting, dwarfing, or incomplete differentiation. A third response is an increased growth of an abnormal and morbid type. This results in overgrowths—enlargements of organs, tissues, or cells—or tumor-like protuberances called galls. Although it is true that the basic responses of the plant are limited, the many variations involving different tissues often permit accurate diagnosis from the symptoms alone.

The Pathogen

The **pathogen** is a biological agent that produces a disease. The number of different kinds of pathogens affecting horticultural plants is truly awesome. Their effect may be as transient as an insect bite or as persistent as a virus infection.

279

The pathogen's association with the plant provides it with nourishment, shelter, support, or some other advantage. Competition between plant and pathogen is part of the natural order of things. Disease is not evil, malicious, or particularly unusual, nor do plants escape disease by virtue of their being "healthy." Many pathogens attack only vigorous, thrifty plants.

The association of living organisms in which one organism derives nourishment or other benefit from another, the host, is known as **parasitism.** The terms *parasite* and *pathogen* are not synonyms. A pathogen is injurious to the host at some stage of its life cycle, whereas a parasite is not necessarily injurious. Most pathogens are, however, parasitic in nature. Many disease-causing organisms may be only incidentally pathogenic; their usual mode of life may consist in living on naturally dead or decayed tissue. They are known as **saprophytes.** Some pathogens have evolved with a specific host plant to such an extent as to be **obligately parasitic;** that is, they can only survive on the living tissue of the host. The host range of some pathogens may be extremely large, or it may be specific enough to include only particular cultivars within a single species.

Viruses

Viruses are small infectious particles made of a core of nucleic acid surrounded by a protein sheath (Fig. 8-2). Their small size (approximately one-millionth of an inch) makes them visible only with the electron microscope. Particles may be short or long rigid rods, flexible threads, or "spherical" (actually 20-sided) forms.

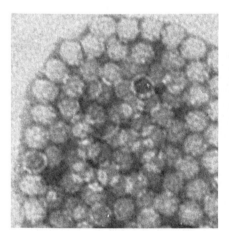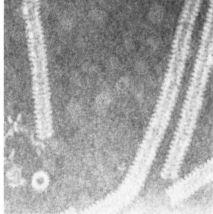

FIGURE 8-2. Plant viruses. *Left,* portions of tobacco-mosaic virus particles. Even the spiral arrangement of the protein sheath is visible in this remarkable electron photomicrograph. The complete virus particle is about 15 × 300 nm. *Right,* the protein subunits are visible in this photomicrograph of the turnip-mosaic virus particles (30 nm in diameter). [Photographs courtesy R. W. Horne, Cambridge, England.]

Viruses are obligate parasites; they reproduce only in living cells of the host, but they may be removed from the organism and remain capable of causing infection. Some viruses remain active in extracted plant juices for many months; the tobacco-mosaic virus remains active for years in dried material. Some can be crystallized and still cause the disease when reintroduced into the plant.

The question of whether viruses are living is a matter of terminology or philosophy. If the living system is defined as a self-duplicating entity capable of reconstructing itself from different component parts, then viruses can be considered alive. But viruses are not complete, living systems, because they are unable to generate the energy required for their multiplication and must exploit the enzyme system of infected, living cells.

The genetic structure of certain viruses that infect bacteria (bacteriophages) has been determined. These viruses mutate and show genetic recombination that is akin to sexual reproduction. Thus the development of various strains is possible. Their genetic material is arranged in linear sequence, and linkage maps (see Chapter 10) can be constructed on the basis of the recombination of characteristics that affect the expression of symptoms. The complete RNA structures of only a very few bacterial viruses are known.

Viruses cause diseases of many organisms, including many important crop plants. Symptoms of viral diseases of plants are usually systemic, and they range from death, through mild stunting with reduced quality and yield, to no obvious affects. Symptoms may include small lesions, a spotted or mosaic pattern of green and yellowish areas, yellowing, stunting, leaf-curling and edge-crinkling, excessive branching, and color-striping or total disappearance of color. The virus is usually named by the description of the major symptom coupled with the name of the host plant where first described. Thus tobacco-mosaic virus retains its name when it infects tomatoes. Viruses are identified by a combination of symptoms, serological tests (with the use of antibodies from warm-blooded animals, usually rabbits), electron microscopy, and chemical analysis.

The effect of viruses on the plant depends upon the sensitivity of the host cultivar and the strain of the virus. Beet curly-top, a virus disease that also affects tomatoes, results in the quick death of the plant. Some viruses produce no obvious symptoms but still cause considerable economic loss by reducing yields and performance. A combination of viruses generally complicates and increases the severity of symptoms.

Viruses are commonly insect-transmitted, many of them by aphids. A few viruses are soil borne, and some of these are transmitted by nematodes or by fungi. Some viruses are transmitted from one generation to the next through the seed. But even when a virus is seed-transmitted, only a fraction of the seedlings become infected. Some highly infectious mosaic viruses may be transmitted by touching healthy leaves after having touched infected leaves. In laboratory assays, viruses are transmitted by rubbing infectious sap on healthy susceptible leaves. Viruses are commonly transmitted through a graft union, and some pass from plant to plant through dodder, a parasitic plant.

FIGURE 8-3. A screenhouse for growing virus-free strawberry plants. The fine screen mesh keeps out aphids, which transmit the virus. These plants are maintained as a "nuclear" source of virus-free plants for nurserymen. A screenhouse for trees is in the background. [Photograph courtesy Purdue University.]

The vegetative propagation of virus-infected plants also propagates the virus. The only satisfactory way of maintaining virus-free stocks of vegetatively propagated plants is by the perpetuation of plants that are determined to be free of the virus. Maintenance of virus-free plant stock is achieved by isolation, roguing of infected plants, and control of insect vectors (Fig. 8-3). As a rule, once a plant contains a virus, little can be done to get rid of it, although heat treatment of the plant is an effective way of inactivating some viruses, and excised shoot tips (particularly if rapidly growing) may be free of virus.

Viroids

A number of virus-like particles implicated as the cause of plant diseases have resisted numerous attempts at identification by conventional techniques. Recently the infectious materials causing potato spindle-tuber disease and citrus-exocortis disease have been shown to be low-molecular-weight RNA (about one-tenth the size of the RNA in the smallest known virus). These infectious agents appear to represent a new class of pathogens now referred to as **viroids,** or **infectious** or **pathogenic RNA.** Viroids may also be implicated as causal agents in some animal diseases.

Mycoplasmas

For many years a number of strange organisms have been known to exist that have the property of passing through filters that trap bacteria (as do viruses), but that live on inorganic media (as do bacteria). In the late 1960s the discovery was made that these organisms, now called **mycoplasmas,** are also responsible for causing plant diseases. Mycoplasmas are now implicated in a number of plant diseases formerly thought to be caused by viruses, particularly in a group called "yellows" diseases, which are principally spread by leafhoppers. Examples of plant diseases known to be caused by mycoplasmas include aster yellows, mulberry dwarf disease, and potato witches' broom.

The size of mycoplasmas is variable. When grown in culture they produce long filaments resembling fungal hyphae, hence their name, which means "fungal-form." However, the filaments may break up into very small, round cells known as **elementary bodies,** ranging in size from 125 to 250 nm. These are the particles that pass through bacterial filters. From the elementary bodies mycoplasmas may form either branching filaments or "large cells" (500 nm) that approach the dimensions of bacteria.

Mycoplasmas, unlike viruses, are definitely lifelike in that they contain enzyme catalysts, energy systems, as well as genetic information in the form of DNA and RNA. Although surrounded by a membrane, they have no exterior cell wall as do most bacteria. This explains their apparent resistance to penicillin, which works on the bacterial cell wall. However mycoplasmas are inhibited by tetracycline compounds (such as chlorotetracycline), and many suspected viral diseases have been proved, through the application of these chemicals, to be caused by myco-plasmas.

Bacteria

Bacteria, one-celled "plants" and the smallest of living "organisms," are respon-sible for many plant diseases. Seven genera of bacteria, none of which form spores, are plant pathogenic. Bacteria as a group are able to enter plants only through natural openings, such as the stomata and lenticels, or through wounds. Insects are important in the transmission of bacterial diseases.

One of the most serious bacterial diseases of plants, and the one in which bacteria were first shown to cause disease, is fireblight, a serious infection of apple and pear caused by *Erwinia amylovora* (Fig. 8-4). Insects disseminate these bacteria, which penetrate the plant either through the nectar-producing glands in the flower, causing a blighting or death of the blossom (blossom blight), or through shoot terminals (shoot blight). The bacterial pathogen is also carried from infected parts to other parts of the tree by rain. The organism survives through winter in older bark lesions called cankers (Fig. 8-5). The disease is extremely serious on pear and has confined commercial production in the United States to the Pacific states and the Great Lakes area. Even in these locations, however, careful control measures must be used. This involves the constant removal of blighted wood, the

A

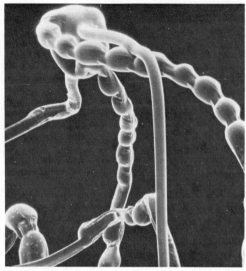

B

C

FIGURE 8-4. Fireblight, which is caused by the bacterium *Erwinia amylovora*, is one of the most devastating diseases of pear and apple. *A*, profuse production of bacterial strands on stem and leaf petioles of 'Bartlett' pear. The aerial strands may be smooth or beaded. *B*, the strands contain large numbers of pathogenic bacteria embedded in a matrix. The strands are wind disseminated, and they appear to play an important role in fireblight epidemics. *C*, Fireblight symptoms on a 4 year-old pear tree. [Photographs *A* and *B* courtesy Harry L. Keil, U.S. Department of Agriculture; photograph *C* courtesy Purdue University.]

use of antibiotic sprays, and cultural practices that discourage rapid, succulent growth.

Symptoms of bacterial diseases include the death of tissues and the formation of galls. The soft rots common in stored fruits and vegetables are caused by pectin-dissolving enzymes produced by the bacteria. Wilts produced by some bacterial diseases (Stewart's wilt of corn, for example) are a result of vascular disturbances, specifically a "plugging" of the vascular system by masses of bacteria.

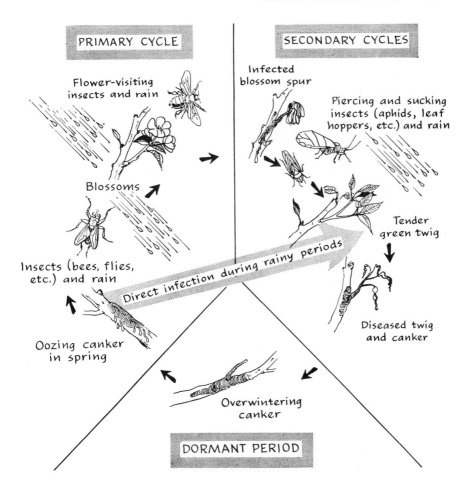

FIGURE 8-5. Life cycle of the bacterium *Erwinia amylovora*, which causes fireblight. [Adapted from Klos, Michigan State University, Extension Folder F-301, 1961.]

Fungi

Fungi, which cause the great majority of plant diseases, are multicelled plants. Except for some primitive types, they are characterized by a branching, thread-like (mycelial) growth. Fungi do not have chlorophyll and hence depend ulti-mately on green plants for their food. They may be saprophytic or parasitic, and many are both. Fungi reproduce by spores, which may be mitotically (asexually) or meiotically (sexually) produced. The life cycle of fungi is typically quite involved and comprises many different stages.

Fungi form three large well-defined classes, **Phycomycetes, Ascomycetes,** and **Basidiomycetes.** Those whose sexual stage is not known (and that may presumably

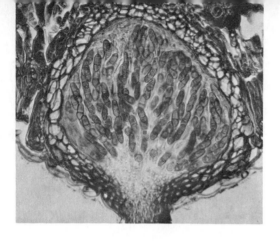

FIGURE **8-6.** The fruiting body (perithecium) of *Venturia inaequalis*, which causes the apple-scab disease. Eight two-celled ascospores are contained in each ascus. [Photograph courtesy S. Maciejowska.]

never form a sexual stage) are lumped together in a more or less artificial class (the mycologist's "trash pile") known as **Fungi Imperfecti** or **Deuteromycetes.** The phycomycetes are primitive fungi whose characteristic feature is the absence of cross walls in the mycelium. Examples of well-known horticultural plant diseases caused by phycomycetes are downy mildew of grapes, seedling damping-off (caused by a number of species), and late blight of potatoes. The ascomycetes, which produce the largest number of plant diseases, are distinguished by their specialized sacs (asci), which contain the sexual spores (Fig. 8-6). Ascomycetes are responsible for the following diseases: apple scab, powdery mildews, brown rot of peaches and plums, and black spot of roses. The basidiomycetes, the "higher fungi," produce a specialized sexual spore-forming structure called a **basidium.** Mushrooms, the culture of which is considered by some as part of the horticultural industry, belong to this group. Diseases caused by basidiomycetes are among the most destructive scourges of crop plants. They are popularly called **smuts** and **rusts** because of the appearance of masses of black or red colored spores on the host plant, as in onion smut, corn smut, and asparagus rust. Species of rust fungi may produce many spore forms (Fig. 8-7). Some rusts require two unrelated plants for the completion of their life history. Thus, different stages of the cedar–apple rust organism (*Gymnosporangium juniperi-virginianae*) infect different hosts.

Fungi are capable of entering plants by themselves, although wounds and natural openings are utilized by many. Fungal disease is spread in a great number of ways. Spores are produced in enormous numbers and are spread by water or air currents. The pathogen that causes the apple-scab disease is capable of forcibly discharging spores into the air. Mycelia growing saprophytically are a factor in the spread of soil-borne fungi. Some fungi are spread by insect vectors, as is the fungus that causes Dutch elm disease; others utilize wounds inflicted by insects for access, as does the fungus responsible for brown rot of plums and peaches.

Symptoms of fungal diseases are extremely varied; all parts of the plant may be affected. Plant-pathogenic fungi generally cause localized injury, although some are responsible for vascular wilts. The presence of visible (to the unaided eye) forms of the fungus, such as mycelial growth, masses of spores, or fruiting bodies, are often an integral part of symptom expression and help identify fungal diseases (Fig. 8-8).

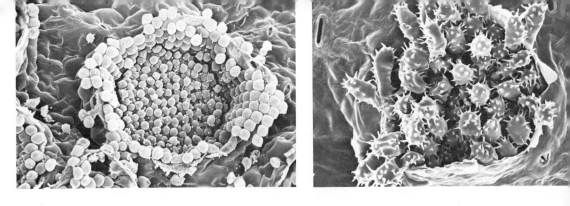

FIGURE 8-7. Fungal Spores of the rust fungus *Puccinia podophylli* on May-apple: *Left*, aecial stage; *right*, telial stage. As many as five spore forms may develop in rusts. These form in the following sequence: *pycniospores* (spermatia)—uninucleate and haploid spores of two mating types; *aeciospores*—binucleate, usually formed on side of leaf opposite the side bearing the pycniospores; *urediniospores*—binucleate, "red" spores; *teliospores*—binucleate resting spores; *basidiospores* (sporidia)—uninucleate products of meiosis. [Scanning electron micrographs courtesy H. P. Rasmussen and G. R. Hooper.]

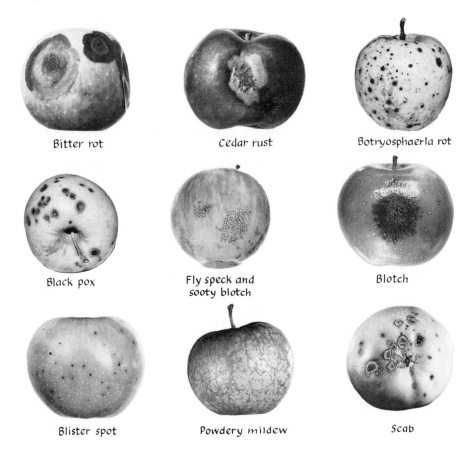

Bitter rot	Cedar rust	Botryosphaerla rot
Black pox	Fly speck and sooty blotch	Blotch
Blister spot	Powdery mildew	Scab

FIGURE 8-8. Fungal diseases of apple. [Photographs courtesy E. G. Sharvelle.]

Nematodes

Nematodes are unsegmented worms (not segmented "earth worms") of the phylum Nematoda (Fig. 8-9). Many species of nematodes are parasitic on plants and animals, including man; the major portion, however, are "free-living" in the soil and represent a large portion of the soil fauna. Although their size is macroscopic, they are small enough (usually less than 3 mm long) to be generally inconspicuous. The importance of nematodes in causing plant disease is becoming increasingly apparent.

The majority of nematodes that attack plants are soil-borne and generally feed on plant roots. They may feed superficially, or they may be partially or completely embedded in the root tissues. A few species of nematodes feed on the aerial portions of the plant (for example, the foliar nematode of chrysanthemum and the seed-gall nematode of bentgrass). Some nematodes are very specialized parasites and attack only a few species of plants; others have a wide host range.

Symptoms of root injury are variable. Those nematodes belonging to the genus *Meloidogyne* produce gall-like overgrowths on roots (Fig. 8-10) and are known as **root-knot** nematodes. These readily observable symptoms allow positive identification of nematode injury and are used as criteria in inspection of planting stock. The infected roots eventually deteriorate, and may afford access to bacterial and fungal rots. The tissues in the gall (especially the vascular tissues) become disorganized; giant cells are often formed. Thus, the above-ground symptoms appear as drought injury (that is, excessive wilting and weak, yellow growth). Other plant-parasitic nematodes, such as those of the genus *Pratylenchus* (meadow nematodes), do not form galls. Both the root damage and the above-ground symptoms resemble those of root rots. Owing to the difficulty of recognizing this pathogen, the meadow nematode may be easily transported in infected planting stock.

1mm

FIGURE 8-9. Nematodes of two genera, representing different morphological forms: *Left,* the swollen female of *Meloidogyne* (root-knot nematode). *Right,* the worm-like female of *Tylenchorhynchus* (stunt nematode). [Photograph by H. H. Lyon, courtesy W. F. Mai.]

FIGURE 8-10. Increasing severity of root-knot lesion (galls) on musk-melon. [Photograph courtesy G. W. Elmstrom and D. L. Hopkins.]

Arthropods

The phylum Arthropoda (literally, "jointed-legged") includes the invertebrate animals having an external skeleton, paired jointed limbs, and a segmented, bilaterally symmetrical body. Arthropoda is an enormous group, and contains about 75% of all known animal species, of which 90% belong to one order, Insecta, the true insects. Almost 700,000 species of true insects are known, and the estimated number of species actually existing in the world ranges from 2 million to 10 million.

Species in two classes of arthropods, Arachnida (spiders and mites) and Insecta (true insects), are the major plant pests. Species of the class Arachnida have four pairs of legs, are wingless, and have a cephalothorax (the head and thorax are fused). Insects have three pairs of legs, almost all are winged (although some wings are rudimentary or degenerate), and all have three body regions, including a separate head. A brief classification of the arthropods showing some of the orders of true insects that attack plants is shown in Table 8-1.

The war over crop plants between people and arthropods is a continuous one. The battle lines are not clearly drawn, however. We are aided by the intense competition between insect species, and predatory insects must be regarded as beneficial. Furthermore, if we were to rid the earth of all insects, we would be much the worse off, since we depend upon insect species for the pollination of many of our crop plants.

Our antagonists have many built-in advantages. These include their small size, which makes them difficult to find; their power of flight; their extremely rapid rate of reproduction; and their specialized structural adaptations, which enable some species to exist in practically any location and to infest almost any plant species. Further, the division of the life cycle of some insects into separate stages is an enormous advantage. This **complete metamorphosis** permits specialized structural adaptation for feeding and reproduction. The life cycle of an insect that undergoes complete metamorphosis consists of four stages: (1) **egg**, (2) **larva**, or

TABLE 8-1. Partial classification of arthropods, including some of the major orders of insects that attack plants.

Class and order	Examples	Typical mouthparts	
		Larvae	Adults
Chilopoda	Centipedes		
Diplopoda	Millipedes		
Crustacea	Crayfish, lobster		
Arachnida	Scorpions, mites, ticks, spiders		
Insecta	True insects		
Gradual metamorphosis			
Orthoptera	Grasshoppers, crickets		chewing
Thysanoptera	Thrips		rasping-sucking
Hemiptera	True bugs		piercing-sucking
Complete metamorphosis			
Hymenoptera	Bees, ants, wasps	chewing	chewing-lapping
Coleoptera	Beetles	chewing	chewing
Lepidoptera	Butterflies, moths	chewing	sucking
Diptera	Two-winged flies and mosquitos	chewing	piercing-sucking-sponging

feeding stage, (3) **pupa,** a quiescent stage in which the larva is transformed into the adult form and, (4) **adult,** or reproductive stage. Other insects, such as grasshoppers, undergo **incomplete metamorphosis,** in which the physical changes from egg to adult are gradual. The intermediate stages are known as **nymphs.**

The larval stage, which often bears no obvious resemblance to the adult stage (for example, the caterpillar and the butterfly), is often the most injurious to crop plants. Thus, many of the descriptive names (tomato hornworm, apple maggot) given to insect pests refer to the larval form. The terms for the larval stages of the insect orders are not too specific. In general, the name **maggot** refers to the larval stage of species of the order Diptera (flies, mosquitos, gnats). **Miners** are dipteran larvae that tunnel within leaves. **Caterpillars** are the larval stages of species of the order Lepidoptera (butterflies and moths). The term **grub** is used with reference to some of the soil-borne larvae of species of the order Coleoptera (beetles), although the name is also applied to any other soil-borne larvae. The larvae of "click" beetles are known as **wireworms.** The larval and adult forms of beetles that infest grains and seeds are referred to as **weevils.** The name **slug** may be applied to any slimy larva, specifically to larvae of the Hymenoptera (bees, ants, and wasps). They should not be confused with true slugs, which belong to the phylum Mollusca, along with snails and clams. **Borers** are larvae, usually of moths or beetles, that tunnel within roots or stems. The term "worm" is also used to refer to insect larvae, but this is a misnomer and should be avoided.

Insects (and mites, which we shall also consider here) injure plants in their attempts to secure food. The damage caused directly and indirectly by insects is enormous and varied. It includes the destruction of plants, either in whole or in part, by chewing insects; the debilitation of plants by sucking insects; the spreading of such plant pathogens as viruses, bacteria, and fungi; and the contamination of plant products by the decomposed bodies or excreta of insects.

Insects feed on plants in two distinct ways. They either tear, bite, or chew portions of the plant (**chewing insects**) or pierce or rasp the plant and suck or lap up the sap (**sucking insects**). Specialized mouth parts have been evolved for these two basic feeding patterns.

The chewing insects, adults or larvae, eat their way through the plant, riddling it with holes and tunnels (Fig. 8-11). Leaf-chewing insects that do not eat the tougher vascular portions may completely skeletonize the leaves; others, less selective, may devour the entire plant. Injury caused by chewing insects that feed externally is seldom confused with anything else. Damage done by chewing insects that girdle the plant, or those that feed internally or on roots, is not as apparent. The internal feeders gain entrance to the plant from eggs deposited in the plant tissue or by eating their way in soon after being hatched on or near the

A

B

FIGURE 8-11. Insect damage: A, cabbage looper on cabbage; B, grasshopper on corn; C, tomato hornworm feeding on tomato fruit. [Photographs by J. C. Allen & Son.]

C

surface of the plant. These internal feeders are almost impossible to control once they have entered the plant. Control consists in destroying them in their external stages. The symptoms of infestation by the peach-tree borer, which may enter the tree soon after hatching, are typical of the symptoms caused by internal insect feeders, namely, a weakened, devitalized growth of the tree and a yellowing of the foliage. A gummy exudate may be observed where the borer has wounded the trunk. Peach or plum trees infested with borers often die in a few years.

Injury caused by sucking insects (aphids, scale insects, leafhoppers, and plant bugs) results in a distinctly different type of symptom—the curling, stunting, and deforming of plant parts, usually the stem terminals. Spotting, yellowing, and a glazed appearance of the leaves are also common symptoms. The small size of red spider mites, which have sucking mouth parts, makes them difficult to see without a magnifying glass. They are often not diagnosed until they form conspicuous webs, by which time their damage may be extremely severe.

Another symptom caused by chewing and sucking insects alike is the formation of overgrowths called galls. These are formed by an abnormal growth of plant tissue in response either to the feeding insect or merely to the presence of eggs deposited in the tissue. These galls, suggestive of cancerous growth, may be quite elaborate and structured. Some appear to do little damage to the plant; others are obviously quite injurious.

Birds, Rodents, Rabbits, and Deer

Among the vertebrates, or backboned animals, birds, rodents, and deer are considered the greatest pests. Birds may become quite troublesome in such fruit crops as grapes, blueberries, and cherries. Birds may do severe damage to grapes; even a few pecked fruits permit the introduction of insects and rot, which may mean ruin of the whole cluster.

Rodents (including mice and moles) and rabbits are among the most serious orchard pests. They feed in winter and early spring and often completely girdle fruit trees, especially apples. Unless bridge grafting is done promptly, even large trees may be killed outright. The tunnels of moles can be a real nuisance in lawns or gardens.

In areas where they are naturally abundant or where their population is unchecked, deer may prove quite damaging to orchard and nursery stock. They are most troublesome during the winter, when their natural browse is in short supply, and when the horticulturist's vigilance is at its low ebb.

The Disease Cycle

The disease cycle includes all the series of sequential changes of the pathogen and the plant in the course of the disease. It involves the life cycle of the plant as well as that of the pathogen.

Life Cycles

The **life history** of an organism includes all of the diverse forms and stages through which it passes. The life history of higher organisms, including that of man, is synonymous with the sexual cycle. Rather than the sequential steps from "womb to tomb," it is more correctly "gametes to fertilized egg to adult to gametes." The life history of lower organisms may be considerably more complex, and is often made up of a number of continuous stages of existence called **life cycles.** This may involve a number of asexual cycles within the sexual cycle. In addition, many lower organisms such as bacteria do not ordinarily pass through the sexual stage but exist as a single continuous form or utilize asexual spores exclusively, as in the members of the fungal class Deuteromycetes (Fungi Imperfecti).

In temperate areas the life history must adjust to the seasonal cycle. Most pathogens pass through the winter in a stage of inactivity or dormancy. This overwintering stage, although usually specific for a particular organism, may be any stage in the life history. With the advent of spring the cycle resumes. The first cycle initiated at this time is known as the **primary cycle.** Subsequent cycles within the year are all known as **secondary cycles.** In areas where the change in seasons is not distinct—that is, where there is neither a sharp temperature differential nor separate periods of wet and dry—the pathogenic cycle may be a continuous secondary cycle. Greenhouse pests often live in a continuous secondary cycle.

In reference to the disease, the life cycle of the pathogen can be divided into two phases; a **pathogenic phase,** in which the pathogen remains associated with the living tissues of the principal host plant; and an **independent phase** not associated with the living plant, in which the organism may be saprophytic, dormant, or pathogenic to another plant. The relative length of these two phases varies greatly. For example, the independent phase of plant viruses lasts only while they are being transported from plant to plant via insects. On the other hand, the independent stage of some pathogens may be very long. Some fungi, such as *Verticillium,* live saprophytically as a natural part of the soil flora and become pathogenic only with the introduction of a suitable host plant. Similarly, some animal pests of plants may have long independent phases. Mouse damage to orchards, for instance, is done usually only in brief periods in late winter.

Pathogenic Phase

The pathogenic phase may be divided into the stages of **inoculation, incubation,** and **infection.** Inoculation consists in the transference of some form of the pathogen (the **inoculum**) to the plant. The pathogen may be transferred under its own power, as are insects, or by some agent of inoculation. The important vectors are wind, water, insects, and man. In gall-forming insects, the adult form is the vector and the egg is the inoculum. **Incubation** includes all activities of the pathogen from the time it actually enters the plant until the plant reacts to it—the **infection** stage. The important stages with respect to control are inocula-

293

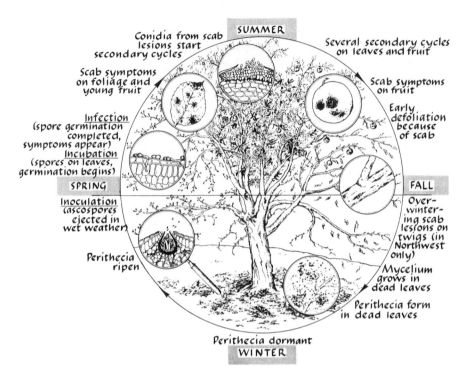

Conidia from scab lesions start secondary cycles

SUMMER

Several secondary cycles on leaves and fruit

Scab symptoms on foliage and young fruit

Scab symptoms on fruit

Infection (spore germination completed, symptoms appear)

Early defoliation because of scab

Incubation (spores on leaves, germination begins)

SPRING

FALL

Inoculation (ascospores ejected in wet weather)

Over-wintering scab lesions on twigs (in Northwest only)

Perithecia ripen

Mycelium grows in dead leaves

Perithecia form in dead leaves

Perithecia dormant

WINTER

FIGURE 8-12. The life history of the apple-scab fungus (*Venturia inaequalis*). [From Pyenson, *Elements of Plant Protection*, Wiley, 1951.]

tion and incubation. By the time the plant reaches the infection stage, the damage has been done.

It is not possible to generalize on the relationship between the pathogenic stages and the life cycle. For example, the primary cycle of the fungus *Venturia inaequalis,* which causes apple scab, is formed when spores (ascospores) are produced from the sexual stage, which overwinters on leaves under the tree (the saprophytic stage). The secondary cycles are formed when asexual spores (conidia) are produced from primary infection. The life history of this pathogen unfolds in a single year and contains several asexual cycles (Fig. 8-12).

The codling moth, which infests apple, pear, and several other fruits, demonstrates a different type of cycle. The life history may occur several times within a single year. The full-grown larval stage overwinters in silken cocoons in trees or on the ground; the pupal stage is formed in midwinter. In the spring the grayish moths emerge and lay their eggs on the upper sides of leaves, twigs, or spurs. The eggs hatch in 6–20 days, and the larvae chew their way into the young fruit. This is the primary cycle. After becoming full grown in 3–5 weeks, the larvae crawl out of the apple, drop to the ground, and spin cocoons. The moths may emerge to produce secondary cycles. There may be two or three generations, some of which may be incomplete, depending upon the latitude.

Principles of Disease Control

> There's small choice in rotten apples.
> SHAKESPEARE, *The Taming*
> *of the Shrew* [I. 1]

The study of pathogenic plant diseases in the United States is usually divided between two disciplines: **plant pathology,** which deals with diseases of viral, bacterial, and fungal origin; and **entomology,** the study of arthropods. (Nematodes may be claimed by both, but mice and other vertebrates are claimed by neither.) The applied phase of these biological sciences is directed toward the **control** of these plant pests. This involves both alleviating the injury and preventing the spread of the pathogen. In Europe and increasingly in the United States the control of plant pests is considered a distinct agricultural discipline called **plant protection.**

There are two approaches to the control of pathogenic plant diseases. One is directed at the pathogen and involves the use of techniques that prevent or restrict the pathogen's invasion of the plant. These techniques interfere at some point with the successful completion of some stage in the disease cycle. The other approach is directed toward the plant's ability to resist or at least tolerate the intrusion of the pathogen. It should be obvious that both methods of control depend upon an intimate knowledge of the disease cycle. To control some diseases successfully, many different methods must be utilized; effective control requires experience, vigilance, and persistence.

The economic value of pest control is the additional benefit obtained by the control measure less the cost of the control. Predicting the economic benefits of any control measure is particularly hazardous, because control, to be of value, must usually be applied in anticipation of the pest. Thus, many control measures can be thought of as a form of disaster insurance. In agriculture, the use of expensive methods of pest control is only feasible with high-value crops. Consequently, horticulture bears a disproportionate share of the cost of pest control.

Legal Control

The separation of the pathogen and the plant may be accomplished by legal methods. For example, **quarantines,** which prohibit the importation of certain plants, may effectively eliminate, or at least slow down, the introduction of new pests. Where outright quarantines are not feasible or practical, subjecting plant shipments to inspection helps to prevent the spread of some pests.

Cultural Control

A number of techniques may be employed to reduce the effective population level of a pathogen. These include the elimination of diseased plants or seeds (**roguing**); the removal of infected parts of the plant (**surgery**); or the removal of plant debris, which may harbor some stage of the pathogen (**sanitation**). Cultural practices that alternate plants unacceptable to the pathogen (**rotation**) may effectively starve it out.

The effective population of the pathogen may be reduced by any method that renders the environment unfavorable to increases in its population, such as draining land to discourage water-loving fungi, pruning to reduce foliage density and to increase the rate of drying, and varying temperature and humidity conditions in the greenhouse.

Physical Control

When individual plants are valuable enough to warrant the expense, physical barriers may be employed to protect the plant from larger pests. (Such barriers are also effective in controlling those microscopic pathogens borne by insect vectors.) Examples are screening to keep out insects or birds, guards to protect trunks of apple trees from injury by mice, and the traditional garden fence to keep out rabbits. In Japan, developing pears are sometimes enclosed individually in bags to protect them from insect injury.

Physical methods may be used to eliminate certain pathogens selectively or to protect the plant against the intrusion of the pathogen. This requires the use of techniques that are differentially effective in destroying or repulsing the pathogen but will not damage the plant. An example of such a method is the use of heat. Immersion in hot water (110–122°F, or about 44–50°C, for 5–25 min) is used as a treatment for destroying pathogens that infest seeds and bulbs. Hot water treatment is used to control some fungal diseases of grains, a bacterial disease (black rot) of crucifers, and nematodes in dormant strawberry plants. Heat-treating strawberry plants has also been effective in inactivating some viruses. Steaming of potting soils is a common control for many soil pests.

The action of a strong stream of water (**syringing**) has some value in reducing the infestation of spider mites in home or greenhouse plantings. The use of firecrackers or noisemakers to discourage birds from small fruit plantations is successful to a limited extent (although often no more effective than the old-fashioned scarecrow). In many areas, however, these are subject to antinoise ordinances.

Traps may also be used to control the pest population. Substances to lure the pest (**attractants**) are incorporated in the trap. A recent example is the use of "blacklight" (light of wavelengths between 340–380 nm) to attract insects to the trap, where they are killed by some toxic solution or by rotating fan blades. The use of traps is generally not efficient for mice, but it has proved successful for rabbits in limited areas.

Chemical Control

The concentration of the horticultural industry has resulted in the concentration of plant pests. As a result the entire industry is now almost completely dependent

FIGURE 8-13. Control of summer diseases of apple with fungicide application. Fruits on right received no spray and show a number of summer diseases, including Brooks spot, black pox, and sooty blotch. Fruits on left were grown under a full-season program of captan sprays. [Photograph courtesy J. R. Shay.]

upon the use of chemical control (Fig. 8-13). Commercial growers of practically all horticultural crops rely on complete "schedules" of chemical application and utilize many different compounds. The agricultural chemical industry is a vigorous one and new materials are continually being released.

BIOLOGICAL ACTION OF CHEMICALS. **Pesticide** is the generic name for all chemicals that control pests. They are usually toxic to the pest at some stage of its life cycle. (**Repellants** are compounds that may not be actively poisonous but make the crop plant unattractive to animal predators by virtue of their odor, taste, or other physical properties. They are included in the legal definition of pesticides.) Pesticides are classed according to the organisms they control; for example, bactericide, fungicide, rodenticide. (Herbicides, chemicals that kill plants, will be discussed separately, later in this chapter.)

Pesticides are generally selective in their action. Thus, chemicals that are fungicidal are usually not insecticidal. But there are exceptions; for example, Bordeaux mixture (100 gal water; 10 lb hydrated lime; 6 lb copper sulfate), which is primarily fungicidal, also has some value as an insect repellant. Pesticides that are toxic to a broad spectrum of organisms, including the crops they are meant to protect, must be used in the absence of the plant, as in preplanting treatments of the soil.

It should be reemphasized that pesticides are not necessarily toxic to all stages of a particular pest, nor is it necessary that they be. Usually, a particular stage of the life cycle is especially vulnerable to chemical attack. This "weak link" may be a germinating fungal spore, the young larval stage of an insect, or the insect vector of a virus disease. For example, it is much more difficult to kill fungi after they have extensively entered the plant. However, materials that will kill or prevent spores from germinating on the plant may prevent further inoculation and, in effect, control the disease.

Pesticides are classified as systemics or nonsystemics. **Systemics** are actually absorbed by the plant, and may be translocated within the plant, rendering the plant itself toxic to the pathogen. Systemics must of course be restricted in their use to nonedible plants unless they break down within the plant before consumption. Much more common are the **nonsystemics,** which merely coat the plant with a toxic substance. This distinction between systemics and nonsystemics is not clearcut, however; many compounds whose main action is on the surface of the plant may be absorbed to some degree.

Insecticides are classified, according to their action, as stomach poisons or contact poisons. **Stomach poisons** are usually used against chewing insects. The poison must be ingested by the insect in order to do its work. Since sucking insects feed on plant juices, they are usually not affected by stomach poisons (unless, of course, they are also systemics). They are usually controlled by **contact poisons,** which kill by penetrating the insect body directly or through breathing or sensory pores.

A great number of compounds, both organic and inorganic, are used to kill plant pests. The inorganic materials are usually metallic salts of such metals as copper, mercury, lead, and arsenic. A few of the organic compounds occur naturally, such as nicotine, pyrethrum, and rotenone. However, most of the present-day organic compounds are now completely synthetic, as are DDT, the organic phosphates, and the carbamates. A list of some of the important pesticides is presented in Table 8-2.

APPLICATION OF PESTICIDES. Chemicals are applied to plants in a number of formulations of the liquid, solid, or gaseous state, and consequently by a variety of methods. The aim is to obtain uniform coverage at a controlled rate. For highly active compounds the rate of application of the active ingredient may be extremely low, thus it is difficult to maintain adequate coverage without special equipment.

It is seldom practical to apply chemicals in their undiluted form. Consequently their distribution is carried out by the use of an inert carrier. The carrier may be a solid, such as talc, in which case the material is applied as a **dust.** Dusts have the advantage of lightness, but there is some question as to the persistence of the material. Moreover, some materials (for example, oils) cannot be applied in this manner.

The disadvantages of dusts are obviated by the use of a liquid carrier, usually water, in which the chemical is dissolved, suspended, or emulsified. The material is applied under pressure in droplets of various sizes (Fig. 8-14), but usually in the form of a **spray** (Fig. 8-15). The disadvantages of spraying lie in the bulk and weight of the water carrier. The trend in spraying is, therefore, to use high concentrations of the active material and achieve dispersion by a blast of air (Fig. 8-16).

Another difficulty in using a water carrier is that plant surfaces, since they are heavily cutinized, are "water repellant," which causes the spray to form droplets

TABLE 8-2. Classification of some important pesticides (excluding herbicides) by chemical composition.

Class of material	Examples	Major type of action
Inorganic		
Arsenic	Lead arsenate	Insecticide
Copper	Copper sulfate	Fungicide, bactericide
Sulfur	Lime sulfur	Fungicide
Organic		
Naturally occurring		
Antibiotics	Streptomycin, cycloheximide	Bactericide
Nicotine	Nicotine sulfate	Insecticide
Pyrethrum		Insecticide
Rotenone		Insecticide
Oils	Diesel oil	Insecticide
Synthetic		
Benzene compounds	Chlorothalonil	Fungicide
Benzimidazoles	Benomyl, thiophanate	Fungicide
Carbamates	Maneb, zineb	Fungicide
	Carbaryl	Insecticide
Carbon tetrachloride		Fumigant
Chlorinated hydrocarbons	DDT, methoxychlor	Insecticide
	DD	Nematocide
Heterocyclic compounds	Captan, folpet	Fungicide
Organophosphates	Malathion	Insecticide, miticide
Pyrethroids		Insecticide
Quinones	Dichlone	Fungicide

on the surface instead of a continuous film. As a result, the deposited chemicals are irregularly dispersed. This may be overcome through the use of **wetting agents**—surface-active chemicals that break the surface tension. Other additives known as **stickers** improve the retention and persistence of the chemical application.

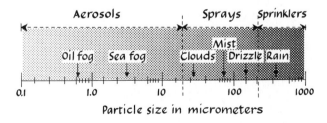

FIGURE 8-14. The drop-size spectrum.

FIGURE 8-15. Application of fungicides to tomatoes with a boom-type sprayer. [Photograph by J. C. Allen & Son.]

FIGURE 8-16. Orchard speed sprayers use a blast of air as the carrier for highly concentrated sprays. [Photograph courtesy Farm Equipment Institute.]

Volatile substances that are pesticidal in the gaseous state are known as **fumigants.** Some substances applied as sprays or dusts may also be used as fumigants; for example, nicotine sulfate may either be applied as a spray or volatilized and applied as a fumigant. Fumigants are, as a rule, extremely toxic to all life, and the problem in their use is to obtain selectivity. Thus, fumigants have been widely used in the absence of plants to control soil pests (Fig. 8-17). Some fumigants may be stored under pressure in the liquid state. Others may be solids at

FIGURE 8-17. Soil fumigation is achieved in a single pass with a tractor-mounted applicator that lays a polyethylene tarp and injects a mixture of methyl bromide and chloropicrin (*top*). The plastic strips are glued together to keep in the fumes. The polyethylene tarp is removed after 48–72 hours (*bottom*). [Photographs courtesy Bernarr J. Hall and Victor Voth.]

normal temperatures, and these are usually volatilized by heat. Such substances as paradichlorobenzene (mothballs) sublimate at normal temperatures and are applied as a solid. Soil fumigants are now being synthesized that will volatilize when wet, and these may be applied to the soil in granular form. Increasing in importance for enclosed areas such as greenhouses are **aerosol bombs,** from which the toxic chemical is dispensed by means of a gaseous carrier. In **fogging,** the active ingredient is dispersed by heat volatilization in kerosene or some other petroleum oil.

PROBLEMS OF CHEMICAL CONTROL. Although chemicals have greatly diminished the problems of pest control, they have by no means solved them. Furthermore, the use of chemical control imposes many new problems.

1. *Residues on edible products.* Severe restrictions are imposed upon the use of chemicals in order to avoid health hazards. The quantitative determination of residue, and the determination of limits of safety for the pesticide operator, as well as for the consumer, must be established for each chemical by the manufacturer. The controversy over residues involves the question of the potential health hazards of minute quantities of chemicals that are toxic at higher dose levels. The solution to the problem consists in choosing pesticides that break down before consumption or are metabolized harmlessly by the body, and in timing their application such that residues may be eliminated entirely. Some chemicals leave soil residues that may interfere with subsequent plant growth.

2. *The technical problems of application.* The timing of application must be extremely precise with regard to the effectiveness of treatment as well as the elimination of residues. Usually the pesticide must be applied before the trouble appears; once the symptoms are obvious, it is often too late for control. The necessity for thoroughness of application in tree crops often requires special pruning procedures. Due to chemical reactions that occur between certain chemicals, compatibilities must be taken into consideration when more than one pesticide is applied at one time.

3. *Spray injury to the plant.* Care must be taken to insure that the cure is not more injurious than the disease. Spray injury can reduce yield as well as spoil the appearance or "finish" of the product. Some russetting of fruit can be traced to cuticle damage by pesticides.

4. *The development of genetic resistance by the pest.* The pest is not a static factor. Natural variation among pests produces types that may be resistant to the pesticide. Because of the enormous rate of reproduction of many pests, the pesticides act as a screening device for the selection of resistant types. When DDT was introduced in the 1940s, many entomologists had hopes of victory in the war against insects. Within ten years, however, mosquitos, house flies, and many other insects had developed a resistance to DDT.

5. *The disturbance of the biological balance.* When one pesticide is used in place of another, pests that formerly appeared to cause little damage may begin to assume major importance: pests controlled by the former pesticide may not be by the new one. For example, the older methods for controlling apple scab utilized lime-sulfur, which was also effective against powdery mildew. But because of its adverse effect on fruit finish, lime-sulfur was replaced by organic fungicides. Although these fungicides have proven very satisfactory for scab control, their use has resulted in severe outbreaks of mildew, which have necessitated additional control measures.

6. *Pollution.* Widespread attention was focused on pesticides as a source of soil and water pollution with the publication of *Silent Spring* by the biologist Rachel Carson in 1962. Carson argued that, although pesticides had been created to rid agriculture of undesirable organisms, the extent of their harm was widespread and often unpublicized, and that the substances themselves should be considered undesirable. Residues of chemicals were found remaining in the soil long after their application, concentrating in some plants, and passing into the groundwater and ultimately into rivers and oceans. Through the 1960s a controversy over the polluting effects of chlorinated hydrocarbons—DDT is the best-known example—and organic phosphates became public. Opponents of the use of DDT, led by ecologists and conservationists, cited disturbing evidence of its dangers and wanted its use outlawed. DDT is an extremely persistent fat-soluble molecule that is greatly concentrated as it is passed along food chains. Concentrations of DDT in the food of fish-eating birds have been responsible for the failure of those birds to produce eggs with shells strong enough to survive the incubation period. The maximum concentration of DDT considered safe in the cow milk consumed by humans is 0.05 ppm, yet breast-milk human mothers produce for their babies has been shown to contain commonly between 0.05 and 0.26 ppm DDT, and concentrations as high as 5 ppm have been reported.

Defenders of DDT pointed out that, over the years, a good many materials far more noxious than DDT had been released into the environment, and that it had not been proven that DDT has the deleterious effects in humans that it has in other organisms. Defenders also claimed that there were irrefutable benefits to DDT use (such as malaria control), that its inexpensiveness and effectiveness made it unique, and that residues had not remained after DDT had been replaced by other materials. The public furor over pesticide-caused pollution resulted, in 1970, in a Federal restriction of DDT to use only on cotton and citrus crops.

More recent developments indicate that the DDT story is not yet closed: some of the nonpersistent substitutes for the persistent pesticide DDT may be extremely poisonous to applicators and handlers. Moreover, in regions where the use of DDT has been suspended, the incidence of malaria appears to have increased. Federal agencies have been reluctant to broaden the initial prohibition, causing some to wonder if DDT has been made an emotional scapegoat for the continuing problem of general pollution.

The problem of pesticides as pollutants is real. The necessity for nonpersistent, narrowly specific chemicals, harmful to pests and harmless to other living things, is clearly evident. Many agriculturists and horticulturists have concluded that complete reliance by pest controllers on chemicals is a dangerous concept and a wistful dream. The present feeling is that **integrated control** is required, with reliance on greater genetic resistance to pests and biological control, as well as judicious use of nonpersistent, narrowly specific chemical materials.

Biological Control

So naturalists observe, a flea
Has smaller fleas that on him prey,
And these have smaller still to bite 'em,
And so proceed *ad infinitum.*

JONATHAN SWIFT

Biological control utilizes direct competition between organisms. Certain insects
feed on other insects (Fig. 8-18). In addition, there are bacterial diseases of insects,
and there are virus diseases of bacteria. This competition between organisms may

FIGURE 8-18. Some insects parasitize other insects. *Top,* cocoons of braconid
wasps emerging from the body of a tomato hornworm. The cocoons contain
the pupal stage of the wasp, which lays its eggs on the body of the hornworm. *Bottom,* the potato-beetle killer (*Perillus bioculatus*) attacking a larva
of the Mexican bean beetle. [Photograph by J. C. Allen & Son.]

be directed toward the control of plant pests by the introduction of a natural parasite or predator of the pest. A recent example is the use of spores of *Bacillus thuringiensis,* a natural pathogen of caterpillars, as a spray material.

An example of successful biological control is the introduction of the vedalia beetle into California from Australia by Albert Koebele in 1888. This beetle feeds upon the eggs and larvae of the cottony-cushion scale, a serious insect pest of citrus. This beetle successfully controlled the scale until the use of DDT became prevalent in the late 1940s. Then, injury to the vedalia beetle, apparently caused by the DDT, upset the biological balance and resulted in the first outbreaks of cottony-cushion scale since 1890.

Biological control requires an organized attack on a pathogen. The use of biological control by individual growers by encouraging or importing insect predators, such as the praying mantis, has had at best only very limited success.

The advantage of biological control is that, once put into effect, it generally works without further human interference. However, the biological balance is often more apparent than real. The upsetting of this delicate balance, as by random environmental fluctuations, is a perfectly natural phenomenon. Commercial horticulture, which by its nature disturbs the natural biological pattern, cannot afford the risk inherent in biological control and at present depends upon chemical control as its main weapon against plant pests. Biological control, however, is an attractive measure, and one that is being given increasing attention.

The control of the screwworm, a severe pest of livestock, by utilizing artificially reared insects made sterile by being exposed to powerful radiation, has opened up a new approach to biologic control. Normal females are "monogamous" and mate only once. When mated to an irradiated male they will produce only sterile eggs. The basis of the control is the relatively low population of reproductive adults during the winter. If the irradiated sterile males released continually outnumber normal males, the number of fertile eggs will continually decrease. (This will be the case regardless of the mating habits of the female.) The greater the proportion of irradiated males over normal males, the faster will be the control. The success of this program suggests its use as a method for controlling plant-attacking insects.

Physiological Alteration of the Host

Plants do not have an antibody mechanism, as animals do, that can be utilized to resist disease. Thus, they cannot be made immune by vaccines. However, the physiology of the plant can be altered to affect the plant's ability to either resist invasion by the pathogen or to overcome the deleterious effects of the pathogen. For example, many vascular wilts caused by the fungus *Verticillium alboatrum* (for example, verticillium wilt of maple) can be compensated for by vigorous growth of the plant. Applications of fertilizer to increase vigor causes the plant literally to outgrow the pathogen. The reverse technique is utilized in fireblight of pears. Because infection and growth of the bacterium causing the disease is extremely rapid in fast-growing, succulent shoots, one method of control is achieved by slowing down rapid growth of the tree by eliminating excessive nitrogen fertilization or extensive pruning. The direct action of inorganic nutrients gives protection in some instances. For example, clubroot of cabbage appears to diminish in severity when the ratio of calcium to potassium in the soil is decreased. The affect of various levels of nutrients on disease resistance has not, however, been intensively investigated.

Genetic Alteration of the Host

The innate ability of the plant to avoid the injurious effects of the pathogen is the ideal method of control. This **genetic resistance** varies from complete absence of injury (**immunity**) through various degrees of **partial resistance.** The lack of resistance is referred to as **susceptibility. Tolerance** is a type of resistance in which the plant suffers infection and some injury, but is able to live with it without serious impairment.

Examples are known of plant resistance to viruses, bacteria, fungi (Fig. 8-19), nematodes, and insects (Figs. 8-20 and 8-21). The nature of plant resistance lies in the structural alterations or biochemical effects that either prevent or discourage intrusion and persistence of a particular pathogen. Some plants have resistance to whole groups of pathogens; others have only a specific resistance to a particular species, or race, of pathogen. Where pathogen and plant are closely adapted to each other, a close relationship exists: the plant evolves a genetic resistance and the organism evolves the genetic ability for violating or overcoming this resistance. The spontaneous origin of new races of pathogens is one of the major problems the plant breeder faces in attempting to incorporate genetic resistance into an improvement program.

The combination of resistance and horticultural quality is one of the main objectives of the plant breeder (see Chapter 10). Resistance may be incorporated

FIGURE 8-19. Resistance of lima bean to downy mildew caused by *Phytophthora phaseoli:* The plants in the middle are 'Early Thorogreen' lima bean, a cultivar susceptible to the mildew. The healthy plants on either side are of the cultivar 'Thaxter.' [Photograph courtesy USDA.]

FIGURE 8-20. Corn earworm larvae cause less feeding damage and develop slower on the resistant sweet corn experimental hybrid 'M4399' (*left*) than on the susceptible 'Iobelle Hybrid' (*right*). [Photograph courtesy of E. V. Wann, USDA.]

FIGURE 8-21. *Top*, muskmelon plants exposed for 8 days to mass infestation by the melon aphid, *Aphis gossypii*. The plant on the left is resistant to the aphids. Adult aphids on such resistant plants (*bottom left*) grew slowly and were stunted as compared with aphids on susceptible plants (*bottom right*). [Photographs courtesy G. W. Bohn, A. N. Kishaba, and H. H. Toba.]

in the whole plant or, when the plant is composed of separate, grafted components, in part of the plant. Thus, resistance to root pests, such as the woolly aphid of apple, may be incorporated into a grafted tree through the use of an aphid-resistant rootstock. Similarly, fireblight-susceptible cultivars of pear are often grafted to a framework of the resistant 'Old Home' cultivar. This prevents an infected limb from destroying the whole tree.

COMPETITION BETWEEN CROP AND WEED

> Now 'tis spring, and weeds are shallow-rooted;
> Suffer them now, and they'll o'er grow the garden,
> And choke the herbs for want of husbandry.
> SHAKESPEARE, 2 *Henry VI* [II. 4]

Weeds as Pests

A weed may be defined as any plant that is undesirable. According to this definition, any crop plant that is out of place may be termed a weed. More typically, however, the term refers to certain naturally occurring, aggressive plants that are injurious to people or to agriculture. The degree of undesirability of weed species varies greatly. Extremely noxious weeds, if left unchecked, may completely dominate crop plants. Crop losses are usually the result of competition for light, water, and mineral nutrients. Weeds are also indirectly responsible for crop losses because they harbor other plant pests, such as viruses, fungi, and insects. In addition, weeds may lower the quality and economic value of crops. A horticultural example is the lowered quality of peppermint oil when the crop is contaminated with weeds. Because of their rank growth and unsightliness, weeds are a perpetual nuisance in turf. They represent a safety hazard along roadsides and railroad right-of-ways, and they often clog irrigation ditches and streams. Finally, poisonous weeds (such as poison ivy and poison oak) directly affect the health and comfort of people and livestock, and the pollen of such plants as ragweed is a source of misery to the millions who suffer from hay fever. The annual cost of weed control in the United States exceeds the combined losses from all other types of plant pest. The fierce competition offered by certain weeds is due to a combination of their prolific reproductive capacity and vigorous, exuberant growth.

Reproductive Capacity of Weeds

In general, the destructive power of weeds is due to their sheer number. Some weeds produce seeds in enormous quantities. Weed species differ greatly, however, in terms of the number of seeds they produce: a single plant of the wild oat produces about 250 seeds, whereas a plant of the tumbleweed produces several million. One study of 181 perennial, biennial, and annual weed species reported an average of more than 20,000 seeds per plant.

The seeds of certain weed species have become structurally adapted for dispersal by wind, water, or animals. For example, many weeds have hard seed coats and can remain viable when passed through the digestive tract of an animal. People are among the chief disseminators of weeds through the shipment of crop seeds and plants. Most of the noxious weed species in North America are native to other parts of the world.

The large number of seeds and their efficient dispersal only partly explain the high reproductive capacity of weeds. (Many crop plants that are also prolific seed producers do not become weed-like.) The high reproductive capacity of weeds is particularly effective because of extended seed viability coupled with delayed germination brought about by dormancy. The failure of a weed seed to germinate may be due either to natural dormancy or to induced dormancy. Natural dormancy was discussed in Chapter 5. **Induced dormancy** is brought about by environmental conditions that limit germination. Seeds buried deep in the soil by tillage may lack either sufficient oxygen or the light stimulus necessary for germination. The viability of weed seed buried in the soil may be extremely long in contrast to the relatively brief viability of the seeds of most crop plants. In the eightieth year of an experiment on buried weed seeds—a study started in 1879 by Professor W. J. Beal at Michigan Agricultural College—three species—evening primrose (*Oenothera biennis*), curly dock (*Rumex crispus*), and moth mullein (*Verbascum blattaria*)—had survivors as viable seed. More than half of the twenty weed species in the experiment had survivors after twenty-five years.

The combination of dormancy, extensive seed viability, and high seed production makes weed control exceedingly difficult. If weed seeds would all germinate at once, their control might be accomplished by a rigorous and intensive eradication program. The high weed-seed population of agricultural soils makes weed control a continuing and integral part of crop culture.

Many weeds reproduce vegetatively as well as by seed. Some of the most pernicious weeds reproduce in this way: Johnson grass and quack grass, by rhizomes; wild morning glory, by roots; and wild garlic, by bulblets. Vegetative reproduction by underground stem modifications or by roots makes control by cultivation particularly difficult.

Weed Growth

The competition between weed and crop adversely affects both plants, but weeds usually win out over the crop plants. However, the exact physiological basis of the growth advantages that enable weeds to do this is not clear. Among the growth characteristics that explain the competitive ability of certain weeds are rapid germination, rapid seedling growth, and a root system that is deeply penetrating yet fibrous at the surface. Furthermore, weeds possess a natural resistance to many of the pests that plague crop plants. Their resistance to heat and cold undoubtedly gives them an added advantage.

The luxuriant growth characteristic of weeds might also have been a character of crop plants, but one that was sacrificed during "domestication" concomitant with the selection of horticultural attributes such as large fruit size. The loss of seed dormancy in most crop plants propagated by seed is an example of the loss of an adaptive trait. Another more reasonable explanation is that, to survive, weeds must be uniquely adapted to their surroundings, whereas crop plants may be grown in locations far removed from the conditions to which they are best adapted. Weed species change much more dramatically than do the crops across the United States. Thus, the study of weed control must be preceded by a knowledge of the natural history and ecology of the particular weed species under consideration.

Weed-Control Methods

The many techniques utilized in the control of weeds may be grouped into physical, biological, and chemical methods. Weed control was given a new impetus with the discovery of 2,4-dichlorophenoxyacetic acid (2,4-D), a selective weed killer, in the middle 1940s. This proved to have far-reaching effects, not only on chemical weed control, but on weed research in general. So far, however, chemicals have not proved to be the ultimate solution to the weed problem. Weeds are well endowed in their struggle for space and survival. Successful control still involves the judicious combination of many methods (Fig. 8-22).

Physical Techniques

Various controls entail the physical destruction of weeds. The pulling or grubbing of weeds is the simplest and most ancient form of weed control. The hoe, the basic hand tool, is still widely used. Various mechanical devices have been developed to automate this process (Fig. 8-23). The basic principle is to cut out, chop up, or cover the weeds and thereby destroy them. For maximum efficiency, cultivation should be carried out when weeds are very small. Weeds that are able to propagate vegetatively by means of underground parts are extremely difficult to control by cultivation and may actually be dispersed in this way.

Cultivation and tillage, the loosening or breaking up of the soil, are such widespread agricultural practices that many have come to believe that the loosening of the soil has beneficial functions other than the control of weeds. Yet a number of experiments have indicated that weed control is, indeed, the primary benefit of cultivation. The other advantages of cultivation, such as increased soil aeration or the conservation of soil moisture by the formation of a soil or dust mulch, may actually be counteracted by the destructive effects of inadvertent root pruning in the surface layer, the most productive portion of the soil. In addition,

FIGURE 8-22. Onion-seed fields occupy land for 6 to 14 months, depending on whether they are seed-planted or bulb-planted, and the onions compete poorly with weeds. Onion-seed producers must contend with three weed populations: winter, spring, and summer annuals. Control is achieved by (1) early spring cultivation, (2) midspring cultivation with an application of a pre-emergence herbicide and (3) an early-summer application of herbicide when fragile seedstalks prevent cultivation. *Left*, an unweeded onion-seed field. *Right*, a field with nearly 100% weed control. [From P. J. Torell and D. F. Franklin, *HortScience* 8(5):419–420, 1973.]

extensive cultivation with heavy machinery often leads to serious compaction of the soil. Although cultivation may conserve moisture by preventing runoff, it also contributes to considerable erosion of loose soil during heavy rains. Nevertheless, cultivation is still the major form of weed control for most crops, and it should therefore be timed to coincide with the time most favorable for efficient weed control.

The control of weed germination by mulching has been recommended from time to time. The use of black polyethylene plastic film has been encouraging in the culture of vegetables. Mulching, however, is too expensive as a means of weed control in commercial plantings, except with high-value crops, but it is a valuable practice for home gardens.

Heat may be used to control weeds: greenhouse or cold-frame soils may be "pasteurized" by steam at 180°F (82°C) for half an hour. Weed seeds and certain "damping-off" organisms are controlled in this manner. Care must be taken to avoid sterilizing the soil, since this will destroy the bacteria involved in the nitrogen cycle. It is a good practice to avoid planting in freshly steamed soils; it is best to allow the balance of microorganisms to be restored first.

Fire has been used to destroy weeds. Flame throwers have been adapted to control weeds in such places as railroad beds, and have been used in weeding cotton and onions. Similarly, burning has been used to dispose of weed trash, but this must be considered a poor practice. The burning over of muck soils to control weeds is a flagrant example of a resource waste.

FIGURE 8-23. Weed control by cultivation. *Top*, a ten-row onion tillivator. *Bottom*, a "power hoe" developed for strawberry cultivation. The operator steers with his feet and directs a movable hoe attachment. [Photographs courtesy Purdue University, and Friday Tractor Co., Hartford, Mich.]

Biological Techniques

The utilization of the natural competition between weeds and other organisms is the basis of biological control. The most spectacular example of biological control involves the introduction of insects that feed specifically on certain weeds. Prickly pear (*Opuntia* species), first introduced as an ornamental into Australia prior to 1839 by the early colonists, escaped cultivation and quickly became a noxious weed. By 1925, the cactus had infested 60 million acres, and the infested area was increasing at the rate of one million acres annually. Finally, a natural enemy of the cactus was imported from Argentina—the moth *Cactoblastis cactorum,* whose larvae bore into the cactus and feed upon it. Ten years after the introduction of the moth, control of the cactus was almost complete. Although there were successive waves of regrowth, they were of diminishing proportions, owing to the successful establishment of the insect.

More recently the weed *Hypericum perforatum* (Saint-John's-wort, or Klamath weed) has been successfully controlled in California by the introduction of the beetle *Chrysolina gemellata.* This beetle, native to France, had proved to be a satisfactory natural enemy of this weed in Australia.

The use of insects to control weeds can be handled only on a large scale, and, because of the problems inherent in such programs, they must be placed under the control of some national agency. Insects are usually used for controlling introduced weeds that are not attacked by native predators. In order for the introduction of insect predators to be successful, the insect must thrive in the new habitat and yet not become a pest to other agricultural crops. Thus, only those insects that are highly selective in their feeding habits can be imported. Care must also be taken to avoid introducing parasites of the imported insect.

Crop competition is an important biological method of weed control. Weed populations can be reduced by proper rotation involving well-adapted crops that can compete with weeds—for example, silage corn and alfalfa. Often it is the kind of tillage used in the rotation that brings weeds under control. For example, cultivation in corn may reduce the grass weeds that become established in small grain crops. Similarly, horticultural practices that facilitate rapid growth and good crop stands will encourage crop competition.

The use of geese to control weeds has had limited success in some horticultural crops. Geese will selectively weed strawberry patches of grass, provided there is enough grass present, but they cannot be used in fruiting patches. Fields must be fenced when geese are used, and careful management is required.

Chemical Techniques

> Buildings and walls were razed to the ground; the plough passed
> over the site, and salt was sown in the furrows made.
> A solemn curse was pronounced that neither house nor crops should
> ever rise again.
> A description of the fall of Carthage (147 B.C.)[1]

Substances such as common salt have been used for centuries to destroy vegetation. However, practical weed control in agriculture depends largely upon the selective destruction of weeds. In the early 1900s a number of compounds were

---

[1]B. L. Hallward, 1965. "The Siege of Carthage." *The Cambridge Ancient History,* Vol. VIII, p. 484. Cambridge University Press.

shown to have selective action in destroying broad-leafed weeds in grain; for example, various copper salts, sulfuric acid, iron sulfate, and sodium arsenite. But interest in these materials waned because of the unreliability of the results and the inadequacy of application equipment. The introduction of 2,4-D and other auxin-like herbicides in the 1940s rapidly transformed chemical weed control into a method of major importance. In rapid succession, many other chemicals were introduced as weed killers. Herbicides have accounted for an increasing percentage of all pesticide sales. Chemical weed killers are widely used on lawns but are not advisable for home gardens because of the difficulty of applying them at the proper rate and the danger of injuring adjacent plants.

SELECTIVITY. **Nonselective herbicides** kill vegetation indiscriminately. A **selective herbicide** is one that, under certain conditions, will kill certain plants and not harm others. Selectivity in herbicides is a relative quality, and it depends to a large extent on the interaction of a number of factors: dosage, timing, method of application, chemical and physical properties of the herbicides, and the genetic and physiological state of the plants.

In order for a herbicide to cause death it must be absorbed by the plant and cause some toxic reaction. Some kill only the area of the plant actually covered (**contact herbicides**—for example, dinitro compounds, oils, and arsenates) while others are translocated within the plant (**noncontact** or **translocated** herbicides—for example, 2,4-D). Selectivity may be brought about by directing the herbicide away from the crop plant (**positional tolerance**) or by inherent morphological differences between tolerant and susceptible plants: the amount and type of waxy cuticle, which results in differential wetting and absorption (for example, dinitro compounds on peas versus weeds); the angle and shape of the leaves (for example, the differences between broad-leafed plants and grasses); or the location of the growing point (for example, protected in grasses, exposed in broad-leafed plants). Diuron is an effective weed killer on grapes because it does not readily reach their deeply growing roots. Selectivity may be achieved as a function of dose. An overdose of 2,4-D will seriously affect all plants, whereas in low doses it will effectively "discriminate" between certain plants (Fig. 8-24).

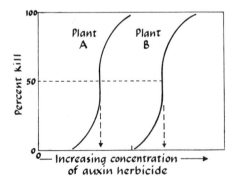

FIGURE 8-24. Selectivity of auxin-like herbicides is a function of dose. Arrows on mortality curves of two species indicate relative concentration of herbicide required to kill 50% of the plants. [From Leopold, *Auxin and Plant Growth*, University of California Press, 1955.]

Physiological distinctions that result in selectivity exist between certain plants. This may be manifested in varieties of the same species. The precise mechanism of physiological selectivity is unknown, since the exact mechanism by which many herbicides kill is still obscure. Some interfere with enzyme systems; others disturb the metabolism of the plant in some manner. Physiological selectivity may be due to differences in the plant's ability to translocate herbicides. The tolerance of carrots to Stoddard solvent is apparently due to the inherent resistance of the cell membrane to penetration.

Plants may show differences in the intensity of the toxic reaction at different stages of growth. Thus, some herbicides may be effective only during a very early stage of plant development, such as seed germination; others may be effective only at some later stage, such as flowering. Combinations of different herbicides are often required to control the many weed species usually present.

TIME AND METHOD OF APPLICATION. **Preplanting treatments** are applied before the crop is planted. Fumigants and other nonselective herbicides achieve selectivity between weed and crop by the timing of application. If preplanting herbicides are nonselective and have residual effects, sufficient time must elapse before the crop can be planted. Selective preplanting herbicides that are now available discriminate between germinating weed seeds and crop seeds or between germinating weed seeds and transplants.

Preemergence treatments are applied after the crop is planted but before it has emerged from the soil (Fig. 8-25). To be effective, the herbicide must have good coverage, remain on the surface of the soil, and be relatively unleached. Timing and soil moisture are very critical. Water is often required to activate the

FIGURE 8-25. Pre-emergence application of herbicide to seed bed. Uniform and thorough coverage is essential. [Photograph courtesy Farm Equipment Institute.]

herbicide, but too much water may leach the herbicide. Because their action is restricted to the soil surface, the herbicides are applied after the crop is planted. Physiological selectivity is often utilized so that the crop will germinate but the weed will not. In addition, selectivity may be enhanced by time. The weed seeds germinate first, since the crop seeds have a slightly delayed germination owing to the time required to imbibe water or to their depth of planting.

Postemergence treatment is applied to the growing crop. Selectivity may be physiological, or it may be due to directing the application away from the crop plant. (Care must be taken to avoid drift of herbicides.) Selectivity may be achieved as a result of plant age—for example, if the herbicide being used is toxic only to germinating seedlings at the dose being applied. When used in this way, herbicides are usually applied immediately after a thorough cultivation, since the ground must be free of germinated weeds.

Herbicides may be applied as liquids, solids, or gases. Specialized equipment has been devised that accurately meters low dosages. This is essential because of the extremely low concentrations required for some highly active substances. The increased use of granular herbicides that are absorbed by roots eliminates the need for heavy, bulky carriers for liquids.

CLASSIFICATION OF HERBICIDES. The number of chemicals known to have herbicidal activity is large and is increasing at a rapid rate. Table 8-3 presents a classification of some important families of herbicides in terms of their chemical

TABLE 8-3. Classification of some important herbicides by chemical composition.

Class of Material	Examples
Inorganic	Ammonium sulphamate
	Sodium arsenite
	Sodium chlorite
	Sulfuric acid
Organic	
Oils	Diesel oils, Stoddard solvent
Phenoxyacetic acids	2,4-D, 2,4,5-T, MCPA
Chlorinated aliphatic acids	TCA, dalapon
Dinitro analines	Benefin, trifluralin
Amide-like compounds	
Acetamides	Alachlor, CDAA, propachlor
Carbamates	CIPC, EPTC, IPC
Triazines	Atrazine, cyanazine, simazine
Ureas	Chloroxuron, diuron, binuron
Substituted phenols	DNBP
Others	Dacthal, glyphosate, nitrofen, paraquat

composition. Although many compounds fit into well-defined groups on this basis, others do not appear to fit into any particular grouping. Lists of herbicides become outdated very quickly, since this technology is expanding at a rapid rate.

Oils such as diesel oil have long been used as contact, nonselective herbicides. Lighter fractions, such as stove oil and Stoddard solvent (used in dry cleaning), have proved to have selective herbicidal action. Stoddard solvent is widely used with carrots and cranberries. Heavy aromatic fractions were found to be superior to diesel oils as contact herbicides, and large numbers of these materials are used as nonselective herbicides.

Phenoxy or auxin-like materials are those substances that have a physiological cation resembling indoleacetic acid (see Chapter 4). The most common of these are 2,4-D, 2,4,5-T, and MCPA. Note how similar their structural formulas are to one another and to that of indoleacetic acid (Fig. 8-26). Various derivatives of these compounds may be achieved by different substitutions and formulations. These affect the herbicidal as well as the physical properties of the molecule. For example, the amine formulation of 2,4-D is less volatile than the ester form. Volatile esters of 2,4-D are hazardous around sensitive crops such as grapes or tomatoes. The auxin herbicides are generally highly selective with respect to dose and are effective at extremely low concentrations. They have a short residual life in the soil and are low toxicity to animals. Selectivity is achieved both by differential absorption and genetic differentiation to dose. At high enough concentrations auxin herbicides are toxic to all plants. The herbicide 2,4-D is used to control weeds in corn and strawberries, and is used as a broadleaf weed killer in turf.

Among the chlorinated aliphatic acids used in weed control are such compounds as TCA (trichloroacetic acid) and dalapon (2,2-dichloropropionic acid). As a group these compounds are more selective against monocots than dicots and are

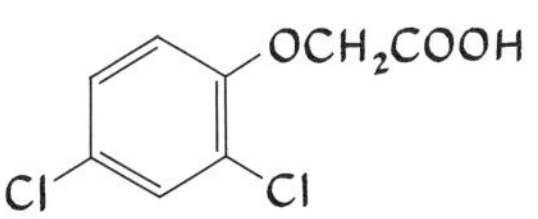

2,4-Dichlorophenoxyacetic acid (2,4-D)

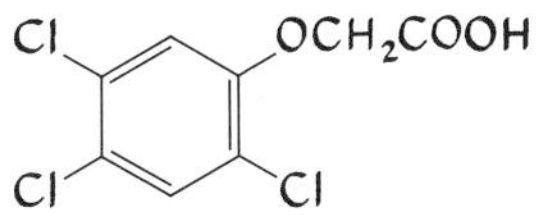

2,4,5-Trichlorophenoxyacetic acid (2,4,5-T)

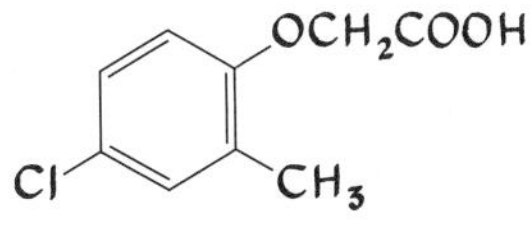

Methylchlorophenoxyacetic acid (MCPA)

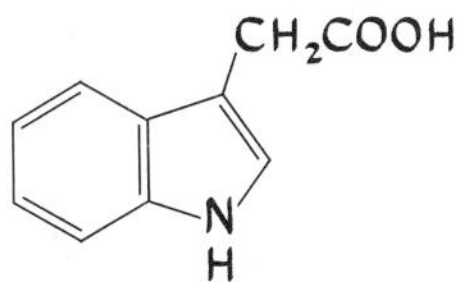

Indoleacetic acid (IAA)

FIGURE 8-26. Structures of auxin-like herbicides compared with that of indoleacetic acid.

thus used in the control of grassy weeds. Dalapon has the widest herbicidal use and is used for the control of both seedling and established perennial grasses.

Amide-like and related compounds contain the amide moiety N—C=O or the related groupings N—C=S or N—C=N. The physiological significance of their chemical structure is not known. Many important herbicides are included in this group, such as the triazines, carbamates, ureas, and acetamides. These are now some of the most important ones in horticulture.

FATE OF HERBICIDES. Herbicides, in order to be successful agricultural tools, must dissipate so that they will not interfere with future use of the land. Their eventual disappearance may result from vaporization, chemical breakdown, biological decomposition, leaching, or adsorption on soil colloids.

POPULATION COMPETITION

Population pressures markedly affect plant performance. As plant population increases per unit area, a point is reached at which each plant begins to compete for certain essential growth factors: nutrients, sunlight, and water. The effect of increasing competition is similar to decreasing the concentration of a growth factor.

Yield, whether it be of root, shoot, flower, fruit, or seed, is usually expressed on a "per unit area" basis rather than on a "per plant" basis. One of the principal reasons is that space and the fixed costs associated with it are usually much more valuable than the costs of individual plants. Thus, the most important consideration is the effect that varying the plant population has on the yield per unit area rather than on the yield per plant. The optimum population, however, is the one which produces the greatest net return to the grower. It should be emphasized that yield must be interpreted in both quantitative and qualitative terms. The value of the total yield is not merely the total bulk, but is related to the quality of the yield (size per unit, color, appearance, culinary properties, and so on). There are two aspects to the problem of yield in horticulture. One is competition between plants as plant density changes (**interplant competition**), the other is competition between parts of the same plant (**intraplant competition**).

Interplant Competition

Yield

The yield per unit area is equal to the yield per plant times the number of plants per unit area. When the population is below the level at which competition between plants occurs, increasing the population will have only an indirect effect

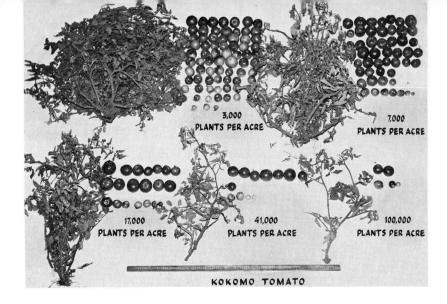

FIGURE 8-27. Increasing plant population drastically reduces plant size and yield per plant in tomato, but the yields per unit area increase with plant population and then level out. [From R. L. Fery and J. Janick, *J. Amer. Soc. Hort. Sci.*, 95:614, 1970.]

on individual plant performance; the yield per unit area will increase in direct proportion to the population increase. As soon as competition between plants occurs, however, the yield per plant will decrease (Fig. 8-27).

Once competition exists, the change in yield per unit area becomes a function of changes in the total weight per plant and in the proportion of the plant that is harvested as yield. A number of studies have shown that, although the yield (of dry matter) per plant decreases, the relationship is such that the total weight per unit area levels out as shown in Figure 8-28. Thus the yield of a particular plant part per unit area is related to its proportion of the total. If part yield is a constant proportion of the total, then yield will level out with the total weight. However, if part yield decreases, then yield will peak at some population, then decline, and density above or below the optimum will adversely affect yield.

For some crops—corn is a good example—the yield of grain (or ears) is a decreasing percentage of total plant weight as competition increases. This means that there is an optimum population where yield is at a maximum. Putting restraints on yield (for example, on the ear size considered to be marketable) accentuates this effect. In many crops, part yield is a relatively stable proportion of total yield under competition. For example, the proportion of a tomato or bean plant that is fruit or seed is relatively constant over low and high populations. Fruit or seed yield then increases with increasing population but at a decreasing rate, eventually leveling out (in other words, yield increases asymptotically toward a limit); and the higher the population the higher the yield, although economic considerations (for example, seed or plant costs) determine the practical limit for the population. However, if we apply restraints on yield quality to this

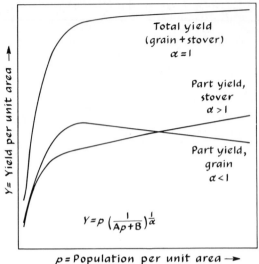

FIGURE 8-28. The relation between yield per unit area and population in corn. Note that total plant yield levels out as population increases. This appears to be true for all crops. The relation between population and the yield of some plant part (such as grain) depends on its proportion to total yield as population changes. In corn grain, yield is a decreasing proportion of total plant yield as population increases. Thus, grain yield increases up to some optimum population (usually between 12,000 and 15,000 plants per acre) and then decreases. In many crops, part yield is a fairly constant proportion of total yield as population changes, and thus part yield levels out as does total plant yield.

illustration—for example, by defining marketable yield in terms of fruit or seed size—yield may be maximum at a finite population.

The effect of population on plant performance thus depends on the part of the plant being measured and the restraints governing quality. Plant parts may respond to increased population pressure after competition by decreasing in size or in number or in both. The decrease in the number of plant parts (for example, the response of potato tubers) may produce a very small corresponding decrease in the size of the remaining parts. These compensating responses to competition may differ with various plant parts, even on the same plant. In corn, for example, an increase in population pressure causes a decrease in ear number, ear size, and kernel number, but the kernel size remains relatively constant.

The optimum size of plant parts (fruit, tuber, bulb, flower) is not necessarily the largest size. (In pickling cucumbers, it is the smallest.) Population competition may be utilized to produce plant parts of a particular size. For example, onion seed grown for "sets" to be replanted is sown at the rate of 70 lb/acre. The crowding produces a small bulb about three-quarters of an inch in diameter,

For those mathematically inclined, the following formulas may be useful:

$$y_t = \frac{1}{Ap + B},\qquad(1)$$

where y_t = total yield per plant, p = number of plants per unit area, and A and B are constants. The linear form of the equation,

$$y_t^{-1} = Ap + B,\qquad(2)$$

indicates that the constants may be derived from two widely spaced populations.

Total yield per unit area (Y_t) is equal to the yield per plant times the number of plants per unit area:

$$Y_t = py_t.\qquad(3)$$

Thus,

$$Y_t = p\left(\frac{1}{Ap + B}\right) = \frac{p}{Ap + B}.\qquad(4)$$

The relation between total yield per plant and yield of a plant part (y_p) can be described by the following relationship:

$$y_t = Ky_p^{\alpha},\qquad(5)$$

where K and α are constants. The linearized form of this equation,

$$\log y_t = \log K + \alpha \log y_p,\qquad(6)$$

indicates it is possible to estimate α from two widely spaced populations.

By substituting equation (5) into equation (1) the yield of plant part (y_p) can be described as follows:

$$y_p = \left(\frac{1}{A'p + B'}\right)^{1/\alpha} \text{ or } y_p^{-\alpha} = A'p + B',\qquad(7)$$

where $A' = KA$ and $B' = KB$. Therefore the yield of a plant part per unit area (Y_p) is described by the equation

$$Y_p = p\left(\frac{1}{A'p + B'}\right)^{1/\alpha}\qquad(8)$$

Note that when $\alpha = 1$, equation (8) is in the same form as (4).

which is ideal for replanting. When onion seed is planted for an edible crop and a large bulb size is desired, the planting rate is only 2–4 lb/acre. Similarly, large increases in pineapple yields have been obtained by close spacing of large-fruited varieties.

Environmental factors such as nutrition or moisture levels may drastically change the level at which the response to crowding occurs. Thus, in tomatoes, the response of a given cultivar differs with the season. Of particular interest is the different type of response that may occur with different genetic types of the same plant. The dwarf corn produced with the *compact* gene appears to demonstrate a different response to competition than normal corn, which means that yields may be constantly increased in proportion to population. The rearrangement of a plant population by rows increases competition in comparison to equidistant plant spacing, but this may be necessary for cultural considerations such as cultivation for weed control and access for spraying and harvesting equipment.

Quality of Yield

Population pressures affect quality factors that must be considered. For example, crowding tomatoes increases the foliage canopy, which tends to protect the fruit from being burned by the sun. On the other hand, excessive crowding of potted chrysanthemums produces undesirably spindly growth.

A high population level has an adverse effect on disease control. Plants grown exceptionally close together produce a dense cover that discourages rapid drying and produces conditions favorable for the growth of many fungi. In addition, the dense cover is impenetrable to spray application and limits chemical disease control. Sometimes dense foliage facilitates the actual spread of diseases by contact, as is true of tobacco mosaic virus in tomato. High population may be utilized as a method of weed control. The increased shade produced by a dense cover of vigorous plants may permit the crop plant to outcompete weeds. This is utilized in turf management.

Intraplant Competition

The relationship between parts of the same plant is an important component of population competition. This almost invariably involves fruit and flower size. Size per unit is an important component of value in practically all horticultural crops. For example, high yields of extremely small fruit may be economically worthless.

Fruit Size

An important factor in fruit size is the leaf-to-fruit ratio. Since leaves are the carbohydrate source nourishing the fruit, fruit size will be related in any given genotype to the amount of leaf area per fruit (Fig. 8-29). The seeds get first call

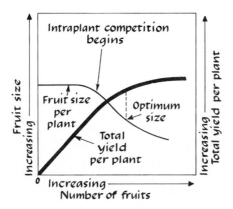

FIGURE 8-29. The relation between the number of leaves per fruit and fruit size in the 'Delicious' apple. [Adapted from data of Magness, Overly, and Luce, Washington Agricultural Experiment Station Bulletin 249, 1931.]

on the carbohydrates produced by the leaves. When the seed requirements are satisfied, the extra carbohydrates become available to the fruit. Only after the requirements of the fruit are satisfied do the extra carbohydrates become available for the vegetative organs.

Leaf area may not be a constant value. Indeterminate plants (plants in which the main axis remains vegetative) have a constantly increasing leaf area as the plant grows. Under a heavy load of maturing fruit, however, many indeterminate plants stop vegetative growth. The result is a constant leaf-to-fruit ratio. In mature, bearing fruit trees the amount of leaves produced in any one season is relatively constant because the amount of new growth tends to be small.

There are two ways to compensate for fruit competition. One is to increase the number or efficiency of the leaves. The other is to reduce the fruit load by blossom or fruit removal—that is, by **thinning** (Fig. 8-30). Increasing leaf growth and efficiency is accomplished by various cultural practices, especially improving the plant's nutritional status. This may be self-defeating, if better nutrition also increases fruit number. Thinning, on the other hand, is an expensive practice confined largely to high-priced fruit crops whose value is largely dependent on size per unit (such as apples, pears, peaches, and plums).

Yield and Fruit Size

Below the levels of intraplant competition, yield is directly related to the number of fruit. Under competition, however, the increase in yield levels off. The practical level for thinning popular peach and apple cultivars occurs at levels where total yield is practically unaffected. This is because these cultivars are potentially

FIGURE 8-30. The effect of thinning on fruit size. The left side of the tree was thinned; the right side of the tree was not. [Photograph courtesy F. H. Emerson.]

large-fruited. Thinning must be early to be effective. The general relationship between the number of fruit, fruit size, and yield in apples and peaches is plotted in Figure 8-31.

Other Considerations

In addition to size, there are other important considerations in fruit competition. In apples, for instance, a relationship exists between quality and fruit load. With an unusually light crop, apples of some cultivars may get too large (more than $3\frac{1}{2}$ inches in diameter), and their storage quality decreases. On the other hand, with fewer leaves per fruit, there is a decrease in fruit sugars (mainly sucrose), which affects the quality of the fruit. A corresponding decrease in anthocyanin pigments per fruit decreases red color, although this effect is relatively slight.

Competition between fruit and vegetative parts may result in severe damage to the plant; in grapes, for example, excessive fruit load renders the plant susceptible to winter injury as a consequence of low sugar accumulation by the vegetative organs. In periods of stress, such as during a drought, fruit is removed from young

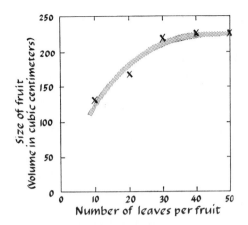

FIGURE 8-31. Idealized relation between fruit size and yield. The total yield per tree is only slightly reduced by thinning popular large-fruited varieties of apples and peaches. However, unthinned fruit may be economically worthless. Some reduction in total yield is desirable in order to keep alternate-bearing cultivars producing on an annual basis.

trees in order to prevent desiccation and subsequent winter injury. Heavy fruiting in tomatoes will increase the plant's susceptibility to foliage diseases.

In fruit trees that produce flower buds in the year previous to fruiting (especially apples and pears), fruit competition may be responsible for **biennial bearing.** The competition for nutrients between a large number of developing fruit apparently prevents the development of fruit buds for the next year's crop. Once this pattern has started it is difficult to stop, for in the off-bearing year an extra abundance of flower buds will form in the absence of fruit competition. Biennial bearing of apples may be controlled by fruit thinning to prevent overbearing.

Flower Size

In plants grown for individual flowers rather than for clusters, unit size is an important component of value, just as it is in fruit. For this reason, flower size is commonly increased by reducing the number of developing flowers per shoot. This is accomplished by **disbudding,** the removal of all buds but one per shoot. Disbudding is a standard practice in the culture of roses, carnations, chrysanthemums, and peonies.

Economics of Population Control

If we grant that the goal of the commercial producer of plants is profit, then the optimum plant population will be the one from which profits are greatest, and populations of plants or of plant parts present a number of factors that must be considered: for example, the costs of plants and planting (or thinning), as well as the costs of cultivation, disease control, and harvesting. From these figures the actual costs of production for each level of population can be determined. Similarly, the returns at each population level can be determined if the yield responses are known. This will be the value of each unit determined from its size and quality multiplied by the total number of units of that value. The profit (or loss) at each population level may be determined by subtracting total costs from total returns.

If we could predict all plant responses, costs, and returns, it would be relatively easy to "program" the problem—that is, to substitute values into a mathematical formula and solve (with an electronic computer, if necessary) the equation. Although this degree of control is not yet possible, it is being approached, as has been illustrated by the cling peach predictions made in California. However, empirical solutions to these problems have been obtained by trial and error through the years. It will be interesting to compare these empirical answers with the mathematically determined solutions when they become available. The obvious advantage of the mathematical approach is its "instantaneous" utilization of changing information concerning prices, costs, and yield responses. It should be pointed out that the same type of analysis is available to "noncommercial" growers. Although monetary values are usually not established for pride or satisfaction, they might well be for purposes of analysis.

Selected References

Agrios, G. N., 1969. *Plant Pathology*. New York: Academic Press. (An excellent text on plant pathology.)

Anderson, H. W., 1956. *Diseases of Fruit Crops*. New York: McGraw-Hill. (The important deciduous temperate fruits are treated. Insects are not covered.)

Anderson, W. P., 1977. *Weed Science: Principles*. St. Paul, Minn.: West Publishing. (A broad perspective of weeds and weed control.)

Bleasdale, J. K. A., 1966. Plant growth and crop yield. *Annals of Applied Biology*, 57:173–182. (A discussion of the effect of crop density on plant performance.)

Chupp, C., and A. F. Sherf, 1960. *Vegetable Diseases and Their Control*. New York: Ronald Press. (A useful treatise on diseases of importance in the United States.)

Crafts, A. S., 1975. *Modern Weed Control*. Berkeley and Los Angeles: University of California Press. (A comprehensive text and manual on the principles of weed control.)

Johnson, W U., and H. H. Lyon, 1976. *Insects That Feed on Trees and Shrubs: An Illustrated Practical Guide*. Ithaca, N.Y.: Cornell University Press. (The best illustrated book on insect pests of ornamentals.)

Kenaga, C. B., 1974. *Principles of Phytopathology*, 2nd ed. Lafayette, Ind.: Balt. (A general introductory text on plant pathology.)

Klingman, G. C., F. M. Ashton, and L. J. Noordhoff, 1975. *Weed Science: Principles and Practices*. New York: Wiley. (An excellent general textbook.)

Metcalf. C. L., W. P. Flint, and R. L. Metcalf, 1962. *Destructive and Useful Insects: Their Habits and Control*, 14th ed. New York: McGraw-Hill. (The bible for economic entomology.)

Pirone, P. P., B. O. Dodge, and H. W. Rickett, 1970. *Diseases and Pests of Ornamental Plants*, 4th ed. New York: Ronald Press. (A thorough treatment.)

Roberts, D. A., 1978. *Fundamentals of Plant-Pest Control*. San Francisco: W. H. Freeman and Company. (A comprehensive introduction to the nature and control of plant pests, stressing an interdisciplinary approach.)

Roberts, D. A., and C. W. Boothroyd, 1972. *Fundamentals of Plant Pathology*. San Francisco: W. H. Freeman and Company. (This fine phytopathology text presents specific diseases in groups according to their physiological impact upon plant welfare.)

U.S. Department of Agriculture, 1952. *Insects* (USDA Yearbook, 1952). (This and the following two publications are broad, popular treatments of aspects of crop protection.)

U.S. Department of Agriculture, 1953. *Plant Diseases* (USDA Yearbook, 1953).

U.S. Department of Agriculture, 1966. *Protecting Our Food* (USDA Yearbook, 1966.)

Walker, J. C., 1969. *Plant Pathology*, 3rd ed. New York: McGraw-Hill. (An excellent introduction, with an organization based on causal agents. There are many other textbooks whose organizations are based on crops.)

Webster, J. M. (editor), 1972. *Economic Nematology*. New York: Academic Press. (A reference text to the important nematode pests of the world's major crops.)

Weed Science Society of America, 1974. *Weed Science Society of America Herbicide Handbook*, 3rd ed. Champaign, Ill. (A list of herbicides, desiccants, plant-growth regulators, various adjuvants, and chemicals of interest to weed scientists.)

Wiley, R. W., and S. B. Heath, 1969. The quantitative relationships between plant population and crop yield. In A. G. Norman (editor), *Advances in Agronomy*, vol. 21, pp. 281–321. New York: Academic Press. (An excellent review that examines the usefulness and biological validity of the different mathematical equations that have been proposed to describe the relationships between plant population and crop yield.)

9

Mechanisms
of Propagation

Yet Nature is made better by no mean
But Nature makes that mean; so over that art
Which you say adds to Nature, is an art
That Nature makes. You see, sweet maid, we marry
A gentler scion to the wildest stock,
And make conceive a bark of baser kind
By bud of nobler race. This is an art
Which does mend Nature—change it rather; but
The art itself is Nature.

SHAKESPEARE, *The Winter's Tale* [IV. 4]

Propagation refers to the controlled perpetuation of plants. The basic objective of plant propagation is twofold: to achieve an increase in number and to preserve the essential characteristics of the plant. There are two essentially different types of propagation: sexual and asexual. **Sexual propagation** is the increase of plants through seeds formed from the union of gametes. **Asexual propagation** is the increase of plants through ordinary cell division and differentiation. The essential feature of asexual propagation is that plants are capable of regenerating missing parts. Thus, a stem cutting initiates roots, a root cutting develops shoot buds, and a leaf cutting initiates both roots and shoots.

REPRODUCTION

Reproduction is the sequence of events resulting in the perpetuation and multiplication of cells and organisms. It involves a controlled self-duplication of the life-controlling mechanism. We now appear to be on the verge of understanding some of the mysteries of reproduction.

Replication of Living Systems

The principle upon which the replication of living material is based is somewhat analogous to that of a machine that forms printing type, such as a linotype. The machine arranges the type matrices in certain sequences to transmit information—that is, in accordance with the text that is to be set in type. These matrices (molds or templates) of letters give form to the cast type. New type is continually formed from new combinations (or sequences) of matrices. In this analogy, the reproducing element is the type matrix; in living systems, the reproducing elements are the nitrogenous bases in deoxyribonucleic acid (DNA).

In the English language, 26 letters of the alphabet produce a dictionary of tens of thousands of words. The same results can be achieved with the two symbols of the Morse code. The four symbols of the genetic alphabet, or code, are the four nitrogenous bases that make up the nucleotides in DNA: adenine, guanine, thymine, and cytosine (Fig. 9-1). If two symbols can express 26 letters, thousands of words, and thus an infinite series of messages, is it not conceivable that four bases can make up the code by which the information for maintaining living matter is transmitted? The way in which this is accomplished is the theme of intense study. The general theory is that the sequence of bases in DNA directs the sequence of amino acids in the synthesis of protein, specifically enzymes. Enzyme specificity is related to its amino acid sequence, and it is not unreasonable to assume that enzyme control can account for the biochemical control of living systems.

Replication of DNA

The replication of DNA appears to be accomplished by a system quite analogous to the template system of the type machine. The template model, however, is based on chemical bonding rather than physical impression. Chemical analysis of DNA shows that, although the relative quantity of the four nitrogenous bases varies, the amount of adenine is always the same as thymine, and the amount of guanine is always the same as cytosine. This suggests that adenine and thymine, and guanine and cytosine, occur in pairs. The significance of this becomes apparent when the structure of DNA is considered. DNA normally occurs in double strands, and each strand is connected by linkages between the nitrogen bases. But because of the configuration of the bases, adenine is always paired with, and opposite to, thymine; and guanine is always paired with, and opposite to, cytosine. Thus if the sequence of one strand is given, the sequence of the opposite strand is fixed. Each strand, then, is the "template" for the other.

The theory of DNA replication follows nicely from its structure. The separation of the strands of DNA is apparently followed by a realignment of complementary bases on each strand (Fig. 9-2).

The structure of DNA accounts both for its ability to replicate and for its ability to synthesize. The actual synthesis of protein is carried out not in the nucleus but

329

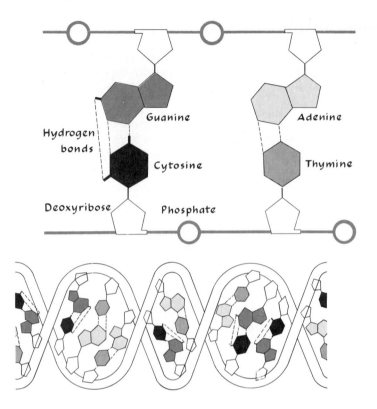

FIGURE 9-1. *Top*, a portion of the DNA molecule, shown untwisted. The molecule consists of a long double chain of nucleotides made up of phosphate-linked dexoyribose sugar groups, each of which bears a side group of one of the four nitrogenous bases. Hydrogen bonds (*broken lines*) link pairs of bases to form the double chain. The bases are always paired as shown, although the sequence varies. *Bottom*, the double-helix form of the DNA molecule. [After Allfrey and Mirsky, "How Cells Make Molecules," copyright © 1961 by Scientific American, Inc., all rights reserved.]

in the cytoplasm. The information from the nucleus must be transferred to the structural cytoplasmic sites of synthesis, called ribosomes. The information is carried by a substance called ribonucleic acid or, RNA. (The components of RNA are almost identical to those of DNA except that the sugar portion of the molecule is slightly different—ribose takes the place of deoxyribose, and the nitrogenous base thymine is replaced by a very similar base called uracil.) The particular form of RNA that carries the genetic instructions, in the form of a complementary copy of the DNA series of bases, is called messenger RNA. Another form of RNA, transfer RNA, brings the amino acids to the ribosomes to construct protein (Fig. 9-3).

Protein synthesis can be achieved outside of the living organism by placing together the component parts in an energy-generating system: ribosomes (from

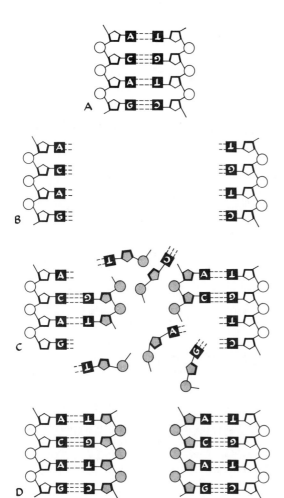

FIGURE 9-2. DNA replication. The linked DNA strands (A) separate in the region undergoing replication (B). Free nucleotides (indicated by shading) pair with their appropriate partners (C), forming two complete DNA molecules (D). [After Crick, "Nucleic Acids," copyright © 1957 by Scientific American, Inc., all rights reserved.]

bacteria), amino acids, and RNA. Very elegant experiments involving the direction of protein synthesis have been made possible through the use of artificial RNA. In this way, the genetic code has been deciphered; that is to say, the sequence of bases in RNA has been associated with the incorporation of a particular amino acid into protein. The message has been shown to be in the form of a three-letter code; that is, a sequence of three bases directs each amino acid. For example, RNA composed only of uracil (UUU) incorporates the amino acid phenylalanine into protein; the sequence UUA incorporates leucine. The sequence of amino acids in the polypeptide (protein unit) of the enzyme is related to the three-letter code sequence of DNA. Thus the four-letter alphabet represented by the four nitrogenous bases is used to make three-letter "words," each corresponding to one of the approximately twenty amino acids. Each sentence spells out the amino acid sequence for one protein. This "sentence" is the gene!

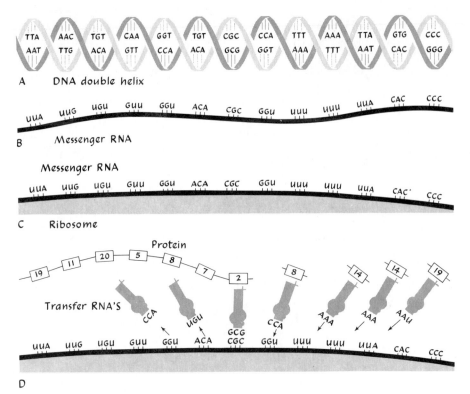

FIGURE 9-3. A schematic representation of protein synthesis as related to the genetic code. The complementary code, here derived from the dark-lettered strand of DNA (A), is transferred into RNA (B) and moves to the ribosome (C), the site of protein synthesis. Amino acids (D, numbered rectangles) are presumably carried to the proper sites on messenger RNA by transfer RNA and are linked to form proteins. [After Nirenberg, "The Genetic Code," copyright © 1963 by Scientific American, Inc., all rights reserved.]

CELL DIVISION

The basic difference between asexual and sexual propagation involves the distinction between the two types of cell division—**mitosis** and **meiosis**—in which the chromosomes are distributed. Chromosomes are linear structures in the nucleus that contain the **genes,** the carriers of genetic information.

Mitosis

In mitosis a synchronized division takes place in which both the chromosomes and the cell divide (Fig. 9-4). The chromosomes duplicate themselves, the members of each pair of duplicate chromosomes moving to opposite ends of the cell. The cell

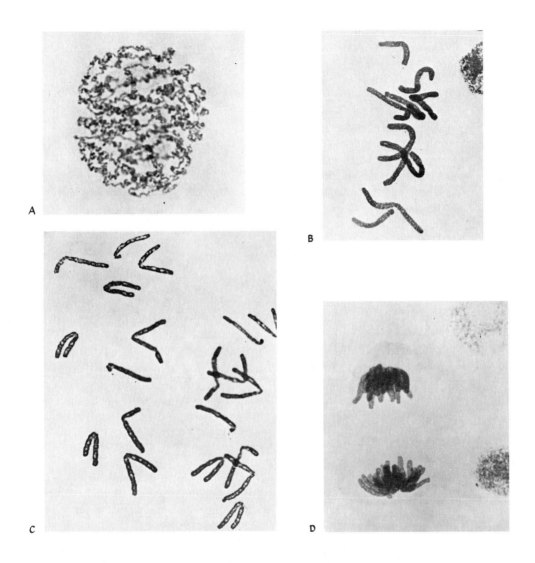

FIGURE 9-4. Mitosis in the California coastal peony. The vegetative cells of this species have 10 chromosomes ($2n = 10$). Although mitosis is continuous, the process is broken down into a number of stages for descriptive purposes. A, prophase: During this stage the nucleus becomes less granular and the linear structure of the chromosomes can readily be discerned. Note the chromosome coils. B, metaphase: The ten chromosomes line up in the equatorial plate, which appear as an "equatorial line" due to the smearing process. The chromosomes have reduplicated, and each appears visibly doubled. C, anaphase: The chromosomes have separated, and approach the poles of the cell. It is possible to pick out the pairs in each group and to match the daughter chromosomes, which have just separated from each other. D, telophase: During this stage the contracted chromosomes are pressed close together at each end of the cell. A wall subsequently forms across the cell, making two "daughter" cells with the same number and kind of chromosomes as exist in the original cell. [Photographs courtesy M. S. Walters and S. W. Brown.]

then divides, and the two resulting daughter cells thus receive exactly the same number and kind of chromosomes. The division of the cell distributes equally the other constituents of the cell.

This kind of cell division, in plants as well as in animals, is called mitotic division. The differentiation of the cells into tissues and organs ordinarily is not related to any chromosomal difference. The fact that each cell contains all the necessary genetic material implies that any cell potentially can give rise to the entire organism. Indeed, the production of an entire plant from a single carrot parenchyma cell has been achieved. The formation of shoot buds from roots, or roots from leaves, is common in plants. The genetic continuity of mitosis insures that a bud on the potato tuber will, under the right conditions, produce an entire potato plant.

Meiosis

In meiosis—a sequence of cell divisions that occurs in the formation of gametes—the number of chromosomes is reduced by half (Fig. 9-5). The chromosomes in the ordinary vegetative cells of higher plants normally occur in pairs. The two chromosomes of a pair (homologues) are morphologically similar and contain the same kind of genes, although each member of the gene pair may not be identical. The combination of different genes is responsible for the genetic variability between living things. The sexual process is one mechanism that provides for the reassortment and recombination of genetic factors so that organisms may be able to survive through time in an ever-changing environment. The reassortment of genetic factors taking place between and within chromosomes is accomplished by meiosis; the recombination is accomplished by fertilization (see Chapter 10).

Meiosis is basically a series of two divisions in which the cells divide twice but the chromosomes divide only once. This results in four cells, each having the haploid number of chromosomes, that is, half the number possessed by the vegetative cells (the diploid number). Each of these four cells potentially generates a gamete. Fertilization, the fusion of two gametes, subsequently restores the diploid number.

The most obvious effect of meiosis is the reduction in chromosome number from the diploid to the haploid number. This is necessary to provide a constant chromosome level when gametes recombine at fertilization. The less obvious but most important effect of meiosis is the reassortment of chromosomes, which are distributed to the gametes.

The basic difference between mitosis and meiosis occurs at the first division. In meiotic division the chromosomes become visibly double, as they do in mitosis, but this is the last time the chromosomes divide, although the cells will divide once more. Unlike mitosis, the two homologous chromosomes, now visibly doubled, pair up (**synapse**) along their length. The attraction between the doubled

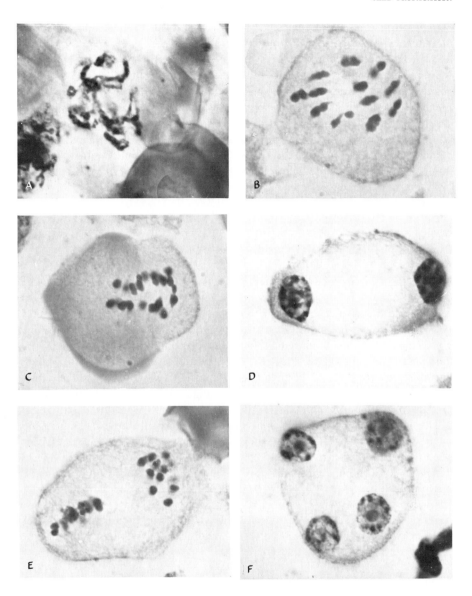

FIGURE 9-5. Meiosis in pepper. This species has 24 chromosomes in the vegetative cells
(2n = 24). Note that there are two divisions in meiosis. In the prophase of the first division (A),
the 24 chromosomes pair up along their length at the same time that they appear doubled. This
can be seen clearly in the bottom of the picture. At the first division (B), 12 pairs of chromosomes are visible. Each chromosome pair consists of two doubled chromosomes (four chromo-
tids). (C), anaphase of the first division. (D), telophase of the first division. (E), metaphase (left)
and prophase (right) of the second division. (F), telophase of the second division. Walls will form
across the cell to produce four pollen grains, each containing 12 chromosomes. Each grain will
thus contain half as many chromosomes as the original cell.

chromosomes changes to repulsion, and each doubled chromosome of a pair moves to opposite poles of the cell, which then divides. As a result, each of the two daughter cells has half as many "whole" chromosomes.

Apparently, when the chromosomes pair at the first division, an actual exchange of segments (**crossover**) occurs between homologous chromosomes. The precise mechanism by which the exchange of chromosome material takes place is still not fully understood. The net result of the first division is thus not only a reduction in chromosome number but a rearrangement of segments between homologous chromosomes. In the second division, which immediately follows the first, the division of doubled chromosomes is similar to that of mitosis. However, owing to the previous crossovers, these two daughter nuclei may not be duplicates of each other.

Consequences of Sexual Reproduction

The genetic consequences of sexual propagation are more fully discussed in Chapter 10, but a brief discussion will be given here. Plants that are continually self-pollinated, such as the tomato or pea, contain essentially similar pairs of genes on homologous chromosomes (that is, they are **homozygous**). Homozygous plants will reduplicate themselves exactly by sexual reproduction, or **breed true.** Plants that tend to crosspollinate, such as the petunia or cucumber, will have many dissimilar pairs of genes on homologous chromosomes (that is, they are **heterozygous**). Sexual reproduction constantly rearranges these genetic factors. Cross-pollinated plants *do not* breed true, but **segregate.**

The problem of reproducing a particular plant exactly thus depends on its natural method of pollination. Seed propagation will duplicate naturally any plant that is highly self-pollinated because such plants tend to be homozygous. A particular cross-pollinated plant can be duplicated exactly only by asexual methods because it is heterozygous. However, a high degree of uniformity in some character may be achieved in the seed propagation of cross-pollinated plants as a result of constant selection. For example, in well-selected petunia cultivars each plant may have the same flower color, although it can be demonstrated that the plants are not identical for all characters.

SEED PROPAGATION

Seed is the most common means of propagation for self-pollinated plants, and is extensively used for many cross-pollinated plants. It is often the only possible or practical method of propagation. There are many advantages in propagating from seed. It is usually the cheapest method of plant propagation. Seeds also offer a convenient method for storing plants. When kept dry and cool, seeds remain viable from harvest to the following planting season. Some seeds remain viable for very long periods: those of Indian lotus (*Nelumbo nucifera*) remain viable for as long as 1000 years; and seeds of Arctic tundra lupine, uncovered after being frozen for an estimated 10,000 years, germinated and produced normal plants.

Another advantage to seed propagation is that it provides a method for starting "disease-free" plants. This is especially important with respect to virus diseases, since it is almost impossible to free an infected plant of viruses. Most virus diseases are usually not seed transmitted. The major disadvantages to seed propagation besides genetic segregation in heterozygous plants, which has already been discussed) is the long time required by some plants to reach maturity from seed. For example, eight years is usually required for pears to fruit from seed. Potatoes do not produce large tubers the first year when grown from seed. Thus, asexual propagation not only provides trueness to type but, in many plants, saves several years.

Seed Origin and Development

The seed is the result of complex growth and developmental events. These include the development of the pollen and embryo sac (male and female gametophyte), pollination, fertilization, and maturation processes.

Pollen Formation

In the anther, **pollen mother cells** undergo meiosis to produce microspores— haploid male spores that, when developed, are known as pollen. The pollen grain can be thought of as a separate plant, the **male gametophyte** (Fig. 9-6). This haploid "plant" producing the male gametes is the evolutionary remnant of the gametophytic generation, which may be well developed in more primitive plants, such as ferns and mosses. In seed plants this stage is very much reduced. The haploid nucleus of the microspore divides mitotically to form the generative nucleus and the tube nucleus. The generative nucleus often appears to have cytoplasm associated with it, resembling a cell within a cell. The generative nucleus is destined to divide mitotically, either in the pollen grain or in the pollen tube, to form two nuclei—the **male gametes.**

Embryo-Sac Formation

There is no common term, unless it be *embryo sac*, that corresponds to the word *pollen*. This is probably because the embryo sac is inconspicuously located inside the enlarged base of the pistil.The **megaspore mother cell,** borne inside a specialized region in the ovary called the ovule, undergoes meiosis to produce four haploid cells. Of the four haploid cells produced, three disintegrate. The remaining cell, the **megaspore,** or female spore, divides mitotically and when mature develops into the **female gametophyte,** or **embryo sac.** This development proceeds in various ways. Most commonly, however, the haploid nucleus undergoes three

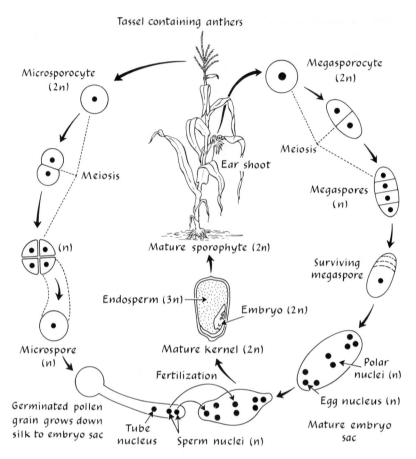

FIGURE 9-6. The life cycle of corn (*Zea mays*), illustrating pollen and embryo-sac formation. [Adapted from Srb, Owen, and Edgar, *General Genetics*, 2nd ed., W. H. Freeman and Company, copyright © 1965.]

successive mitotic divisions. The eight resulting nuclei form a cell membrane around them and arrange into three groups within the embryo sac. The middle of the three cells at the micropylar end develops into the female gamete, the **egg cell.** The two **polar cells** at the center of the embryo sac will eventually be part of the endosperm. The function of the two cells (**synergids**) that accompany the egg cell and the three cells (**antipodals**) at the other end of the embryo sac is obscure.

Pollination

Pollination is the transfer of pollen from the anther to the stigma. The transfer of pollen to any flower on the same plant or clone is **self-pollination** (or **selfing**); the transfer of pollen to a flower on a different plant is **cross-pollination.** Self-pollination is usually accomplished by gravity or by the actual contact of the shedding anther with the sticky stigmatic surface. In cross-pollination, wind and insects are the important agents of pollen transfer. Most plants both self- and cross-pollinate naturally to varying extents, depending upon structural features of the flower or upon genetic incompatibility. Plants are referred to as self-pollinated when the amount of cross-pollination is less than about 4%, as cross-pollinated when cross-pollination is predominant. Those plants somewhere in between are referred to as self- and cross-pollinated.

Natural self-pollination is achieved through functional and structural features of the flower. Flowers that lend themselves to this mode of pollination are perfect; that is, they contain both stamens and pistils. In some plants, such as the violet, the shedding of pollen before the flower is open (**cleistogamy**) insures self-pollination. A common structural feature of self-pollinating plants consists of a pistil growing through a sheath or ring of anthers, as in the tomato.

Cross-pollination, typical of many horticultural plants, is brought about in many different ways. The most basic is the separation of stamens and pistils into separate flowers on the same plant, as in corn and cucumbers (**monoecious** plants). The extreme form of this is the separation of staminate and pistillate flowers on different plants, as in spinach, asparagus, and holly (**dioecious** plants). However, many perfect-flowered plants cross-pollinate. This is achieved by anatomical or physiological features of the flower that prevent self-pollination. For example, differential maturation of stamens and pistils will prevent natural selfing. The structural features of the flower that insure cross-pollination are often related to pollen transfer by insects. Among the special adaptations that aid in insect pollination are petal color, odor, and the presence of nectar. In some species an intimate interdependence exists between the plant and the insect. For example, pollination of the Smyrna fig, which produces only female flowers, is carried out by the *Blastophoga* fig wasp, which develops only in the wild, inedible caprifig (see page 517).

Pollen incompatibility (self-sterility) is a physiological mechanism that prevents self-fertilization. A genetic factor (or factors) serves to prevent pollen tubes produced by the plant from growing in the style of the same plant. Incompatibility factors prevent self-pollination in such crops as cabbage, tobacco, petunia, and apple.

Fertilization

The pollen grain, after landing on the stigmatic surface of the pistil, absorbs water and other substances, such as sugars, and forms a tube. The tube literally grows down the style to the embryo sac. The pollen tube penetrates the embryo sac, where one male gamete unites with the egg to form the **zygote.** After mitotic division, the zygote becomes the **embryo** of the resultant seed. The other male gamete fuses with the two polar nuclei and forms the **endosperm.** This complete process is referred to as **double fertilization.**

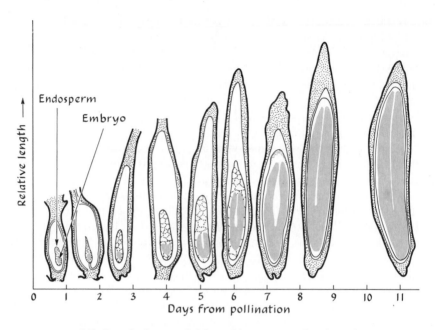

FIGURE 9-7. Growth of one-seeded fruit of lettuce. Note that the early growth of endosperm precedes that of the embryo. [Adapted from Hartmann and Kester, *Plant Propagation: Principles and Practices*, Prentice-Hall, 1959; after Jones.]

Seed Maturation

Fertilization initiates rapid growth of the ovary and subsequent development of the seed. Usually the ovary will not develop unless it contains viable and growing seed. Common exceptions (**parthenocarpy**) are seedless grapes and oranges. In these the developing seed breaks down at an early point in its development. The growth of the fruit has been discussed in Chapter 5.

The growth pattern of a developing lettuce seed is presented in Figure 9-7. Note that the development of the embryo is preceded by endosperm growth. When rapid growth of the embryo begins, it does so at the expense of the endosperm. The amount of endosperm at maturity varies with different plants. The endosperm may be completely absorbed at seed maturity, in which case stored food may be in the embryo itself, as in the cotyledons of bean seeds.

Seed Production and Handling

Although seeds of some vegetables (for example, tomato and watermelon) are grown in the same locations used for crop production, most seed crops are grown in specialized locations. (The principal horticultural seed-producing areas of the United States are in the western states, principally Idaho and California.) The limitation may be imposed by specific flowering requirements, such as cold induction or photoperiod. In addition, there are specialized requirements for commercial seed production: low moisture at harvest to permit proper maturation of the seed and to reduce the incidence of fungal and bacterial disease; and isolation, which is required to prevent contamination in cross-pollinating species. For example, sweet-corn seed cannot be produced in the Midwest because the abundance of pollen from field corn or popcorn would result in undesirable hybrids. Very little tree and shrub seed is grown commercially. This seed is collected from natural stands, nurseries, and arboreta. Seed from fruit trees, the seedlings of which are often used as rootstocks, is obtained as a by-product of the fruit-processing industry.

A considerable amount of hand labor is required in the harvesting of flower and vegetable seed. This is particularly true of species in which the seed head, or pods, shatter easily, or of plants on which the seed matures gradually, such that at any one time there may be mature seed pods, flowers, and flower buds. Examples are aquilegia, delphinium, salvia, petunia, and pansy. One great advantage of hand picking is that the cleaning operations are greatly simplified. With other crops the entire plant is cut and placed on sheets of canvas or in windrows to dry and is then threshed. This procedure is used with carnation, centaurea, daisy, gypsophila, hollyhock, larkspur, French marigold, lychnis, phlox, scabiosa, snapdragon, verbena, sweetpea, nasturtium, and morning glory.

In many crops the seed must be extracted or milled from the fruits and then cleaned. The separation of the seed from fleshy fruits, such as the tomato, is accomplished through fermentation of the macerated pulp. Seed is removed from dried pods or seed heads by milling. Cleaning is facilitated by differences in size, density, and shape of the seed in comparison with the plant debris or other seeds that may have been harvested incidentally. Screens may be used to separate large particles from the seed. Small, light fragments are blown from the seed by passing an air stream through the seed as it passes from one screen to another or as it is passed across a porous bench or against an inclined plane. The heavier seed remains at the base while lighter material moves up the plane. Seeds or particles that are of the same density as the crop seed but of a different shape can be removed on an "indent machine." A wheel covered with indentations is passed through a mass of seed, and each "indent" picks up a seed. The size and shape of the indentation is determined by the crop being cleaned. Some seeds, particularly beans and peas, can be separated on the basis of color. Single seeds are picked up by suction through perforations on a hollow wheel and then are passed through a photoelectric cell. If the cell detects a seed of the wrong color, a device releases the vacuum and ejects the seed. Throughout the milling and cleaning operations, extreme care must be taken in the adjustment of the machinery. Seeds which are chipped or damaged may be less viable or may produce abnormal seedlings.

Seed Storage

The storage life of seeds varies greatly with the species. With any species the longevity is greatly affected by storage conditions. Most seeds retain the highest viability in a relative humidity of 4–6%, although those of some plants (for example, silver maple and citrus) lose their viability under low moisture. The best temperatures for storing many seeds have been shown to be in the range of 0–32°F (-18–0°C). Actual storage conditions required for seed depends ultimately on the species and on the length of storage time desired. For most seeds, temperatures of 32–50°F (0–10°C) and a relative humidity of 50–65% are adequate to maintain full viability for at least one year.

Germination

Germination, the series of events from dormant seed to growing seedling, is dependent upon seed viability, the breaking of dormancy, and suitable environmental conditions. Because germinating seeds and young seedlings are vulnerable to certain diseases (for example, damping-off disease), protection must be provided.

Viability

Seed viability refers to the percentage of seed that will complete germination, the speed of germination, and the resulting vigor of the seedlings. The viability of seed lots can be determined by standardized testing procedures. Probably the most significant measure of viability is the **germination percentage,** the percentage of seed of the species tested that produces normal seedlings under optimum germinating conditions (Fig. 9-8). Germinating tests are usually performed on moistened absorbent paper under rigidly controlled environmental conditions (Fig. 9-9). The length of the test varies, for some species are notoriously slow to germinate. Perhaps the greatest problem is in distinguishing between dormant and nonviable seed. Seed dormancy must be overcome to obtain a reliable test. A rapid chemical test involving tetrazolium (2,3,5-triphenyltetrazolium chloride) makes it possible to evaluate viability in nongerminating dormant seed. Living cells treated with the chemical turn red, whereas nonliving cells show no color.

Breaking Dormancy

Breaking dormancy and creating a suitable growing environment are necessary to initiate the germination process. There are many treatments that can be used to break dormancy, the appropriate ones being determined by the particular type of dormancy (see Chapter 5). They include scarification, dry storage, stratification, embryo culture, or various combinations of these treatments, along with suitable environment control.

FIGURE 9-8. Germination of corn seed is being tested by the "rolled towel" technique. One hundred seeds are placed on moist paper toweling, which is sealed in wax paper, rolled, and stored at a standardized temperature. After seven days, the percentage of germinated seeds is determined. [Photograph by J. C. Allen & Son.]

FIGURE 9-9. Seed germinator with complete environmental control, including control of light. This germinator was designed for the Indiana State Seed Laboratory. [Photograph by J. C. Allen & Son.]

SCARIFICATION. The germination of seed that contains an impervious seed coat may be promoted by **scarification**—the alteration of the seed coat to render it permeable to gases and water. This is accomplished by a number of techniques, mechanical methods involving abrasive action being the most common. The action of hot water (170–212°F, or 77–100°C) is effective for the seed of honey locust. Some seeds are best scarified by the corrosive chemical action of sulfuric acid.

DRY STORAGE AND STRATIFICATION. Seed that will not germinate immediately after harvest requires **dry storage** for a period of days or months. The physiological basis of this type of dormancy is not clear but it has been associated with the evolution of volatile germination inhibitors. The afterripening of some seeds requires a period of moist storage known as **stratification.**

Cold-stratification—the afterripening of dormant embryos by storing them at high moisture and low temperature—is a prerequisite for the uniform germination of many temperate-zone species, such as apple, pear, and redbud. The cold-stratification of apple and pear seed involves storing the moist aerated seed at around 32°F (0°C). The germination percentage increases with time until the third month of treatment. The stratification medium consists of moist soil, sand, and peat, or such synthetic substances as vermiculite. An effective means of preventing the loss of moisture and of providing an adequate exchange of oxygen and carbon dioxide consists in sealing the seed in polyethylene bags containing a moist blotting paper.

Warm-stratification—moist storage above approximately 45°F (7°C)—promotes germination in some species as a result of microbial decomposition of the seed covering. Seed of such plants as viburnum, which exhibits two different types of dormancy (**double dormancy**), are first **warm-** and then **cold-stratified.** For redbud seed, a combination of scarification and cold-stratification is used.

EMBRYO CULTURE. The aseptic growth of excised embryos (often with associated parts, as placenta tissue) in artificial media is known as **embryo culture.** This specialized technique is used to facilitate seed germination in some species. For example, the embryos of many early-ripening peaches (such as 'Mayflower') are not sufficiently mature to germinate when the fruit is ripe. A serious impediment to the breeding of early-ripening peaches, this problem can be overcome, however, by excising the embryo from the pit and culturing it under aseptic conditions in media that provide certain nutrients. Tukey's Solution is one such medium. This technique is also used to produce viable seedlings from interspecific crosses that produce defective endosperms. A number of new types of crucifers from

Tukey's Solution for culturing mature and relatively immature embryos

Stock chemical	Relative amounts
KCl	5
$CaSO_4$	1.25
$MgSO_4$	1.25
$Ca_3(PO_4)_2$	1.25
$Fe_3(PO_4)_2$	1.25
KNO_3	1

Note: Use 1.5 g of mixture per liter of water.

interspecific crosses have been developed with the aid of embryo culture. Embryo culture is also used to circumvent dormancy caused by inhibitory substances associated with the seed coat. It is for this reason that the technique is used with viburnum.

The routine germination of orchid seed involves culture in artificial media. Orchid seed, which is almost microscopic, consists of a very simple, undifferentiated embryo and contains no reserve food. The stage between germination and the developing seedling is called a **protocorm.**

Knudson's Solution B for growing orchid seedlings

Chemical	mg/liter water
$Ca(NO_3)_2 \cdot 4H_2O$	1000
$(NH_4)_2SO_4$	500
KH_2PO_4	250
$MgSO_4 \cdot 7H_2O$	250
$FePO_4 \cdot 4H_2O$	50
Agar	17,500
Sucrose	20,000

Environmental Factors Affecting Germination

The germination of seed that does not require afterripening, or of seed that has had this requirement satisfied, depends upon external environmental factors—namely, water, favorable temperature, oxygen, and, sometimes, light.

The amount of water required for germination varies somewhat from species to species. For example, celery seed requires that soil moisture be near field capacity, whereas tomato seed will germinate with soil moisture just above the permanent wilting point. For most seed, excessive wetness is harmful, since it prevents aeration and promotes disease. Moisture must be maintained during germination, however, lest the germinating seedling dry out. Shading to conserve moisture is recommended until germination is complete. The use of glass over seed flats conserves moisture, but care should be taken to prevent the seeds from getting too hot.

The effect of temperature upon germination varies by species and is related somewhat to the temperature requirement for optimum growth of the mature plant (Table 9-1). In general, the germination rate increases as temperature increases, although the highest germination percentage may be at a relatively low temperature. An alternating temperature is usually more favorable than a constant temperature. Because of its critical role in respiration, oxygen is necessary for seed germination in all plants except some water-loving species (for example,

TABLE 9-1. Soil temperatures for vegetable seed germination.

Minimum					
32°F (0°C)	40°F (4°C)		50°F (10°C)	60°F (16°C)	
Endive	Beet	Parsley	Asparagus	Lima bean	Okra
Lettuce	Broccoli	Pea	Sweet corn	Snap bean	Pepper
Onion	Cabbage	Radish	Tomato	Cucumber	Pumpkin
Parsnip	Carrot	Swiss chard		Eggplant	Squash
Spinach	Cauliflower	Turnip		Muskmelon	Watermelon
	Celery				

Optimum					
70°F (21°C)	75°F (24°C)	80°F (27°C)	85°F (29°C)		95°F (35°C)
Celery	Asparagus	Lima bean	Snap bean	Pepper	Cucumber
Parsnip	Endive	Carrot	Beet	Radish	Muskmelon
Spinach	Lettuce	Cauliflower	Broccoli	Sweet corn	Okra
	Pea	Onion	Cabbage	Swiss chard	Pumpkin
		Parsley	Eggplant	Tomato	Squash
				Turnip	Watermelon

Maximum					
75°F (24°C)	85°F (29°C)	95°F (35°C)		105°F (41°C)	
Celery	Lima bean	Asparagus	Eggplant	Cucumber	Squash
Endive	Parsnip	Snap bean	Onion	Muskmelon	Sweet corn
Lettuce	Pea	Beet	Parsley	Okra	Turnip
Spinach		Broccoli	Pepper	Pumpkin	Watermelon
		Cabbage	Radish		
		Carrot	Swiss chard		
		Cauliflower	Tomato		

Source: Hartmann and Kester, 1959, *Plant Propagation: Principles and Practices* (Prentice-Hall).

rice and cattails). The maintenance of proper drainage and tilth in seed beds promotes rapid germination, largely as a result of good aeration (Fig. 9-10).

The effect of light in stimulating or inhibiting the germination of some seed (discussed in Chapter 5) is a reversible red–far red phenomenon. To produce good stocky plants ample light must be supplied during early seedling growth.

The action of certain salts has been shown to influence germination. At concentrations of 0.1–0.2%, potassium nitrate will increase germination in a number of plants, and for this reason it is used in seed testing. Seed treatment with salts has recently been used to achieve rapid field germination. In general, however, high concentrations of salts brought about by overfertilization inhibit germination.

FIGURE 9-10. Good seedbed preparation promotes rapid seed germination. [Photograph courtesy Ford Motor Co., Tractor and Implement Division.]

Disease Control

Disease is a critical factor in the germination process. This is especially true for seed that must be stratified or that requires an extensive period of time for germination. The control of these diseases is an integral part of the technology of seed propagation.

The major diseases of germinating seeds are grouped under a single name: **damping off.** These diseases are caused by a number of separate fungi, mainly species of *Pythium, Rhizoctonia,* and *Phytophthora.* Damping off is expressed either by the failure of the seedling to emerge or by the death of the seedling shortly after emergence. A common symptom is the girdling of young seedling stems at the soil surface. Damping off usually occurs only in young, succulent seedlings during or shortly after germination, but older plants may be affected in severe cases. Damping off can be severe in both greenhouse and field soils, and it is often a limiting factor in the success of seed propagation. Protection of seedlings from damping off and other diseases involves both the direct control of the pathogens and the regulation of environmental conditions such that they favor the rapid growth of the plant rather than the growth of the pathogens.

SEED AND SOIL TREATMENT. A number of seed treatments are available either to eliminate the pathogens from the seed or to provide protection to the seedling when planted in infested soil. These consist in coating the seed with a suitable fungicide, such as mercuric chloride, cuprous oxide, or calcium hypochlorite. A common seed treatment is a 5-minute dip in a 10% solution of Clorox (which is a 5.25% solution of sodium hypochlorite). A number of compounds intended for seed treatment are available commercially. The treatment of seeds in hot water—122°F (50°C) for 15–30 minutes—has been used for seed-borne diseases of vegetables (for example, alternaria of onion). Such treatment must be precise, however, or the seed may be seriously injured.

Soil may be treated by applying fungicides to the upper surface or by applying heat. Raising the soil temperature to 180°F (82°C) for 30 minutes ("pasteurization") is always recommended for potting soils to control weeds and nematodes, as well as damping-off organisms. Complete sterilization of soil interferes with the availability of nutrients and should be avoided. Sphagnum moss has proved satisfactory as a germination and stratification medium for some seeds because inhibitors and low pH prevent the growth of many of the damping-off organisms. The use of such sterile media as sand, vermiculite, or perlite may be desirable for seed germination. However, care must be taken to avoid recontamination of sterilized soil. The absence of natural predators (bacteria and other fungi) may result in great damage if such soil becomes infested with a pathogen.

CONTROL OF THE ENVIRONMENT. Any environmental effect that encourages rapid plant growth more than it does the buildup of pathogens is effective in the control of such seedling diseases as damping off, because older seedlings appear to resist attack. The temperatures most favorable to damping-off fungi are approximately 70–85°F (21–29°C). Thus, damping off tends to be severe when cool-season crops are germinated at temperatures that are too high, and vice versa. For best control, germinating temperatures should be optimum for the crop. This principle can be utilized in the control of damping off in the field by regulating planting dates. Good viable seed and rapid seedling growth are important. Many of the fungi responsible for damping off are water-loving,[1] and are encouraged in wet soils.

---

[1] The fungi of class Phycomycetes, to which both *Pythium* and *Phytophthora* belong, are often referred to as "water molds."

Cloudy weather and periods of poor drying encourage the damping-off complex; consequently, frequent and shallow watering should be avoided after planting.

Sanitation to reduce the buildup of organisms responsible for damping off should be practiced in the greenhouse, where this trouble is a perpetual problem. This involves eliminating plant refuse, disinfecting the walks and the potting area, keeping unsterilized soil out of the potting area, and general cleanliness.

Planting

Seed may be sown in a permanent location (**direct seeding**) or it may be planted first in some container from which the young plants can be transplanted once or many times before permanent planting. The growing of **transplants** makes it possible to provide precise environmental control during the critical stages of germination and early seedling growth. Many ornamentals are grown from transplants, as are vegetables for early production.

Direct Seeding

Plants that are difficult to transplant, or those for which the individual value of the plant does not justify the trouble and expense that transplanting entails, are grown by direct seeding. Many of the common vegetables (for example, beans, sweet corn, and radishes) are always grown by direct seeding (Fig. 9-11). Although direct

FIGURE 9-11. Direct seeding with an eight-row planter. [Photograph by J. C. Allen & Son.]

seeding requires much less labor and trouble than transplanting, one of its limitations is weed control. However, the recent advances in chemical weed control have made the direct seeding of some crops, such as tomato, economically feasible.

Precision spacing is important in direct seeding to prevent the need for extensive thinning or replanting. Because this is difficult to accomplish with small seeds, attempts have been made to "pelletize" such seeds by coating them with some suitable material, usually clay with additives. The increased size of the seed facilitates planting, and the coating material may be treated with fertilizer to encourage rapid seedling growth. Pelleting has been somewhat successful with lettuce, but with present materials it is of doubtful value for most crops, since the pelletizing materials can reduce or retard germination. But this is an area of seed technology that can be expected to change dramatically.

Transplants

The growth of transplants is a specialized part of seed propagation. Seed may be germinated first in seedling flats containing specially prepared media, and the seedlings transplanted later to suitable containers. If the seedlings transplant with difficulty (as do those of the cucurbits) the seed may be planted directly into individual containers, such as small plastic pots or 3-inch veneer plant bands. The germination media in the seedling flats is usually sand or a sand-soil mixture. Sand has the advantage of being well drained and relatively easy to keep free of disease-producing fungi. Germination media are lacking in nutrients, but this is not important as long as the seeds are transplanted soon after emergence. Supplemental feedings with nutrient solutions can be provided. The depth of planting in seedling flats depends on the size of the seed. As a rough approximation, the depth of planting should be 1–2 times the largest seed diameter. Very fine seed may simply be sprinkled over the surface of the soil.

Seedlings grown in flats should be transplanted as soon as they are large enough to handle. The transplanting operation must be done carefully to prevent injury. (In many plants, however, the destruction of the tap root results in a more fibrous root system, which may be advantageous.) The transplanting operation is best made with the soil medium just wet enough to be impressionable, but not wet enough to be sticky. A "dibbler" is useful for making the planting hole.

A number of containers, made of various materials, are available for transplanting—for example, flats, pots, and bands. Containers made of a decomposible organic material such as peat are proving of value, especially for retail flower transplants. In their manufacture, peat pots are treated with a fungicide to prevent decomposition by mold and with nitrogenous fertilizer to overcome nitrogen deficiencies commonly associated with the use of organic materials. Wooden plant bands are best soaked in nitrogenous fertilizer for the same reason.

Field transplanting is a part of both seed propagation and vegetative propagation. Transplants may be planted along with the soil in which they were grown, or

FIGURE 9-12. A tomato transplanting operation. Starter solution is being added to the transplanter. [Photograph by J. C. Allen & Son.]

they may be "bare-rooted." The transplanting of bare-rooted plants of many crops (tomato, strawberry, and many kinds of nursery stock) is well adapted to mechanization (Fig. 9-12). Transplanting machines are often equipped to apply water and starter solution. Bare-root nursery plants are usually covered with mud before transplanting to prevent drying.

VEGETATIVE PROPAGATION

Vegetative propagation involves nonsexual reproduction through the regeneration of tissues or plant parts. In many cases this process is a completely natural one; in others it is more or less artificial, depending on the degree of human interference and regulation. There are many methods of vegetative propagation, the best one to use depending on the plant and the objectives of the propagator. The advantages of vegetative propagation are readily apparent. Heterozygous material may be perpetuated without alteration. In addition, vegetative propagation may be easier and faster than seed propagation, as seed-dormancy problems may be completely eliminated and the juvenile stage reduced. Vegetative propagation also makes it possible to perpetuate clones that do not produce viable seed or that do not produce seed at all—for example, 'Washington Navel' orange, 'Gros Michel' banana, and 'Thompson Seedless' grape.

The various methods of vegetative propagation are summarized in the following list.

1. Utilization of apomictic seed (as with citrus)
2. Utilization of specialized vegetative structures
   Runners (strawberry)
   Bulbs (tulip)
   Corms (gladiolus)
   Rhizomes (iris)
   Offshoots (daylily)
   Stem tubers (potato)
   Tuberous roots (sweetpotato)
3. Induction of adventitious roots or shoots
   Layerage (regeneration from vegetative part while still attached to the plant)
   Cutting propagation (regeneration from vegetative part detached from the plant)
4. Graftage (the joining of plant parts by means of tissue regeneration)
5. Tissue culture

Utilization of Apomictic Seed

Apomixis refers to the development of seeds without the completion of the sexual process. It is therefore a form of nonsexual or vegetative reproduction. The most significant type of apomixis is that in which the complete meiotic cycle is eliminated. The seed is formed directly from a diploid cell, which may either be the nonreduced megaspore mother cell or some cell from the maternal ovular tissue. As a result of apomixis, a heterozygous cross-pollinating plant will appear to **breed true.**

Although apomixis is widespread within the plant kingdom, it is not commonly used as a means of asexual propagation. It is utilized in the propagation of Kentucky bluegrass, citrus, and mango. These species, however, are only partially apomictic, and the seed they produce is derived from both the sexual process and apomixis.

Utilization of Specialized Vegetative Structures

The natural increase of many plants is achieved through specialized vegetative structures. These modified roots or stems are often also food-storage organs (bulbs, corms, and tubers), although in some plants they function primarily as natural vegetative extensions, as do runners. These organs enable the plant to survive adverse conditions, such as the cold period in temperate climates or the dry period in tropical climates, and provide the plant with a means of spreading. These specialized structures renew both the plant and themselves through adventitious roots and shoots, and they are commonly utilized by people as a means of propagation. When these structures subdivide naturally, the process is called **separation;** when they must be cut, the process is called **division.**

Stem Modifications

BULBS. **Bulbs** are shortened stems with thick, fleshy leaf scales (Fig. 9-13). In addition to their development at the central growing point, bulbs produce buds at the axils of their leaf scales that form miniature bulbs, or **bulblets.** When grown to full size, bulblets are known as **offsets.**

The development of bulbs from initiation to flowering size takes a single season in the onion, but most bulbs, such as those of the daffodil and hyacinth, continue to grow from the center, becoming larger each year while continually producing new offsets. The asexual propagation of bulb-forming plants is commonly achieved through the development of scale buds. Various stages of development may be utilized from the individual scales, offsets, or the enlarging mature bulb itself. In hyacinth propagation the bulb is commonly wounded to encourage the formation of adventitious bulblets. The bulblets develop into usable size in 2–4 years.

CORMS. Although they resemble bulbs, **corms** do not contain fleshy leaves, but are solid structures consisting of stem tissues, complete with nodes and internodes. The gladiolus, crocus, and water chestnut (*Eleocharis tuberosa*) are examples of corm-forming plants. In large, mature corms, one or more of the upper buds develops into a flowering shoot. The corm is expended in flower production, and the base of the shoot forms a new corm above the old. By season's end, one or

FIGURE 9-13. The structure of a tulip bulb. [Adapted from Hartmann and Kester, *Plant Propagation: Principles and Practices,* Prentice-Hall, 1959; after Mulder and Luyten.]

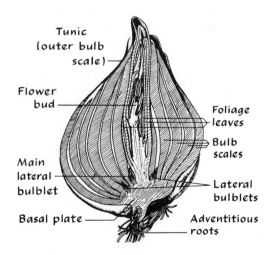

more new corms may have developed in this manner. **Cormels,** or miniature corms, are fleshy buds that develop between the old and new corms. Cormels do not increase in size when planted, but produce larger corms from the base of the new stem axis. These require 1–2 years to reach flowering size. This is the usual method of propagating corm-producing plants. Corms may also be increased by division, but this is not commonly practiced because of disease problems.

RUNNERS. **Runners** are specialized aerial stems that develop from the leaf axils at the base or crown of a plant having rosette stems (Fig. 9-14). They provide a means of natural increase and spread. Among the plants propagated by runners are the strawberry, strawberry geranium (*Saxifraga sarmentosa*), and bugle weed (*Ajuga*). The commercial propagation of strawberries is done through runner-plant production. Leaf clusters, which root easily, are formed at the second node of the runner. These rooted plants may in turn produce new runners. Runnering is photoperiod sensitive, being commonly initiated under a day length of 12 hours or longer. Dormant plants are dug by machines in the fall or in the spring. The yield in plants per mother plant varies with the variety, but under optimum conditions it may be as high as 200:1. A field increase of 20:1 or 30:1 is common. Some species of strawberries are nonrunnering, and many of the everbearing varieties usually form relatively few runners. These plants may be vegetatively propagated by crown divisions, but the increase in plants is much lower than in those with runners.

FIGURE 9-14. Runnering in the strawberry geranium (*Saxifraga sarmentosa*). [Photograph courtesy E. R. Honeywell.]

FIGURE 9-15. Root system and rhizomes of the tawny day-lily, *Hemerocallis fulva*. [Photograph courtesy G. M. Fosler.]

RHIZOMES. Horizontal cylindrical stems growing underground are called **rhizomes** (Fig. 9-15). Rhizomes contain nodes and internodes of various lengths, and they readily produce adventitious roots. Rhizomes may be thick and fleshy (iris) or slender and elongated (Kentucky bluegrass). Growth proceeds from the terminal bud or through lateral shoots. In many plants the older portion of the rhizome dies out. If new growth proceeds from branching, then the new plants eventually become separated. Rhizomatous plants are easily propagated by cutting the rhizome into several pieces, as long as each piece contains a vegetative bud.

TUBERS. Fleshy portions of underground rhizomes (of which the potato is probably the best known example) are known as **tubers.** The potato is propagated by planting either the whole tuber or pieces containing at least one "eye," or bud. If the whole tuber is planted, the terminal eye commonly inhibits the other buds, but this apical dominance is destroyed when the tuber is cut. Commonly the "seed pieces" are kept at 1–2 ounces (25–50 g) to provide sufficient food for the young plant. The seed pieces may be cured for a while to heal the cut surface (see Chapter 11). Chemical treatments are used to prevent disease.

OFFSHOOTS. In many plants lateral shoots develop from the stem, which, when rooted, serve to duplicate the plant (Fig. 9-16). These have been referred to in horticultural terminology as **offsets, suckers, crown divisions, ratoons,** or **slips,** depending on the species. Lateral shoots may be referred to collectively as **offshoots.** The increase of bulbs and corms by offsets is a similar phenomenon. Propagation of plants that produce offshoots is easily made by division. In temperate climates, rooted offshoots of outdoor perennials may be divided either in fall or spring.

FIGURE 9-16. Pineapple may be propagated from slips—leafy shoots originating from axillary buds borne on the base of the fruit stalk. They may also be grown from the crown that issues from the top of the pineapple or from suckers that grow lower down on the stem. [Photograph courtesy Dole Corp.]

Root Modifications

TUBEROUS ROOTS. Roots as well as stems may be structurally modified into propagative and food-storage organs. Fleshy, swollen roots that store food materials are known as **tuberous roots** (Fig. 9-17). Shoot buds are readily formed adventitiously. Tuberous roots of some species may contain shoot buds at the "stem end" as part of their structure.

The sweetpotato is commonly propagated from the formation of rooted adventitious shoots called slips. In the dahlia, the roots are divided, but each tuberous root must incorporate a bud from the crown. In the tuberous begonia, the primary tap root develops into an enlarged tuberous root with buds at the stem end. This root can be propagated by division, but each section must contain a bud.

SUCKERS. Shoots, that arise adventitiously from roots are called **suckers,** although the term has been commonly used (less precisely, perhaps) to refer to shoots originating from stem tissue. The red raspberry, for example, is propagated by suckers abundantly produced from horizontal roots. In the red raspberry, suckering may be stimulated by extensive pruning. The rooted suckers are usually dug during the period of plant dormancy.

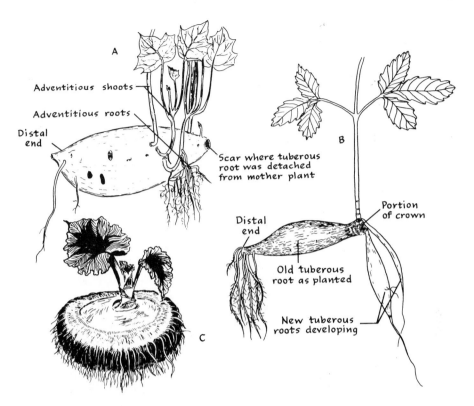

A
Adventitious shoots
Adventitious roots
Distal end
Scar where tuberous root was detached from mother plant
B
Portion of crown
Distal end
Old tuberous root as planted
New tuberous roots developing
C

FIGURE 9-17. Tuberous roots of sweetpotato (*A*), dahlia (*B*), and tuberous begonia (*C*). The tuberous root of the sweetpotato and dahlia disintegrate in the production of the new plant. The tuberous root of the begonia enlarges each year. [From Hartmann and Kester, *Plant Propagation: Principles and Practices*, Prentice-Hall, 1959.]

Induction of Adventitious Roots and Shoots

The regeneration of structural parts in the propagation of many plants is accomplished by the artificial induction of adventitious roots and shoots. When the regenerated vegetative part is attached to the plant, the process is called **layerage;** when the regenerated vegetative part is detached from the plant of origin, the process is called **cutting propagation.** These two processes, although technically different, are part of the same phenomenon—namely, the ability of vegetative plant parts to develop into a complete plant.

Layerage is often a natural process. In the black raspberry, the drooping stem tips tend to root when in contact with the soil; in strawberries, the runners form natural layers. Because the regenerated stem is still attached and nourished by the parent plant, the timing and techniques of layerage are not as critical as in cutting propagation, in which the vegetative part to be regenerated is severed from the parent plant. Rooting may be facilitated by such practices as wounding, girdling, etiolation, and disorientation of the stem, which affects the movement and accumulation of the carbohydrates and auxin needed to stimulate root initiation.

Layerage is a simple and effective means of propagation that can be practiced in the field. It is especially suited to the amateur because of the high degree of success possible with only a minimum of specialized facilities. However, it is relatively slow, offers less flexibility than techniques of cutting propagation, and requires much hand labor. For these reasons, layerage is not adaptable to large-scale nursery practices, and it is normally used only for plants that are most naturally adapted to this method of propagation, or in which propagation by cuttings is difficult. Different types of layerage are illustrated in Figure 9-18.

Cutting propagation is one of the most important means of vegetative propagation. The term "cutting" refers to any detached vegetative plant part that can be expected to regenerate the missing part (or parts) to form a complete plant. Cuttings are commonly classified by plant part (root, stem, leaf, leaf bud). In **stem cuttings** or **leaf-bud cuttings** a new root system must be initiated; in **root cuttings** a new shoot must be initiated; and in **leaf cuttings** both roots and shoots must be initiated.

Anatomical Basis of Adventitious Roots and Shoots

The formation of adventitious roots can be divided into two phases. One is initiation, which is characterized by cell division and the differentiation of certain cells into a root initial. The second phase is growth, in which the root initial expands by a combination of cell division and elongation. Although the two processes usually occur in sequence, in some plants, such as the willow, the time between initiation and development is well separated.

Root initials are formed adjacent to vascular tissue. In herbaceous plants that lack a cambium, the root initials are formed near the vascular bundles close to the phloem. Thus, roots will appear in rows along the stem, corresponding to the major vascular bundles. In woody plants, initiation commonly occurs in the phloem tissue, usually at a point corresponding to the entrance of a vascular ray.

The production of both adventitious roots and shoots from leaf cuttings commonly originates in secondary meristematic tissues—cells that have differentiated but later resume meristematic activity. In *Kalanchoë* (*Bryophyllum*) new plantlets form during leaf development from meristematic regions on the leaf edges (Fig. 9-19).

Adventitious roots and shoots may be derived from different kinds of tissue; for example, in leaf cuttings of African violet the roots are initiated from cells between the vascular bundles, whereas shoots are initiated from cells of the epidermis or cortex. In the sweetpotato, on the other hand, roots and shoots may be derived from callus tissue formed on the cut surface.

The formation of complete plants from pieces of root is dependent upon both the initiation of adventitious shoots and the extension of new root growth.

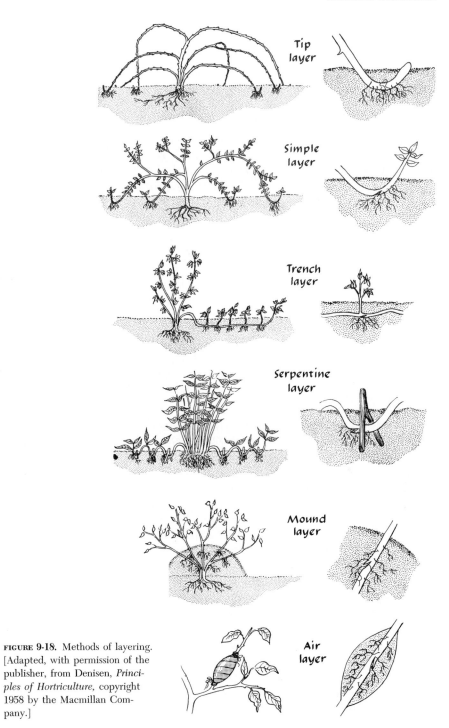

Tip
layer

Simple
layer

Trench
layer

Serpentine
layer

Mound
layer

Air
layer

FIGURE 9-18. Methods of layering. [Adapted, with permission of the publisher, from Denisen, *Principles of Horticulture*, copyright 1958 by the Macmillan Company.]

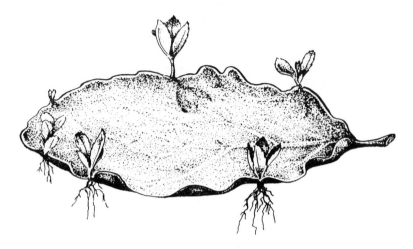

FIGURE 9-19. In *Kalanchoë*, new plants arise from meristem located in notches at the leaf edge. Shoot and root primordia are present in adult leaves. [From Mahlstede and Haber, *Plant Propagation*, Wiley, 1957.]

Adventitious shoot buds develop from cells of the phloem parenchyma and from rays. New roots originate from older tissues through latent root initials, although new root initials may arise adventitiously from the vascular cambial region.

Physiological Basis of Rooting

The variables that enable a stem to root depend upon the plant and its treatment. Some insight into the physiological basis of rooting has been developed from studies on **easy-** and **difficult-to-root** plants. The capacity for a stem to root has been shown to be due to an interaction of inherent factors present in the stem cells as well as to transportable substances produced in leaves and buds. Among these transportable substances are auxins, carbohydrates, nitrogenous compounds, vitamins, and various unidentified compounds. Substances that interact with auxins to affect rooting may be referred to as **rooting cofactors.** In addition, such environmental factors as light, temperature, humidity, and the availability of oxygen play an important role in the process. The physiological factors involved in rooting are only beginning to be understood; it is still not possible to effect rooting in many plants, including blue spruce, rubber tree, and oak.

Auxin levels are closely associated with adventitious rooting of stem cuttings, although the precise relationship is not clear. The normal rooting of stems appears to be triggered by the accumulation of auxins at the base of a cutting. The increase in rooting produced by the application of indoleacetic acid or auxin derivatives supports this concept (see Chapter 7). However, it is certain that auxins are only a part of the stimulus, for rooting of many difficult-to-root cuttings is not improved by auxins alone. Other specific factors that either stimulate rooting (as does catechol) or inhibit it have been isolated. More such factors can be expected to be found.

The presence of leaves and buds exerts a powerful influence on the rooting of stem cuttings. In many plants the effect of buds is due primarily to their role as a source of auxins, whereas the rooting stimulation provided by leaves is related in part to carbohydrate production. But in many plants the effect of leaves and buds can be shown to be due to additional transportable cofactors that complement both carbohydrate and auxin application (Fig. 9-20).

An important component of the capacity for a stem to root is the nutritional status of the plant. In general, high carbohydrate levels are associated with vigorous root growth. The nitrogen levels of the plant will affect the number of roots produced. Although a low nitrogen level will increase the number of roots produced, an outright deficiency will inhibit rooting.

The accumulation of auxins as well as the accumulation of carbohydrates explains in part the effectiveness of ringing and wounding in stimulating rooting. In addition, wounding stimulates root initiation by some unknown process. Callusing of the wounded surface also increases the efficiency of water absorption. This wounding effect is utilized to increase the absorption of applied auxins.

The effectiveness of stem rooting varies with the stage of development and the age of the plant, the type and location of stem, and the time of year. Owing to the great variation between species, precise conclusions concerning the relationship of these factors to rooting cannot be made. In general, rooting ability is associated with the juvenile stage of growth. Such plants as English ivy, apple, and many conifers become very difficult to root when they reach the mature stage. Mature, difficult-to-root plants may be made easy to root by a reversion to the juvenile stage. Generally, adventitious shoots from the base of mature plants tend to assume juvenile characteristics. In mature plants that become more difficult to root with increasing age, these adventitious shoots may be induced by severe pruning. A form of layerage called **stooling** maintains the juvenile stage of growth

FIGURE 9-20. The rooting of a cutting is dependent on auxin, carbohydrates, and the presence of rooting cofactors. The cofactors interact with auxin in trigger rooting.

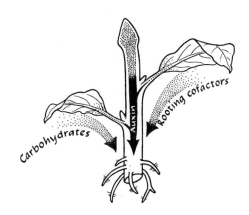

by continued pruning at the base of the plant. The stem bases are mounded with soil to facilitate rooting.

The ability of a stem to root is also affected by its position on the plant; lateral shoots tend to root better than terminal shoots. Vegetative shoots also tend to root better than flowering shoots. These differences may be related in part to auxin levels and the amount of stored food.

Cuttings vary in their ability to root, depending upon the type of stem tissue from which they are derived. Cuttings may be made from succulent nonlignified growth (**softwood cuttings**)[2] or from wood up to several years old (**hardwood cuttings**). Although almost all types of cuttings of easy-to-root plants root readily, softwood cuttings of deciduous, woody plants taken in the spring or summer generally root more easily than hardwood cuttings obtained in the winter. However, dormant hardwood cuttings are used when possible because of the ease with which they can be shipped and handled. Dormant cuttings must be stored until the shoot's rest period is broken, although rooting is less affected by dormancy. The time that softwood cuttings are taken varies greatly with different plant species. Softwood cuttings of azalea root best in the early spring; for other broadleafed evergreens, the optimum time for rooting softwood cuttings may be from spring to late fall.

Environmental Factors Affecting Rooting

HUMIDITY. The death of the stem as a result of desiccation before rooting is achieved is one of the primary causes of failures in propagation by cuttings. The lack of roots prevents sufficient water intake, although the intact leaves and new shoot growth continue to lose water by transpiration. Leaves or portions of the leaf may be removed to prevent excessive transpiration. However, this practice is not desirable because the presence of leaves encourages rooting. The use of mist (Fig. 9-21) maintains high humidity and also reduced leaf temperature by maintaining a water film on the leaf. This makes it possible for the cuttings to be exposed to light of greater intensity, so that photosynthesis can proceed at a high rate. The use of automatic controls to produce an intermittent mist is desirable because it eliminates the accumulation of excess water, which may be harmful to many plants, and because it makes it possible to maintain higher temperatures in the rooting medium.

TEMPERATURE. The use of bottom heat to maintain the temperature of the rooting medium at about 75°F (24°C) facilitates rooting by stimulating cell division in the rooting area. The aerial portion may be kept cool to reduce transpiration and

---

[2] Softwood cuttings of plants that are normally nonwoody are sometimes referred to as **herbaceous cuttings**, whereas those of woody plants prior to lignification are known as **greenwood cuttings**.

FIGURE 9-21. An in-bench mist installation. The supply pipe runs along the bottom of the bench. The polyethylene wind barrier eliminates the problem of drift. The deflection-type nozzle produces a mist by directing a fine stream of water against a flat surface. An intermittent mist, commonly 4 seconds on and 56 seconds off, is controlled by a timer. [Photograph courtesy Purdue University.]

respiration. Daytime air temperatures of 70–80°F (21–27°C) and night temperatures of 60–70°F (16–21°C) are optimum for the rooting of most species.

LIGHT. Light in itself appears to inhibit root initiation (or, conversely, the lack of light encourages it). Softwood and herbaceous cuttings indirectly respond to light because of its role in the synthesis of carbohydrates. However, deciduous hardwood cuttings that contain sufficient stored food, and to which artificial auxins can be supplied, root best in the dark. The role of light in inducing rooting thus varies with the plant and with the method of propagation. The reason that the absence of light favors root initiation in stem tissues is not clear. Root promotion may be achieved by the use of opaque coverings that etiolate the stem. Etiolation probably affects the accumulation of auxins and other substances that are unstable in light.

ROOTING MEDIA. Rooting media must provide sufficient moisture and oxygen and must be relatively disease free. It is not necessary that a rooting medium be a source of nutrients until a root system is established. The medium may have an effect on the percentage of cuttings rooted and on the type of roots formed. Various mixes containing soil, sand, peat, and artificial inorganic substances such as vermiculite (expanded mica) and perlite (expanded volcanic lava) have been widely used. Perlite used alone or in combination with peat moss has proven especially effective because of its good water-holding properties, drainage, and freedom from root-rotting diseases (Fig. 9-22). Sand or water alone may be satisfactory for some easy-to-root cuttings (Fig. 9-23). When water is used alone, improved results are achieved with aeration.

FIGURE 9-22. Rooted cuttings of woody ornamentals planted in a mixture of perlite and peat. Fertilizer must be added. They will be marketed as container-grown stock. [Photograph courtesy Perlite Institute.]

FIGURE 9-23. *Left*, a home propagator made up of a large pot filled with sand. The hole of the smaller inner pot is plugged with cork and kept filled with water. Uniform moisture is maintained by seepage from the small pot's porous sides. *Right*, cuttings that are easy to root, such as coleus, may be propagated in water. [Photograph courtesy E. R. Honeywell.]

Graftage

Graftage involves the joining together of plant parts by means of tissue regeneration, in which the resulting combination of parts achieves physical union and grows as a single plant. The part of the combination that provides the root is called the **stock;** the added piece is called the **scion.** The stock may be a piece of root or an entire plant. When the scion consists of a single bud only, the process is referred to as **budding.** (Budding and propagation by cuttings represent the most important commercial methods of asexual propagation.) When the graft combination is made up of more than two parts, the middle piece is referred to as an **interstock,** body stock, or interpiece.

There are two basic kinds of grafts: **approach** and **detached-scion.** In the approach graft the "scion" and "stock" are each connected to a growing root system. In the detached-scion graft, the kind most commonly used, only the stock is rooted, and the scion is severed from any root connection. The approach graft is used when it is difficult to obtain a union by ordinary procedures. The roots of the scion act as a "nurse" until union is achieved, at which time the scion is severed from its own roots.

The Graft Union

The **graft union** is the basis of graftage. It is formed from the intermingling and interlocking of callus tissue produced by the stock and scion cambium in response to wounding. The cambium, the meristematic tissue between the xylem (wood) and phloem (bark), is continuous in perennial woody dicots. Monocotyledonous plants with diffuse cambium cannot be grafted. Callus tissue is composed of parenchymatous cells. Under the influence of the existing cambium, the callus tissue differentiates new cambium tissue. This new cambium redifferentiates xylem and phloem to form a living, growing connection between stock and scion (Fig. 9-24). The basic technique of grafting consists in placing the cambial tissue of stock and scion in intimate association such that the resulting callus tissue produced from stock and scion interlocks to form a continuous connection. A snug fit is often obtained by utilizing the natural tension of the stock and/or scion. Tape, rubber bands, or nails may also be used to facilitate contact. Various types of budding and grafting are shown in Figure 9-25. Natural grafts may be formed as a result of the close intertwining of roots or stems.

Although there is usually no actual interchange of cells through the graft union, the connection is such that many viruses and hormones pass through unhindered. This principle is utilized in virus identification. Plants containing a suspected virus, but which may not show obvious morphological symptoms, are grafted to a plant that is sensitive and will show the symptoms of the virus. This process is known as **indexing** (Fig. 9-26).

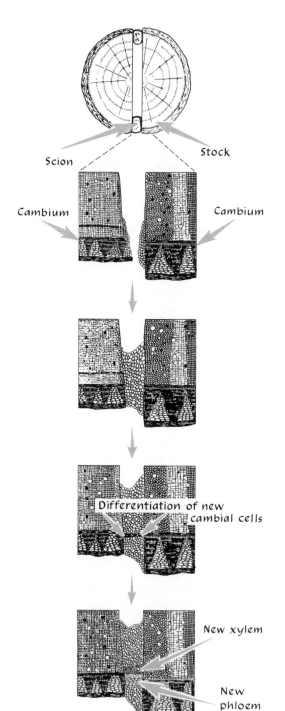

Scion

Stock

Cambium

Cambium

Differentiation of new
cambial cells

New xylem

New
phloem

FIGURE 9-24. Developmental sequence during the healing of a cleft-graft union. The graft union is formed from the redifferentiation of the callus tissue under the influence of the stock cambium. [From Hartmann and Kester, *Plant Propagation: Principles and Practices,* Prentice-Hall, 1959.]

Graft Incompatibility

Owing to the lack of any antibody mechanism, plants have a greater tolerance to grafting than do animals. The ability of two plants to form a successful graft combination is related in large part to their natural relationships. The inherent failure of two plants to form a successful union—**graft incompatibility**—may be structural as well as it is physiological. Graft incompatibility may be the cause of a high percentage of grafting failures; poor, weak, or abnormal growth of the scion; overgrowths at the graft union; or poor mechanical strength of the union, which in extreme cases results in a clean break at the graft. Incompatibility may be manifested immediately or it may be delayed for several years. In some cases, incompatibility has been traced to a virus contributed by one of the graft components, which was itself virus tolerant. In such cases incompatibility is due to the virus sensitivity of the other component. If the sensitive component is the root stock, the entire tree may be adversely affected.

In general, grafting within a species results in compatible unions. Grafts made between species of the same genus or species of closely related genera vary in their degree of compatibility. For example, many but not all pears (*Pyrus* species) may be successfully grafted on quince stock (*Cydonia oblonga*), but the reverse combination, quince scions on pear stock, is not successful. Incompatibility may be bridged by an interpiece composed of a variety compatible with both components. A few exceptional graft combinations involving species of different families have been experimentally produced in herbaceous plants.

Factors Influencing Grafting Success

In addition to the inherent compatibility of the plants, there are a number of factors that affect successful "take" in grafting. Skillful grafting or budding techniques are, of course, necessary. Success is then dependent upon environmental factors, which promote callus formation. In general, callus formation is optimum at about 80–85°F (26.5–29.5°C). After the grafting of dormant material (bench grafting), the completed graft is best stored under warm and moist conditions for a week or two to stimulate callus formation.

It is very important that high moisture conditions be maintained to prevent the scion from drying out. Waxing the tissue after grafting serves to prevent desiccation of the delicate callus tissue. Special waxes are available that consist of various formulations of resins, beeswax, paraffin, and linseed oil. Bench grafts should be plunged in moist peat to prevent desiccation during the period of healing at warm temperature. The use of plastic films has proved successful in conserving moisture. Oxygen has been shown to be required for callus formation in some plants (for example, the grape). Waxing should not be used with such plants because of its effect in limiting the oxygen supply.

The percentage of "take" in grafting is often improved if the stock is in a vigorous state of growth. The scion, however, should be dormant to prevent

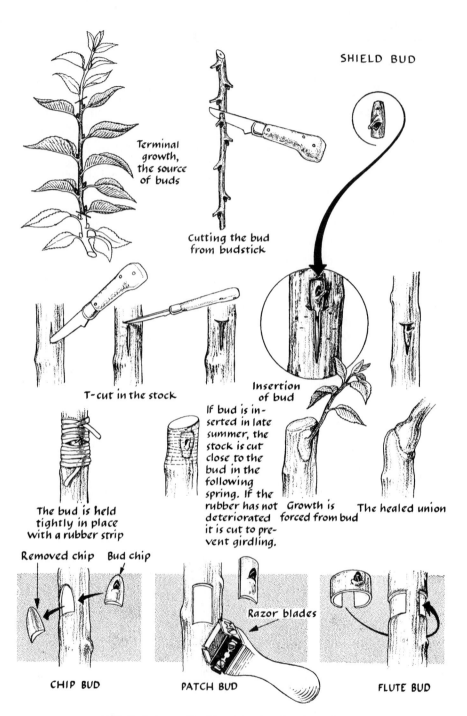

SHIELD BUD

Terminal growth, the source of buds

Cutting the bud from budstick

T-cut in the stock

Insertion of bud

If bud is inserted in late summer, the stock is cut close to the bud in the following spring. If the rubber has not deteriorated it is cut to prevent girdling.

The bud is held tightly in place with a rubber strip

Growth is forced from bud

The healed union

Removed chip   Bud chip

CHIP BUD

Razor blades

PATCH BUD

FLUTE BUD

FIGURE 9-25. Techniques of budding (*above*) and grafting (*facing page*).

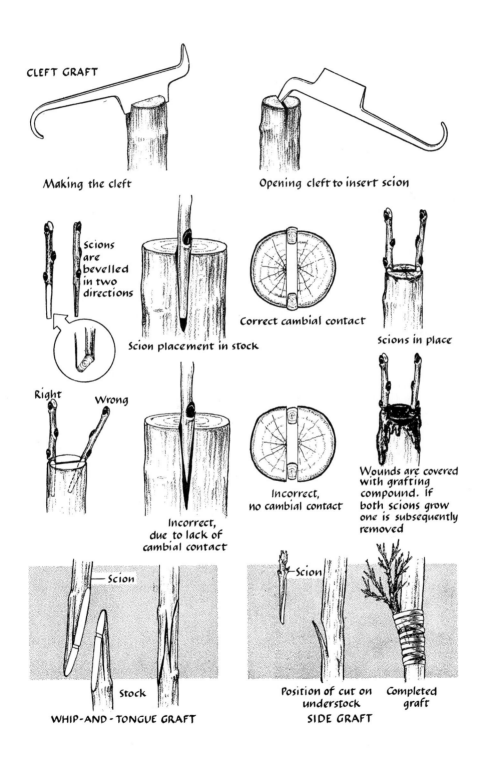

CLEFT GRAFT

Making the cleft

Opening cleft to insert scion

Scions are bevelled in two directions

Right

Wrong

Scion placement in stock

Incorrect, due to lack of cambial contact

Correct cambial contact

Incorrect, no cambial contact

Scions in place

Wounds are covered with grafting compound. If both scions grow one is subsequently removed

Scion

Stock

WHIP-AND-TONGUE GRAFT

Scion

Position of cut on understock
SIDE GRAFT

Completed graft

369

FIGURE 9-26. Indexing strawberries by means of the leaf graft. *Left,* the middle blade of the trifoliate leaf of a strawberry is cut to a wedge and inserted into the split petiole in place of the removed blade of the sensitive indicator, *Fragaria vesca.* The blade is usually held in place with a self-sticking latex tape. *Right,* after two months the grafted leaf from each of two different plants is still alive. The older leaves have been removed to show the new growth. The normal regrowth on the left indicates that the inserted leaf was free of any viruses to which the indicator plant is sensitive. The stunted, deformed growth on the plant at the right indicates transmission of a virus from the excised leaf to the indicator plant. [Photograph courtesy Purdue University.]

premature growth and subsequent desiccation. Growing trees are often grafted with dormant scions from refrigerated storage. In summer budding, irrigation should be supplied before budding in order to invigorate the stock. The leaf buds of most temperate woody plants, when inserted in late summer, remain naturally dormant until the following spring.

Grafting of Established Trees

In addition to its role as a method of propagation, grafting is useful in many instances for cultivar change, repair, or invigoration of older established trees. Grafting to affect growth is discussed in Chapter 7.

TOPWORKING. Owing to the long time required for growth of many fruit or nut crops, it is often desirable to utilize the existing framework and root system of a larger, established tree. It becomes possible to rework a tree and rapidly bring an improved cultivar into production. The regrafting of scions onto a large growing tree to utilize an existing framework is called **topworking** (Fig. 9-27). A cleft graft is usually used.

In topworking, the length of the scion inserted may have an effect on subse-

FIGURE 9-27. One-year-old olive scions growing on a 30-year-old stump. [Photograph courtesy USDA.]

quent growth. The few buds from short scions all tend to produce vigorous vegetative growth. Bud break on long scions is confined to the terminal buds; the basal buds often grow into fruiting spurs. This technique is utilized to bring new cultivars into early bearing for observation.

INARCHING, BRIDGING, AND BRACING. **Inarching** is a form of "repair" grafting that involves reinforcing the existing root system of an already growing tree. It is used when the existing root system is weak and must be replaced to save the tree. A number of seedlings or rooted cuttings are planted around the tree and are then approach-grafted to the stem. The grafts are usually held in place by nailing.

Bridge grafting, as the name infers, is a means of "bridging" an intact root and stem when the connection between them has been damaged. Bridge grafting can save a tree from death when the stem has been girdled by mice (Fig. 9-28).

Bracing the framework of a tree can be accomplished by means of grafting. The technique involves the twisting of young branches around each other to encourage the formation of a natural graft.

FIGURE 9-28. Five steps in the development of a bridge graft. [From Eaton, Ontario Department of Agriculture Publication 439, 1961.]

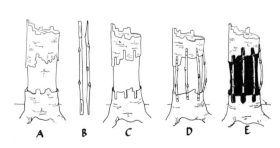

A   B   C   D   E

Tissue Culture

Tissue culture (or **in vitro culture**) refers to the aseptic growth of cells, tissues, or organs in artificial media. Although the culture of plant cells and tissues has long been a tool of the physiologist, these techniques are now increasingly used as a means of rapid plant propagation. Plant propagation by tissue culture (also known as **micropropagation**) involves a sequence of steps, each of which may necessitate a specific set of conditions. Three distinct steps are usually involved: (1) establishment of an aseptic culture, (2) multiplication of the propagule, and (3) preparation and establishment of the propagule for an independent existence by hardening and acclimation.

The first step requires surgical removal of a suitable tissue or organ. While almost every plant organ has been successfully cultured, the most suitable material often differs from species to species. Shoot tips are extensively used, but sections of stems, leaves, petals, and cotyledons have also been cultured, as have shoot apical meristems (the actual growing point) and nucellar tissues.

The probability of isolating tissues free from contamination by microbes and viruses is inversely proportional to the size of the **explant** (the piece of tissue taken). Explants taken from the true shoot meristem consist only of the apical meristem dome, which is less than 0.1 mm in height. These are usually too small for use in practical propagation, but they are used whenever it is especially important to establish a tissue culture entirely free of pathogens, including viruses and viroids. However, shoot tip pieces as tall as 1 mm have been used successfully in the establishment of pathogen-free cultures. These shoot tips establish more quickly and growth is faster.

Although such environmental factors as light can be critical to the success of a tissue culture, growth and differentiation of the explant is dependent primarily on the culture medium. Various formulas have been developed that make use of combinations of inorganic salts, sugars, vitamins, amino acids, purine bases, such organic complexes as coconut "water" or yeast extract, and growth regulators (Table 9–2). One of the key factors controlling morphogenesis is the auxin–cytokinin balance. A high cytokinin:auxin ratio promotes shoot formation while a high auxin:cytokinin ratio promotes root formation (see Fig. 4-13). Species and tissues differ in the critical balance required as well as in their reactions to different auxins and cytokinins. A critical mass of tissue is usually required before differentiation can proceed.

Actual propagation after the explant is established can be achieved by one of three pathways: (1) asexual formation of embryos (**embryogenesis**); (2) proliferation of axillary shoots; and (3) formation of adventitious shoots.

Embryogenesis

Embryogenesis cannot yet be induced in all species, and its applicability to commercial propagation is confined to relatively few plant families. This process may proceed directly from the cultured explant or indirectly from a **callus**

TABLE 9-2. Inorganic constituents of various media for culture of tissues, organs, and embryos (in mg/liter).

Compound	Knop medium	White medium	Heller medium	Murashige and Skoog medium°
KNO_3	200	80		1900
$Ca(NO_3)_2 \cdot 4H_2O$	800	200		
$NaNO_3$			600	
NH_4NO_3				1650
KH_2PO_4	200			170
$NaH_2PO_4 \cdot 4H_2O$		17	125	
Na_2SO_4		200		
$MgSO_4 \cdot 7H_2O$	200	360	250	370
KCl		65	750	
$CaCl_2 \cdot 2H_2O$			75	440

Source: Hartmann and Kester, 1968, Plant Propagation: Principles and Practices, 2nd ed.
° Also contains $MnSO_4 \cdot 4H_2O$, 22.3; H_3BO_3, 6.2; $ZnSO_4 \cdot 4H_2O$, 8.6; KI, 0.83; $Na_2MoO_4 \cdot 2H_2O$, 0.25; $CuSO_4 \cdot 5H_2O$, 0.025; and $CoCl_2 \cdot 6H_2O$, 0.025; all in mg/liter.
Also add Na_2EDTA, 37.3, plus $FeSO_4 \cdot 7H_2O$, 27.8 (as 5 ml/liter of stock solution of $FeSO_4 \cdot 7H_2O$, 5.57 g, plus 7.45 g Na_2EDTA per liter of water).

intermediary, unorganized growth induced from the cut surface of plant tissues. In certain plants, pollen grains (or microspores) have been induced to proliferate embryos directly, or from callus growth, to yield haploid plants (see p. 413).

During this remarkable process, embryo development proceeds through small globular, heart-shaped, and torpedo-shaped stages and mimics the development of normal embryos from true seeds. The asexually formed embryos may be produced in very large numbers: literally hundreds of embryos may be formed from the relatively small volume of tissue in a culture tube. Each embryo is potentially a separate plant. Embryogenesis was first observed in carrot, but it also occurs in other plants, such as *Kalanchoë*, asparagus, poppy, and citrus (Fig. 9-29).

The clonal increase of the members of some orchid genera has proven very efficient with tissue-culture techniques utilizing a system that appears to be related to embryogenesis. This is potentially very valuable because the usual method of vegetative propagation of orchids (by offshoots) is very slow. In some genera (*Cymbidium, Cattleya,* and *Dendrobium*) the culture of shoot tips results in the development of bulblets that resemble protocorms (orchid plants in the "embryonic" stage between germination and developing seedlings). These protocorm-like bulblets may be multiplied by cutting them into quarters. Vegetative propagules of orchid produced in this manner are called **mericlones.** Vegetative increase may be extremely rapid, with the number of propagules quadrupling every 10 days, but such rapid propagation is not possible with all orchids. Species of *Vanda* and *Phaleonopsis* do not form such protocorm-like bodies.

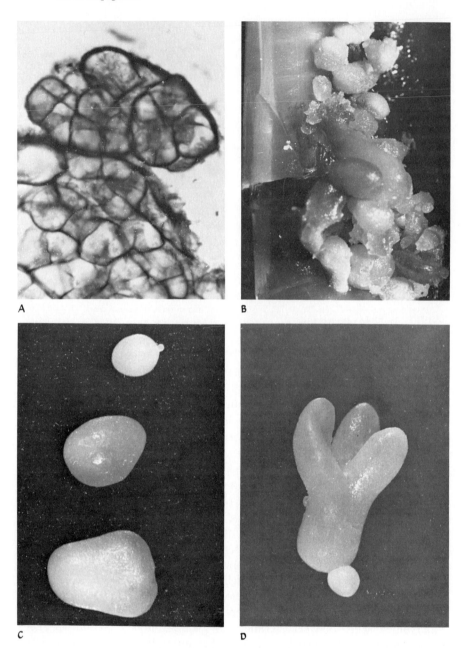

FIGURE 9-29. *In vitro* embryogenesis in orange from nucellus explants. *A*, embryos and embryogenic callus. *B*, cross section of callus showing cluster of developing proembryos. States of embryo development: (*C*) globular, elongated, heart-shaped, and (*D*) cotyledonary. Note a globular embryo bud at the base of the most advanced stage. [From J. Kochba and P. Spregel-Roy, *Hort-Science* 12:110–114, 1977.]

FIGURE 9-30. *Top*, proliferation of axillary shoots in carnation after 5 weeks of growth in a liquid medium containing 2.0 mg/liter kinetin and 0.02 mg/liter NAA with a single replacement of the medium after 3 weeks. *Bottom*, close-up showing single shoots and clumps. [From E. D. Earle and R. W. Langhans, *HortScience* 10:608, 1975.]

Proliferation of Axillary Buds

Enhancement of the development of axillary buds has wide applicability to tissue-culture propagation. Although the process is the slowest of the three, the tissues are under the normal control of the apical meristem and are not likely to produce genetically deviant plants—especially not those of the sort caused by such chromosome irregularities as chromosome doubling. Shoot tips are usually proliferated with high cytokinin:auxin ratios; 2iP, kinetin, and BA are the cytokinins usually used. The proliferated shoot must be separated and induced to form roots for the propagation function to be complete. Carnations (Fig. 9-30) and many tropical foliage plants are typically propagated by axillary-bud proliferation.

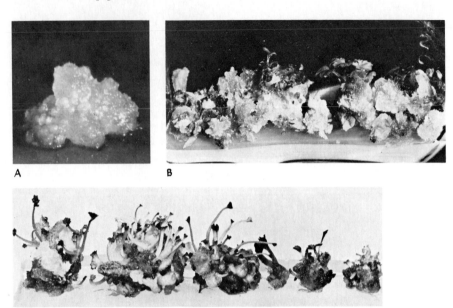

FIGURE 9-31. Adventitious shoot formation from callus culture of geranium. *A*, undifferentiated callus; *B*, various stages in callus differentiation; *C*, shoot development; *D*, root development; *E*, a plantlet two inches long derived from callus culture. [Photographs courtesy A. C. Hildebrandt.]

Formation of Adventitious Shoots

The formation of shoots from callus (or other tissues, such as leaf discs) is a very rapid way to propagate certain plants. However, the process of forming adventitious organs (**organogenesis**) is associated with a higher incidence of genetically aberrant plants than is the proliferation of axillary buds. Although the process is controlled by the cytokinin–auxin balance, organogenesis has yet to be demonstrated in many plants, especially many woody species. Intense effort is now being applied to the problem of developing organogenetic propagation methods for all horticulturally important species. The induction of roots from callus without the induction of shoots is common, and root initiation appears to be a physiologically simpler process than shoot initiation. Of course, when shoots are induced they must be rooted to complete the propagation cycle. Boston fern and geraniums (Fig. 9-31) are among the species propagated by inducing the formation of adventitious shoots.

Plant tissue culture has had a significant impact on clonal multiplication of floricultural crops, including foliage plants and potted plants. The potential rate of increase per year may be a million-fold greater than with conventional methods. Other uses of tissue culture include the production, isolation, and maintenance of disease-free plants (Fig. 9-32) and the storage of germplasm. The use of tissue culture for plant improvement will be discussed in the following chapter.

FIGURE 9-32. The long-term maintenance of "pathogen-free" plants is one of the promising uses of tissue culture technology. Disease-free stawberry plantlets (*top*) have been stored in the dark at 4°C for as long as 6 years. *Bottom*, the same plantlets after transplanting. [From R. H. Mullin and D. E. Schlegel, *HortScience* 11:100–101, 1976.]

Selected References

Gamborg, O. L., and L. R. Welter (editors), 1975. *Plant Tissue Culture Methods.* Saskatoon, Sask.: Prairie Regional Laboratory, National Research Council of Canada. (A useful manual of laboratory methods in tissue culture.)

Hartmann, H. T., and D. E. Kester, 1975. *Plant Propagation: Principles and Practices,* 3rd ed. Englewood Cliffs, N.J.: Prentice-Hall. (The most valuable and authoritative book on the subject.)

Hawthorn, L. R., and L. H. Pollard, 1954. *Vegetable and Flower Seed Production.* New York: Blakiston. (The technology of seed production.)

Kozlowski, T. T. (editor), 1972. *Seed Biology.* New York: Academic Press. (A complete, three-volume reference on seeds.)

Mayer, A. M., and A. Poljakoff-Mayber, 1975. *The Germination of Seeds,* 2nd ed. New York: Pergamon Press. (A monograph on the physiology of seed germination.)

Murashige, T., 1977. Plant propagation through tissue cultures. *Annual Review of Plant Physiology,* 25:136–166. (An up-to-date review of micropropagation.)

Thomas, E., and M. R. Davey, 1975. *From Single Cells to Plants.* New York: Springer-Verlag. (An outline of tissue-culture procedures.)

U.S. Department of Agriculture, 1961. *Seeds* (USDA Yearbook, 1961). (The story of seeds.)

U.S. Forest Service, 1974. *Seeds of Woody Plants in the United States* (USDA Agricultural Handbook 450). (A manual on all aspects of the seeds of woody plants and their proper handling.)

10

Plant Improvement

From fairest creatures we desire increase,
That thereby beauty's rose might never die,
But as the riper should by time decrease,
His tender heir might bear his memory:

SHAKESPEARE, Sonnet 1

Almost all of the edible plants in cultivation today were domesticated before the advent of written history. Many ornamentals, such as the rose and lily, have been in cultivation for thousands of years. Plant improvement has been continuous during this time as a result of deliberate differential reproduction of certain plants by people of many cultures. This process of **selection** over the years has been extremely effective; most of our cultivated plants no longer even remotely resemble their wild ancestors. **Plant breeding,** the systematic improvement of plants, is an innovation of the last century. In the past 75 years, the study of the mechanisms of heredity—**genetics**—has placed plant breeding on a firm theoretical basis. Plant breeding has become a specialized technology, and it is responsible for a large part of the current progress in horticulture.

Since human needs and standards change, the job of the plant breeder is a continuous one. For example, mechanical harvesting becomes possible only with plants (or plant parts) that are ready for harvest all at once. It demands that plants be structurally adapted to the machine. The raw material for these and other changes may be found in the tremendous variation that exists in cultivated plants and their wild relatives. The incorporation of these alterations into plants adapted to specific geographical areas demands not only a knowledge of the theoretical basis of heredity but mastery of the art and skill necessary for the discovery and perpetuation of small but fundamental differences in plants.

THE GENETIC BASIS OF PLANT IMPROVEMENT

Variety is more than the spice of life; it is the very essence of it. Differences between horticultural plants range from the obvious (water-lilies versus water-

379

melons) to the almost imperceptible variation that might exist between two apple trees of the same clonal cultivar growing side by side. Variation can be of two types, **environmental** and **genetic.** Genetic differences are due to the hereditary makeup of the organisms. This variation can be traced to differences in **genes,** the fundamental units that determine heredity. Environmental variation can be demonstrated by comparing organisms of identical genetic makeup grown in different environments. Similarly, differences between plants grown in identical environments must be due to genetic differences. The range of environmental variation is enormous, but its boundaries are determined by genetic makeup. (It is difficult to conceive of any environmental condition that will transform water-lilies to watermelons.) The genetic makeup of an organism is referred to as its **genotype.** The net outward appearance of the organism is its **phenotype.** In a stricter sense, the phenotype is the interaction product of a genotype and its environment.

To ask whether genetic variation or environmental variation is the more important is meaningless. The pertinent question is, *Which genotype is best suited to a particular set of environmental conditions?* And, given a particular genotype, *Which environmental conditions will permit the optimum phenotypic expression of that genotype?*

The fundamental discovery that the genotype is inherited as discrete units was first published in 1865 by Gregor Mendel, an Austrian abbot, whose discovery grew out of experimental work carried out with the garden pea. Environmental variation, however, is not transmitted, as was first firmly established by the Danish geneticist and breeder Wilhelm Johannsen in 1903, from research on seed size in the common bean. Johannsen demonstrated that inheritance of phenotypic variation is only possible when it is a result of genetic differences.

Genes and Gene Action

The chromosomes contain the code-controlling mechanism that coordinates the physiological activities of the cell and, consequently, the organism. The study of many organisms has indicated that the information provided by the chromosomes directs the formation of enzymes that affect biochemical reactions. The unit of the chromosome conferring a single enzymatic effect is called the **gene.** Structurally, the gene appears to be a portion of the deoxyribonucleic acid (DNA) molecule, the nonprotein portion of the chromosome; the genetic code inherent in the DNA molecule effects enzyme synthesis (see Chapter 9). Considerable information has accumulated concerning gene-mediated biochemical pathways.

Gene action can be demonstrated with flower pigmentation. Two anthocyanin pigments distinguished by differing degrees of redness have been shown to be chemically differentiated by a single hydroxyl group. The synthesis of these pigments proceeds in a stepwise fashion, as shown in Figure 10-1. The addition of a single hydroxyl group changes the color of the plant's petals from bright red to

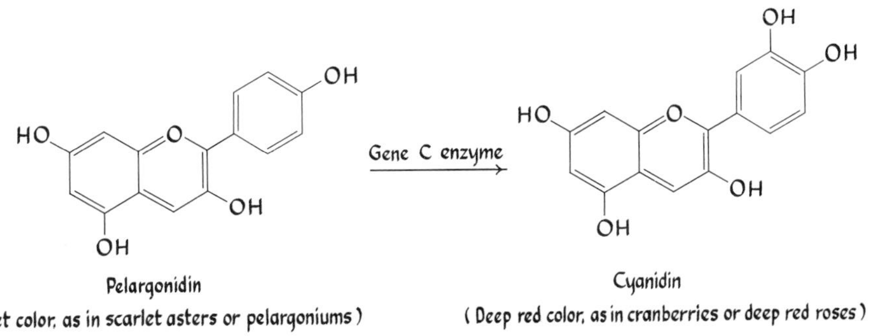

Pelargonidin
(Scarlet color, as in scarlet asters or pelargoniums)

Gene C enzyme

Cyanidin
(Deep red color, as in cranberries or deep red roses)

FIGURE 10-1. The conversion of pelargonidin to cyanidin involves the addition of a single hydroxyl group and is controlled by a single gene.

bluish red. This biochemical step is gene controlled. Thus, the petals of a plant that contains a gene (C) controlling the transformation of pelargonidin to cyanidin will be bluish red. A plant having only an altered version of this gene (c) incapable of adding the hydroxyl group will have bright red petals. Alternate forms of a particular gene are referred to as **alleles** (C and c are alleles). The change in the gene's internal structure, which gives rise to new alleles, is known as **mutation.** Mutations are changes in genetic information and are ultimately responsible for the inherent variation in all living things.

Each plant species contains a characteristic number of chromosomes known as the 2n, or somatic, number (2n = 12 in spinach, 14 in pea, 16 in onion, 18 in cabbage, 20 in corn, 22 in watermelon, 24 in tomato, and so on). The reproductive cells, or **gametes,** contain the haploid number (n) of chromosomes. Chromosomes occur in pairs in the vegetative cells, which therefore contain the somatic number. In meiosis one chromosome of each pair is distributed to the gametes, reducing the diploid number by one-half. Fertilization subsequently restores the diploid number in the zygote, the fertilized egg. Thus, in somatic cells of diploid plants, each gene is present twice. A particular gene (for example, C and its allele c) may be present in any one of three combinations—CC, Cc, or cc. A plant containing two identical genes, CC or cc, is **homozygous** for that allele. When the alleles are different, as with Cc, the plant is **heterozygous.**

What is the difference in outward expression (phenotype) when the genetic constitution (genotype) is CC, Cc, or cc? With reference to our example of petal color, the assumption can be made that, if the single allele c is completely nonfunctional in regard to hydroxylation of pelargonidin, two alleles cc should not be any more efficient. It can be shown experimentally that flowers of cc plants are bright red, and chemical analysis of their petals yields no cyanidin. The difference between plants that are homozygous (CC) and those that are heterozygous (Cc) cannot be predicted. The allele may be efficient enough to produce sufficient enzyme such that plants with only one functioning allele (Cc) cannot be distinguished phenotypically from those having two functioning alleles (CC). When this

381

is the case it would appear that the allele C dominates c. The condition in which heterozygous plants Cc are indistinguishable from homozygous CC plants is termed **dominance.** Thus C is referred to as a **dominant,** c as a **recessive.** If the heterozygote Cc is intermediate in phenotype between the two homozygous types CC and cc, dominance is said to be **incomplete.**

Assume the existence of a completely dominant gene A with a recessive allele a: A heterozygous plant Aa will produce two kinds of gametes (A and a) in equal proportions. A cross of these alleles, designated as $Aa \times Aa$, will produce three kinds of progeny in a predictable ratio if all gametic and zygotic types are equally viable. It can readily be seen that the possible types of zygotes resulting from all combinations of the two kinds of gametes A and a will be AA, Aa, aA, and aa (or $1AA$, $2Aa$, and $1aa$—see Fig. 10-2). If we assume dominance, AA plants are indistinguishable from Aa plants. Thus the phenotypic ratio becomes $3A\_\_$ to $1aa$.

The genotypes AA and Aa may be distinguished by a genetic test. The genotype AA will produce a single type of gamete (A). The genotype Aa will produce two kinds of gametes (A and a). By crossing plants of the phenotype A with themselves, or with the double recessive aa, these two genotypes may be separated on the basis of their progeny:

$$AA \text{ selfed } (AA \times AA) \longrightarrow \text{all } AA \ (AA \text{ plants ``breed true''}),$$

in contrast with

$$Aa \text{ selfed } (Aa \times Aa) \longrightarrow 3A\_\_ : 1aa \ (Aa \text{ plants segregate});$$

or,

$$AA \times aa \longrightarrow \text{all } Aa,$$

in contrast with

$$Aa \times aa \longrightarrow 1Aa : 1aa.$$

Generations are designated by specialized terminology. The first cross (usually referring to homozygous genotypes that differ with respect to a particular character) is the P_1, or the **parental generation.** The progeny of such a cross is the first filial generation (always referred to in writing and speaking as the F_1). If the parents are homozygous for all genes concerned, the F_1 is heterozygous and nonsegregating. The F_2 (second filial generation) results from intercrossing or selfing F_1 plants:

$$P_1 \qquad AA \times aa$$
$$\downarrow$$
$$F_1 \qquad Aa$$
$$\downarrow$$
$$F_2 \qquad 1AA : 2Aa : 1aa.$$

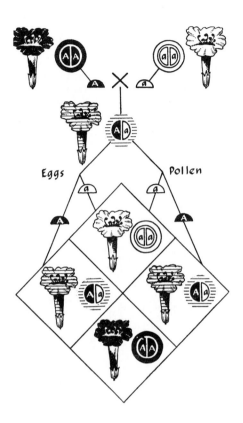

Eggs

Pollen

The F_2 is typically the segregating generation, which, if large enough, could theoretically include every possible genotype. Further generations are known as F_3, F_4, F_5, and so on. A cross of the F_1 with one of its parents is known as a **backcross.**

Multigenic Inheritance

Our discussion of the inheritance of a single gene difference can be expanded for examples involving two or more genes. The assortment of one pair of chromosomes at meiosis has no effect on the assortment of the other pairs; consequently, genes on separate chromosomes segregate independently of each other. Thus, a plant heterozygous for two gene pairs (*Aa* and *Bb*) will produce four different types of gametes (*AB*, *Ab*, *aB*, and *ab*) in equal proportions. The self progeny of the *AaBb* plant will segregate into phenotypic ratios of 3:1 for each factor, if we assume complete dominance. The combined phenotypic ratio will be $9A\_\_B\_\_:3A\_\_bb:3aaB\_\_:1aabb$. This cross is diagrammed in Figure 10-3.

When two genes affect the same biochemical pathway the phenotypic ratio gives some idea of the gene action involved. For example, two genes, both of

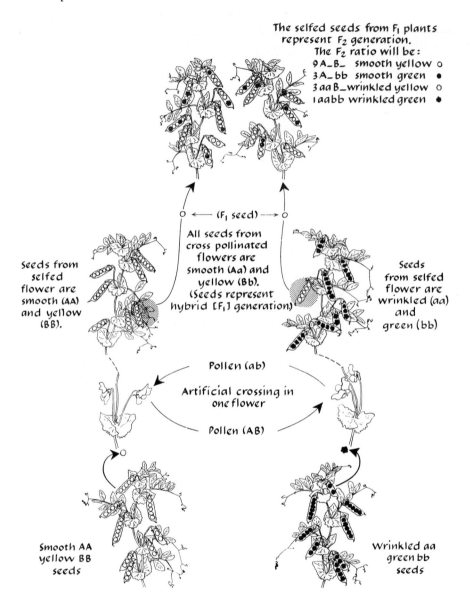

The selfed seeds from F₁ plants
represent F₂ generation.
The F₂ ratio will be:
9 A_B_ smooth yellow ○
3 A_bb smooth green ●
3 aa B_wrinkled yellow ○
1 aabb wrinkled green ●

○ ← (F₁ seed) → ○

All seeds from
cross pollinated
flowers are
smooth (Aa) and
yellow (Bb).
(Seeds represent
hybrid [F₁] generation)

Seeds from
selfed
flower are
smooth (AA)
and yellow
(BB).

Seeds
from selfed
flower are
wrinkled (aa)
and
green (bb)

Pollen (ab)

Artificial crossing in
one flower

Pollen (AB)

Smooth AA
yellow BB
seeds

Wrinkled aa
green bb
seeds

FIGURE 10-3. The independent assortment of two genes in peas. [After "The Gene" by N. H. Horowitz. Copyright © 1956 by Scientific American, Inc. All rights reserved.]

which show complete dominance, and both of which interact to effect a single character, will transform the 9:3:3:1 ratio into a ratio of 9:7. Either gene, when homozygous recessive, will block some essential step, as shown in Figure 10-4. This phenomenon of gene interaction is called **epistasis**. Different types of gene

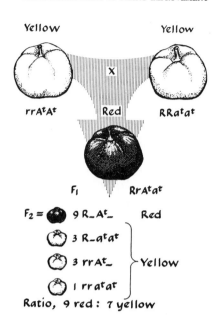

In the tomato, dominant genes R and A^t (apricot) are both required to synthesize lycopene, the red pigment in the fruit.

rr – Depletes all pigment
a^{t}a^t– Depletes lycopene

Genotype

R_ A^t_	Red
R_ a^{t}a^t	Yellow
rr A^t_	Yellow
rr a^{t}a^t	Yellow

(Because of differences in amount of β carotene, which produces yellow color, yellow genotypes can be differentiated by chemical analysis.)

FIGURE 10-4. Epistasis in the tomato.

action produce various F$_2$ ratios as a result of the consolidation of various genotypic classes having similar phenotypes.

Quantitative Inheritance

Characters that are continuous from one extreme to the other, such as size, yield, or quality, are of utmost significance in plant improvement. The inheritance of continuous or **quantitative** characters is an extension of the genetic principles briefly discussed for noncontinuous or **qualitative** traits. It can usually be shown that there are many genes that will each contribute a small increment of effect to modify the quantitative character. The individual effects of quantitative genes (also known as multiple factors, polygenes, and modifiers) are extremely difficult to ascertain because of the large environmental effects present in relation to the small contribution of the individual genes. As a result it is not possible to determine the precise pattern of gene action. The special techniques used to analyze the inheritance of quantitative characters involve mathematical and statistical analysis of the variation through many generations. These techniques make it possible to predict the number of genes differing in a particular cross, as well as their average contribution and average dominance. These methods rely, however, on particular assumptions (for example, the absence of epistasis) that are difficult to prove precisely.

An example of the pattern of quantitative inheritance is shown in Figure 10-5.

A cross between two homozygous lines of corn, one short-eared, and the other long-eared, produced an F_1 population whose mean ear length was between those of the parental lines. The parental and F_1 variation is presumably all environmental. The distribution of the F_2 population includes the homozygous parental types. The F_2 can be shown mathematically to have more variation than either the F_1 or the lines of the parental generation. This increased variation in the F_2 is a result of genetic segregation.

Except for a deficiency of parental types, the data as presented are similar to the data that would be expected from a single partially dominant gene, if we assume enough environmental variation to obscure the differences between the three F_2 classes (AA, Aa, aa). However, the major differences between a single-gene difference and a many-gene difference can be shown by selfing F_2 plants to produce F_3 lines. If we assume a single-gene difference, there will be only three genotypes in the F_2 (the genotypic ratio being $1AA:2Aa:1aa$). Two of these would be true breeding genotypes (AA and aa), duplicating the parental types. The other genotype (Aa) would segregate to produce an F_3 distribution identical to the F_2. Note that each parental type would be duplicated in one-fourth of the F_2 population. It turns out that there are many more than three F_2 genotypes in this F_2!

As the number of genes differentiating the parents with respect to ear length

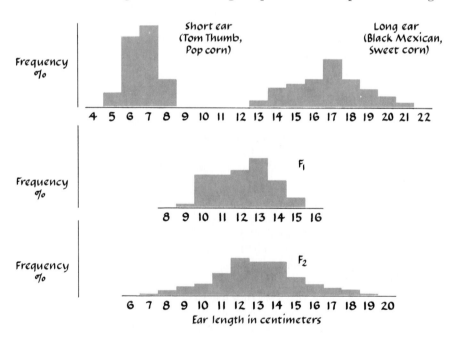

FIGURE 10-5. Distribution of ear length in a popcorn–sweet corn cross. [Adapted from Srb, Owen, and Edgar, *General Genetics*, 2nd ed., W. H. Freeman and Company, copyright © 1965; after Emerson and East.]

increases from one to n, the number of F_2 genotypes increases from 3 to 3^n. On the other hand, the number of genotypes resembling the parental types decreases from $2(\frac{1}{4})$ of the F_2 population to $2(\frac{1}{4})^n$. If we assume that only 10 genes differentiate ear length, there are 3^{10}(or 59,049) different genotypes in the F_2. To recover every genotype, a minimum population of 4^{10}(or 1,048,576) plants, would have to be grown. In this population there would be only one plant duplicating each of the parental genotypes.

The genetic value of a seed-propagated plant is the average performance of its progeny. Thus, the basic problem in the improvement of quantitatively inherited characters lies in distinguishing genetic from environmental effects. The plants that have the best appearance may not have the most desirable genotype. The great economic importance of quantitative characters demands that the breeder of horticultural plants be familiar with the basic principles governing their inheritance.

Linkage

Since plants contain thousands of genes but only a limited number of chromosomes, it is apparent that each chromosome must contain many genes. Genes located on the same chromosome are not randomly assorted but tend to be inherited together. This condition is referred to as **linkage.** There is a relationship between the physical closeness of genes and the intensity or strength of their linkage. The assortment of genes on the same chromosome is related to an actual exchange of chromosome material in meiosis.

By special techniques it is possible to locate genes on their respective chromosomes. Since the genes occur in a linear sequence on the chromosomes, it is possible to determine their relationship to each other by their linkage strengths. In this way the topology of the chromosomes may be mapped, as shown for the tomato in Figure 10-6. The linkage map may be constructed only for genes for which mutant forms are known.

Linkage may occur between genes controlling qualitative and quantitative characters. In the peach there is a particularly undesirable association between small fruit size (a quantitative character controlled by many genes) and the single gene for smooth skin that distinguishes the nectarine from the peach. The breaking of this apparent linkage is a prerequisite for the breeding of large nectarines. Large populations are needed to obtain the rare but desirable combination that produces large nectarines from crosses between large peaches and small nectarines. The smooth-skin gene also appears to be related to susceptibility to brown rot. Another interesting association in the peach is that between rubbery flesh (versus melting flesh) and clingstone (versus the freestone) character. Similarly, the glandless varieties of peaches and nectarines are highly susceptible to mildew. Unless recombinant types are obtained, it is difficult to prove whether such associations are actually genetic linkages or whether they are due to some physiological relationship.

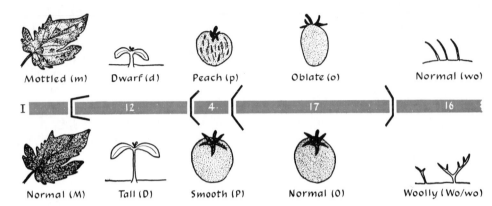

FIGURE 10-6. Map of chromosome 2 (linkage group I) of tomato. [From Butler, *J. Hered.* 43:25–35, 1952.]

Hybrid Vigor

Hybrid vigor, or **heterosis,** refers to the increase in vigor shown by certain crosses as compared to that of either parent. The classic example of hybrid vigor is the mule, an interspecific cross of a mare and a jackass. The effect in plants has been described as similar to that caused by the addition of a balanced fertilizer. Hybrid vigor is often associated with an increased number of parts in **indeterminate** plants (whose main axis remains vegetative and in which flowers form on axillary buds—for example, cucumber) and with increased size of parts in **determinate** plants (whose main axis terminates in a floral bud—for example, sweet corn). In perennial plants the vigor persists year after year.

The term *hybrid* is applied loosely; any organism resulting from genetically dissimilar gametes is technically a hybrid. Thus, a plant that is heterozygous for a single-factor pair is a genetic hybrid. In horticulture and botany the word hybrid is often incorrectly used to refer specifically to the result of crosses made between species. The word in the expression "hybrid corn" refers to a particular combination of inbred lines.

Genetic Consequences of Inbreeding

The phenomenon of heterosis is intimately associated with the decline in vigor brought about in some plants by continued crossing of closely related individuals (**inbreeding**). The genetic consequences of inbreeding are best explained with **selfing** (crossing a bisexual plant with itself), an extreme example of inbreeding.

Selfing has no effect on homozygous loci:

$$AA \times AA \longrightarrow AA,$$
$$aa \times aa \longrightarrow aa.$$

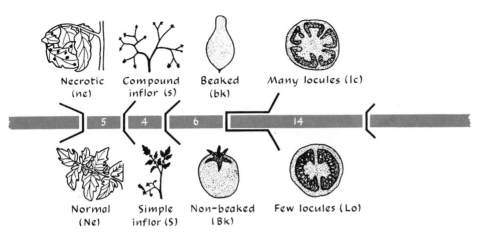

As shown previously, however, the selfing of a plant that is heterozygous for a single gene (Aa) produces progeny of which half are homozygous (AA and aa). This phenomenon may be generalized for any number of genes. With each generation of selfing, 50 percent of the heterozygous genes become homozygous (Fig. 10-7). Continued selfing will ultimately produce plants that are homozygous for all genes. Lines of such plants are known as **inbred** or **pure** lines. Inbreeding can now be defined as any breeding system that increases the homozygous state.

The effect of inbreeding depends on the degree of heterozygosity and on the method of pollination of the plant. Homozygous plants (naturally cross- or self-pollinated) are unaffected by inbreeding. Nevertheless, heterozygous cross-pollinated plants tend to show a loss of vigor upon inbreeding that closely parallels the increase in homozygosity. There are, however, great differences between species and lines. Heterozygous individuals of self-pollinated plants can be obtained by artificial crosses between different homozygous lines. When these heterozygous lines are selfed, the characteristic decline in vigor is usually absent.

The crossing of unrelated inbred lines of cross-pollinated crops restores the vigor lost by inbreeding (Fig. 10-8). The progenies of these crosses are referred to as F_1 hybrids. In maize some F_1 hybrids have shown increases in yield as high as 25% above the yield of the original open-pollinated cultivars from which the inbred lines were derived. This increased vigor is apparently due to the genotypic control, which is made possible by the selection of inbreds. A vigorous, uniform F_1 hybrid line assures an improvement of the F_1 over the segregating open-pollinated cultivars. This lack of variation in F_1 hybrids, although disadvantageous under certain conditions, is usually desirable in an age such as this that values mechanization and prepackaging.

Crossing various cultivars of self-pollinated crops (which in reality are also inbred lines) increases vigor in isolated cases, but the magnitude of the increase is very much less than it is in cross-pollinated crops.

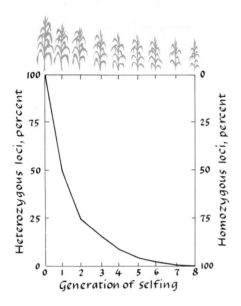

FIGURE 10-7. Selfing reduces the number of heterzygous genes by half in each generation. By the sixth generation, more than 95% of the genes are expected to be homozygous. Most cross-pollinated plants show a decrease in vigor as homozygosity is reached.

Genetic Explanation of Heterosis

Although the incorporation of heterosis is a most significant feature in plant improvement, its genetic basis is not completely established. The genetic explanation must reconcile both inbreeding depression and the apparent stimulation of vigor associated with heterozygosity. The two major theories proposed are now referred to as the **dominance** and **over-dominance** hypotheses.

The dominance hypothesis, first proposed by D. F. Jones in 1917, is based on the assumption that cross-pollinated plants contain a large number of recessive genes concealed by their heterozygous condition. As the recessive condition is largely due to an absence of some essential gene function, these concealed recessives are largely deleterious. Upon inbreeding they become homozygous, resulting in a decline in vigor. Crossing inbred lines from diverse sources restores vigor. The recessives of each inbred are "covered up" by their dominant or partially dominant alleles, carried by the other inbred, much as the wearing of two moth-eaten bathing suits might better protect the modesty of a bather than the wearing of only one. In successful combinations the inbreds mutually complement each others' deficiencies. In its simplest case this situation may be expressed by the cross

$$aaBB \times AAbb \longrightarrow AaBb.$$

The hybrid *AaBb* is more vigorous than either of its homozygous parents because it contains both dominant alleles, whereas each of the parents is deficient either in gene *A* or in gene *B*.

According to the dominance hypothesis, the vigor of a hybrid is due to the large

FIGURE 10-8. Hybrid vigor in the onion from crossing two inbred lines. [Photograph courtesy K. W. Johnson.]

number of dominant genes it contains. The high heterozygosity is merely a concomitant and nonessential part of hybrid vigor. The theory assumes that, were it possible to obtain each heterozygous gene in the homozygous dominant condition, the vigor of this genotype would be as great or greater than that of the heterozygous type. The failure to obtain true-breeding homozygous lines that are as vigorous as F_1 hybrids is due to the great number of genes involved and perhaps to linkages between deleterious and beneficial alleles.

The concept of inbreeding depression lies in the quantitative level of deleterious genes in the population, a direct function of the natural method of pollination and the phenomenon of dominance. The lack of a significant inbreeding depression and of heterosis in naturally self-pollinating crops is due to the absence of any deleterious "hidden" recessives. Any deleterious recessives become quickly exposed by the natural inbreeding mechanism of self-pollinating crops and are eliminated by the forces of natural selection.

The dominance hypothesis has not been an entirely satisfactory explanation to some geneticists. This is due largely to the high estimates of recessive alleles in the open-pollinated population required for the phenomenon and to other mathematical considerations. The overdominance hypothesis assumes that heterozygosity *by itself* may be advantageous. The reduction of vigor upon inbreeding is a direct result of the attainment of homozygosity. This theory assumes that the heterozygous condition *Aa* is superior to either homozygous condition (*AA* or *aa*). This has been termed overdominance. There are two interpretations of overdominance at the gene level. One assumes that both alleles are physiologically active. The combined activity of both alleles in the heterozygous condition produces the heterotic effect. The alternative explanation assumes that the recessive allele is inactive but that dominance is incomplete. However, the reduced amount of essential gene product is optimum; the amount produced by the homozygous types is either deficient or inhibitory. There are experimental verifications, although rare, of both types of gene action.

It is difficult to distinguish between the dominance and overdominance hypotheses. The essential features of the hybrid effect are explained by both. The correct explanation lies in determining primary action of individual quantitative genes. In general, however, the dominance hypothesis is thought to be the most likely explanation of hybrid vigor.

Polyploidy

The condition in which an organism has more than the normal $2n$ number of chromosomes in its somatic cells is referred to as **polyploidy.** Although many plants normally have only two complete sets of chromosomes (the diploid number), other multiples of chromosome sets may occur. Plants containing various numbers of sets are referred to as follows:

Number of chromosome sets	Name of type
1	monoploid (haploid)
2	diploid
3	triploid
4	tetraploid
5	pentaploid
6	hexaploid
7	septaploid
8	octoploid

In addition to the variation of whole sets of chromosomes (**euploidy**) there is variation of the number of chromosomes within a set (**aneuploidy**). An example of aneuploid variation is shown in Figure 10-9.

Many plants can be shown to be naturally polyploid. Some cultivars of apple and pear are triploid; tart cherries are tetraploid; cultivated strawberries are octoploid. In this regard the English convention of referring to the basic number of chromosomes in a set as the x number avoids confusion. Thus, the basic number of chromosomes in the genus *Fragaria*, to which the strawberry belongs, is 7 ($x = 7$). The chromosome number in the vegetative cells of the cultivated strawberry is 56 (8 sets of 7 chromosomes), or $2n = 56 = 8x$. The chromosome numbers in the gametes, the haploid or n number, is $n = 28 = 4x$. The concept of n is halfness, whereas the concept of x is oneness.

Polyploids may occur spontaneously or may be artificially induced by special treatment. For example, the formation of callus tissue by wounding is effective in inducing tetraploidy in the tomato. An alkaloid called **colchicine,** derived from the autumn crocus, has the remarkable property of doubling the chromosome number in a wide variety of plants when applied in concentrations of 0.1–0.3%. This drug interferes with spindle formation in mitosis; the chromosomes divide,

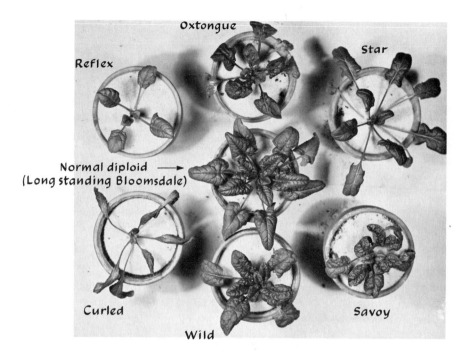

FIGURE 10-9. The six trisomics of spinach. The spinach plant in the center is a diploid having 12 chromosomes (six pairs) in the vegatative cells. The six plants surrounding it all have 13 chromosomes. Each trisomic has a different extra chromosome that confers a characteristic appearance.

but the cell does not. Tetraploids are produced routinely with the use of colchicine. Seed treatment (usually for 24 hours) produces tetraploidy in about 5% of the surviving seedlings. The drug may also be applied to the growing points of seedlings or large plants by a variety of methods.

An artificially induced tetraploid may usually be distinguished from a normal diploid by its larger, thicker leaves and organs (Fig. 10-10), and somewhat slower, coarser growth. Cell size is larger, and fertility is often reduced. Larger pollen size is usually a reliable indicator of tetraploids. Actual chromosome counts are necessary, however, for positive identification. The morphological changes associated with chromosome doubling vary within and between species.

Chromosome Behavior of Polyploids

Tetraploids, unlike diploids, have four chromosomes of each type. The chromosome behavior of tetraploids depends on their pairing relationships at meiosis. If they pair two by two (bivalent pairing) they will separate normally. However, if all four chromosomes pair together (quadrivalent pairing) there is a possibility of a 3–1 distribution leading to gametes with unbalanced chromosome numbers. This

FIGURE 10-10. Increased cluster size as a result of colchicine-induced polyploidy in the 'Portland' grape. *Left*, diploid; *right*, tetraploid. [Photograph courtesy H. Dermen.]

type of chromosome separation is largely responsible for the reduced fertility of tetraploids.

The cross of a tetraploid with a diploid produces a triploid:

$$2x \text{ gamete} + x \text{ gamete} = 3x \text{ zygote.}$$

Chromosome pairing commonly involves the three chromosomes of each type (trivalent pairing). When the chromosomes separate at the first division of meiosis, two chromosomes go to one end of the cell, and the other chromosome goes to the other. Since the assortment of each trivalent is independent, gametes with chromosome numbers anywhere between x and $2x$ may be formed. Consider spinach, for example, in which $x = 6$. In a triploid ($3x = 18$) the possible chromosome assortments at anaphase 1 of meiosis are 6–12, 7–11, 8–10, and 9–9. The seven types of gametes in terms of chromosome number (6,7,8,9,10,11,12) are produced in a binomial distribution and occur in the frequency of $1:6:15:20:15:6:1$. Only $\frac{2}{64}$ of the gametes are x or $2x$. As the chromosome number of plants increases, the frequency of x or $2x$ gametes in triploids becomes very small. Gametophytes having unbalanced numbers of chromosomes are either nonviable or are at a great selective disadvantage. Consequently, triploids are commonly quite sterile. They may be propagated asexually (as are triploid apples and pears) or produced anew each year from tetraploid-diploid crosses (as are seedless watermelons).

Genetics of Polyploids

The single-gene ratios in diploid organisms are based on the assortment of two alleles per gene. In tetraploids, however, four alleles are present at each locus. There are two types of homozygous genotypes (*AAAA* and *aaaa*) and three kinds of heterozygous genotypes (*AAAa, AAaa,* and *Aaaa*). The genetic ratios from crosses involving these types differ from those of diploids. The *AAaa* heterozygote will produce gametes in a ratio of 1*AA*:4*Aa*:1*aa*. If *A* is completely dominant and produces the *A__* phenotype even with three doses of *a* (*Aaaa*), the progeny of *AAaa* selfed will produce a phenotypic ratio of 35*A__*:1*aa*. This ratio, however, is affected by the location of the gene on the chromosome. Thus, tetraploidy tends to muddle the genetic picture. Recessive genes appear hidden, for they are expressed less frequently. As a result of this, genetic analysis of polyploid plants is exceedingly complex.

Use of Polyploid Breeding

Polyploidy is used in a number of important ways in breeding. Although the indiscriminate induction of tetraploidy seldom leads to anything of immediate value, manipulations of ploidy have become a valuable tool in breeding. In some plants the large size associated with higher ploidy is of value; for example, there are now a number of very attractive tetraploid snapdragon cultivars. Most lilies are bred on the tetraploid level because of the thicker petals and larger flowers associated with the increased number of chromosomes.

There has been great interest in the use of polyploidy as a means of extending variability by creating new species or by transferring genes from other species. Induced polyploidy makes it possible to overcome sterility associated with interspecific hybrids. This sterility is a result of the inability of chromosomes to pair at meiosis. Fertility is restored by doubling the chromosome number of the sterile hybrid or by crossing autotetraploids of each species. Using capital letters to designate the chromosome complement (**genome**), we can represent this as follows:

$$AA \times CC \xrightarrow{} AC \text{ (sterile hybrid)} \xrightarrow{\text{chromosome doubling}} AACC \text{ (fertile amphidiploid)}$$

$$AA \xrightarrow{\text{chromosome doubling}} AAAA$$
$$CC \xrightarrow{\text{chromosome doubling}} CCCC$$
$$AAAA \times CCCC \xrightarrow{} AACC \text{ (fertile amphidiploid)}$$

The resulting **amphidiploid** is an allopolyploid made up of the entire somatic complement of each species (Fig. 10-11). Fertility results from bivalent pairing within each genome.

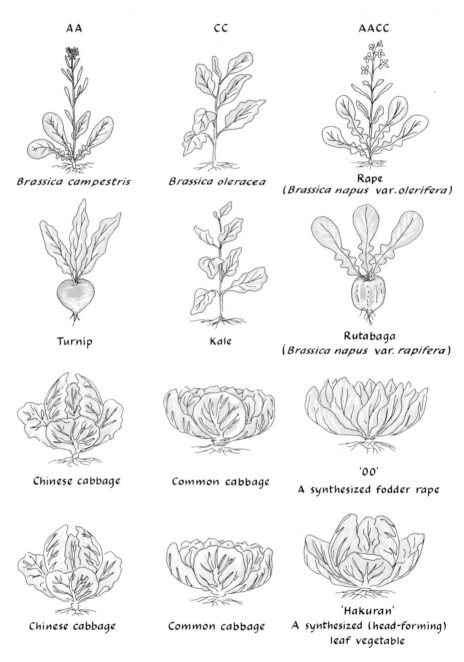

AA

CC

AACC

Brassica campestris

Brassica oleracea

Rape
(*Brassica napus var. olerifera*)

Turnip

Kale

Rutabaga
(*Brassica napus var. rapifera*)

Chinese cabbage

Common cabbage

'00'
A synthesized fodder rape

Chinese cabbage

Common cabbage

'Hakuran'
A synthesized (head-forming)
leaf vegetable

FIGURE 10-11. Natural and artificial amphidiploids in *Brassica*.
[Courtesy H. Kihara, *Seiken Zihô* 20:1–14, 1968.]

In some crops the triploid condition is desirable because of increased vigor (Fig. 10-12) and fruit size (pears, apples) or to take advantage of reduced fertility, such as in the banana or in the seedless watermelon (Fig. 10-13).

Polyploidy has not proved to be a panacea in plant breeding. It is, however, another in the arsenal of weapons the breeder can use to change plants to better

FIGURE 10-12. Diploid, triploid, and tetraploid spinach. In spinach and sugar beets, the triploid condition appears to be the most vigorous.

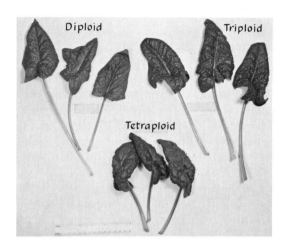

FIGURE 10-13. The seedless watermelon is the result of a cross between a tetraploid and a diploid. The triploid fruit matures, but, as a result of sterility, most of the seeds do not develop normally. The small underdeveloped seeds are similar in texture to seeds of cucumber in the edible stage. [Photograph courtesy Purdue University.]

suit his needs. Polyploid manipulation has greatly enlarged the genetic base from which the breeder may draw. These methods undoubtedly will contribute much to future crop improvement in horticulture.

BREEDING METHODS

Sources of Variation

Genetic variability is the raw material of the plant breeder. The richest source of genetic variability for a particular species has been shown to be its geographical **center of origin** (Fig. 10-14). This is also a **center of diversity** where a pool of genes exists for exploitation by the plant breeder. The incorporation of this genetic variability, when in the form of closely related species, may involve such special techniques as the manipulation of chromosome numbers or the artificial culture of embryos. In the tomato, valuable genes for resistance to fusarium wilt, tomato-leaf mold, and gray leaf spot have come from a related species that grows wild in Peru, *Lycopersicon pimpinellifolium*.

For many cultivated plants a wealth of material is available from varieties already under cultivation. A complete collection of such widely grown plants as the apple, tomato, or rose must be worldwide in scope (Fig. 10-15). The New

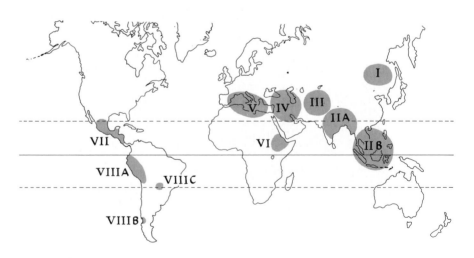

FIGURE 10-14. Centers of origin of important cultivated plants. See key on opposite page.

Crops Research Branch of the Agricultural Research Service (USDA) facilitates the exploration and introduction of genetic variability for the improvement of plants and the establishment of new crops. Societies have been organized around many plants (especially ornamentals), and often assist in the dissemination of plant material. Examples in the United States are the African Violet Society and the Tomato Breeders Cooperative. Breeders are often good sources of plant material.

KEY TO THE CENTERS OF ORIGIN OF IMPORTANT CULTIVATED PLANTS.

Old World
   I. *China*—The mountainous and adjacent lowlands of central and western China represent the earliest and largest center for the origin of cultivated plants and world agriculture. Native plants include millet, buckwheat, soybean and many other legumes, bamboo, crucifers, onion, lettuce, eggplant, various cucurbits, pear, cherry, quince, citrus, persimmon, sugar cane, cinnamon, and tea.

   II. *South Asia*—This area has two subcenters. The principal one, Hindustan (IIA), is considered the center for rice, sugar cane, many legumes, and such tropical fruits as mango, orange, lemon, and tangerine. The other, Indo-Malaya (IIB), is the center for banana, coconut, sugar cane, clove, nutmeg, black pepper, and manila hemp.

   III. *Central Asia*—This region is most important as the center of origin of common wheat. Native plants include pea, broad bean, lentil, hemp, cotton, carrot, radish, garlic, spinach, pistachio, apricot, pear, and apple.

   IV. *Asia Minor*—At least nine species of wheat, as well as rye, are native. This is the center for many subtropical and temperate fruits (cherry, pomegranate, walnut, quince, almond, and fig) and forages (alfalfa, Persian clover, and vetch).

   V. *Mediterranean*—This is the home of the olive and many cultivated vegetables and forages. The effect of early and continuous civilizations is indicated in the improvement of crops that originated in Asian centers.

   VI. *Abyssinia* (Ethiopia)—Wheats and barleys are especially rich; sesame, castor bean, coffee, and okra are indigenous.

New World
   VII. *Central America*—Extremely varied native plants, including maize, bean, squash, chayote, sweetpotato, pepper, upland cotton, prickly pear, papaya, agave, and cacao.

   VIII. *South America*—A number of South American centers are noted. The West-Central area (VIIIA)—now Ecuador, Peru, and Bolivia—is the center of origin of many potato species, tomato, lima bean, pumpkin, pepper, coca, Egyptian cotton, and tobacco. The island of Chiloe in southern Chile (VIIIB) is thought to be the source of the potato. The peanut and the pineapple originated in the semi-arid region of Brazil (VIIIC), and manioc and the rubber tree in the tropical Amazon region.

[After N. V. Vavilov, 1951. *The Origin, Variation, Immunity and Breeding of Cultivated Plants*, translated by K. Starr Chester. New York: The Ronald Press.]

FIGURE 10-15. A collection of English cultivars of tomatoes selected for greenhouse production. [Photograph courtesy Purdue University.]

Sports

Mutations of spontaneous origin, although rare, contribute to genetic variability. These changes are referred to in horticultural terminology as **sports.** Desirable mutations occurring in adapted, asexually propagated plants may result in an immediate improvement such as the color sports in many apple varieties and tree types in coffee plants.

Mutations occurring somatically may involve only a sector of tissues, resulting in a **chimera** (Fig. 10-16). (The original chimera, a mythical beast, was part lion, part goat, and part dragon.) Chimeras, when vegetatively propagated, tend to be unstable. This is because buds may be formed from tissues with or without the mutation. For example, the 'White Sim' carnation, a sport of the 'Red Sim' cultivar, is a chimera. Buds that form from internal tissues will produce red flowers. Upon close inspection, the normal 'White Sim' blooms also show small islands of red tissue. It is important to know if a sported plant is stable before undertaking extensive propagation.

From a study of the performance of various types of natural variants under different environmental conditions, it may be possible to make an immediate improvement merely by selecting the best individuals. However, with established crops it becomes more and more difficult to make improvements in this manner.

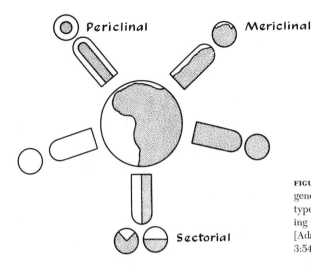

Periclinal

Mericlinal

Sectorial

FIGURE 10-16. Plant chimeras contain genetically different tissues. Various types may be derived from buds arising from a sectored chimeral stem. [Adapted from Jones, *Bot. Rev.* 3:545–562, 1937.]

Usually, the individual desirable characters are present in many plants, no one of which contains all the desirable attributes, in which case it becomes necessary to attempt to recombine the characters by hybridization (sexual crossing) in the hope of obtaining plants having as many of the desirable features as possible. When these characters are spread over many cultivars and lines, the process necessitates making many crosses.

The desired characters are too often found only in plants that may be horticulturally unsatisfactory from other standpoints. Resistance to the apple scab disease is found in species of apple with almost inedible fruit less than 1 inch in diameter. In addition, there are a great many characters to be considered. For example, a successful strawberry cultivar must satisfy (1) the growers, with respect to yield, season of ripening, and disease resistance; (2) the consumers, with respect to appearance and quality; (3) the nurserymen producing the plants; and (4) the shippers, handlers, and processors.

The incorporation of these various characteristics, most of which are quantitatively determined, involves many series of crosses carried over many generations. With each generation the amount of plant material tends to pyramid into unmanageable proportions unless rigid selection is maintained. Breeding becomes a problem in biological engineering. The main effort of the plant breeder is directed toward the recombination of desired characters in as efficient a manner as possible in terms of land, time, and labor.

Artificial Creation of Variability

One of the limitations to plant improvement is the dependence upon naturally occurring variation. The artificial induction of mutation by radiation (X-rays, gamma rays, thermal neutrons) and by chemical means (ethyl dimethyl sulfate) provides a method for creating changes that have not occurred naturally. Since such induced changes are completely random and are largely harmful, refined techniques for screening are needed. For example, large populations of plants can easily be screened in the seedling stage for resistance to a disease if the susceptible plants died soon after inoculation. A promising application of irradiation may lie in its use in obtaining beneficial changes in asexually propagated material. In this way, otherwise adapted material may be screened in the search for a single desirable change in the hope of eliminating further generations of breeding and extensive testing.

Genetic Structure of Crop Cultivars

The method of pollination has a profound effect on the genetic constitution of the plant. The amount of cross-pollination varies from essentially none in plants such as the garden pea to 100% in dioecious and self-incompatible plants. Two main groups are recognized; naturally **self-pollinated** plants, in which cross pollination is less than 4%, and naturally **cross-pollinated** plants, in which cross pollination exceeds 40%. The intermediate types are usually considered along with the cross-pollinated group.

Self-pollinated plants are ordinarily homozygous for practically all genes. The exceptions are a result of chance cross-pollination and mutation. However, any heterozygosity is quickly eliminated as a consequence of natural inbreeding. Thus, although a hybrid between two divergent types will be heterozygous for many genes, by the F_6 to F_{10} generations of natural selfing, the population will consist of many different individuals, each of which are for all practical purposes completely homozygous. The basic problem in the improvement of self-pollinated plants lies in selecting the best genotype from the unlimited number of different genotypes. Once a satisfactory homozygous plant is selected the problem of genetic maintenance is small as compared to that of cross-pollinated plants.

The genes in naturally cross-pollinated, seed-propagated plants are recombined constantly from generation to generation. A cross-pollinated cultivar is typified not by any one plant but by a population of plants. The problem of improving cross-pollinated plants consists in somehow maintaining visible uniformity while avoiding the decline in vigor associated with homozygosity. In general, procedures that result in severe inbreeding cause a loss of vigor. Once a desirable population is achieved there is still the perpetual problem of maintenance. A unique method of producing uniformity and at the same time maintaining heterozygosity has been the process of producing F_1 hybrids. At present, however, technical difficulties prevent this breeding method from being used with all cross-pollinated crops.

A large number of cross-pollinated crops (including apple, rose, and gladiolus) are normally vegetatively propagated. Here improvement depends on the selection of a single desirable genotype. The problem of genetic maintenance is solved by the elimination of the sexual process. In improving these plants by recombination, inbreeding must also be avoided. Hybridization between unrelated plants is usually made in order to obtain a vigorous population within which selection may be practiced.

Selection

The problem of the breeder confronted with a population containing many diverse genotypes is to recognize and save only the most desirable types. This process of selection differs depending on the method of reproduction of the plant.

Selection in Self-pollinated Crops

Two fundamentally different types of populations of self-pollinated crops exist. One is a mixture of different homozygous lines as found in a collection of cultivars. Here selection consists of determining the best genotype by testing. As each cultivar is homozygous, the problem of genetic maintenance is eliminated. The best genotype can be expected to be duplicated from its selfed seed. The other type of population is a mixture of different heterozygous genotypes, as found in the F_2 population of a cross between different homozygous varieties. As discussed previously, some completely homozygous types are theoretically expected, but their number greatly decreases as the number of genes differentiating the original cross increases. The problem of improvement is now twofold: to select the best genotype and to transform this genotype as closely as possible into a homozygous line.

The genetic value of a self-pollinated, seed-propagated plant is the average performance of its progeny. The task of selecting the best genotype from an F_2 population is difficult because of the problem of distinguishing between genetic and environmental variation. Selection of the most desirable genotypes is accomplished on the basis of progeny performance. This process, which is known as **pedigree selection,** is based on the assumption that the best homozygous genotype will be derived from the heterozygous plant that produces the most desirable progeny.

Assume that selfed seed from a number of F_2 plants are planted out to produce F_3 lines. Since each selfing brings about homozygosity in a 50% increase each generation, the variability within F_3 lines will be half as great as the variability between F_2 plants. That is, the plants within a particular F_3 line will resemble each other more so than will the aggregate of all F_3 lines. For example, some lines will be uniformly large and some uniformly small. By choosing between a number of lines rather than choosing a single F_2 plant there is less chance of confusing genetic and environmental effects.

The selection process may then be repeated. What appear to be the best plants in the best F_3 lines are planted out, and selection is made again on a "pedigree" basis between F_4 lines. By the sixth to tenth generation the lines derived from single plant selections will be homozygous for more than 95% of their genes. Such lines are, for practical purposes, considered to be true breeding. If one of these lines is of superior type, it may be "named," and it is then considered a new horticultural cultivar.

The problem of straight pedigree selection is the extreme expense it entails. Fairly extensive records must be maintained, and unless rigid selection is maintained the program soon "mushrooms" to extremely unwieldy proportions.

An alternate method of selection is known as **mass selection.** In this technique, what appear to be the best plants are selected and maintained in bulk without testing the progenies separately. This may be accomplished by eliminating (**roguing out**) the undesirable plants in each generation and harvesting the remaining plants. This can be done mechanically for some crops, such as by screening for seed size or harvesting at a particular date to select for a particular season of ripening. After 6–10 generations, the population will consist of a heterogeneous mixture of "somewhat selected" homozygous genotypes. The progeny of any plant can be expected to form the basis of a new cultivar. The problem now is to determine the best genotype by testing. This method will be successful if the better genotypes are retained in the mass selection process either by judicious selection or merely by virtue of large numbers.

The processes of mass selection and pedigree selection can be combined. For example, F_2 plants can be "pedigree selected" on the basis of F_3 line performance. Thus the greatest genetic differences are exploited early, and the obviously undesirable types are eliminated. The best F_3 lines may then be bulked and carried on by mass selection until homozygosity is reached, at which time "pedigree selection" is resumed.

Selection in Cross-pollinated Crops

Pedigree selection depends on genotype evaluation by inbreeding. Straight pedigree selection is undesirable as a method of improving cross-pollinated plants, in which inbreeding leads to loss of vigor, unless some procedure is set up to combine inbred lines and restore vigor. Mass selection enables a cross-breeding population to become relatively uniform for certain visible characters and to conserve enough variability to maintain vigor. Inbreeding is avoided by natural interpollination.

Mass selection often leads to a rapid improvement of cross-pollinated plants. However, it sometimes becomes difficult to increase this "genetic gain" after a certain point. To obtain more control of the genotypes that make up the cross-pollinating population, pedigree and mass selection may be combined. This is accomplished by selfing individual plants and pedigree-selecting them on the basis of their progeny. The selected lines are then allowed to interpollinate to restore heterozygosity. The process may be repeated for a number of cycles. There are several variations of this procedure. When the character evaluated is a fruit or a seed, selection must be made after pollination. In this case the "female" parent is pedigree-selected, and the "male" parent is mass-selected.

In a very real sense, F_1 hybrids are an extreme form of this method. Here the inbreeding process is carried on to homozygosity, and vigor is restored by combining two inbreds (or four in a double cross). The selection of the inbred combination results in uniform, genetically controlled hybrids.

Selection of Asexually Propagated Crops

In asexually propagated plants, selection is straightforward, since any genotype may be perpetuated intact. The problem is one of testing—that is, of determining the best genotype. If the most desirable selection is still unsatisfactory and further improvement is necessary, the selection of the best genotype to be used for further crossing is more difficult because the best-performing selection does not always make the best parent. A sample of genotypes must be selected and tested as parents on the basis of their progeny performance.

Utilization of Hybrid Vigor

The production of F_1 hybrids as a source of seed-propagated, cross-pollinated plants depends upon a number of factors. The first requirement is that it must be possible to obtain and maintain inbred lines. Some inbred lines become so weak that they are difficult to maintain. Inbreeding itself may be very difficult with some plants.

A number of crops resist self-fertilization because of incompatibility mechanisms. Self-incompatibility is usually controlled by alleles at a single gene locus. When pollen contains the same incompatibility allele that is present in the style, tube growth is arrested. In a few crops, such as cabbage, incompatibility may be circumvented by pollinating before the flower opens (bud pollination). Homozygous lines may be produced in a single step, avoiding the long process of inbreeding. Monoploids are produced in certain crops, such as corn, in a very low but predictable frequency. They apparently result from a disruption in the double fertilization process. One male gamete combines with the two polar nuclei to form the triploid endosperm, but the fertilization of the egg nucleus does not occur. Monoploids may be recognized in the seedling stage by the use of special marker genes. Many of these monoploid lines double spontaneously to produce "instant" homozygous diploid types.

The second requirement for producing hybrids is that it must be possible to make an efficient cross between the inbred lines. This may be exceedingly difficult in perfect-flowered plants, unless each hand pollination yields a great number of seeds. One method used in overcoming this is the incorporation of **male sterility,** which transforms one inbred into a "female" line (Fig. 10-17). Another technique is the use of the self-incompatibility mechanism. If this method is used, special breeding techniques are required to perpetuate the male-sterile line.

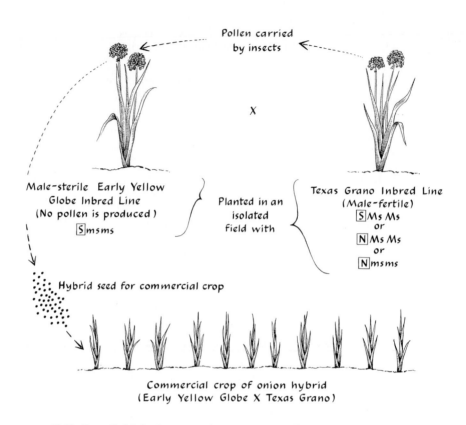

Pollen carried
by insects

X

Male-sterile Early Yellow
Globe Inbred Line
(No pollen is produced)
$\boxed{S}msms$

Planted in an
isolated
field with

Texas Grano Inbred Line
(Male-fertile)
$\boxed{S}Ms\,Ms$
or
$\boxed{N}Ms\,Ms$
or
$\boxed{N}msms$

Hybrid seed for commercial crop

Commercial crop of onion hybrid
(Early Yellow Globe X Texas Grano)

FIGURE 10-17. Controlled hybridization in the onion. Genetically determined male sterility transforms one inbred to a female line. Male sterility in onions results from interaction between a
cytoplasmic factor $\boxed{s}$ and recessive nuclear gene *ms* in the homozygous state. Plants with $\boxed{N}$
(normal) cytoplasm or with gene *Ms*, either homozygous or heterozygous, produce viable pollen.
[Adapted from Srb, Owen, and Edgar, *General Genetics*, 2nd ed., W. H. Freeman and Company, copyright © 1965.]

The third requirement is an economic one. The improvement must be great
enough to warrant the extra expense of hybrid seed.

The process of producing F_1 hybrids by the inbreeding process has not proved
practical in cross-pollinated plants that are vegetatively propagated, although it
is often suggested as a breeding method. Each clonal variety is essentially a
hybrid, which, if duplicated asexually, need not be duplicated by crossing two
unique inbred lines. The great advantage of F_1 hybrids in cross-pollinated plants is
their *uniformly* high vigor. However, the process of vegetative propagation
assures absolute genetic uniformity.

Hybrid seed is now commercially available for the following vegetables:
broccoli, Brussels sprouts, cabbage, carrot, cauliflower, Chinese cabbage, cucumber, eggplant, muskmelon, onion, pepper, spinach, squash, sweet corn, radish,
tomato, and turnip. F_1 hybrid flowers include *Ageratum*, begonia, *Calceolaria*,

FIGURE 10-18. 'Fresno,' 'Tioga,' and 'Torrey' are three strawberry cultivars obtained from crossing 'Lassen' with a California selection '42.8-16.' [Photograph courtesy Royce S. Bringhurst and Victor Voth.]

cyclamen, geranium (*Pelargonium*), gloxinia, Iceland poppy, *Impatiens,* marigold (*Tagetes*), pansy, petunia, African violet, *Salpiglossis,* snapdragon, and zinnia. Vigorous hybrid genotypes can be selected from segregating populations instead of creating them from inbred lines (Fig. 10-18).

As heterosis appears to be involved to a lesser extent in self-pollinated crops, the creation of F_1 hybrids (actually cultivar crosses) is not usually an important breeding method for these plants. However, since cultivar crosses offer a quick means of producing a particular combination, this method has found a limited place in some self-pollinated crops, such as greenhouse tomatoes.

Although the maximum amount of heterosis is obtained by crossing two inbreds, a number of other less heterotic combinations may be made. The various kinds of crosses that are referred to as hybrids in the trade are designated as follows.

Type of cross	Name of hybrid
inbred × inbred	single cross
F_1 hybrid × inbred	three-way cross
F_1 hybrid × F_1 hybrid	double cross
inbred × cultivar	top cross

Since many inbreds are relatively weak, the double cross is an attempt to produce seed on a vigorous plant. It has been used very widely in field corn (Fig. 10-19). The top cross has been used in spinach, and the three-way cross is popular in sweet corn.

The genetic improvement of inbreds is a specialized type of breeding. Basically, they are treated as self-pollinated plants. However, the success of an inbred is often difficult to determine phenotypically but is related to its success as a parent.

DETASSELING SYSTEM

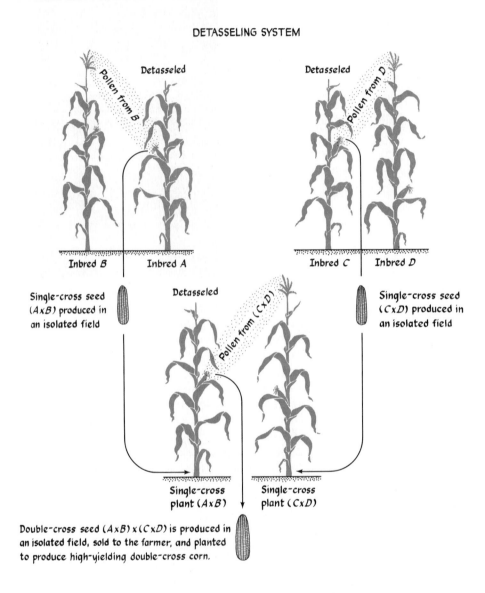

Single-cross seed
(A x B) produced in
an isolated field

Single-cross seed
(C x D) produced in
an isolated field

Double-cross seed (A x B) x (C x D) is produced in
an isolated field, sold to the farmer, and planted
to produce high-yielding double-cross corn.

FIGURE 10-19. The production of single-cross and double-cross hybrid corn. The detasseling process (above) can be eliminated by manipulating male sterility (opposite page) determined by an interaction between a cytoplasmic factor $\boxed{\text{s}}$ and a double recessive nuclear gene *rf*. Only the combination $\boxed{\text{s}}$ *rf rf* is male-sterile. Plants with $\boxed{\text{N}}$ (normal) cytoplasm or with the pollen restoring allele *Rf*, either homozygous or heterozygous, produce viable pollen. The cytoplasmic

CYTOPLASMIC MALE-STERILE SYSTEM

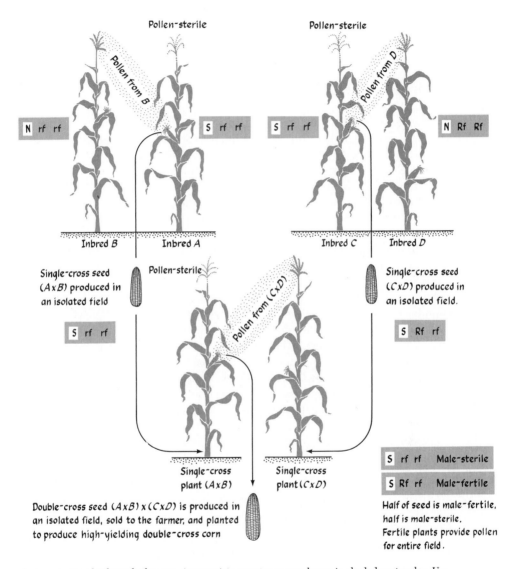

Pollen-sterile

Pollen-sterile

Pollen from B

Pollen from D

N rf rf

S rf rf

S rf rf

N Rf Rf

Inbred B   Inbred A

Inbred C   Inbred D

Single-cross seed
(A×B) produced in
an isolated field

Pollen-sterile

Pollen from (C×D)

Single-cross seed
(C×D) produced in
an isolated field.

S rf rf

S Rf rf

Single-cross
plant (A×B)

Single-cross
plant (C×D)

| S rf rf | Male-sterile |
| S Rf rf | Male-fertile |

Double-cross seed (A×B) × (C×D) is produced in
an isolated field, sold to the farmer, and planted
to produce high-yielding double-cross corn

Half of seed is male-fertile,
half is male-sterile.
Fertile plants provide pollen
for entire field.

factor passes only through the egg. Appropriate genotypes are shown in shaded rectangles. Unfortunately an association of the common male-sterile cytoplasm (T-cytoplasm, from Texas) and susceptibility to Southern corn blight caused by *Helminthosporium maydis* has almost eliminated the commercial use of the cytoplasmic male-sterile system after the disastrous epidemic of 1970. [From Janick et al., Plant Science, 2nd ed., W. H. Freeman and Company, copyright © 1974.]

This ability of an inbred to produce good hybrids is known as **combining ability.** Combining ability is inherited, but the only way to select for it is to test it directly. Combining ability may be selected for early in the inbreeding process by crossing the plant to a tester at the same time it is selfed. The selfed seed is set aside for further inbreeding until after the results of the crossing trial are evaluated.

The Backcross Method

The backcross has been defined as a cross of an F_1 plant with one of its parents. The backcross method is a breeding technique for transferring single characters, readily identifiable, from one cultivar to another. Characters controlled by single genes, such as certain types of disease resistance, male sterility, and so on, are most easily transferred by this method. By repeated backcrossing of the hybrid to the parent cultivar that carries the most desirable characters (**the recurrent parent**), but selecting in each generation for a single character from the other parent (**the nonrecurrent parent**), a genotype will be eventually obtained that has all the genes of the recurrent parent except those affecting the selected character from the nonrecurrent parent. In the following explanatory discussion, assume that the "cultivars" are of self-pollinated plants.

As regards the consequences of the backcross method, two sets of genes must be considered. One set is small, consisting only of the genes from the nonrecurrent parent that are to be transferred. The other set consists of a large group of genes from the recurrent parent, which we do not wish to lose. The genes of the recurrent parent can be shown to be transferred by an increment of 50% in each generation of backcrossing. By selecting plants at random in each generation, we can expect the genes of the recurrent parent not only to be incorporated in the hybrid but also to become homozygous. This can be shown easily with respect to a single gene.

Assume a cross involving a single gene pair, $AA \times aa \longrightarrow Aa$. The two types of backcrosses are

$$Aa \times AA \longrightarrow 1Aa:1AA,$$
$$Aa \times aa \longrightarrow 1Aa:1aa.$$

Compare these backcrosses with selfing:

$$Aa \times Aa \longrightarrow 1AA:2Aa:1aa.$$

Note that, in backcrossing and selfing, half of the genes become homozygous. This can be generalized for any number of genes. The rate of return to homozygosity in backcrossing is the same as in selfing, but *all of the homozygous genes resemble the recurrent parent.* If backcrossing to the same homozygous type is continued for 6–10 generations, more than 95% of genes of the hybrid will be identical to those of the recurrent parent and will be in the homozygous condition.

If the gene to be transferred is dominant, the procedure is straightforward. After the last backcross, the gene may be made homozygous by selfing. If the character to be transferred is recessive, special testing must be carried on in each generation to be sure the plant selected for backcrossing is heterozygous and thus contains the desired recessive.

The backcross method has found its greatest usefulness in self-pollinated crops and in improving inbred lines of cross-pollinated plants. It is not as useful in cross-pollinated crops because backcrossing is equivalent to selfing in achieving homozygosity. This may be overcome, however, by using a large number of selections of the recurrent line. In this way, inbreeding may be avoided (Fig. 10-20).

The main advantage of the backcross method is its predictability. The improved cultivar is often indistinguishable from the old one with the exception of the added character. As a result, extensive testing may be avoided. But the backcross method cannot be expected to improve the cultivar any more than the addition of the single selected character. Moreover, unless this character is inherited relatively simply, the method is difficult to use.

Tissue-Culture Methods

Recently, great excitement has been generated about the possibility of using tissue-culture techniques for crop improvement. As we have seen in this chapter, crop improvement is based largely on the creation of variable populations from which individual genotypes are isolated. In traditional plant-breeding systems, genetic variation in the population is stored and released through gene assortment and recombination brought about by meiosis and fertilization. Because most crop improvement is based on the sexual cycle, genetic progress has been limited by the way the sexual cycle can be controlled and manipulated by the plant breeder.

It now appears that the sexual system for releasing variability may be short-circuited by imposing selection on populations of cells rather than on populations of whole plants. For the technique to be successful, it must be possible to regenerate plants from individual cells. This has been achieved with a great number of horticultural plants (see pp. 372–377). The advantage of this system is that extremely large populations of cells (literally in the billions) can be grown in a very small space, such as a few culture tubes or flasks. The problem is to be able to select for variants at the cellular level that will be expressed as desirable characters in the growing plant.

Variation in cell populations occurs naturally as a result of spontaneous mutation, but it can also be induced by means of various mutagenic agents. Selection can be achieved by exposing cell or tissue cultures to environmental stresses, such as extremes in heat or cold, or to high concentrations of pathogenic toxins or herbicides: susceptible cells die while only resistant ones survive to grow and reproduce. Many research studies are now under way to test the practicality and usefulness of this system.

A

B

C

D

E

F

FIGURE 10-20. Backcross bleeding for scab resistance in the apple. A, the fruits of *Malus flori-bunda* '821,' which is heterozygous for a dominant gene that confers resistance to apple scab. B, selection derived from cross of *M. floribunda* '821' × 'Rome Beauty.' Subsequent backcrosses to different large-fruited types (C and D) produce resistant seedlings with increased fruit size and quality. The fruits in the basket (E) are the 'Prima' apple, the first release of the scab-resistant apple breeding program cooperatively conducted by Purdue University, Rutgers University, and the University of Illinois. 'Prima' has been successfully introduced in the United States as an early apple both for home gardens and for commercial use. Its use is expanding in Eastern and Western Europe, with the most extensive plantings in Yugoslavia. F, resistant seedlings are selected in the seedling stage. Susceptible seedlings are killed by the disease.

Tissue culture offers the possibility of restructuring the genetic organization of an organism through special techniques developed by molecular biologists. This so-called "genetic engineering," which includes cell hybridization, organelle transfer, genetic transformation, and haploid production, offers potentially great opportunities for horticulturists.

Plant-cell hybridization has been shown to be feasible with protoplasts. Protoplasts can be isolated from plant cells by digesting cell walls with appropriate enzymes. A small but significant percentage of protoplasts grown in suspension culture show genetic fusion (Fig. 10-21). This means that, if protoplasts of two genetic backgrounds are mixed, some of the fused protoplasts will be genetic "hybrids." One method to isolate hybrid protoplasts from those that are not fused, and from those that are fused with protoplasts of the same genotype, is to use a medium in which only hybrid products will be able to grow. Mature plant hybrids have been produced from cellular hybridization in tobacco and petunia. The refinement of this process may make it possible to achieve wide crosses that are not possible with traditional methods of sexual hybridization. Various organelles that are capable of regeneration with their own DNA (such as chloroplasts) may also be transferred from one species to another in this way.

Gene transformation has been achieved in plants by extracting DNA from one plant and incorporating it into another through the use of bacterial viruses (bacteriophages) as transfer vectors. This technique is now only experimental, but it opens up the possibility of transferring to plant cells genes that have been created by isolating messenger RNA and producing DNA transcripts (gene copies).

Finally, pollen of various plants can be cultured to produce haploid cells, or even haploid "embryoids." Haploid cell lines of plants are extremely valuable for mutation breeding because, since dominance is eliminated, all mutated recessive genes will be immediately expressed. Further, doubling such haploids produces "instant" homozygous lines, a process that may require many years with ordinary sexual inbreeding. Haploid cells could also be used to produce "hybrids" directly by protoplast fusion.

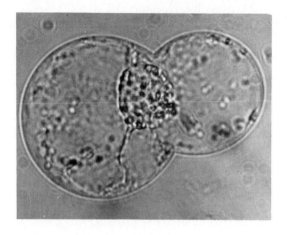

FIGURE 10-21. Initial stage of fusion between two protoplasts of tobacco. [From T. Murashige, *HortScience* 12:127–130, 1977.]

Controlled Hybridization

Plant breeding is concerned largely with the control of the sexual process in plants. Before two plants can be crossed they must be induced to flower simultaneously. Because this may not always occur naturally, various environmental factors, such as photoperiod, temperature, and nutrition, may have to be manipulated to achieve synchronization of flowering. In some varieties of sweetpotato, flowering can be induced only with difficulty; therefore, they must be grafted to species of morning glory and then subjected to specialized photoperiods and temperatures.

The storage and shipping of pollen facilitates artificial hybridization and, in some cases, saves many years. Pollen is quite variable in its ability to remain viable. For example, the pollen of cucurbits remains viable only for about three hours. Apple pollen stored under low humidity and cool temperatures (30–36°F, or −1–2°C) may retain its viability for more than two years. A new technique of freeze-dehydration is a promising method for pollen storage.

The basic procedure in making artificial crosses involves the following practices: (1) avoidance of contamination before artificial pollination, (2) application of the pollen, and (3) protection of the pollinated flower from subsequent contamination.

In perfect-flowered plants contamination by selfing is avoided by the process of **emasculation.** This consists in removing the anthers before they have begun to shed pollen. It is done before the flower opens, and the technique varies with the flower structure (Fig. 10-22). In monoecious plants (those with separate pistillate and staminate flowers) the pistillate flowers must be protected before pollination from wind or insect pollination by suitable protective devices, such as paper bags, glassine bags, or cheesecloth nets. In insect-pollinated plants the petals are often removed to discourage insect visitation.

The pollen may be applied by one of several methods. Often the anthers are

FIGURE 10-22. Preparing a rose bud for cross-pollination. The petals are cut at the base (*left*) and at the tip (*center*). Pollen is being supplied to an emasculated flower with a camel-hair brush (*right*). [Photographs courtesy USDA.]

collected the day before pollination and dried. The pollen is then applied with a camel's-hair brush, the fingers, a blackened matchstick, or the dried flower itself, used as a brush. The pollen should be applied in large amounts. In the greenhouse, the pollination of dioecious plants, such as spinach, may be controlled by isolating the pistillate and staminate plants to be crossed. The pollination of onions is carried out by enclosing their flower heads in mesh cages containing blow flies (Fig. 10-23).

After pollination the flower may need to be protected from contamination by cross-pollination. This is usually accomplished by enclosing the flower in a paper or glassine bag (Fig. 10-24). If the petals have been left, as in peas, they may by themselves serve as the protective device with the aid of cellophane tape. When it is only necessary to keep away insects, cheesecloth is usually sufficient. Care must be used in selecting protective devices that do not act as heat traps and thus prevent successful fruit set. In citrus and in apples, insects will not visit the flowers if the petals are removed, thus protection of emasculated flowers is not necessary.

FIGURE 10-23. Pollen transfer in cross of male-fertile × male-sterile onions is achieved by caged blow flies. [Photograph courtesy K. W. Johnson.]

FIGURE 10-24. A piece of ordinary soda straw protects pistil of daylily from contamination before and after pollination. [Photograph courtesy G. M. Fosler.]

Maintaining Genetic Improvement

Once a genetic gain has been achieved, vigilance is still required to maintain improvement. Mutation, natural crossing, contamination, and, in vegetatively propagated material, diseases, especially those caused by viruses, tend to cause the deterioration or "running out" of cultivars. Genetic deterioration may be reduced to a minimum by continued selection and careful propagation. In an effort to control the purity of seed many semipublic organizations and seed associations have been formed. In order to protect their reputations, seed companies and nurseries must strive constantly to keep their stocks free of contamination.

Control of Cultivars

To encourage improvement of plants, United States patent laws protect the right of the originator to control the increase of new forms of certain vegetatively propagated materials (potatoes are an exception). Only the right of vegetative propagation is protected by patent; the right to propagate by seed is not. Many cultivars of gladiolus, rose, apple, peach, and pear are patented. In this way, the breeder, whether a private individual, a group, or a public agency, may justly profit from his discoveries, just as the inventor or chemist may be rewarded by his invention, discovery, or new process.

Although seed-propagated crops cannot be patented, they are now protected in the United States by the Plant Variety Protection Act, which became law in December 1970. (Six processing crops were excluded from this act: tomato, green pepper, celery, carrot, okra, and cucumber.) The establishment of "Breeder's Rights" is expected to have important implications for the seed industry.

It has always been possible to protect seed-propagated crops indirectly through copyright of the name, but this did not prevent renaming. However, seed-propagated hybrids can be biologically controlled by the breeder. For example, control of the unique sets of inbreds protects the hybrid even more effectively than does legal control. In hybrids between cultivars of self-pollinated plants, such as tomato, merely keeping secret the names of the two cultivars that are used to make the hybrid from the hundreds available effectively controls its production.

Selected References

Allard, R. W., 1960. *Principles of Plant Breeding*. New York: Wiley. (An outstanding presentation of the genetic foundation of plant breeding.)

Crane, M. B., and W. H. C. Lawrence, 1947. *The Genetics of Garden Plants*. London: Macmillan. (An important compilation.)

Ferwerda, F., and F. Wit (editors), 1969. *Outlines of Perennial Crop Breeding in the Tropics* (Miscellaneous Papers 4, Landbouwhogeschool). Wageningen: H. Veenman. (A review of tropical tree breeding including many tropical fruits.)

Frey, F. J. (editor), 1966. *Plant Breeding*. Ames, Iowa: Iowa State University Press. (A collection of articles on plant breeding with the emphasis on general methods rather than specific crops.)

Janick, J., and J. N. Moore (editors), 1975. *Advances in Fruit Breeding*. West Lafayette, Ind.: Purdue University Press. (A monograph on fruit breeding by crop, each chapter written by a distinguished fruit breeder. Crops covered include apples, pears, strawberries, brambles, grapes, blueberries and cranberries, currants and gooseberries, minor temperate fruits, peaches, plums, cherries, apricots, almonds, pecans and hickories, walnuts, filberts, chestnuts, citrus, avocados, and figs.)

Reinert, J., and Y. P. S. Bajaj (editors), 1977. *Applied and Fundamental Aspects of Plant Cell, Tissue and Organ Culture*. New York: Springer-Verlag. (A collection of chapters by various authorities. Includes discussion of the uses of tissue culture for genetic engineering and crop improvement.)

U.S. Department of Agriculture, 1937. *Better Plants and Animals II* (USDA Yearbook, 1937). (A valuable reference for horticultural plant improvement. Together with the 1936 yearbook, this work is an historical landmark in genetics and breeding of agricultural crops in the United States.)

11

Marketing

I cannot do't without compters. Let me see what am I to buy for our sheep-shearing feast? Three pound of sugar, five pound of currants, rice—what will this sister of mine do with rice? But my father hath made her mistress of the feast, and she lays it on. She hath made me four and twenty nosegays for the shearers (three-man song-men all, and very good ones), but they are most of them means and bases; but one Puritan amongst them, and he sings psalms to hornpipes. I must have saffron to color the warden pies; mace; dates, none—that's out of my note; nutmegs, seven; a race or two of ginger, but that I may beg; four pounds of pruins, and as many of raisins o' th' sun.

SHAKESPEARE, *The Winter's Tale* [IV. 3]

HORTICULTURE AFTER HARVEST

The path of horticultural products from the growing plant to the consumer is complex. **Marketing** may be defined broadly as the activities that direct the flow of goods from producer to consumer—that is, the operations and transactions involved in their movement, storage, processing, and distribution. These operations have been developed into a highly specialized technology. Many production operations are a direct part of marketing. When the grower selects a particular cultivar to plant, he is very often making a marketing decision. Production decisions that determine ultimate quality, such as methods of disease control, may ultimately become as much a part of marketing as decisions to store or to sell, or to prepack or to bulk ship. Many production and marketing operations are connected and interrelated. For example, the harvesting, grading, and packing of lettuce is combined in one field operation.

The unique aspects of marketing of horticultural products are due to their perishable nature. The horticultural product is still alive after harvest, but subject to deterioration. The ultimate quality of the product for the consumer, which is what determines its real economic value, hinges upon how it is treated after harvest.

418

MARKETING FUNCTIONS

Marketing increases the value of horticultural products through the application of **marketing functions.** These functions have been categorized as exchange, physical, and facilitating. The **exchange functions** are those activities related to the transfer of title to goods—that is, buying and selling. Buying in its broadest sense includes the seeking out and assembling of sources of supply. This may involve many transactions and many people in various marketing operations. Similarly, selling includes not only the transfer of goods; it also includes merchandising (packaging, advertising, and promotion). **Physical functions** include storage, transportation, and processing. **Facilitating functions,** which make possible the orderly performance of the exchange and physical functions, include grading, financing, risk bearing, and communication (market information). The total cost of the marketing functions, what is known as the **marketing margin,** may greatly exceed the entire cost of production (Fig. 11-1). These are the charges of the much maligned middlemen of agriculture.

This chapter is largely devoted to a discussion of technological problems associated with marketing operations. The detailed structure and economic ramifications of marketing, however, are beyond the scope of this book.

HARVESTING

The close relationship of harvesting to subsequent marketing operations makes its discussion a proper introduction to marketing technology. Harvesting is one of the crucial features in the horticultural operation. It is often the costliest part of production, and its timing has a direct bearing on the final quality of the product. For example, maturity has a great influence on the subsequent storage behavior of many crops. Technological advances in harvesting have become an outstanding example of the current trend of substituting capital for labor in horticultural operations.

Predicting Harvest Dates

A number of factors must be considered with regard to the timing of harvesting: maintaining orderly production operations that make possible the maximum utilization of equipment and labor; setting up an orderly marketing sequence; and the ultimate quality and appearance of the product. For many crops, harvesting must proceed within a certain narrow time interval, for quality is a fleeting and elusive factor. Often there are a number of conflicting factors to be considered. For example, in apple harvest, storage quality is adversely affected by delaying maturity, yet, on the other hand, red color tends to increase with time. Determining the optimum time of harvest is not a simple, straightforward decision.

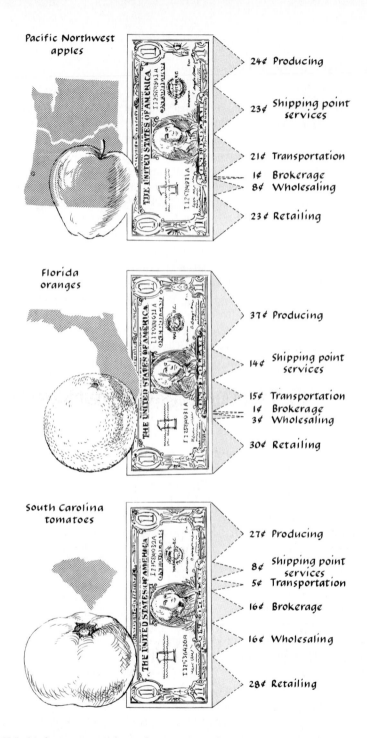

Pacific Northwest apples

24¢ Producing
23¢ Shipping point services
21¢ Transportation
1¢ Brokerage
8¢ Wholesaling
23¢ Retailing

Florida oranges

37¢ Producing
14¢ Shipping point services
15¢ Transportation
1¢ Brokerage
3¢ Wholesaling
30¢ Retailing

South Carolina tomatoes

27¢ Producing
8¢ Shipping point services
5¢ Transportation
16¢ Brokerage
16¢ Wholesaling
28¢ Retailing

FIGURE 11-1. Marketing margins for apples, oranges, and tomatoes. Note that two-thirds to three-quarters of the consumer's dollar is spent for marketing. [Data from USDA.]

The factors affecting the harvest date of a crop depend upon the genetic nature of the cultivar, its planting or blooming date, and the environmental factors that exist during the growing season. Successive harvest dates of sweet corn may be obtained by planting cultivars with different maturity dates. Similarly, in tree crops, there are different cultivars that mature at different dates, and planting a well-chosen assortment of such cultivars can make for an orderly succession of crops throughout the season. In the past, tremendous market gluts of peaches have been produced as a result of extensive plantings of one cultivar, 'Elberta.' This has been alleviated by breeding programs that have produced groups of cultivars that ripen in an orderly sequence over almost a three-month period.

The harvest dates of annual crops may be altered by changing the sequence of planting dates. However, a delay in planting date does not necessarily cause the same delay in harvest date. The time required to reach the harvestable stage varies widely from crop to crop and is based largely on the crop's temperature requirements for growth. When temperatures are below the minimum required for growth, a long delay in planting may not affect the harvest date.

In some crops the time required to reach the harvestable stage may be expressed with temperature-time values called **heat units** by calculating time in relation to temperatures above a certain minimum. For example, if the minimum temperature for growth of a particular crop is 50°F (10°C), then a day with an average temperature of 68°F (20°C) would provide 18 degree-days F (10 degree-days C) of heat units. A day with an average temperature of 41°F (5°C) would provide 0 degree-days of heat units. The harvest date can be ascertained by an accounting of accumulated heat units. In Wisconsin it has been shown that 1200–1250 degree-days F are required for the 'Alaska' cultivar of pea to mature if planted in the early spring. Under increasing spring temperatures it takes longer to accumulate heat units in the early part of the season than later. Assuming that all temperatures above a minimum have similar effects on growth, there would be a decreasing interval between planting and harvest dates as the season progresses.

The heat-unit system has a number of limitations. For example, soil temperatures more accurately indicate early growth than do air temperatures. Differences in day–night temperature shifts, day-length effects, and the differential effects of temperature on various stages of plant growth also affect the results. In addition, temperatures above a minimum may not have a similar effect on growth, but, within limits, may act exponentially, approximately doubling many physiological processes with every 18°F (10°C) rise in temperature. The precise determination of harvest date by the accumulation of temperature data depends upon a knowledge of the general climate of an area and upon experience with harvest dates based upon planting dates for each crop.

For many long-season fruit crops, such as apples, the most reliable index for calculating maturity is the number of days from flowering. This value, determined from an analysis of previous data, is surprisingly consistent. It would indicate that other environmental factors besides temperature are important in determining fruit maturity.

A number of physiological criteria are used to determine harvest date. In pears the proper harvest date for each cultivar is determined by a combination of criteria, among which are pressure testing of the fruit, ground color of the skin, seed color, and percentage of soluble solids (basically sugar) in the flesh. In some crops, other criteria are also utilized. These include the formation of abscission zones separating fruit from stem (muskmelons), the visible stage of development (such as degree of bud tightness in roses), color (tomatoes), sugar–acid balance (blueberries), and even the way the fruit sounds when thumped (watermelons). A combination of methods is often used to determine the correct degree of maturity.

Harvesting Methods

Hand harvesting is still the only practical method for many high value crops that are either sensitive to bruising or that must be selectively picked (strawberries and apples, for example). The mechanization of harvesting has proceeded in stages. Thus, we may speak of completely mechanized, semimechanized, and non-mechanized harvesting. For example, in Arizona muskmelons are harvested by hand, but they are then placed directly on conveyors, where they are graded, and they are packed in the field on moving machines. Similar operations are used for cabbage, peppers, cauliflower, and broccoli.

The harvesting of many crops is completely mechanized, although it is done in a number of separate steps. For example, when grown for processing, pumpkins and squash are raked into windrows in one operation and are loaded by conveyor in another.

For other crops, harvesting is completely mechanized, and all operations are performed by a single machine in a single step. Thus, potatoes are harvested by a combine that digs, removes vines, and loads tubers. A somewhat similar operation is used to harvest radishes, beets, carrots, and floral bulb crops. Asparagus, pickling cucumbers, snap beans (Fig. 11-2), sweet corn (Fig. 11-3), and tomatoes are mechanically harvested; canning peas are now picked, vined, and shelled in the field.

An evolutionary pattern is evident in the development of mechanical harvesters and associated growing systems. For example, although late-maturing large-vined cultivars of vegetable crops were planted once in wide rows and hand harvested several times, early-maturing dwarf cultivars are grown now in close spaced rows for a single destructive harvest. Similarly, harvesting methods proceed from hand harvesting, to mechanical aids for hand harvesting (conveyors), to destructive machine harvest following hand harvest, to a single destructive machine harvest. When machine harvesting is first implemented, the crop is usually collected in the same small containers previously used for hand harvesting; sometimes packing and sorting operations are an integral part of the harvest. In more advanced operations the crop is handled in bulk and transported to a central location for washing, grading, and packing or processing.

FIGURE 11-2. Mechanical harvest of snap beans in Wisconsin. *Top*, each picker harvests 8–10 acres of snap beans per day. *Bottom left*, a pallet box containing about 700 pounds of snap beans being lowered to the ground. *Bottom right*, tractors with hydraulic lifts loading pallets onto a truck, which is dispatched to the canning plant. [Photographs courtesy National Canner's Association.]

FIGURE 11-3. Mechanical harvest of potato (*top*) and sweet corn (*bottom*). [Photograph at bottom by J. C. Allen & Son; photograph at top courtesy Purdue University.]

Fruit crops present a somewhat different problem in that the plant itself must not be destroyed. Cranberries were the first fruits to be harvested by machine; the plants are essentially combed to remove the berries. Equipment has been developed for shaking plums, clingstone peaches, blueberries, and nuts onto canvas sheets. Mechanical harvestors for grape, raspberry, blackberry, and strawberry are also commercially available.

In general, machine harvesting must proceed hand in hand with breeding to develop machine-compatible plants. Such plants must usually have fruits that are highly uniform, that ripen all at the same time, and that are firm enough to resist excessive bruising. When fruit-harvesting machines are perfected, however, they often do a more satisfactory job than hand harvesting. Machine-harvested potatoes show a decrease in defects due to bruising. In many vegetable crops, machine harvesting is facilitated by the planting of F_1 hybrids, which tend to be more uniform in size and shape.

Postharvest Alterations

The desirability, or quality, of horticultural products may be determined by many different things, depending on the product. For processing tomatoes, desirability implies freedom from cracking and good internal red color. In the 'Delicious' apple, however, desirability connotes a particular flavor, shape, and pattern of external red color. Desirability of nursery stock refers to viability, form, and size, plus the proper proportion and packaging of roots and soil.

For the sake of precision, the general term **quality** is best divorced from the meanings suggested by the terms **condition** and **appearance.** In food crops the word quality is best used with reference to palatability. This implies a pleasing combination of flavor and texture; flavor resulting from taste and smell, and texture perceived as "mouth feel." Quality can also properly be used to describe nonedible products such as seed or nursery stock. Used in this sense, the term refers to the physiological state of the material, connoting high viability and trueness to type. **Condition** is the presence of, or freedom from, disease, injury, or physiological disorders. Although we associate good condition with high quality, this is not always the case. **Appearance** refers to the visible attributes of the product. It includes color, conformation, and size. Unfortunately, appearance is not always a reliable index of quality.

A number of physiological and biochemical processes occur in the harvested, nonprocessed horticultural product that contribute to change (Table 11-1). This may result in the deterioration of quality, condition, and appearance in some crops, and in the improvement of quality and appearance in those crops that complete ripening after harvest. For most commodities the objective is to maintain the product as close to harvest condition as possible. The product must be maintained in the living state because death causes irreversible biochemical changes. These may involve gross deterioration and drastic differences in flavor, texture, and appearance.

TABLE 11-1. Changes that occur in harvested produce.

Change	Process	Examples and significance
Water loss	Transpiration; evaporation	Unattractive appearance, texture changes, weight loss, shriveling
Carbohydrate conversion	Enzymatic	Starch to sugar: detrimental in potatoes, beneficial in bananas and pears
		Sugar to starch: detrimental in sweet corn and most edible crops
Change in flavor	Enzymatic	Usually detrimental, but beneficial in persimmons, pears, and bananas
Softening	Pectic enzymes	Usually detrimental
	Water loss	Beneficial in pears, bananas
Change in color	Pigment synthesis or destruction	May be detrimental or beneficial
Toughening	Fiber development	Detrimental in celery
Change in vitamin content	Enzymatic	May be gain (vitamin A) or loss (vitamin C)
Sprouting, rooting, or elongation	Growth and development	Detrimental in potatoes, onions, and asparagus
Decay and rot	Pathological; physiological	Detrimental

GRADING

The inherent variability of horticultural products at harvest and their differences in value make it necessary to grade them according to some objective standard. Grading is the basis of long-distance trade. It permits the description of products in terms that are understandable both to buyer and to seller. Without a system of grading, all products would have to be individually inspected. Grading thus adds tangible value to horticultural products.

Grading has two distinct functions. The first is to eliminate completely all obviously unsatisfactory items. This is extremely important in packaging, inasmuch as diseases spread rapidly in the enclosed environments afforded by packages. Furthermore, one poor item visibly detracts, out of all proportion to its bulk, from the appearance of a large sample. The second function of grading is consolidation. Products may be consolidated by cultivar, size, appearance, defects, and, where possible, quality. Because of inherent differences, certain crops (for example, apple and pear) are grouped and identified by cultivar. Within a cultivar, size is the most obvious gradable factor (Fig. 11-4). Various size characteristics may be used: stem length (roses), stem diameter (trees), spread (shrubs), weight (watermelons), and diameter (most fruits). Grading for appearance may be based on the absence of defects, conformation, and the amount and intensity of

FIGURE 11-4. This fruit grader sorts fruit by weight. The fiber cartons are tray-packed. [Photograph courtesy FMC Corp.]

color. Various techniques have been used for the objective evaluation of palatability, including determinations of oil content (avocados), firmness (tomatoes), sugar–acid ratio (oranges), specific gravity (potatoes), and various maturity criteria.

One of the factors that determines the quality grade of a processed product is the quality of the raw product. Raw product grades for processing must take into account particular demands of the industry. Some factors assume greater importance, whereas others assume less. For example, apple skin color is of small importance, since the skins are removed in processing, but large fruit size is extremely important because of the limitations of mechanical peelers. There are a number of quality features that have special importance to processing—for example, total solids (tomatoes), tenderness (peas and corn), and sugar content (grapes for wine). Defects, texture, and internal color become very important.

Inspection and grading is backed by Federal law. The basic piece of legislation relating to inspection and grading is Title II of Public Law 733, known as the Agricultural Marketing Act of 1946, which directs the Secretary of Agriculture

> to develop and improve standards of quality. . . . To inspect, certify, and identify the class, quality, quantity, and condition of agricultural products when shipped or received in interstate commerce, under such rules and regulations as the Secretary of Agriculture may prescribe . . . to the end that agricultural products may be marketed, to the best advantage, that trading may be facilitated, and that consumers may be able to obtain the quality product which they desire. . . .

Most federal standards for horticultural products are permissive; that is, they are officially recommended but are not compulsory. But if federal grades are used to describe a product, they must be complied with. Mandatory grades do exist for some crops under certain conditions, however, as in apples and pears intended for export. Most of the grade standards for fruits and vegetables are designed for wholesale trading, and are not directly carried over into retail trading.

Federal grading regulations are complemented by state laws. This has created confusion and disorder where state regulations differ widely from each other and from federal standards. The inspection of fresh fruits and vegetables at shipping points is accomplished through the combined efforts of federal and state agencies.

MARKET PREPARATION

Most horticultural crops require special preparation after harvest. They are usually cleaned, trimmed, or specially treated in some manner. Root crops must be cleaned to remove adhering soil and debris. Potatoes grown in muck soil are at a serious disadvantage unless the extremely fine black soil is removed. Washing or brushing of fruit is done to enhance its appearance. The outer leaves of lettuce and cabbage are routinely removed, as are the tops of beets and carrots. Leaves of many florist crops, such as chrysanthemums and snapdragons, are stripped by the grower at harvest. Strawberry plants are "cleaned" by removing the dead leaves and runners.

Many products are waxed to prevent water loss and to improve appearance (turnips, citrus fruits, and cucumbers). Waxing is now a standard practice for dormant nursery stock, especially rose plants. Bananas are dipped in solutions of copper sulfate to control storage and shipping rots.

Conserving Life of Cut Flowers

> When I have pluck'd thy rose, I cannot give it vital growth again,
> It needs must wither. I'll smell it on the tree.
>
> SHAKESPEARE, *Othello* [V. 2]

The life of cut flowers is markedly shorter than that of the same flower left on the growing plant. The reasons for the short "vase life" of most flowers is not completely understood, but water stress accentuated by vascular blockage, the high respiratory activity of petals and the subsequent depletion of metabolites, and the extreme sensitivity of flowers to ethylene are key factors.

The traditional way to conserve the life of cut flowers has been through refrigeration, which slows down metabolism and respiration and minimizes transpiration. Unfortunately, the vase life of flowers after cold storage may be very brief, and in many cases refrigerated flowers do not undergo normal development, especially if harvested too early. Because of the high perishability of flowers and their high value, much effort has been expended recently to find ways to maximize cut-flower longevity by means of postharvest treatments.

Pulsing

The use of immature flowers or unopened buds greatly facilitates packing and shipment, but they must retain the ability to open normally. Pulsing (or "loading") refers to short-term treatment of cut flowers with sugar solutions before shipment. This treatment makes it possible to ship immature gladiolus, carnation, and chrysanthemum flowers and have them open normally. Typically, the stems of the cut flowers are immersed in solutions of sucrose (5–40%) for about 20 hours. The sugar accumulates in the cells of the petals, enhancing their water-absorbing capacity and allowing them, therefore, to maintain their turgidity even under subsequent water stress. The increased sugar content of pulsed flowers also allows for the maintenance of the high respiratory activity of the petal cells.

Improving Water Relations

Various treatments that maximize water uptake and reduce excess transpiration will extend the vase life of cut flowers. Maximum water uptake may be increased by sharp, diagonal cuts (to avoid crushing) of the stem each time plants are removed from the holding solutions; by using warm solutions (110°F, or about 43°C) when the stems are first immersed; and by the avoidance of microbial buildup in solutions through the use of microbicides or acidifying agents. Diagonal cuts increase the surface area of the stem and prevent the cut end from resting on the bottom of the container. The warm water has a lower viscosity, which increases its uptake, and it is better able to dissolve air bubbles inside the water-conducting tissues. Water loss may be minimized by removing excessive foliage, by maintaining high humidity in shipping containers, and by avoiding high temperatures and excessive air movement.

Vascular blockage has been identified as a major cause of reduced water uptake in cut flowers. Cut roses often die prematurely due to "bent neck," or water deficiency exacerbated by vascular blockage. The precise cause of vascular blockage is unclear. Because it seems to be increased by bacterial toxins, bactericides such as HQC (8-hydroxyquinoline citrate) are added to cut-flower solutions to reduce the disorder. A combination of HQC and sugar is the basis of most conserving solutions.

Elimination of Contaminants

Flowers in general are extremely sensitive to various gaseous toxins and water pollutants. Flowers are exquisitely sensitive to ethylene, for example, and should not be stored with fruits or vegetables that release it. A consumer who stores a floral bouquet in a refrigerator containing fruit may reduce instead of lengthen the life of the flowers. Recently, silver ions have been shown to counteract the effect of ethylene, and silver nitrate is therefore often added to cut-flower solutions. The use of filters to remove ethylene and other contaminants from the air in flower-storage areas can increase floral life considerably. Water quality is also important: water high in salts, fluorides, and other pollutants must be avoided when preparing cut-flower solutions.

Components of Conserving Solutions

For prolonging the vase life of flowers, growers, wholesalers, retailers, and consumers can use cut-flower solutions. These solutions increase the availability of energy, reduce microbial build-up and vascular blockage, and provide for maximum water uptake by providing the flowers with increased water-absorbing capacity. The development of cut-flower solutions is a relatively recent innovation in horticulture. Although there are numerous formulations and many physiological differences between species, the following are classes of ingredients common to most solutions.

ENERGY SOURCE. Energy may be supplied to the flowers in the form of a metabolizable sugar, such as sucrose. Most solutions contain 1–5% sugar. Another effect of sugar is to improve the flower's ability to withstand water stress by increasing the water-absorbing capacity (osmotic potential) of its petal cells.

MICROBICIDE. HQC at 200–600 ppm is commonly added to most keeping solutions. Sodium benzoate at 100 ppm may also be used as a microbicide, and citric acid at 75–200 ppm may be used to impart antibacterial properties to the solution by lowering its pH.

ANTI-ETHYLENE AGENT. Silver ions, which help to protect flowers against the effects of ethylene, may be supplied by adding silver nitrate to the solution at the rate of 25–50 ppm.

WATER. Water of high quality has been shown to be essential to prolonging the vase life of cut flowers. Total dissolved solids of more than 100 ppm, or fluoride levels in excess of 3 or 4 ppm, will shorten flower life. For this reason, distilled or deionized water may be better for cut flowers than tap water.

Curing

Curing is a postharvest treatment used to prolong the storage life of certain commodities, such as potatoes, sweetpotatoes, and bulb crops. Curing involves exposing the product to particular temperature or humidity conditions. Potatoes are cured at 55–60°F (about 13–16°C) during the first two or three weeks of storage. The humidity is kept very high during this period to prevent shrinkage due to the loss of water. Because the tubers are dormant, they do not sprout. Curing brings about the healing of cuts and bruises incurred at harvest. Two physiological processes are involved: **suberization,** the deposition of a waxy material by the tuber, which produces a fairly effective barrier to the loss of moisture; and **active cell division** of the periderm, which produces a new "hide." After the curing period, temperatures are lowered for prolonged storage to prevent sprouting. A similar treatment is undertaken in the processing of sweet-potatoes for storage, although higher curing temperatures are used (around 80°F, or 27°C).

Bulb curing involves the removal of water from the outer scales and neck. This was formerly done in the field, but now the trend is toward the use of artificial drying in storage. Proper curing prevents rot by eliminating from the surface of the bulb those environmental factors favorable to the growth of harmful micro-organisms. Unlike the curing of potatoes, the curing of bulbs involves no active cell division.

Packaging

Packaging affords protection, convenience, economy, and appeal. The changes in packaging that have occurred in this century have had a great impact on the horticultural industry. Before World War I, a tremendous variety of wooden containers was used (barrels, boxes, crates, and baskets of various shapes and dimensions). By World War II, a certain amount of standardization had been achieved, and the wooden bushel basket and box became the principal packages used for a great variety of vegetables and fruit. These containers originated as a grower pack, but they were also used by the retailer. The wooden bushel basket has proved unsuitable for many crops. It is bulky, relatively heavy, and costly. It is also expensive to fill, since it must be faced. Nor is its shape particularly suitable for packing: stacking them wastes space and badly bruises the contents, especially the top layer. Furthermore, the bushel is no longer a convenient retail package, since its use necessitates completely repacking the contents into the refrigerated display cases now commonly in use.

The materials used for packing differ with the product. Wooden packages are now being used for only a few products, such as melons. Light-weight fiber cartons have replaced wood for most packages (Fig. 11-5); some are specially coated to give them wet strength. Plastics, paper, and cloth are now commonly used as packaging materials.

The term **prepackaging** has been used for the process of putting produce in a consumer unit package at some point before it is put on display in the retail store. The rise of the self-service supermarkets has irrevocably altered packaging of horticultural commodities, although perishable commodities were the last to be

FIGURE 11-5. Different kinds of packages for apples. [Photograph courtesy Purdue University.]

prepackaged. By 1970, almost one-half of all fruits and vegetables were prepackaged before they reached the retail store.

The prepackaging of fruit and vegetables has been made possible by the creation of satisfactory films. Polyethylene has proved to be extremely versatile. It is strong, transparent, moisture-resistant, and can be made permeable for gas exchange. The natural appeal of products in transparent bags with a minimum of printing has proved to be a selling attraction. At present, carrots and radishes are nearly always prepackaged. Apples are commonly retailed in five-pound "poly" bags packaged in fiber cartons. Polyethylene films have had a great influence on the marketing of perennial plants by eliminating the need for heavy moisture carriers such as peat. In general, specialized films are being created to suit individual products. Other materials such as cardboard, paper, cloth, and plastics have proven useful for particular commodities.

The prepackaging of produce has put a premium on uniformity. This has eliminated customer handling and has prevented much in-store damage. Prepackaging has not in any way eliminated refrigeration, though it has in many cases increased shelf life by improving moisture retention. In general, prepackaging has improved quality levels.

The question of who should do the prepackaging has not yet been settled. There is controversy over whether prepackaging is best handled at the shipping point or at the distribution point. Although different crops have been handled in different ways, there is a trend with many commodities for grower packaging. As producers have increased in size, associations of buyers have dealt directly with growers and have specified package requirements. The producer, if large enough, often finds it profitable to assume the packaging operations.

PRECOOLING

Precooling refers to the rapid removal of heat from freshly harvested fruits and vegetables in order to slow ripening and reduce deterioration prior to storage or shipment. The rate of deterioration depends on many factors: temperature, the natural respiration rate of the crop, the moisture content, the presence of natural protective barriers to water loss, and the presence of decay organisms. The major effect of precooling consists in reducing the respiration rate. Precooling also slows deterioration and rot by retarding the growth of decay organisms; and it reduces wilting and shriveling, since transpiration and evaporation occur more slowly at low temperatures. The internal temperature of a horticultural product (such as a peach) harvested on a hot day may be $20°F$ ($11°C$) higher than the air temperature. The removal of field heat, to reduce the temperature of the harvested product to $32–40°F$ (roughly $0–4°C$), must be as rapid as possible; consequently, a great deal of energy is required. Hence, the harvesting of many perishable crops is now done at night or early in the morning to avoid excessive field heat. With the field heat removed, considerably less energy is required to maintain low temperatures, inasmuch as the respiration rate at temperatures of $32–40°F$ is relatively slight. The special techniques developed to precool vegetables and fruits are **contact icing, hydrocooling, vacuum cooling,** and **air cooling.**

Contact icing refers to the use of crushed ice placed in or on the package to effect cooling. The major advantage of ice is that it does not dry out the food in cooling it. Another advantage is that produce may be shipped immediately after treatment, since cooling takes place in transit. Although icing requires relatively small outlays of special equipment, a large weight of ice must be shipped. Furthermore, after the ice has melted, the package is left only partly full ("slack pack"). When lettuce was precooled by icing, a packed crate would contain 60 lb of lettuce and 30 lb of ice. The use of ice spread mechanically over the produce after loading (**top icing**) has eliminated the slack pack, but cooling is not as efficient. Contact icing is now being replaced by hydrocooling and vacuum cooling.

Hydrocooling is the cooling of fruits and vegetables with water (Fig. 11-6). The water (usually iced) flows through the packed containers and absorbs heat directly from the produce. The high efficiency of the system is due to the large heat capacity and high rate of heat transfer of water. The time required is related to the thickness of the individual product, as well as the internal temperature at the beginning of the cooling process, in relation to the desired temperature drop. Fungicides, such as calcium hypochlorite at concentrations of 50–100 ppm, are used to prevent the spread of decay organisms. One of the chief advantages of

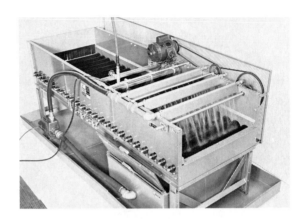

FIGURE 11-6. A hydrocooler in which the fruit may be brushed as it is cooled. [Photograph courtesy FMC Corp.]

hydrocooling is that it prevents the loss of moisture during the process. The crops most commonly hydrocooled are peaches, sweet corn, and celery. The system requires application to large volumes of produce in order to operate efficiently.

Vacuum cooling utilizes the rapid evaporation (actually boiling) of water at reduced pressures to effect cooling. At roughly 0.089 lb/in^2 (which is $\frac{1}{165}$ of the normal atmospheric pressure of 14.7 lb/in^2) water will boil at 32°F (0°C). This rapid evaporation takes up heat directly from the crop. The crop must have a large surface area in proportion to its volume to cool quickly without excessive drying of the surface layers. Lettuce and other leafy crops are ideally suited to vacuum cooling (Fig. 11-7). Sweet corn may be vacuum cooled if it is moistened beforehand. The expensive equipment required for vacuum cooling makes it feasible only for large growers or organizations, although portable vacuum coolers are in use.

Air cooling, the basis of most refrigeration systems, depends upon the movement of cold air. For rapid cooling, special equipment is needed to effect the high-volume air circulation required. The air must be as cold as possible, but it must not freeze the produce. To prevent drying, produce must be removed from the air blast as soon as it is cooled. Solanaceous fruit crops, beans, and berries can be efficiently air cooled. Leafy crops, however, cannot be satisfactorily precooled by this method because of desiccation.

TRANSPORTATION

Marketing depends upon short-haul movement and handling by the growers and upon large-scale, long-haul transportation facilities. One of the great revolutions in horticultural marketing has been the bulk handling of produce. Great savings are made possible by the synchronization of harvesting and hauling operations to minimize handling. The harvesting container has increased in size to conform to

FIGURE 11-7. Lettuce harvested in the Salinas Valley of California is picked directly into cartons (*top*), vacuum cooled (*middle*), and transferred in pallets (*bottom*). [Photograph courtesy T. W. Whitaker, USDA.]

machine power rather than muscle power. For example, the bushel field crate is being replaced as the harvesting unit. Automatic potato and onion harvesters unload directly into truckbeds. In hand-picking operations, as in those for apple and pear, the pickers unload into 30-bushel pallet boxes, which may be stored directly or shipped to processing plants, where they are automatically unloaded (Fig. 11-8). Industrial management techniques in packing and storage layout have been utilized to facilitate bulk handling. This has made centralized grading,

FIGURE 11-8. An automatic pallet unloader in a canning factory. [Photograph courtesy Gerber Products Co.]

storage, and packing operations particularly efficient. Thus, a large portion of the fruit production in the Northwest is handled through the centralized facilities of grower organizations.

Long-haul shipment of horticultural products is effected principally by truck, railroad, and boat. Special railroad cars have been constructed to transport potatoes in bulk directly from the harvesting operation. This has materially decreased bruising, an important cause of trouble in potato storage. Air shipment of certain high-value perishables, such as flowers and strawberries, is increasing. Long-haul transportation contributes storage as well as movement, and it necessitates specialized packaging, packing arrangements, and environmental controls for the reduction of in-transit spoilage.

STORAGE AND PRESERVATION

The demand for most horticultural products is continuous. Their seasonal production and their rapid deterioration after harvest make preservation and storage essential in order to insure an extended supply. Extending the supply of any horticultural commodity requires retarding the natural physiological deterioration that occurs inherently in living systems as well as preventing decay by microorganisms. What method to use depends on the product, its use, and the required storage time. The imposed limitation is that the product be acceptable after storage. If stored plant parts are to be used for reproduction (seeds, bulbs, whole plants), the overriding factor is viability. For fresh fruits, vegetables, and flowers, the maintenance of acceptable quality is dependent upon preserving the natural, living state, although reproductive viability is not necessary. For the extended storage of certain food crops, a number of processes have been developed that stop the life functions but maintain edible quality, even though the product may be materially altered from its harvest condition. Changes in form brought about by food processing may result in new products that effectively extend the utility of the original product, rather than simply preserving it. Changing tomatoes to ketchup is an example of this.

Storage of Perishables

The storage of perishable plant products in their natural state may be accomplished by means of environmental control. The means of prolonging the life of a product are the slowing of respiration in order to retard microbial activity and the prevention of excessive water loss. Respiration and microbial activity may be regulated by the regulation of temperature and the control of oxygen and carbon dioxide levels. Water loss may be prevented by controlling humidity.

The storage life of produce depends not only on storage conditions but on the natural rate of respiration. This in turn depends upon the species, the plant part, its maturity level, and its degree of dormancy. Bruising and decay must be considered because of their adverse pathological and physiological effects (including increased respiration and discoloration).

Temperature Control

For best results the storage temperature must be held constant, since the optimum temperature range for many products is narrow. For example, most apple cultivars keep best and longest when held at a constant temperature of 30–32°F (-1–0°C). Temperature increases of as little as 5°F (3°C) will hasten ripening in proportion to the duration of the increase.

The evolution of heat by respiration must be considered as a factor in storage-temperature control. The heat evolved may be calculated from the respiration rate, which increases with temperature (Table 11-2). The respiration rate of produce varies considerably. The respiration rate of spinach is high enough that, even if it were stored at 40°F (4.4°C), its temperature could increase to more than 100°F (38°C) in 5 days if no heat were allowed to escape. This can be a real problem in shipment if shipping packages are not adequately ventilated.

The lower limits of storage temperature must be determined individually for each crop. In general, most products suffer some injury when frozen, although the degree of injury differs with the product. Some commodities are severely damaged by even slight freezing, whereas others less susceptible may undergo a number of successive freezes and thaws without perceptible permanent injury. Storage temperatures are usually kept above the point at which the commodity freezes, although strawberry plants are often stored frozen.

TABLE 11-2. The relation between temperature and heat evolved due to increasing respiration.

Temperature		Heat (in Btu) evolved per ton of fruit per 24 hours	
(°F)	(°C)	Lemons	Strawberries
32	0.0	480–900	2730–3800
40	4.4	620–1890	3610–6750
50	10.0	1610–3670	7480–13,090
60	15.6	2310–4950	15,640–20,280
70	21.1	4050–5570	22,510–30,160
80	26.7	4530–5490	37,220–64,440

SOURCE: *USDA Agricultural Handbook 66.*

In many commodities, injury can occur at temperatures considerably above freezing. This injury is known as **chilling injury** in contrast to **freezing injury.** Chilling injury interferes with the ripening sequence in tomatoes and bananas. Chilling injury affects the appearance of produce, and it may result in actual breakdown of tissues.

Oxygen and Carbon Dioxide Levels

In addition to being temperature dependent, respiration is also directly affected by the oxygen and carbon dioxide levels. The atmosphere normally contains about 78% nitrogen, 21% oxygen, 0.03% carbon dioxide, and small percentages of several inert gases. Since respiration is an oxidation process, a reduction in the amount of oxygen reduces the respiration rate. Although slight variations in the amount of carbon dioxide show little effect on respiration, high concentrations may inhibit or prevent respiration and may also act to inhibit the ethylene production that stimulates ripening (as will be discussed in a subsequent section on organic volatiles).

The effect of oxygen and carbon dioxide levels on respiration is utilized in the storage of fruit. Carbon dioxide may be increased in storages by carbon dioxide generators. However, in a gas-tight room filled with freshly harvested "living" apples, the respiration of the fruit will consume the oxygen and at the same time give up equal concentrations of carbon dioxide. When the oxygen is exhausted the carbon dioxide level will have reached the original level of the oxygen—that is, 21%. At this point anaerobic respiration begins, and alcohol is formed in the fruit. However, if fresh air is introduced before this occurs, the amounts of oxygen and carbon dioxide may be kept at compensating levels. The carbon dioxide concentration may be further decreased by passing storage air through a "lime scrubber," usually freshly hydrated lime. This may eliminate the detrimental effect of the high concentration of carbon dioxide observed with some apple cultivars.

If respiration rate is reduced through the control of oxygen and carbon dioxide levels, then storage temperatures may be kept higher than normally required. In the "controlled-atmosphere" storage of apples, storage temperatures may be maintained at 37–45°F (3–7°C) rather than 30–32°F (−1 to 0°C), eliminating disorders of certain kinds associated with low temperatures.

The principle of modified atmosphere may be utilized by using sealed film liners that are differentially permeable to carbon dioxide and oxygen. This creates a microenvironment one bushel in size (Fig. 11-9). Polyethylene film is about five times more permeable to carbon dioxide than to oxygen. When apples are placed in sealed polyethylene film liners, the oxygen in the liner is reduced and the carbon dioxide is increased as a result of respiration. The final concentration depends on the storage temperature and the permeability of the film. In general, increasing the film thickness decreases permeability, which results in lower oxygen and higher carbon dioxide concentrations. Film density also influences permeability; for example, a film of density 0.928 transmits oxygen and carbon dioxide at one half the rate of film of density 0.910.

Film liners have made possible an improved method of storing some cultivars of apple. 'Golden Delicious' shows excellent results, although 'Grimes Golden' does very poorly. The high humidity inside the liner is an asset, but temperature must still be closely controlled.

The principle of a modified atmosphere is now extensively used in commercial packaging of fruits, both for home consumption and for shipping by rail, air, truck, and boat.

Organic Volatiles

A number of organic volatiles are produced in ripening fruit. Among these are acids, alcohols, esters, aldehydes, and ketones. The relative proportion of these substances varies with the species and with the degree of maturity, and is reflected in differences in aroma and flavor. In general, the influence of these compounds on

FIGURE 11-9. Polyethylene film liners create a modified-atmosphere storage having a 1-bushel capacity. Film liners are available also for 30-bushel pallet boxes. [Photograph courtesy Purdue University.]

respiration is slight. An exception to this statement is ethylene, an organic volatile produced by many fruits during the ripening phase. Ethylene is an unsaturated hydrocarbon (C_2H_6) that is nonpoisonous and has a sweetish odor. It is generally produced in greater amount than the other organic volatiles, usually accounting for two-thirds of the total carbon lost in volatile form. In apples it may be present in an amount 10–50 times that of other volatiles.

Ethylene gas has a profound effect on the ripening phase of many fruits, although its exact physiological role is not clear. Ethylene applied externally to immature fruits influences their respiration and ripening, but it has no effect on ripe fruits. A certain minimum concentration and minimum time of exposure triggers an irreversible stimulation of respiration. In honeydew melon, 40 ppm is sufficient to stimulate ripening. A 24-hour exposure is sufficient if temperatures are high enough.

Ethylene gas has been used commercially to effect other processes. For example, oranges may be "degreened" by ethylene gas. This involves destruction of chlorophyll in the peel. (This effect was first observed when kerosene heaters were used in shipment. Apparently, some ethylene is released by the incomplete combustion of kerosene.) Ethylene gas is not produced naturally by citrus fruits, but it may be produced from a *Penicillium* mold that is often found with them. Ethylene is used commercially in banana-ripening rooms to produce uniform ripening or to accelerate ripening. At high concentrations, ethylene has been used to defoliate rose bushes. It is also used incidentally in dehusking walnuts, inhibiting potato sprouting, and inducing flowering in pineapple.

Excessive ethylene gas adversely affects growing plants. Thus, neither plants nor cut flowers can be stored with apples; as mentioned previously, scion wood stored in apple storages may show severe damage as a result of bark peeling and splitting.

Humidity Control

The control of humidity is directly related to the keeping quality of many horticultural products. In general, low humidity is likely to result in desiccation and wilting. On the other hand, high humidity favors the development of decay, especially if temperatures are too high. Humidity control has become an important feature of modern storage facilities.

The amount of moisture in the air can be expressed as absolute or relative humidity. **Absolute humidity** is the amount of moisture per cubic foot of air. It is expressed as grains of water (one grain equals $\frac{1}{7000}$ pound). However, the amount of water vapor that can be held in a given space decreases with decreasing temperature. **Relative humidity** is the water-vapor content of the air expressed as a percentage of the amount it is capable of absorbing at the same temperature. The ability of a storage to dry out the products it contains is related to relative humidity and temperature. This "drying power" of air is proportional to the water-vapor deficit below saturation. At high temperatures small differences in

439

the relative humidity represent large differences in drying power. At low temperatures the reverse is true. Relative humidity may be increased in storage either by adding moisture, as by the use of fine mist, or by lowering the temperature. However, in refrigerated storages, if the differences in temperature between the refrigeration coils and the room is very large, water will condense as ice on the coils and will effectively lower the humidity. This may be avoided by keeping the coil temperature within 2–4°F (1–2°C) of the room temperature. Thus, the refrigerant system must be large enough to maintain proper temperature. Humidity control becomes more difficult as the rate of air circulation increases.

For the storage of leafy vegetables and root crops the optimum relative humidity is 90–95%, but for most fruits and vegetables a relative humidity of 85–90% is best. Most seed is best stored at relative humidities of 4–8%.

Ripening

Although most horticultural products ripen on the plant, there are a number of commodities that only ripen to optimum quality when detached from the plant (for example, bananas, pears, and avocados). Pears are picked "green" and, although stored at 30–31°F (about 1°C), they must be ripened at 60–65°F (about 16–18°C) for optimum quality. The length of the storage period depends on the cultivar. When ripened on the plant, bananas become mealy, lack flavor, and are subject to splitting and subsequent decay. When green, bananas can be stored at 56°F (about 13°C). They are best ripened at temperatures of about 64°F (18°C) with 90–95% relative humidity until the fruit is colored. At this point the humidity can be reduced to 85%. Ripening can be hastened by holding initial temperatures at not more than 70°F (21°C) for the first 18–24 hours and at 66°F (19°C) thereafter. Prolonged high temperatures increase deterioration and decay.

Tomatoes ripen both on and off of the plant, but since they do not ship or store satisfactorily when ripe, the fruit may be picked in the "mature" green state and ripened artificially. Tomatoes will ripen at temperatures above 55°F (13°C), preferably at 60–65°F (roughly 16–18°C), although the ultimate quality is often less than that of vine-ripened fruit.

Storage Types

In **common storage** the temperature of the atmosphere is utilized, which makes this kind of facility suitable only where temperatures are naturally low enough. Temperature is regulated by insulation and natural circulation. The most primitive type takes advantage of the reduced temperature fluctuations of the soil. Thus, in the fall, trenches or mounds of earth may be used for storing vegetables or plant material. Caves and unheated cellars provide more usable room, but above-ground structures, properly insulated and provided with sufficient ventilation, may be satisfactory in cooler climates. During warm weather, cooling is accomplished by the intake and circulation of the cool night air. Humidity may be kept high with earthen floors. Although common storage is cheap, the lack of precise temperature and humidity control often makes it economically unsound for many horticultural crops.

In **cold storage,** temperature and humidity are regulated by refrigeration. Many of the present structures are converted common storages, but large structures with better insulation and convenience features are now being constructed especially for storage purposes. The basic refrigeration and ventilation system involves forced air circulation. The structure must be sufficiently insulated to conserve power.

Controlled atmosphere storage involves the regulation of oxygen and carbon dioxide levels as well as the regulation of temperature. These storages are divided into rooms that are sealed in order that all gaseous exchange can be controlled. The rooms are closed after fruit is stored and remain sealed until fruit is removed. Temperature, humidity, and gas concentrations are controlled automatically.

A new system called **low-pressure storage** (LPS), also referred to as **hypobaric storage,** is now under development. This system uses low temperature and subatmospheric pressure (about one-fifth normal), which produce a low partial pressure of oxygen and the other gases in the atmosphere. The system also employs rapid air exchange. Volatiles (such as ethylene) are removed by the air exchange and the reduced pressure. Relative humidity can be controlled within the system.

LPS, by reducing the partial pressure of oxygen as well as ethylene, acts as a modified-atmosphere storage in the control of respiration, ripening, and senescence. Because this is accomplished by reducing pressure, the system does not need to be continually sealed. It is therefore feasible for portable truck-trailer units as well as large stationary storages. At present, LPS seems most useful for ornamentals because of their high price per volume of storage space required, their sensitivity to ethylene, and their amenability to being stored in atmospheres with low partial pressures of oxygen. The use of LPS vans to transport ornamentals could have important effects on the floriculture industry.

Food Processing

Relatively long-term preservation of food may be achieved by physical and chemical processes that sterilize the food or render it incapable of supporting the growth of microorganisms. These processes include drying, canning, freezing, fermentation and pickling, raising the sugar concentration, and irradiation.

Drying

Drying is one of the most ancient methods of food preservation. The process consists of removing water from the tissues, which results in a highly concentrated material of enduring quality (Fig. 11-10). The natural deterioration of the product

FIGURE 11-10. Dehydrated snap beans and the reconstituted product. [Photograph courtesy Quartermaster Food and Container Institute.]

by respiration is stopped because of enzyme inactivation. The lack of free water protects the dried products from decay by microorganisms.

Horticultural products may be naturally dried (**sun drying**) or artificially dried (**dehydration**). Sun drying is relatively inexpensive in locations where summers are sufficiently warm and dry. Dehydration, although a more expensive process, has a number of advantages: the process can be carried on independent of climate, drying time is reduced, and quality may be improved. The yield of dried fruit from a dehydrator is slightly higher than from sun drying because sugar is not lost as a result of continued respiration and yeast fermentation. Furthermore, sun drying requires considerable land and presents sanitation problems.

Dehydration is typically accomplished by hot-air drying. The air conducts the heat to the food and carries away the liberated water vapor. Many types of equipment are used for fruits and vegetables. After being sorted, washed, peeled, and trimmed, fruits to be dehydrated may be treated with sulfur dioxide fumes, which act as a bleaching agent in lighter colored fruits and as a chemical aid to preservation. Safe drying temperatures are near 140°F (60°C). The moisture content of fruits is reduced to 15–25%. In the dehydration of vegetables, enzyme systems are first inactivated by heating in boiling water or steam (**blanching**). Many vegetables are also more stable if given a sulfur dioxide treatment. For satisfactory storage, the moisture content must be reduced to 4% because of the lower sugar content of vegetables as compared to fruits.

A greater quantity of fruit is preserved in the world by drying than by any other method of preservation. Among the important dried fruits are raisin (Fig. 11-11), prune, apricot, date, fig, banana, peach, apple, and pear. In contrast, the quantity of dried vegetables on the market is relatively small. Potatoes are the most important dried vegetable. Most successful dried vegetable products are used as flavoring ingredients (for example, onion, celery, parsley, and their powders). Some dehydrated vegetables are sold in soup mixes; other are used in remanufacturing canned products, such as juices, soups, and stews.

A recent technological development called **freeze-dehydration** may increase the use of drying preservation. Through the use of a high vacuum, quick-frozen

FIGURE 11-11. Sun drying grapes in California. [Photograph courtesy USDA.]

food can be dehydrated. The quality of the reconstituted product is much higher than that of foods dehydrated by ordinary methods. The storage period of dried materials is extended at cool temperatures. At high humidities, mold growth may be a problem.

Canning

Canning is a method of preservation that consists in heat sterilizing food in an air-tight container. Heating destroys the human-pathogenic and food-spoilage microorganisms, and it inactivates the enzymes that would otherwise decompose the food during storage (Fig. 11-12). The sealed container prevents reinfection of the food after it has been sterilized, and it prevents gaseous exchange. There is almost no limit to the size of the container (Fig. 11-13).

The application of sufficient heat to sterilize food and inactivate enzymes results in alterations in the color, flavor, texture, and nutritive value of foods. Quality is therefore the limiting factor in canning. The quality of canned products may be increased by prompt dispatch of high quality raw products through the processing plant, and by proper attention to processing procedures. This involves an understanding of the precise relationship between processing time and temperature control. In general, the reduction in processing time, brought about by increasing temperatures, increases the quality of the product. The precise time and temperature required for each commodity is based largely on the natural acidity of the food. There are basically two groups: a **low-acid** group (pH 4.5–7.0), which includes most vegetables, and an **acid** group (pH less than 4.5), which includes fruits, berries, tomatoes, and fermented and pickled foods. Low-acid foods require relatively severe heat treatment, since they can support the growth of *Clostridium botulinum*, a bacterium that causes a serious form of food poisoning. A millionth of a gram of the toxin produced by this organism is sufficient to kill a normal adult. Thus, all foods capable of sustaining growth of this organism are processed on the assumption that the organism is present and must be destroyed. Since this organism is extremely heat resistant, the high temperatures required often reduce the quality of the food.

The storage life of the canned product decreases as the temperature increases. For extended storage (more than five years), storage temperatures should be below 50°F (10°C). At storage temperatures above 120°F (roughly 50°C), certain heat-loving bacteria (**thermophiles**), which are not ordinarily all killed by the sterilization process, will continue to grow, causing spoilage. In humid regions, and especially in coastal areas, where air-borne salt particles are plentiful, storage life is limited by the life of the container, which may be shortened by salt-induced corrosion of metal cans or of the metal lids on glass containers.

Freezing

Freezing protects food from spoiling because most microorganisms cannot grow at temperatures below $32°F$ ($0°C$). The freezing process stops most enzymatic activity and is not in itself destructive to nutrients. Some enzymatic activity in certain products must be stopped by heat treatment (blanching) in order to keep full flavor and color intact during storage.

The rate of freezing is an important factor in the quality of the thawed product. With slow freezing, relatively large ice crystals develop, which fracture the tissue cells; then, upon thawing, the foods lose cellular fluid, which gives them a soft texture. If freezing is rapid, many small, fine crystals are formed (about one-hundredth the size of those formed during slow freezing). Because these crystals are tightly packed, fewer cells are ruptured.

The basic principle of rapid freezing is the speedy removal of heat from food by various methods, including cold air blasts, direct immersion in a cooling medium, contact with refrigerated plates, and liquid air, nitrogen, or carbon dioxide. Freezing in still air is the slowest method. As living plant cells contain much water, most plant foods freeze between $25°$ and $31°F$ (-3.9 and $-0.6°C$). The temperature of the food undergoing freezing remains relatively constant at its freezing point until it is almost completely frozen. Quick freezing is a process in which the water in food passes through the zone of maximum ice-crystal formation in thirty minutes or less. This usually requires refrigerant temperatures of $-20°$ to $-40°F$ ($-29°$ to $-14°C$).

The success of freezing as a method of preservation depends on the continuous application of the process. Some nutrient loss and deterioration in color, texture, and flavor may occur during frozen storage, depending on the temperature. Best results are obtained at a temperature of $-10°F$ ($-23°C$), although most storages are kept near $0°F$ ($-18°C$). Temperatures must be kept uniform. Frozen foods must be packaged to protect them from dehydration during freezing and subsequent storage. Sublimation of ice occurs in unprotected food, resulting in a freezing disorder called **freezer burn,** which irreversibly alters the color, texture, flavor, and acceptability of frozen foods.

Fermentation and Pickling

Although microorganisms are commonly associated with decay, microbial action may be utilized in food preservation. The action of certain bacteria and yeasts in decomposing carbohydrates anaerobically is known as **fermentation.** The decomposition may be accomplished by a number of different organisms, the end products varying with the particular organism. These end products include carbon dioxide and water (complete oxidation), acids (partial oxidation), alcohols (alcoholic fermentation), lactic acid (lactic fermentation), and others. When built up to sufficient concentrations, some of these fermentation products create conditions unfavorable to microorganisms, including the original one. They act as preservatives by retarding enzymatic deterioration, and they impart flavors that are regarded as desirable. Fermentation may be controlled by conditions that favor the growth of one type of organism. This is done through the regulation of pH, oxygen availability, and temperature, and through the use of salt.

Fermentation is an important processing method for some horticultural crops. The fermentation of the juice of grapes or other fruits produces wines. The further fermentation of alcohol to acetic acid is the basis of vinegar production. When used in combination with salting, fermentation is called **pickling.** This term is used especially for cucumbers, but it also applies to other commodities, such as olives and many vegetables (including onions, tomatoes, beans, cauliflower, cabbage, and watermelon rind). Fermented cabbage (sauerkraut) is the result of a number of distinct fermentation processes. Pickling may be accomplished without the direct use of microorganisms by placing food in organic acids (for example, vinegar or citric acid). For extended storage of pickled products, the enzyme systems must be inactivated. This is usually done by canning.

Sugar Concentrates

Acid fruits, concentrated until at least 65% of their weight is soluble solids, may be preserved with mild heat treatment if protected from air. The high concentration of sugars, and low water content, preserves the food. Depending on the recipe used to make it, the product may be called jelly (made from fruit juice), jam (made from concentrated fruit), preserves (made from whole fruit), fruit butter (a semisolid, smooth product made from high concentrations of fruit), or marmalade (made from citrus fruit and rind). The making of candied or glacéed fruits involves the slow impregnation of tissues with sugar. Storage of these products at high temperatures reduces their quality: their appearance is spoiled by nonenzymatic browning caused by reactions of organic acids with reducing sugars.

Chemical Preservation

In addition to such natural preservatives as salt, vinegar, and spices, there are a number of chemicals that, when added to food, prevent or retard deterioration. These are usually used in conjunction with other methods of preservation. Some of the chemical preservatives used in food preparation are inorganic agents, such as sulfur dioxide and chlorine; others are organic agents, such as benzoic acid, certain fatty acids (including sorbic acid), ethylene and propylene oxides (fumigants), and various antibiotics. Each has a special use in the preservation of fruits and vegetables. For example, sulfur dioxide is widely used in the drying of fruits and vegetables. Chlorine compounds are used in hydrocooling and in processing.

Potassium sorbate and sodium benzoate are useful in preventing growth of yeasts and molds in such fruit products as fresh cider. Ethylene oxide has been used in the sterilization of spices and flavoring compounds.

Chemical preservatives have a legitimate place in food processing. They are not, however, in the best interests of the public if they are used to deceive or if there is any danger inherent in their use.

Radiation Preservation

Radiation is not presently used to preserve horticultural foods, the early enthusiasm with regard to its possible usefulness notwithstanding (Table 11-3). Some of the uses that had been suggested were the extension of storage life of perishables (for example, strawberries) by low-level irradiation, the inhibition of such growth processes as sprouting in potatoes (Fig. 11-14), and disinfestation of Hawaiian fruit fly in papaya. However, the results of extensive studies indicate that irradiation has adverse effects, including the production of off-flavors and severe injury to delicate tissues. Further, the beneficial effects expected could be obtained by cheaper, more effective means.

Radiation has also been suggested to have potential use in certain unit operations in the food industry, such as the aging of wine, sweetening of peas, and

TABLE 11-3. Radiation effects on fruits and vegetables.

Dose absorbed (rad)[°]	Chemical and physiological alterations	Potential horticultural applications
10^1	Interferes with sensitive enzyme systems (auxin, DNA)	
10^2		Treat seed potatoes for storage at room temperature
10^3	Injures many growth processes	Inhibit sprout of potatoes
10^4	Terminates growth processes	Inhibit sprout of bulbs
		Accelerate pear ripening
10^5	Destroys 90–99% of microorganisms	Sterilize fruit and vegetable surfaces to increase shelf life (lengthens storage of strawberries and soft fruits)
10^6	Hydrolysis of carbohydrates	Sterilize fruits
	Termination of respiration	Sterilize vegetables
	Destruction of enzymes, proteins, viruses	Sweeten peas
		Tenderize asparagus
10^7	Complete tissue deterioration	

SOURCE: Data courtesy N. R. Desrosier.

[°] 1 Rad = 100 ergs/g of moist tissue. The energy of radiation supplied must be below 2.2 mev (million electron volts) to prevent induced radiation. Cobalt 60 emits radiation below this energy level.

Control   1,250r   5,000r

20,000r   80,000 r   106,250r

FIGURE 11-14. Inhibition of potato sprouting with irradiation. A dose of 7500 rads permits some cultivars of potatoes to be stored at 50°F (10°C) for two years. [Photograph courtesy Quartermaster Food and Container Institute.]

tenderizing of asparagus. None of these uses has proved practical. While there has been no direct evidence of toxic effects, there is evidence for destruction of vitamins and other nutrients similar to that produced by heat treatments.

Irradiation as a method of food preservation has not lived up to early expectations. The early hopes of irradiation as a substitute for refrigeration, especially in developing countries, has been dispelled. Irradiation cannot substitute for refrigeration and at best can only supplement it.

DISTRIBUTION

Marketing channels are those agencies that handle a commodity along its course from producer to consumer—the actual physical route, as well as the route of title. In general, horticultural commodities travel from producer to wholesaler to retailer to consumer. The concentration of goods by wholesalers and the corresponding dispersion into retail outlets are diagrammed in Figure 11-15. Storage and preservation tend to equalize uneven production with relatively constant demand. The exact marketing channels differ with each commodity and change in

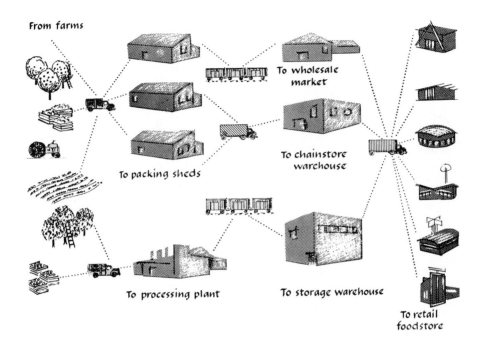

From farms

To wholesale market

To chainstore warehouse

To packing sheds

To processing plant

To storage warehouse

To retail foodstore

FIGURE 11-15. Principal marketing channels for fruits and vegetables. [Adapted from USDA.]

pattern over the years. As examples, the principal marketing channels for nursery crops in 1949 and floriculture crops in 1963 are shown in Figure 11-16.

Wholesaling

The wholesaling of horticultural crops varies with the commodity. The transactions involved in wholesaling are performed by **middlemen** who specialize in the sale of goods, moving the commodity from producer to consumer.

Merchant middlemen act as representatives for buyers or sellers; they are called **commission men** when they actually handle the product, and **brokers** when their relation to the product is less direct.

Speculators take title to products but merely attempt to profit from market fluctuations. Their effect on the market is mainly on pricing structure. Potato "futures" and frozen orange juice concentrate are the chief horticultural commodities traded by speculators in the United States.

A number of organizations, such as **flower exchanges** and **fruit auctions,** are involved in the wholesaling process. They establish procedures at the market and may contribute such services as storage or grading. They operate from fees and assessments paid by those who use the facilities.

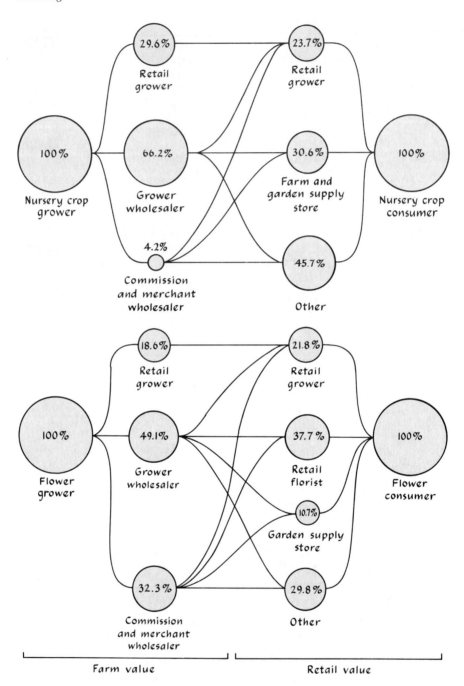

FIGURE 11-16. Marketing channels for horticultural crops based on data for selected years. *Top*, marketing channels for nursery crops (based on 1949 data). *Bottom*, marketing channels for floriculture crops (based on 1963 data). [Courtesy USDA.]

Some processors of horticultural products also perform wholesaling functions by acting as their own buying and selling agents. Large processors advertise their own products to build up brand loyalties among consumers. On the other hand, smaller processors distribute their food through food brokers and commission agents. At the present time, processors are large customers for other forms of processed food products. For example, canners are becoming large buyers of frozen foods. Thus, the assortment of marketing channels utilized by processors is becoming quite complex.

Retailing and Merchandising

Retailing, the final outlet for horticultural goods, is the most expensive marketing operation. It accounts for almost half of the total marketing cost for fruit and vegetables. In recent years great innovations at the retail level have created a revolution in marketing that has been felt throughout the horticultural industry—namely, the nearly complete shift to the self-service supermarket. Mass selling, although adaptable to processed products, was not possible with the older methods of retailing fresh fruits and vegetables. The self-service idea has created the packaging concept, which has emphasized better grading and increased standardization. This has also affected ornamentals. At present, nursery stock and perennial plants are being sold in this manner.

The self-service trend, and the wide variety and form of products available to the consumer, have resulted in competition that is influenced greatly by promotion and advertising. Different commodities compete for quantity and quality of space in the supermarket. Moreover, certain commodities from one region compete with the same commodities from other regions. This struggle has encouraged producers to organize on the basis of crop and region to facilitate the marketing of their products. Probably the best example of this kind of organization is the California Fruit Growers Exchange, which markets citrus fruits under the Sunkist label. Smaller groups, such as the Michigan Blueberry Association, the Idaho Potato Growers, and Ocean Spray, Inc. (a cranberry-growers cooperative), follow this form. The combined resources of many producers make it possible to promote their crop and, at the same time, guarantee a more standardized product.

The marketing of flowers is an exception to this retailing trend. Because of the services they provide (mainly in the form of arrangements), florists have retained their identity as retail units. However, retail florists have organized through such groups as Florist's Telegraph Delivery, Telegraph Delivery Service, and Teleflorists.

Integration

The marketing channels may be integrated horizontally or vertically. Vertical integration is the combination under one management of firms that control two or more steps in the production, handling, processing, distribution, or sales of a commodity. A large canning company such as Dole Pineapple is a good example of vertical integration. Horizontal integration is the combination of firms for the performance of a similar function; a co-op that markets fruits or vegetables from a given area is a good example. Chain stores illustrate both types. They are vertically integrated, in that they control their own wholesaling organization—they are not as yet involved in production—and their organization of retail units constitutes horizontal integration. The degree of integration in marketing does not eliminate the main thoroughfares of the marketing channels, although it may streamline them. All marketing operations, such as grading, storage, and packaging, must still be performed.

Selected References

Cargill, B. F., and G. E. Rossmiller (editors), 1969. *Fruit and Vegetable Harvest Mechanization: Technological Implications* (Rural Manpower Center Report No. 16). East Lansing, Mich.: Michigan State University.

Darrah, L. B., 1971. *Food Marketing*, rev. ed. New York: The Ronald Press. (An excellent introductory text on marketing and food distribution.)

Desrosier, N. W., and J. N. Desrosier, 1977. *The Technology of Food Preservation*, 4th ed. Westport, Conn.: Avi. (The fundamentals of preservation of fresh and processed foods.)

Kohls, R. C., and W. D. Downey, 1972. *Marketing of Agricultural Products*, 4th ed. New York: Macmillan. (An introduction to agricultural marketing.)

Pfahl, P. B., 1977. *The Retail Florist Business*, 3rd ed. Danville, Ill.: Interstate. (A text and manual on the retail florist operator.)

Ryall, A. L., and W. J. Lipton, 1972 and 1974. *Handling, Transportation and Storage of Fruits and Vegetables* (Vol. 1, *Vegetables and Melons*, 1972; Vol. 2, *Fruits and Nuts*, 1974). Westport, Conn.: Avi. (A complete two-volume manual for post-harvest handling and storage of fresh fruits and vegetables.)

USDA Agricultural Marketing Service. *United States Standards*. (Standards and grades are issued and revised periodically by the USDA.)

III

THE INDUSTRY
OF HORTICULTURE

Fruit orchard. [Courtesy International Harvester Corporation.]

12

Horticultural Geography

Now I will believe
That there are unicorns; that in Arabia
There is one tree, the phoenix' throne, one
phoenix
At this hour reigning there.

SHAKESPEARE, *The Tempest* [III. 3]

The commercial production of horticultural crops is not evenly distributed over agricultural regions, but tends to be concentrated in limited areas of the world. Horticultural geography deals with the distribution of the industry and is concerned with the environmental, economic, and social factors that determine its location and development.

CLIMATE

The physical environment of plants includes many factors whose actions and interactions must be considered. Climate, the summation of an area's weather, includes temperature, moisture, and light effects. It largely determines which plants can be grown in the region and where and when they should be planted. Thus, vegetation is one of the obvious determinants of the differences in appearance between climatic regions. A map of the earth's climatic regions is also an approximation of natural vegetation. Climate, as the fundamental force in the environment, also shapes the soil and, to a lesser extent, the configuration of the earth's surface.

The pattern of climate is a result of the circulation of the atmosphere. Solar radiation falling more directly on tropical than on polar regions warms the equatorial air, causing it to flow poleward. The resultant pressure produces a return ground flow of cold polar air. The flow pattern does not follow a simple direct line from the pole to the equator but is deflected as a consequence of (1) the easterly rotational spin of the earth, (2) the seasonal effect, (3) the differential cooling of land and water masses, (4) the altitude and the configuration of the land, and (5) the storms and winds resulting from the interactions of cold and warm air masses. Other extraterrestial factors, such as sun-spot activity, may affect the weather, but they are poorly understood.

MICROCLIMATE

Microclimate refers to the climate of a "small area." The climate at the ground may differ considerably from that at thirty feet above the ground. These climatic differences are of vital importance to people and their agriculture. We become aware of the microclimate as we drive in and out of pockets of fog on a chilly morning. The orchardist who loses the crop on the lower half of his trees as a result of frost becomes painfully aware of microclimate.

Whereas **location** refers to a geographic and climatic area, the term **site** implies microclimatic influences within a specific location. The ultimate success of horticultural enterprises depends to a great extent on proper location and site. Microclimatic variations are due to exposure, slope, vegetation, and the thermal capacity and conductive characteristics of the soil. These will be discussed along with the principal climatic elements: temperature, moisture, and light.

CLIMATIC ELEMENTS

Temperature

The temperature at any point on the earth's surface depends on the geographic ordinates of latitude and altitude, season and time of day, and the mediating influence of microclimate. The major factor that determines temperature is the amount of solar radiation received, which depends upon both the intensity and the duration of radiation. The more vertical the sun's rays, the less atmosphere they must penetrate. In addition, vertical rays provide a greater concentration of energy per unit area than do the oblique rays that reach the poles. The duration of solar radiation is determined both by day length, which imposes an absolute limit on the amount of solar energy received, and by the variable effects of cloud cover. Furthermore, there is a decrease in temperature with an increase in elevation. This averages $3.6°F$ ($2.0°C$) for every 1000 feet of elevation, approximately 1000 times the rate of temperature change with latitude. The reason for this is that much of the atmospheric thermal energy is received directly from the earth's surface and only indirectly from the sun. In addition, the lower tropospheric air contains more water vapor and dust and is therefore a more efficient absorber of terrestrial radiation, which explains the presence of snow-capped mountains near the equator, such as the famous Mt. Kilimanjaro.

The variation in temperature reported at the earth's surface is enormous, with a range of more than 200°F (111°C), from a record −96°F (−71°C) at Verkhoyansk, Siberia, to 136°F (58°C) at Azizia, Libya. The mean annual temperatures range from an estimated −22°F (−30°C) at the South Pole (elevation 8000 ft) to a record 86°F (30°C) at Massawa, Eritrea, Africa. In annual crops the important temperature values are the mean and extreme temperatures as well as the duration of temperatures conducive to plant growth (the length of the "growing season"). Perennial plants are affected by the temperature values during the whole year. Both seasonal variation and average temperature must be considered in relation to plant growth. Peaches, for example, require a long growing season and warm summer temperatures. Their northern distribution is limited by their degree of hardiness; temperatures below −12°F (−24°C) cause flower-bud injury. Their southern distribution is limited by their chilling requirements. A map of the expected minimum temperatures of the horticulturally important areas of the continental United States and Canada is presented in Figure 12-1.

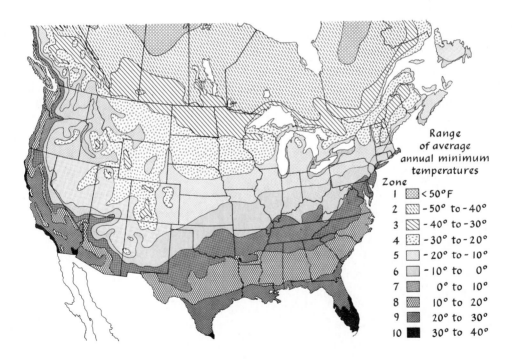

FIGURE 12-1. Zones of plant hardiness in the United States and Canada. [From USDA Miscellaneous Publication 814.]

Frost

> We see hoary-headed frosts
> Fall in the fresh lap of the crimson rose,
> And on old Hiems' thin and icy crown
> An odorous chaplet of sweet summer buds
> Is, as in mockery, set. . . .
>
> SHAKESPEARE, *A Midsummer Night's Dream* [II. 1]

Frost is a thin layer of ice crystals deposited on soil and plant surfaces as a result of freezing temperatures. Two types of weather conditions produce freezing temperatures: (1) rapid radiational cooling, which results in a **radiational frost,** and (2) the introduction of a cold air mass with a temperature below 32°F (0°C), which produces an **air-mass freeze.** Frost is common under conditions of clear calm skies; a freeze, on the other hand, may occur with an overcast sky, and usually when there is considerable wind. Both frost and freeze involve temperatures at or below the freezing point. Because of its local nature, frost may occur when the so-called official temperature (taken relatively nearby, and usually at a height of six feet) is above freezing. Glaze damage (Fig. 12-2) can occur when relative humidity is high and temperatures drop below freezing. The weight of the ice coating on tree limbs causes them to break off.

The earth radiates energy, as do all bodies. At night the earth receives no solar radiation, which means there is a net loss of radiation. Frost occurs when the loss of heat from the ground permits the temperature to drop to or below the freezing point.

Frost is of vital horticultural concern. It defines the season for annual crops in the middle latitudes and is potentially extremely destructive to perennial fruit crops, which flower in the early spring. The danger of frost is usually not due to actual tree damage but to injury of temperature-sensitive flower parts (Fig. 12-3). The year's crop is thus at the mercy of spring frost.

FIGURE 12-2. Glaze damage: limbs have been broken off by the weight of their ice coating. [Photograph courtesy USDA.]

FIGURE 12-3. Frost is a serious hazard to the orchard in bloom. [Photograph by J. C. Allen & Son.]

Frost Conditions

Conditions favorable for radiational frost are those that are conducive to rapid and prolonged surface cooling. For example, the introduction of low-temperature polar air followed by clear, dry, calm nights will facilitate upward radiation. The presence of cloudy, humid air causes reradiation back to the earth and thus prevents frost conditions. The main effect of fogging in the control of frost is the creation of an artificial reradiating surface, although there is some release of heat as the fog droplets cool. Strong evaporation after rainfall, especially on plant cover with a large surface area, will increase cooling. The absence of wind leaves the coldest air undisturbed and next to the ground.

Frost conditions are usually quite variable, since local conditions modify radiation. Site is an important factor in frost. For example, slopes are protected by "air drainage." Since cold air is denser than warm air, it moves downhill. The circulation pattern of the displaced warm air produces a relatively warm area, or **thermal belt,** on the slope itself. It is this microclimatic factor that makes slopes good sites for fruit plantings. Low areas, on the other hand, collect cold air and become "frost pockets." This results in a phenomenon referred to as **temperature inversion,** in which air temperatures increase with altitude (Fig. 12-4). Artificial

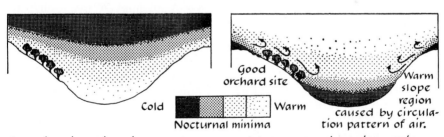

Good orchard site

Warm slope region caused by circulation pattern of air.

Cold

Warm

Nocturnal minima

Except in an inversion, air temperature decreases with height. (Darker tones indicate cooler air.)

A temperature inversion sets in when cool air is trapped under warm layer. The normal temperature gradient is reversed in the inversion layer.

FIGURE 12-4. Schematic representation of the orgin of a thermal belt as a result of a temperature inversion on a hillside. [Adapted from Geiger, *Climate Near the Ground*, Harvard University Press, 1957.]

FIGURE 12-5. Wind machines protect citrus from frost during temperature inversions by mixing the warmer upper air into the colder air below. [Photograph courtesy Food Machinery and Chemical Corp.]

wind machines prevent frost damage by mixing in the upper warmer air (Fig. 12-5).

The frost protection afforded by large bodies of water is due to the high specific heat of water as compared to that of the land; water absorbs and gives up heat slowly. Solar radiation penetrates water more deeply than land, and the continuous internal movement of water results in heat distribution throughout the water mass. Thus, large bodies of water become heat reservoirs in the fall and cold reservoirs in the spring. Because of their great heat capacity, large bodies of water moderate temperatures. In the winter and early spring the influence of water keeps temperatures moderately low and prevents premature plant growth. In the late spring it may provide enough heat to prevent frost. Similarly, in the fall there is a warming influence that tends to delay the advent of frost conditions. This temperature lag is felt mainly along the windward side of large bodies of water.

Many factors affect frost. Anything that prevents the accumulation of radiation during the day will increase frost. For example, vegetation that shades the soil will reduce the amount of heat stored in the day. Thus, a sodded or mulched area is more liable to become frosted than one under "clean cultivation." The necessity for controlling frost is one of the main reasons why peaches are not grown under permanent sod. The exposure of the slope also affects the amount of radiation received. In the Northern Hemisphere, southerly slopes receive considerably more radiation than northerly slopes.

Heat from the lower layers of the earth moves up by conduction. Consequently, the conductivity of the soil will affect frost at the surface. Frost on muck is a serious hazard because organic soils tend to be poor conductors of heat as compared with mineral soils. More important than soil type in the occurrence of frost is the amount of soil moisture. By replacing air (a poor conductor) with water (a better conductor) the danger of frost can be reduced. Thus, frost damage may be prevented in muck areas by flooding.

The **white frost** commonly seen in the morning results from frozen dew (Fig. 12-6). Its occurrence depends on the **dewpoint**—the temperature at which relative humidity reaches 100%. When the air temperature is below the dewpoint but above 32°F (0°C), water vapor condenses in the form of dew. White frost occurs when the air temperature is below the dewpoint and below 32°F. If the humidity is low, frost damage may occur when the air temperature is below 32°F but above the dewpoint. This is known as a **black frost** because the only visible indication of it comes when the vegetation turns dark due to cold injury.

The change in state from water to ice results in the release of energy, the **heat of fusion.** Consequently, if temperatures do not get too low, the freezing of water or the occurrence of a white frost actually protects vegetation from lower temperatures. This phenomenon is exploited in the use of sprinkler irrigation as a method of frost protection. The ice forming on the plant releases heat and acts as a protective buffer against cold injury!

Prevention of Frost Injury

The prevention of frost injury may involve three strategies: (1) escape, (2) reduction of heat loss, and (3) addition of heat.

Escaping spring frost by late planting is possible with annual crops, but in locations with short growing seasons this strategy sometimes backfires because the crops may be exposed to early fall frost. In perennial plantings, spring frost can be escaped with treatments designed to delay blooming. Recently this has been accomplished experimentally by overhead misting: evaporative cooling delays the onset of bud break by as much as two weeks. However, some problems have been encountered with this method, including reduced fruit set. Escaping spring frost by choosing the right location and site remains the preferred method in the fruit industry.

FIGURE 12-6. Frost is frozen dew. Hoar-frost results when sublimation occurs over several hours, leaving heavy deposits of ice crystals in the form of scales, needles, feathers, and fans. [Photograph by J. C. Allen & Son.]

The reduction of heat loss is accomplished through the use of such devices as hot caps, plastic tunnels, and cold frames. A recent practice is the use of stable foams to insulate plants against frost. Such foams are nontoxic combinations of surfactants, stabilizers, and protein materials (gelatin) selected for durability, low cost, and insulating properties. Heat may be obtained from the earth by improving the conduction characteristics of the soil, or it may be obtained from the air by disturbing the temperature inversion. Heat may be added indirectly, by sprinkler irrigation or by techniques that increase the daytime absorption of insolation, or directly, with heaters.

Heat may be added using various kinds of heaters and heating systems (Fig. 12-7) to ward off cold temperatures, especially in high-value fruit crops. A recent innovation is the use of solid-fuel candles. Made of solid petroleum wax, they are about eight inches in diameter and burn with a low flame for eight hours. Two such heaters beneath a grapefruit tree will raise the average air temperature within the canopy of the tree by about 7°F (4°C).

Heaters are commonly used in regions where citrus crops are subject to occasional frost. If growers have a warning from the United States Weather Bureau that there will be a frost on a particular night, they fill their heaters with enough oil to burn throughout the anticipated danger period and place them in their orchards in strategic locations. Before the cold front moves in, the heaters are lighted. The radiant fraction of the thermal output of heaters provides the

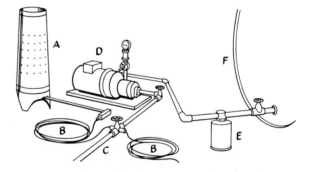

FIGURE 12-7. A popular central-distribution system for heating orchards. Fuel oil, liquefied petroleum gas, or natural gas is supplied to the burners from a central fuel source via an underground trunk-line network. *top*, lightweight portable burners (A) are attached to flexible feeder lines (B), which, in turn, are attached to a trunk line (C). Trunk lines are attached to an oil pump (D) A filter (E) traps impurities from the central storage tank (F). Valves control oil flow and burning rate. *Bottom*, heaters in action in a Midwestern apple orchard. [Courtesy Spot Heaters, Inc.]

most protection, although part of the convection heat is also useful, particularly under conditions of good inversion where wind machines can redirect some of the warm layer overhead back into the orchard. Although the smoke from smoky heaters ("smudge pots") may be of some value as a radiation "blanket" (if atmospheric conditions permit its accumulation in an inversion ceiling during the night), when more than one night of protection against radiation frost is required this blanket may be costly, for the smoke will reduce the incoming solar radiation during the succeeding day. "Smudging" is illegal in California citrus-growing areas because of the air pollution that it causes.

Moisture

In discussions of climate, moisture refers both to precipitation (rain and snow) and to atmospheric humidity. Rainfall is directly related to the circulation pattern of the atmosphere. It results from the cooling of warm humid air forced upward by the convergence of air currents. Total annual precipitation may vary enormously from one place to another. Because of topographical variation, a marked difference in rainfall may occur between points that are relatively close together. This may result in the close proximity of desert and rain forest. A map of average annual precipitation is shown in Figure 12-8.

Of greater agricultural significance than average precipitation is the **effective precipitation,** the water that is not lost by runoff or evaporation and that is consequently available to plants. The percentage of precipitation that is "effective" is higher where temperatures are low. Thus, cool areas require less rain for plants than do warm areas. The natural vegetation of an area is the most satisfactory measure of effective precipitation.

The extremes of precipitation result in drought and flood. Drought may be predictable in some areas on a seasonal basis; in others, it appears to recur on a longer cycle. Unless irrigation is practical, horticultural crop production is effectively restricted in drought areas. Although methods exist for the efficient utilization of existing water, there is no substitute for sufficient water in intensive crop production. Horticultural crops require a plentiful supply of water as compared to agronomic crops. Flood, on the other hand, is equally injurious (Fig. 12-9). Except for such water-loving plants as cranberries, which are grown in boggy areas, an excess of water results in extensive damage to horticultural crops. Much of this is due to root damage from lack of soil oxygen. Excessive moisture also results in disease problems, since high humidity favors the growth of many pathogenic fungi. Many crops (including tomato, peach, apple, and cherry) show abnormal splitting and cracking of the fruit as a consequence of excess moisture during periods of fruit ripening. Flooding can be overcome by proper drainage in some locations. Tiling and proper soil management do much to alleviate the problem.

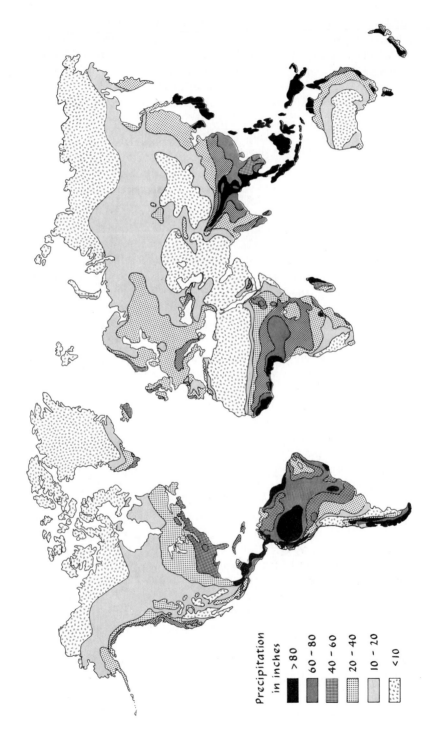

Precipitation
in inches

>80
60 – 80
40 – 60
20 – 40
10 – 20
<10

FIGURE 12-9. Spring flooding and erosion. A folk proverb states: "In dry years farmers complain of starvation; in wet years farmers starve." [Photograph by J. C. Allen & Son.]

Light

The quantity of light is an important climatic feature, and it is a significant part of the plant environment. Day length is the most obvious difference between climates. In the tropics, the day is close to twelve hours long throughout the year. At the polar regions, the summer day length goes up to 24 hours. The world distribution of plant species is greatly determined by their photoperiodic response. It is useless to attempt to grow long-day plants for flowers or fruit in the tropics, for they will remain vegetative indefinitely. In addition to day length, the quantity of light is affected by such atmospheric conditions as the number of sunny versus cloudy days during the growing season. The amount of sunlight greatly affects quality in many crops (for example, grapes and apples).

SOIL AND CLIMATE

The chemical, physical, and biological changes that determine the soil profile are all affected by long-term climatic conditions. The major differences between soils are the result of climate operating through soil-forming processes and vegetation. Thus, climate affects soil formation directly through weathering processes and indirectly through vegetation.

The distribution of the major soil groups can be interpreted through broad climatic types, based on weathering and vegetation (Fig. 12-10). The major zonal soil types, grouped on a climatic basis, are (1) the strongly leached, red soils

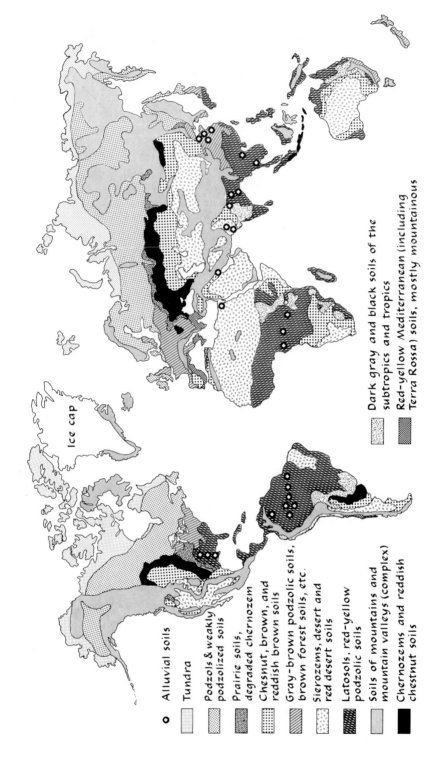

Legend:

○ Alluvial soils
Tundra
Podzols & weakly podzolized soils
Prairie soils, degraded chernozem
Chesnut, brown, and reddish brown soils
Gray-brown podzolic soils, brown forest soils, etc.
Sierozems, desert and red desert soils
Latosols, red-yellow podzolic soils
Soils of mountains and mountain valleys (complex)
Chernozems and reddish chestnut soils

Dark gray and black soils of the subtropics and tropics
Red-yellow Mediterranean (including Terra Rossa) soils, mostly mountainous

Ice cap

FIGURE 12-10. The major soil groups of the world. [Courtesy USDA.]

supporting tropical rain forests in areas of hot, humid climate; (2) the unleached light-colored soils in areas of hot desert climate; (3) the dark-colored soils supporting native tall grasses in areas of subhumid, temperate climate; and (4) the acid, light-colored soils supporting coniferous forests in areas of cool, moist climate. Unfortunately, the climates best suited for plant growth do not always coincide with the areas of naturally fertile soils.

CROP ECOLOGY

Ecology is the study of life forms in relation to their physical and biological environment. Plant ecology deals largely with natural plant communities, and is concerned with the causes responsible for the course and pattern of plant development, succession, and distribution. It is concerned with the relations between climate, soil, and biological interaction.

By definition, horticulture deals with cultivated plants. However, the crop-plant community is subject to many of the same ecological responses as is the natural-plant community. These plant responses determine to a great extent the ability of a region to support successfully a particular crop, and they define the specific problems of land use and crop management.

The climatic environment is a powerful determinant of plant development. It is the extremes of temperature (Table 12-1), rainfall, and light, rather than their yearly means, that determine the status and define the limitations of agriculture. The inappropriate utilization of agriculture in areas of marginal climate results in more "poor years" than "good years." Unfortunately, the occurrence of unusual periods of "good" weather often results in overextension of an unsuited agriculture, with disastrous consequences when the more normal pattern resumes.

The effect of climate upon quality and appearance of horticultural products plays an important role in the location of the horticultural industry. For example, high light intensity favors maximum development of red color in apples. The prominence of the central valleys of Washington State as apple-producing areas is due to their dry climate, which is brought about by a favorable combination of altitude and sheltering mountain ranges that results in bright, cloudless summers.

CLIMATIC REGIONS

There are a variety of ways of classifying climatic regions. We are all familiar with the climatic classification by temperature into tropical, temperate, and polar zones. A more useful classification includes both temperature and moisture to account for seasonal patterns. In this regard the natural vegetation may be used as a meteorological instrument to integrate the climatic elements (Fig. 12-11). Using

TABLE 12-1. Classification of common fruit crops by temperature requirements.° Cold-tolerance increases as the crop listing descends.

Tropical	Subtropical	Temperate	
		Mild winter	Severe winter
Coconut			
Banana			
Cacao			
Mango			
Pineapple			
Papaya			
	Coffee		
	Date		
	Fig		
	Avocado		
	Citrus		
	Olive		
	Pomegranate		
		Almond	
		Blackberry	
		Grape (European)	
		Persimmon (Japanese)	
		Quince	
		Peach	
		Cherry	
		Apricot (blossoms tender)	
		Strawberry ⎫ (very hardy	
		Blueberry ⎭ under snow)	
		Raspberry	
		Cranberry	
			Pear
			Plum
			Grape (American)
			Currant
			Apple
Low-temperature sensitive	Slightly frost tolerant	Tender	Winter-hardy
Noncold requiring		Cold requiring	

° Variation in tolerance depends to a large extent on species, variety, plant part, and stage of growth.

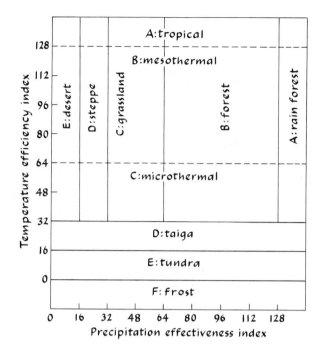

FIGURE 12-11. Temperature and humidity provinces based on natural vegetation. [After Thornthwaite, *Geogr. Rev.* 21:649, 1931.]

precipitation effectiveness and temperate efficiency,[1] Thornthwaite has divided climate into regions associated with a characteristic vegetation—for example, grassland, forest, and rainforest (Fig. 12-12).

A widely used method of classifying climates is the Köppen system, which is based on annual and monthly means of temperature and precipitation. The temperature and precipitation boundary lines are selected by natural vegetation and crop responses rather than by atmospheric circulation patterns (Fig. 12-13). The classification discussed here is derived from Trewartha's modification of the Köppen classification.

Tropical Rainy Climates

The region of tropical rainy climates lies in an irregular belt 20–40° wide that straddles the equator. Its most typical feature is the absence of winter. The difference between the average daytime temperature and the average nighttime temperature exceeds the annual range of 24-hour temperature averages. The temperature boundaries of this region have been arbitrarily placed at the average minimum monthly temperature of 64°F (18°C). This coincides with the poleward limit of plants requiring continuing high temperature, as various tropical palms.

---

[1]Precipitation effectiveness is the summation of monthly precipitation divided by evaporation. Temperature efficiency is the summation of monthly mean temperature minus 32°F divided by 4.

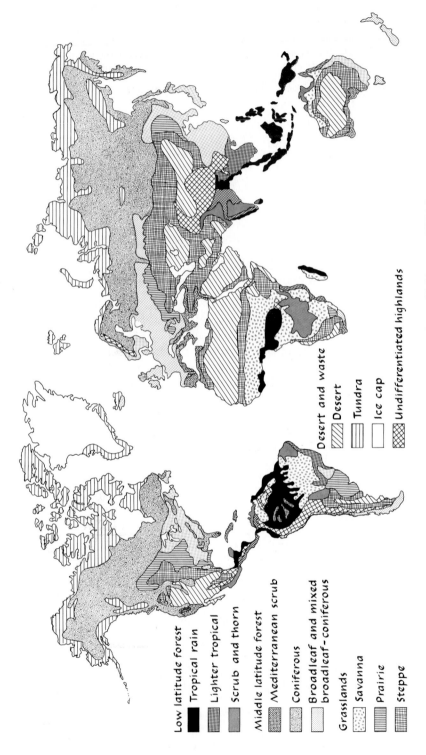

Low latitude forest
- ■ Tropical rain
- ▦ Lighter tropical
- ▨ Scrub and thorn

Middle latitude forest
- ▨ Mediterranean scrub
- ▩ Coniferous
- ▨ Broadleaf and mixed broadleaf-coniferous

Grasslands
- ▨ Savanna
- ▥ Prairie
- ▥ Steppe

Desert and waste
- ▨ Desert
- ▥ Tundra
- □ Ice cap
- ▨ Undifferentiated highlands

FIGURE 12-12. Natural vegetation throughout the world. [Courtesy USDA.]

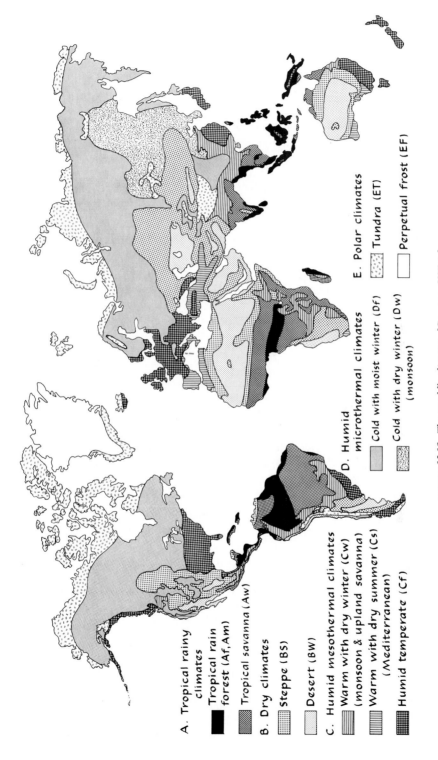

A. Tropical rainy climates
    ■ Tropical rain forest (Af, Am)
    ▨ Tropical savanna (Aw)

B. Dry climates
    ▨ Steppe (BS)
    ▨ Desert (BW)

C. Humid mesothermal climates
    ▥ Warm with dry winter (Cw) (monsoon & upland savanna)
    ▤ Warm with dry summer (Cs) (Mediterranean)
    ▦ Humid temperate (Cf)

D. Humid microthermal climates
    ▨ Cold with moist winter (Df)
    ▨ Cold with dry winter (Dw) (monsoon)

E. Polar climates
    ▨ Tundra (ET)
    □ Perpetual frost (EF)

FIGURE 12-13. The world's climates. [Courtesy USDA.]

Although rainfall is abundant (rarely less than 30 inches per year), its distribution serves to produce two subclimatic types. One type has ample rainfall throughout the year, and although there are seasonal differences, no less than $2\frac{1}{2}$ inches of rain falls in any month. The other subclimate is distinguished by distinct wet and dry seasons.

Tropical soils, which are chiefly **lateritic** soils (red soils rich in iron and aluminum hydroxides), are notoriously unproductive when continually cultivated. The rapid destruction of organic matter in such soils results in a soil structure that deteriorates under intensive crop use. Their low native fertility, a result of constant leaching, makes them responsive to fertilizer application.

In the continually rainy areas of the tropics the natural vegetation is lush and varied. No other area can compare with the tropics in the diversity of plant species. The natural vegetation is broad-leaved and evergreen and has no dormancy requirement. In the seasonally wet and dry areas of the tropics the dominant vegetation is coarse grass, which goes dormant under dry periods. A tropical grassland is called a **savanna,** and the climate that produces it is called a **savanna climate.** The number of trees to be found in such places is determined by the duration of the wet season.

The chief horticultural plants of the tropics grow continuously and have no dormancy requirement. As a group they are quite sensitive to any extended periods without moisture. However, agriculture is generally more extensive in the wet-dry areas of the tropics because continuous rain is unsuitable for many crops. Wet-dry areas are used widely, without irrigation, for grazing. Although many tropical horticultural crops are known principally in their native areas—mango, yam (*Dioscorea*), papaya—many important crops of worldwide significance must be grown in tropical or subtropical climates exclusively. The separation of plant agriculture into horticulture and agronomy is less meaningful in the tropics. Although sugar and rice are considered agronomic crops, the distinction of such crops as palms (for oils), jute, and rubber as horticultural is justifiable because of their intensive culture.

Dry Climates

Dry climates are those in which the potential for evaporation from the soil surface and for vegetative transpiration exceeds annual rainfall. Rainfall is scarce and unpredictable, and there is often a water deficiency. Because evaporation and transpiration increase with temperature, the boundaries of dry climates are affected both by rainfall and by temperature. The dry climates are roughly distributed on either side of the tropical rainy climates.

The dry climates are subdivided on the basis of moisture and temperature into **arid (desert)** and **semiarid (steppe).** Steppe is a transitional zone between humid and desert climates. These dry climates may also be subdivided into tropical and subtropical regions on the basis of minimum temperatures. Thus, there are four general regions—hot and cold desert, and hot and cold steppe.

Although natural vegetation is meager in areas with dry climates, the abundance of such areas (26% of the earth's surface) makes them an important agricultural resource. Because the steppe areas support low-growing, shallow-rooted grasses, their chief agricultural use is for grazing.

Since the soils of dry climates are subject to minimum leaching, they are high in nutrients. Although desert soils tend to be too sandy and otherwise unsuitable for agriculture, steppe soils are superior. The grass roots provide considerable organic matter. On steppe soils, irrigation can transform the land, making it ideal for horticulture. The irrigated lands of our Southwest provide a splendid example of this. There, such subtropical horticultural crops as melons, figs, and citrus are produced in the warmer season, and such temperate-climate vegetables as carrots and lettuce are grown in the cooler season. The agriculture of this area, dependant as it is on irrigation, resembles the oasis agriculture of naturally moist areas (deltas, flood plains, and alluvial fans) within dry climates.

Humid Temperate Climates

The climate of the middle latitudes is characterized by a distinct seasonal rhythm; winter, spring, summer, and autumn have real meaning. The chief factor influencing plant dormancy is low temperature, rather than drought. The humid temperate climates are subdivided into mild-winter (mesothermal) and severe-winter (microthermal) climates. This distinction is based on an average minimum monthly temperature of $32°F (0°C)$. This can be empirically found on the basis of a durable winter snow cover.

Mild-Winter Climates

The mild-wintered portions of the temperate zones are found in the lower latitudes and in marine locations on the westerly side of continents. Within the mild-wintered temperate zone there are three important climates: **Mediterranean,** (2) **humid subtropical,** and (3) **marine.**

MEDITERRANEAN CLIMATE. The Mediterranean climate is a subtropical, dry-summer climate. It represents a most important horticultural region. It is characterized by dry, warm-to-hot summers and mild winters with adequate rainfall. Fog is common and provides some summer moisture. The inland areas have hotter summers than the coastal areas. Because of its delightfully mild winters, bright sunny weather, and strong horticultural associations, it is one of the best known climates. It is found in the Mediterranean region, central and coastal California, central Chile, the southern tip of South Africa, and parts of southern Australia. Although these areas constitute less than 2% of the land, they provide a large portion of the world's horticultural products, especially subtropical fruits (citrus, figs, dates, grapes, and olives).

The Mediterranean soils are variable. The natural vegetation consists of mixed trees and stunted woody shrubs. The leaves of the trees and shrubs are typically thick and glossy, designed to prevent excessive transpiration. The growing season lasts practically the whole year. However, frosts do occur during the winter months. These may be extremely dangerous to horticultural crops that are marginal for the area, such as out-of-season vegetables and citrus. Although the annual rainfall is moderate (15–20 inches), the effective precipitation is high, since the bulk of the rain comes in the cool winter, making much of it available for plant growth. Summers are distinctly dry. This has resulted in a double dormancy in native vegetation: one due to low winter temperatures, the other to summer droughts. The dry heat of the summer is ideal for fruit drying. It is here that this method of fruit preservation originated, and this area still provides dried figs, dates, plums (prunes), and grapes (raisins).

Flower-seed production is an important part of the horticultural industry in this climate in California. Under irrigation, an extensive fruit and vegetable canning industry has developed. Cool-season crops are shipped almost exclusively from California during the winter months. For example, the Imperial Valley provides most of the United States' winter supply of lettuce.

HUMID SUBTROPICAL CLIMATES. The humid subtropical climates are typically on the eastern side of continents (for example, the cotton belt of the United States). The annual precipitation is usually abundant (30–65 inches) and well distributed over the entire year. The summers are hot and muggy. Although winters are generally mild, freezing temperatures occasionally occur. There is, however, considerable variation; in the United States the South experiences more severe winters than do the humid subtropical areas of South America, Australia, or China.

The natural vegetation includes forest (mixed conifers and broad-leaved trees, both deciduous and evergreen) and grassland, depending on precipitation. The soils are deep, owing to continuing weathering, but they are typically low in fertility as a result of constant leaching. The grassland soils are more productive.

The humid subtropics are rich vegetable-producing lands. For example, the long growing season and high summer temperatures of the southeastern United States are suited for the growth of such crops as sweetpotatoes, dried beans, and melons. The relatively mild spring and fall allow two crops of such garden vegetables as spinach, mustard greens, radishes, and snap beans. Tomatoes, peppers, and celery are grown as spring crops. A large industry based on the growing of vegetables and flower transplants for shipment to northern states is located in this area. Only those fruit crops having a low winter chilling requirement can be grown. Thus, this area is more important for growing peaches than it is for apples. In sites where the danger of spring frost is minimal, peaches are an important crop. This is particularly true of the Gulf states. Strawberries and blueberries, which have a wide range of adaptation, are common horticultural crops. The nursery industry is expanding along the northern edge of this area.

MARINE. Marine climates are found on the western side of middle latitude continents and extend into the high latitudes owing to the ocean's influence. The extent to which these climates penetrate depends on topographical features. In North and South America and in Scandinavia, areas with marine climates are narrow because mountain ranges closely parallel the coast. In western Europe extensive lowlands serve to carry the ocean's influence inland.

In contrast to Mediterranean and humid subtropical climates, the marine summers are cool. The warmest months typically average 65–70°F (18–21°C). Winters are exceptionally mild considering the latitude, the average minimum monthly temperature being above freezing. The growing season is long. Frosts are frequent, however, and winter is generally long enough to produce a dormant season.

There is adequate rainfall in all seasons, although precipitation varies greatly, from more than 100 inches in the Pacific Northwest to 20–30 inches in parts of western Europe, which is surprising considering the large number of rainy days. Sunshine is limited; fog and mist are abundant. About half the days each year are cloudy.

The natural vegetation is forest, deciduous in Europe and coniferous in the Pacific Northwest. The soils are variable though generally deep. Gray forest soils of this area have excellent structure, and although their natural fertility is moderate, it is higher than that of other forest soils. In addition, there are areas of organic soils (peat-bogs), areas of thin, stony, glacially deposited soils (Scandinavia), and sandy coastal plains (Germany and Denmark).

Horticulture is not the predominant agriculture in marine climates, but it is extensive. Apples and pears are well suited to marine climates, as is the ubiquitous strawberry. Cool-season vegetables—peas, lettuce, and crucifers—do especially well. There is not enough heat, however, for outdoor plantings of some crops, such as melons. For example, cucumbers and grapes are commonly grown under glass in Holland, Belgium, and England. The long spring extends the flowering season. Bulbous flowers (such as tulips, daffodils, and hyacinth) grow well.

Severe-Winter Climates

The severe-winter climates of the humid temperate regions are characterized by summers that are distinct but shorter than those of the mild-winter climates. These climates, which are located in the interior and poleward areas of North America and Eurasia, are dominated by large land masses. They are bounded in the north by **polar** climates. The climate is divided into areas of warm summers (U.S. corn belt), cool summers (Great Lakes area, western and central Russia), and subarctic (northern Canada, Alaska, Siberia). The insulation provided by permanent deep snow cover serves to prevent excessively low soil temperatures (although winter air temperatures become extremely low in the higher latitudes and vary greatly). Precipitation is generally greatest in the summer, owing to the influx of warm, humid air from the south; the winters, by contrast, are influenced by cold polar air. This is significant in regard to plant growth. Although yearly precipitation is moderate, ample moisture is usually available during the growing season. Largely because of this relatively high effective precipitation, the cool- and warm-summer climates support an abundant agriculture.

The natural vegetation consists of forests (in which conifers predominate) at the northernmost margins, and tall-grass prairies in the sub-humid interiors. The gray-brown soils of deciduous forests are good, but the highly acid **podzol** soils of the cooler, more northerly coniferous forests are poor. The prairie soils are excellent in structure and in fertility, which makes these lands particularly suitable for corn and wheat despite the somewhat lower and less reliable rainfall. Although prairie soils are limited in extent, they make up the finest agricultural lands in North America. Glaciation has molded many of the topographical features in this climate (for example, the rocky New England hills).

The warm-summer climates are ideal for such horticultural crops as melons and tomatoes, which require high temperatures. The cool-summer climates more typically support extensive plantings of hardy fruits, such as apples and pears, although cherries and peaches can be produced in protected sites, such as those found in the Great Lakes area. Cool-summer climates are ideal for cool-season vegetables, such as potatoes and peas. Sweet corn is an important canning crop in the northern part of the United States, where cool summer temperatures tend to prolong the period of good quality for harvesting. The nursery industry has taken advantage of the deep prairie soils; bulb crops, strawberry plants, and nursery shrubs are grown extensively in this region. For example, Iowa, located in the heart of the corn belt, is an important center for mail-order nursery businesses specializing in deciduous plants. These ship well when fully dormant.

The subarctic regions of the severe-winter temperate climates support only a limited horticulture. The long summer days compensate for the cool temperatures and lower light intensity. Here, summer frosts may be a serious hazard. The frost-free season may last only 50–75 days. The important horticultural crops are root crops (such as potatoes, turnips, beets, carrots, and parsnips) and crucifers (such as cabbage and cauliflower). As a result of breeding programs in Alaska, some strawberry cultivars adapted to subarctic climates are now available.

The polar climates are found in the high latitudes and at high altitudes. The boundary is set by a mean annual temperature of 32°F (0°C), with the average warmest monthly temperature being 50°F (10°C). There is no horticultural activity in polar climates.

ECONOMIC FACTORS

In addition to such environmental factors as climate and soil, a number of economic considerations strongly affect the complexion of the horticultural industry. These include land costs, availability of labor, distance to market, and the nature of available transportation facilities. The general level of the economy is also an important factor. A highly developed industrial economy is able to afford an abundance of horticultural products. In the United States an increase in the standard of living results in an increase in the consumption of fresh fruits and leafy vegetables. In addition, the perishable nature of many horticultural crops necessitates advanced technology for movement, storage, and processing. In agricultural societies, most horticultural products are grown for local consumption, and a significant proportion of the products are actually home grown. As a result, the total variety and value of horticultural products is less than in highly developed economies.

Horticultural crops differ as to their adaptability. In general, long-season crops, biennials, and perennials must be restricted to areas where weather remains favorable for extended periods of time; many months are required for long-season annuals, and many years are required for tree-fruit production. This greatly restricts their location, especially when markets are highly competitive. On the other hand, annual crops that mature in a comparatively short time often appear to have a wider range of adaptability. This is also true for biennial plants (such as carrots, onions, and cabbage) that are grown as annual crops. Actually, it is their adaptability *as crops* that is large: they may produce a marketable crop in a short time in any one of a number of diverse climates, but it may be necessary for seed to be produced elsewhere under more favorable conditions.

The wide adaptability of annual horticultural crops makes them responsive to local economic advantages. Increased quality and local market preference account for some of this. With technological advances in the long-distance movement of horticultural crops these local advantages decrease in importance. The trend is for annual crop production to be situated in optimum climatic locations, especially those favorable to off-season production. Nevertheless, the great cities are still ringed by relatively small market gardens that grow fresh produce for local consumption, although their number is decreasing. With the increase in processing and the increase in centralization of the marketing operations of chain stores, the demand for the products of local growers has dwindled to an insignificant level in many areas.

Land Costs

Land cost includes not only land prices but local taxes as well. Horticultural crop production is often intensive enough to justify the use of expensive agricultural land, but it cannot long survive in urban areas. The horticultural enterprises originating on the outskirts of large cities may be literally overrun by urban expansion. High urban taxes soon make any agricultural operation unprofitable. Although "selling out" may result in an immediate profit to individual land owners, the horticultural industry is destroyed, and valuable agricultural land is often lost forever. The disappearance of potato farming on Long Island, New

York, and of the citrus industry around Los Angeles, California, are examples of such losses.

Labor Supply

Many horticultural operations require an abundant supply of labor at some point, usually during harvest. Because of the extensive need for hand work in the past, the horticultural industries have been dependent upon a plentiful supply of low-cost labor. In many parts of the United States, migrant workers move from one horticultural area to another, harvesting crop after crop as the season progresses. In some areas, workers are brought in from other countries. The organization of this labor, or Federal and State legislation requiring minimum wages and standards, increases labor costs and, therefore, the cost of production. This need for transient labor has created great problems and has encouraged the automation of such operations as transplanting, weed control, and harvesting. In two years, for example, the California tomato processing industry changed from total hand picking to 85% mechanized harvesting. Although there is a definite trend toward mechanization, many horticultural operations will continue to depend upon large seasonal labor resources, at least for the foreseeable future.

The recent move of the greenhouse industry away from large centers of population to more rural locations is due to rising labor costs as well as to increased taxation. Cheap labor, however, is not a dependable resource. The solution to extensive labor requirements undoubtedly lies in labor-saving devices and technological innovations.

Market Advantage

Historically, the commercial horticultural industry originated close to the large centers of population. The perishable nature of most horticultural products gave a distinct advantage to market proximity by virtue of the close relation of proximity to quality. As transportation and storage facilities have improved, this advantage has steadily diminished. Rapid, refrigerated transportation facilities operating over great distances have equalized the quality differential. The integration of railroad and motor-truck systems have further increased transportation efficiency. Recent advances in the air shipment of such high-priced horticultural products as flowers and strawberries have reduced the advantages of market proximity to growers of these highly perishable commodities.

Transportation costs have remained relatively constant over the years, and do not, therefore, account for the decrease in the industry's dependence on market proximity. Other factors that determine the distribution pattern are market price, season of shipment, and cost of production. Transportation costs, however, are still a significant cost factor (Table 12-2); but when the reduced cost of production of

TABLE 12-2. Transportation costs for 'Delicious' apples, 1955–1957.

State of origin	City marketed	Retail price per pound (cents)	Transportation cost per pound (cents)	Transportation cost as a percentage of retail price
Washington	Los Angeles	20.5	1.4	7
Washington	Chicago	20.6	2.3	11
Washington	New York	20.0	2.8	14
New York (Hudson Valley)	New York	15.7	0.5	3

SOURCE: Data from Agricultural Marketing Service, USDA.

a different region offsets excessive transportation costs, the industry is bound to move.

One of the chief factors in market advantage has been out-of-season production. Thus, early strawberry production in Louisiana and winter production of vegetables in warmer southern and western climates have created important horticultural industries, despite poor soils and the danger of frosts.

The consolidation of marketing operations by large food chains, which deal directly with growers, favors large operations located in areas with the best climates for production. For example, fewer than 200 buying concerns handle 60% of the nation's fruit. The trend is toward the centralization of the industry in response to climatic factors.

CULTURAL FACTORS

In addition to environmental and economic factors, there are a number of cultural factors that have played a significant role in the distribution of the horticultural industry. In many cases the development of the industry in the United States can be traced along with the history of certain national groups. For example, the greenhouse industry was brought to America primarily by the Dutch immigrants, which is reflected today in the large percentage of people of Dutch descent who still engage in the commercial production of vegetables under glass in the Midwest. This is also true of the onion-set industry located around Chicago. Similarly, the Japanese flower and vegetable growers are prominent in West Coast horticulture.

Many local horticultural enterprises of considerable scale relate to a particular family or firm. A single grower in Wisconsin produces a significant percentage of the horseradish used in the United States. The success of the Hill Nurseries, the largest rose center in the United States, has brought a large floricultural industry to Indiana, although there are better adapted areas in terms of light and temperature. The carnation industry in New England traces its origin to a few private conservatories in wealthy estates.

Market preferences have greatly affected the horticultural industry. For example, collards and mustard greens are typical Southern dishes. The migration of people from the South to Chicago has resulted in these crops being grown by market gardeners located in the Chicago area. Similarly, the large number of Chinese restaurants in the East has stimulated specialized vegetable enterprises in Long Island to serve this trade. In general, these unique markets tend to decrease in importance in the United States as the mobility of the population increases. However, national habits and food preferences change relatively slowly.

Selected References

Cox, G. W., and M. D. Atkins, 1979. *Agricultural Ecology: An Analysis of World Food Production Systems.* San Francisco: W. H. Freeman and Company. (An authoritative but readable treatment of the subject as it applies to agricultural systems.)

Geiger, R., 1965. *The Climate Near the Ground.* Cambridge, Mass.: Harvard University Press. (A revised translation—from the fourth German edition—of an outstanding work on microclimate.)

Gleason, H. A., and A. Cronquist, 1964. *The Natural Geography of Plants.* New York: Columbia University Press. (An interesting text on plant geography.)

Klages, K. H. W., 1949. *Ecological Crop Geography.* New York: Macmillan. (A discussion of the factors determining the distribution of crop plants.)

Trewartha, G. T., 1968. *An Introduction to Climate,* 4th ed. New York: McGraw-Hill. (An elementary but comprehensive text.)

U.S. Department of Agriculture, 1941. *Climate and Man* (USDA Yearbook, 1941). (A broad, popular treatment.)

Wilsie, C. P., 1962. *Crop Adaptation and Distribution.* San Francisco: W. H. Freeman and Company. (A general treatment of the principles of crop ecology.)

13

Horticultural
Production Systems

Her vine, the merry cheerer of the heart,
Unpruned dies; her hedges even-pleach'd,
Like prisoners wildly overgrown with hair,
Put forth disorder'd twigs. . . .
And all our vineyards, fallows, meads, and hedges,
Defective in their natures, grow to wildness.
Even so our houses, and ourselves, and children,
Have lost, or do not learn for want of time,
The sciences that should become our country. . . .

SHAKESPEARE, Henry V [V. 2]

A list of the horticultural production systems existing in the world today is also a list of all the production systems that have been developed in the course of the history of plant agriculture. Which of these systems is utilized in a particular area is determined by the crop, the population density, the cultural and technological level of the society, and the forces of tradition. The variation is very great—from primitive planting of a mixed garden to intensive cultivation of a single crop. In New Guinea, primitive gardens are cultivated by isolated tribal people with a stone-age culture, who, primitive in the use of tools yet sophisticated in the lore of plant material, practice a lengthy rotating system of production in the natural forest based on clearing, burning, production, and abandonment, uninfluenced by the money economy. Banana production in Central America constitutes a highly capitalized plantation system that is in the hands of two large North American corporations whose land holdings are extensive and who have their own internal transportation systems and shipping facilities; the technical personnel come from the United States, but the labor force is made up of local people, many of whom live in company towns. The greenhouse vegetable industry in Holland is based on two crops, tomatoes and cucumbers; it is highly mechanized and the growing operations are conducted by small family businesses whose marketing throughout northern Europe is handled by relatively large cooperatives.

The economic complexity of horticultural crop production is in large part related to the general agricultural situation. At present the limit to agricultural progress has been set by the slow rate of economic development in many parts of the world; the striking increase in productivity that has been achieved through advances in agricultural science has not benefited all countries. In much of the world, such increases will depend not so much upon improved technical knowledge as upon elimination of both rural poverty and the insecurity of farm tenure, and upon increased availability of reasonable credit. In many areas farm prices are so unstable that they discourage the investment required for intensive mechanization of agriculture. In others, marketing channels are inadequate for handling perishable crops.

Within the past decade, however, the production of many horticultural crops has continued to show an upward climb (Table 13-1). This reflects not only the increase in world population but also the increase in world trade and the expansion of processing industries. Lengthening the marketing season by cultivar breeding and selection and by improving the technology of distribution has increased the availability of horticultural crops throughout the year. The following text of a broadcast of the British Broadcasting Company, aimed at the English housewife, illustrates this nicely.[1]

Fruit

If you're looking for luxury this week-end this is undoubtedly the time to buy a pineapple. Those superb large pineapples from the Azores—normally very expensive—are being sold in a number of shops at less than their usual price. There's no catch in it—they are of excellent quality; it just happens that the market price has fallen. Look for bright color on the skin and fresh looking foliage.

If the price is still above your budget, or if you can't find any in your local shops, there are plenty of smaller, far cheaper, pineapples, from South Africa and Jamaica. Their flavour is not as full as that of the Azores ones, but for a fraction of the price they're pretty good all the same.

The first of the blood oranges from Spain and Morocco will already be in a few shops this week-end. Their flavour is pleasant—they're very rarely sharp—and the bright red mixed with the orange colour is attractive.

Israeli oranges, too, are excellent now. And I see the Manchester report is recommending for value large oval oranges imported from Turkey. Another best buy listed in nearly all areas is bananas.

There has not been a heavy crop of South African peaches this year; although there's been a relatively big arrival this week. But you can't expect any low prices. Gaviota plums too—they're the yellowish-orange ones—are not cheap but they're large juicy fruit.

South African grapes—black and white—are now in most areas.

Here's something unusual for this time of year; there have been fresh arrivals of remarkably good quality chestnuts from Italy.

---

[1]Extract from "Shopping List," broadcast by the BBC, February 13, 1959.

TABLE 13-1. World production totals, for vegetables, fruits, and nuts (in thousands of metric tons), 1971–1976.

Economy, Region	1971	1972	1973	1974	1975	1976
Vegetables (including melons)						
Developed	88,559	87,186	90,071	91,437	94,044	91,369
North America	23,322	23,469	24,479	25,049	26,784	24,786
Western Europe	47,010	45,318	47,677	48,610	48,730	47,450
Oceania	1,304	1,220	1,179	1,300	1,309	1,346
Other	16,923	17,179	16,737	16,479	17,221	17,786
Developing	91,946	94,420	94,644	98,898	102,528	105,017
Africa	9,258	9,315	9,193	9,870	10,416	10,770
Latin America	12,449	12,846	13,303	13,542	13,693	13,843
Near East	23,004	24,126	23,221	25,379	27,008	27,783
Asia and Far East	46,985	47,877	48,664	49,839	51,136	52,341
Other	250	256	262	269	274	280
Centrally Planned	106,268	109,405	115,382	114,967	112,695	115,392
Asia	63,306	64,627	66,449	68,031	69,878	71,352
USSR and Eastern Europe	42,964	44,778	48,933	46,936	42,816	44,040
World totals	286,773	291,010	300,097	305,303	309,267	311,777
Fruits						
Developed	91,866	89,342	104,947	100,278	101,045	98,489
North America	22,441	20,343	24,022	23,733	25,309	24,729
Western Europe	57,045	55,205	67,135	63,065	61,753	59,195
Oceania	2,276	2,503	2,279	2,113	2,349	2,352
Other	10,104	11,292	11,512	11,367	11,634	12,212
Developing	111,177	113,109	114,790	122,831	125,279	130,001
Africa	21,310	21,358	22,357	22,787	22,873	23,727
Latin America	46,787	47,941	47,330	52,374	52,821	56,490
Near East	14,061	14,221	14,347	15,141	15,404	15,869
Asia and Far East	27,998	28,550	29,696	31,448	33,086	32,784
Other	1,020	1,039	1,062	1,081	1,095	1,130
Centrally Planned	28,135	25,681	30,919	28,817	31,236	32,946
Asia	7,294	7,319	7,658	7,842	7,771	7,893
USSR and Eastern Europe	20,841	18,362	23,261	20,976	23,465	25,054
World totals	231,178	228,132	250,656	251,927	257,560	261,436
Nuts						
Developed	1,311	1,339	1,445	1,458	1,454	1,519
North America	374	318	422	389	457	437
Western Europe	882	960	955	1,005	931	1,011
Oceania	2	2	2	2	1	1
Other	53	60	67	62	65	69
Developing	1,367	1,438	1,512	1,528	1,596	1,540
Africa	485	478	540	551	486	480
Latin America	144	150	145	120	138	121
Near East	484	544	558	591	694	653
Asia and Far East	251	263	265	263	274	283
Other	3	3	3	3	3	4
Centrally Planned	554	540	584	559	590	602
Asia	258	265	267	276	284	290
USSR and Eastern Europe	295	276	317	283	306	312
World totals	3,232	3,317	3,541	3,545	3,640	3,662

SOURCE: *FAO Production Yearbook 1976*, Vol. 30, 1977.

Vegetables

The Bristol report says there are improved supplies of leeks. Also well worth buying are carrots and all the cabbage family, including savoys and spring greens. There are more parsnips now and generally their quality has improved.

I've been getting some well flavoured, reasonably priced tomatoes. And I see the report from Glasgow agrees that Canary tomatoes are in good condition now. But you need to see that they are firm.

There are some home-grown cucumbers in a few shops, and if you think they're expensive, I'm told they are not as high in price as usual at this time of year.

The report on Wednesday was of lower prices of home-grown potatoes. There is also a bewildering variety of imported ones, the latest arrivals coming from the Canary Islands; these aren't cheap, but they do include some which can be scraped.

In contrast to the world prices of many of the staple agricultural crops, the world prices of nearly all horticultural crops are free from artificial controls. However, in many countries there are stringent import controls that provide protection for domestic horticultural production.

SUBSISTENCE HORTICULTURE: SHIFTING CULTIVATION

A primitive production system known as **shifting cultivation** or **swidden** is practiced by relatively isolated tribal people in many areas of the humid tropics, particularly the tropical highlands of Africa, Indonesia, and the Americas. Shifting cultivation is basically a system of subsistence agriculture characterized by cutting and burning the natural forest, long periods of fallow, lack of tools and draft animals, low population density, and low consumption. Shifting cultivation is usually considered to represent an early stage in the development of cultivation systems—between food gathering and permanent cultivation—but it is often practiced on land with fragile soils ill suited to permanent intensive agriculture, and under such conditions has become a highly developed (though still primitive) system.

The system requires a minimum of work: after the forest has been cleared and burned (hence the sobriquet **slash-and-burn**), little soil preparation is required, although in some locations fences may be necessary. The "garden" is a mixed planting of a diversity of species, including annuals and perennials, but generally only such short-term crops as yam, taro, banana, sweetpotato, and cassava. A garden in full growth appears to be a random assortment of crops of various heights and shapes. Typically it is kept free of weeds. As crops are harvested, they are replanted until, after two or three years, the garden is abandoned and the area reverts to forest.

Attitudes toward shifting agriculture vary. Many agriculturists view the system negatively, describing it as wasteful, extravagant, and an impediment to development. This is because of the obvious waste of timber, the use of burning (a distrusted technique in many parts of the world), the extensive use of land, and a skeptical attitude about the effectiveness of a system that makes little use of technology and is employed by primitive people. On the other hand, anthropologists interested in agriculture have romanticized shifting agriculture in ecological terms. Indeed, at its optimum development shifting cultivation is remarkably in tune with the environment: the lack of tools and draft animals eliminates soil compaction, the mixed planting and permanent cover reduce erosion, the gardens mimic the natural tropical forest in their multistoried complexity and diversity. The nutrients locked in the organic-matter cycle are temporarily released by burning, and the relatively short cropping period allows the land to revert to forest even on steep slopes. The soil structure is restored during the long fallow, which may last as long as 25 years.

The different points of view of agriculturists and anthropologists reflect their different disciplines. The anthropologist is likely to be impressed by a system in which a highly evolved social structure with complex rituals and taboos stabilizes the relation of people to their environment. For instance, in New Guinea some societies are kept in balance by engaging in symbolic warfare that rarely takes human life. There, shifting cultivation, combined with pig raising, which serves both as a means of stockpiling protein and as a symbol for status and ritual, is a remarkable example of adaptation to the forest environment. The natural forest is made to work for people rather than against them, being transformed into a forest-garden. If energy input is measured, rather than yield per acre, the system is remarkably efficient. (The division of labor in which men occasionally clear and burn while the women garden may be another feature of no little attraction to the men of the tribe!)

The carrying capacity of shifting cultivation may be only 100 people per square mile. A breakdown of the ecologically balanced subsistence economy is brought about if population is increased beyond this limit. But such increases often result as new people migrate into an area and impose a civilization that, by comparison, is characterized by less self-sufficiency, a greater number of necessities, and the replacement of society-stabilizing traditions with alternating periods of tranquility and upheaval, often with attendant violence. To feed the increased population the fallow interval must be decreased and shifting cultivation must be expanded to less humid areas where the deciduous forest recovers slowly from cutting and fire is more difficult to control. An increase in intensive use of land, which is usually accompanied by a shift to monoculture, destroys the ecological equilibrium and results in large-scale deforestation and increased erosion. The equilibrium of the tropical environment is particularly delicate and fragile, as the damage done to the forests of Vietnam during the recent war has clearly shown. Deforested hills in Indonesia, now taken over by cogon grass (*Imperata cylindrica*) and likened to a green desert, attest to this. Under such conditions, a highly developed, sophisticated, primitive society becomes a poorly developed, impoverished parody of civilization. At this point the agriculturist becomes disenchanted with the remnant of shifting cultivation—and with ample justification. The anthropologist also loses interest, his enthusiasm being inversely proportional to the similarity of any culture to his own!

Because of its low carrying capacity, shifting cultivation is incompatible with the civilizations of the world's developed areas. It is probable that this system in its pure form will fade into extinction. Yet shifting cultivation has contributed much to the developed world in terms of germ-plasm preservation, the discovery of crop plants (many yet unappreciated), and, most of all, the conservation of natural areas. Further, the system has suggested ways to utilize the humid tropics wisely through fallowing and the development of a natural forest horticulture. Agricultural schemes for these areas have often been carried out in ways best suited to temperate areas (clean cultivation is an example), often with disastrous results.

PLANTATION HORTICULTURE

Plantation horticulture, which is practiced almost exclusively in tropical areas, is characterized by large-scale, centrally managed, commercial production of export crops. At least partial processing accompanies production (for example, oil extraction, drying, or canning). The plantation is a combination agricultural-industrial enterprise tied to the industrial economics of temperate areas rather than to local conditions. Labor, usually hired, is intensive, although this is changing as the new agricultural revolution replaces hands and backs with machines and chemicals. The plantation system tends to be monocultural with continuous year-round production. Management is usually by foreign experts, in the past often North American or European. Typically the plantation is a corporate structure oriented to the creation of profits. High capitalization is a distinguishing feature, and this often involves highly developed internal and external transportation systems (ranging from plantation railroads to refrigerated steamship lines).

The origins of the plantation system can be found in European incursions into tropical America in the sixteenth century. Large-scale production of crops in the tropics for export to temperate countries reached an early pinnacle in the sugar trade from the West Indies. The system ran on a combination of black slavery and extravagant debt financed by London merchants and speculators, and in its heyday in the seventeenth century a saying went: "As rich as a West Indian planter." Croesus had been replaced! The success of the plantation system made the sugar islands of the Caribbean extremely valuable pieces of real estate. Although the boom collapsed, in part because of its own weight but also because of the replacement of tropical cane sugar by temperate beet sugar, the system had proved its capacity for great profits. In the subsequent centuries the plantation system rapidly spread to the rest of the tropical world with other tropical crops, most of which fit a horticultural definition by most standards. These include palm oil in Africa, rubber in Sumatra and Malaya, tea in Ceylon, coconut (for copra) in the Philippines, bananas in Central America, and pineapples in Hawaii.

Under the colonial system, in which "flag followed trade," the plantations were protected by the military and legal apparatus of the mother country, who profited in a number of ways: profit from the enterprise filtered back to the investors in the form of dividends; the mother country profited by supplying manufactured goods, and—more important—controlled the trade processes both ways. One of the causes of the American Revolution was the British legislation that inhibited the growth of American industry and ensured that all goods would travel to the Colonies in British ships.

Protection of plantation property led to the development of colonial empires; puppet regimes passed quickly to outright foreign rule, a political pattern that had deleterious effects on tropical regions. The stress placed on monoculture of export crops retarded the establishment of diversified agriculture as well as industrial development, and resulted in many agricultural countries lacking self-sufficiency in food production. The Caribbean sugar islands at the height of sugar production in the seventeenth century depended on North American flour and Newfoundland codfish for food, and that condition still exists. In addition, the colonial superstructure purposefully refused to develop a native managerial class. This was especially true in the Dutch East Indies (now Indonesia), where two parts of a dual economy—plantations based on sugar and rubber, and subsistence farming based on wet (paddy) rice—operated almost independently.

At its best the plantation system is technically efficient and makes good use of resources. However, because of the high capitalization required and the general poverty of the tropical world, plantations typically have been financed and managed by aliens, sometimes with extreme disregard of the welfare of the local people. Even when management is benign, a good case can be made to show that the development of the tropical country is retarded or hampered by external control.

In spite of the added wealth that it provided to the tropics, the plantation system ran counter to national purpose and pride. In the rush toward self-government after World War II, the plantation system was allowed to deteriorate or was junked outright. It is unfortunate that its social deficiencies brought about the disintegration of what was, in large part, an efficient production system. In many areas plantation agriculture represents the best use of the land. Where the idea that plantations are no more than manifestations of colonialism can be dispelled, it is likely that the system will be maintained. This can come about either through training local people for executive positions, or through changes in land tenure and decentralization of processing and shipping operations.

In response to the rise of nationalism and the popular appeal of "agrarian reform," in some areas the former land holdings revert to local ownership or lease while the industrial sector of the operation remains centrally managed. Technical assistance then must be provided by agricultural "extension" agents for the system to remain efficient.

The plantation system is by no means a dead institution. The economies of many developing countries are dependent upon tropical agriculture for foreign exchange revenues, and in many, a single tropical tree crop constitutes more than half of the total commercial exports. Indications are that the plantation system is in fact moving into temperate areas, where it is often known as **corporate farming.** Spurred by tax advantages, many corporations in the United States have acquired large holdings of nut and citrus orchards. Mechanization and the unionization of farm labor are rapidly removing the differences between urban and rural industry.

The pineapple industry in Hawaii is an example of a highly developed modern plantation system (Fig. 13-1). On 75,000 acres, four companies produce a processed product with an estimated annual value of $125,000,000. All operations from growing to processing are highly mechanized, and many machines have been developed expressly for the industry. The industry operates a centralized research program, and labor is unionized and paid on industry scales. Although increasing labor costs in Hawaii have caused growers to establish plantations in the Philippines, technological progress is drastically reducing the labor input per ton of processed fruit. The fruit is still hand harvested, but self-propelled conveyors now carry it to the packing shed, and the use of growth regulators to cause all the fruit to ripen at the same time will soon make possible complete mechanical harvesting that should restore the competitive position of Hawaiian pineapple.

Although the structure of banana production varies with location, large corporate plantation systems predominate, especially in Central America. Medium-sized holdings (estates) are typical in Ecuador, Colombia, and the Dominican Republic, with small holdings predominant in Jamaica, Mexico, the Caribbean, the Canary Islands, West Africa, and the Pacific Islands.

Typically, the banana-plantation systems of Central America, especially Guatemala, Honduras, Costa Rica, and Panama, are vertically integrated. Private firms own the fruit from planting to its sale in the distribution channels of the consuming country. Two North American companies, United Fruit and Standard Fruit, account for 80 percent of production in Honduras and 100 percent in Guatemala. Production is organized in divisions ranging in size from 10,000 to 30,000 acres; these are divided into unit farms of 800 to 1,000 acres, each with its own operating facilities and labor force. The operations are heavily capitalized. The Honduras Division of Standard Fruit, which embraces 12,500 acres, employs 6,000 workers, owns 300 miles of railroads with more than 900 cars, and has an annual budget of $15 million. The farms are irrigated, and pests are controlled by airplane application of pesticides. Shipping and marketing facilities are company-operated, as are plants for the manufacture of corrugated boxes. Harvested stems are transported by overhead monorail through the washing area, and selected fruit is cut from the stem in clusters (hands). The hands are graded, packed in trays or corrugated boxes, transported to the docks on the company-owned railroad, and loaded on refrigerated ships either owned or chartered by the company.

FIGURE 13-1. Pineapple production in Hawaii. *Top,* a plantation on Oahu, with Diamond Head crater and Honolulu in the distance. *Bottom left,* harvesting with a conveyor system. *Bottom right,* unloading and washing fruit before processing. [Courtesy Dole Corp.]

COMMERCIAL HORTICULTURE

The commercial production of fruit and vegetables in both temperate and tropical regions is largely conducted by relatively small units, usually family-operated enterprises. The intensive nature of horticultural production results in less extensive land holdings but higher capitalization than in other segments of plant agriculture. The trend, however, is toward larger units and even greater capitalization as mechanization increases.

Because of the differences in planting, growing, and harvesting different kinds of crops, commercial horticulture is best discussed in traditional commodity groupings.

Orcharding

By weight, world production of fruit approaches that of some of the world's staple agricultural crops (see Table 14-1, p. 505). Fruit production for human consumption began with the harvesting of wild stands, a procedure still followed with a few crops in some localities. The casual back-yard cultivation of trees and vines, which is still widely practiced throughout the world, is the forerunner of the modern fruit industry. Today the commercial production of fruit is a specialized portion of the horticultural industry known as **orcharding** (Fig. 13-2). Orcharding is typically based on long-lived perennials, many of which do not bear fruit until several years after they are planted, and in consequence orcharding has a number of unique characteristics. In marked contrast to commercial vegetable gardens, which are located as close as possible to the markets of urban centers, commercial fruit production is concentrated in those areas best adapted to the growth of a given crop. Thus the major production areas for apples and pears in North America are to be found in only a few favorable locations, notably the Pacific Northwest and the Great Lakes region. Similarly, citrus production is heavily concentrated in California and especially Florida (Polk County in Florida produces as much as all of California). Such concentration in the production of orchard crops is found throughout the world.

Typically it is not easy to get into or out of the fruit business. Heavy capital expenditures are required, as are financial reserves for the long periods during which the orchardist must wait for newly planted trees to come into bearing. In addition to expenditures for ordinary production equipment, great expenditures are necessary for grading, packaging, and storage facilities. Thus many commercial orchards have been family owned for generations. In contrast to temperate vegetable production, orcharding is a full-year operation; the nongrowing season is a time of considerable activity, particularly pruning and training.

Although the details of fruit growing differ with species and climates, there are a number of common problems and practices. There must be careful selection of a succession of cultivars suitable both for adaptability and marketing. With some fruits (apples and pears, for example) the choice of cultivars must also take cross-pollination into consideration. Clonal propagation is the rule for almost all fruit crops, and most producing units are compound plants consisting of at least two genetically different parts, rootstock and scion. Choice of geographic location and site are important to avoid critical temperatures at flowering and low-temperature injury in winter. Soil type, suitability of drainage and irrigation, and natural soil fertility must also be considered.

The particular planting system is usually a compromise between plant requirements for light (in relation to quality and yield) and management needs

FIGURE 13-2. Apple orchards in the United States. *Top,* an apple orchard in Oregon's Hood River Valley. *Bottom,* an orchard owned by Dixie Orchard Company, Vincennes, Indiana. This large enterprise manages more than 1,000 acres of apples and peaches. [Upper photograph courtesy Oregon State Highway Travel Division; lower photograph courtesy U.S. Soil Conservation Service.]

(movement throughout the orchard for spraying and harvesting). Most fruit crops require extensive pruning and training to keep plants within bounds, to maintain structural strength, and to influence growth and fruitfulness. Extensive labor for pruning ensures that the fruit growing operation will be a year-round endeavor. Fertilization must maintain a balance between tree growth, productiveness, and fruit quality. Thinning of flowers or fruits is often necessary to control fruit size and the overbearing that may result in alternate bearing, poor quality, and even winter injury.

Because orchards take a long time to become established, pests also have ample time in which to become established. Pest control, therefore, is a critical part of the operation. For apple production in the humid areas it is not uncommon to have from 12 to 15, even as many as 20, sprays for insect and disease control; even in winter, pests such as mice and deer must be contended with.

Finally, harvesting procedures assume key importance in crops that have a short storage life. Mechanical harvesting devices and bulk handling are causing great changes in an industry traditionally dependent on high labor inputs during harvest (Fig. 13-3). The life of perishable fruits can be extended through controlled temperature, humidity, and atmosphere, and storage is now an integral part of many orchard operations. Grading and packaging operations may be handled either at the orchard or at some central location.

With the trend toward concentration of the industry in fewer locations, greater reliance on long-term storage, and more exacting marketing requirements, there is a trend also toward larger marketing associations. Thus most of the avocados in California are sold as Calavo brand, the name of a cooperative marketing association; much of California's citrus is sold with the Sunkist label; and Washington apples move through a number of cooperative packing and marketing associations. In the Northeast many fruit growers produce almost exclusively under processing contracts; 'Concord' grapes are handled almost exclusively through grower-processor cooperatives (such as Welch's). In Michigan the large increase in capitalization available for controlled-atmosphere storage has encouraged the formation of small associations of growers around storages and marketing facilities.

At the same time, small growers outside of the major fruit-production areas who have withstood the exodus to more favored locations still find local markets a profitable means of selling their crops—sales on the farm, at roadside markets, and, increasingly, through "pick-your-own" operations.

FIGURE 13-3. Mechanical harvesting: *left*, wine grapes; *right*, dates. [Left photograph courtesy Ontario Ministry of Agriculture and Food; right photograph courtesy USDA.]

Vegetable Growing

The vegetable industry is characterized by its flexibility. Because most vegetables are grown as annuals, shifts in cultivar and crop can be readily made. In the past, a large part of the vegetable enterprise was diversified, and no great long-term investments were required. It was always relatively easy to go into or out of the vegetable business. This is becoming less true as irrigation, specialized equipment, and storage and packing facilities become integral parts of commercial vegetable production. The vegetable industry in the United States may be subdivided into three main categories: home gardening, market gardening, and truck gardening. In addition to these, there are a number of small, specialized parts of the vegetable industry, including plant,growing, greenhouse forcing, seed production, and mushroom culture.

Home gardening, which involves the production of vegetables for home consumption, is still the most important source of vegetables in many countries. It is still a considerable source of vegetable production in the rural United States, although gardening at home appears to be almost more important as an outlet for recreation than as a source of food. In times of national peril, home gardens may become an important part of a country's food supply. **Market gardening** developed from local gardens and involves intensive production of many kinds of vegetables near large centers of population (Fig. 13-4). Market gardening is disappearing in the United States as a result of increasing land costs and improvement in food distribution. The large-scale production of vegetables, commonly less diversified than market-garden production, is known as **truck farming**

FIGURE 13-4. An irrigated lettuce farm in Hawaii. [Photograph by J. C. Allen & Son.]

FIGURE 13-5. Association of climbing beans and maize in field plantings. Beans and maize also are dietary complements because corn protein is deficient in lysine while bean protein is deficient in the sulfur-containing amino acids cystine and methionine. In Latin America, corn (in the form of tortillas) and beans are staple foods that are often eaten together, thus providing better nutrition than either component provides alone. [Photograph courtesy C. A. Francis.]

(from the French *troquer,* to barter). Truck farming, which is based on suitable season, climate, and soil rather than market proximity, has become the most important kind of modern vegetable industry. The produce of truck farming is used either for fresh market consumption or for processing. The rise of the processing industry has made many individual vegetable crops part of the general rotation of farm crops. For example, tomatoes and sweet corn have become important cash crops in many midwestern general or grain farms.

Commercial vegetable production in temperate climates is based on **single cropping,** in which one species is planted *en masse* with only yearly rotations. In tropical areas, however, the traditional pattern of vegetable gardening is **multiple cropping,** which is characterized by simultaneous or relay planting of two or more crops in the same field. The association of beans and maize in South American agriculture is an ancient example of multiple cropping that is still practiced (Fig. 13-5). Multiple cropping predominates on small farms, which often produce cereal grains, legumes, starchy roots, and plantains. The system relies on diversity of plant species, closed cycling of nutrients, reduced pest incidence, erosion control, low but stable yield, and high inputs of human energy for cultivation and weed control. With proper management, however, total yields per unit area can be very high.

Multiple cropping provides small farmers with extra sources of income and greater diversity in their diets. However, the intensity of production requires much hand labor, which is not compatible with many modern farming practices. Recently, multiple cropping has received increased attention from horticultural and agricultural researchers, and studies suggest that multiple-cropping systems can be compatible with increased technological inputs to exploit the increased productivity inherent in such systems.

Of the several specialty operations in vegetable production, greenhouse production of vegetable crops is perhaps the most important. It is a minor but progressive part of the vegetable industry. In the United States greenhouse vegetable production based on tomatoes and lettuce is centered in the Midwest. In the Low Countries of Europe a relatively large greenhouse industry exists with tomatoes and cucumbers as the major crops.

Another specialty industry is mushroom culture. The industry was originally developed in areas with natural caves. Mushrooms are saprophytic and therefore do not require light, but they do require strict temperature and humidity control. At present the production of mushrooms is a highly capitalized operation utilizing specially constructed houses, although natural caves are still used in some areas. Mushrooms have traditionally been grown on composted horse manure, although synthetic compost formulae have been developed from corn cobs, hay, and inorganic fertilizer. In the United States, cultivation is almost completely confined to *Agaricus campestris,* a basidiomycete, although a number of other species belonging to the classes Basidiomycetes and Ascomycetes are commercially produced in various parts of the world.

Ornamentals

The production of ornamental plants can be divided on the basis of crop into **floriculture** (cut flowers, potted plants, and greenery), the growing of **landscape plants** (bulbs, rose plants, ground cover, trees, and shrubs), and the growing of **turf** (seed and sod). Although flowering bulbs and flower seed are important components of agricultural production in the Low Countries of Europe, ornamentals are relatively insignificant in world trade. World statistics on production are unavailable, but in the United States complete statistics have been obtained in connection with the 1890, 1930, 1950, 1960, and 1970 agricultural censuses.

Floriculture

Floriculture has long been an important part of the horticulture industry, especially in Europe and Japan, and it accounts for about half of the nonfood horticultural industry in the United States. The wholesale value of florist crops in 1976 was about one billion dollars; the retail value was estimated to be 3.6 billion dollars. Because cut flowers and potted plants are mainly grown in plant-growing structures in temperate climates, floriculture is largely a greenhouse industry (Fig.

13-6), but there is also outdoor culture of many flowers. Floriculture is a competitive and highly technical business requiring knowledge, skill, and large amounts of capital. Greenhouse floriculture is, in many respects, the most sophisticated kind of plant agriculture. The industry is usually very specialized by crop, and the grower must provide precise environmental control and be vigilant in the constant struggle against pests and other calamities. Exact scheduling must be arranged because most floral crops are subject to seasonal demand (poinsettias at Christmas, lilies at Easter, roses and carnations on Mother's Day, and so on). Because the product is perishable, market channels must function smoothly to avoid losses.

The floriculture industry consists mainly of the grower, who typically mass produces flowers for the wholesale market, and the retail florist, who markets to the public and contributes such services as arrangements and delivery. The growing operation is typically family-owned and highly intensive. The average outdoor producing unit in the middle 1960s was less than five acres, and greenhouse production per farm averaged 13,000 square feet. Nevertheless, as in all modern agriculture, the size of the growing unit is increasing. There is a movement away from eastern urban areas of high taxes and labor costs to locations having lower tax rates and a rural labor pool as well as more favorable climate (milder temperatures and more sunlight). As a result the United States floriculture industry is moving west and south. The development of air freight has emphasized interregional and international competition. Flowers can now be shipped great distances by air and arrive in fresh condition to compete with locally grown products. International air-cargo shipment of flowers is increasing.

Flower growing is based on precise environmental control. This involves structures incorporating glass, plastic, lath or shade cloth, and artificial heating and cooling devices. The rising costs of energy have placed heavy burdens on greenhouse floriculture. For example, the prices of oil, gas, coal, and electricity have increased two to five times from 1973 to 1978. Higher fuel costs have been somewhat offset by a rise in the prices that the growers receive, but there has been concern that greenhouse growers may price themselves out of the market. More dangerous has been the threat individual growers have faced from reduced fuel allotments and even curtailments during recurring energy crises. As a result, increased attention has been paid to energy conservation in the greenhouse industry. Alternatives include the use of greenhouses that are more efficient in capturing solar heat, the use of geothermal heat and waste heat from power plants, and the implementation of energy conservation through such greenhouse modifications as double-wall coverings, thermal blankets (plastic coverings over benches to reduce the volume heated at night), and increased attention to heating and cooling practices.

The production of plants under protected cultivation is constantly becoming technically more complex. Greenhouse soils are usually compounded from a variety of ingredients, sterilized to control weeds, disease organisms, and insects and nematodes. Fertility and pH are precisely controlled through the addition of the appropriate chemicals and amendments. Some growers are enriching the atmosphere in their greenhouses with carbon dioxide. Water regimes are automatically regulated, even with pot culture. In greenhouses, temperatures are regulated by steam heating and evaporative cooling (mist and fan-and-pad). Day length is regulated by a combination of shade cloth and illumination to extend the day or interrupt the dark period. The use of artificial light to increase photosynthesis is usually not economically feasible, except in special situations with seedlings. Pest control relies on a combination of practices involving sanitation, the use of disease-free plants, and an arsenal of organic pesticides. The use of chemicals to control growth is increasing rapidly. The growth retardants in particular are finding much promise in the industry.

Flower growing is highly competitive. With erosion of profit margins, only the large, well-managed organizations can survive economically, and these keep up with technical complexity and high capital requirements by developing highly specialized businesses (Fig. 13-7). The successful grower must be a highly trained, skilled entrepreneur.

The retail florist usually but not always operates a business separate from that of

FIGURE 13-7. A container-filling machine designed for use in the floriculture and landscape industries.

the grower. He typically buys wholesale cut flowers, plants, greenery and associated products, ribbon, and potware, and transforms these into floral arrangements of various types—corsages, table decorations, wreaths, set pieces, and even floral floats. The telephone (and the telegraph before it) enables the retail florist industry to organize to provide nationwide delivery of flowers. This has linked the florist with a number of broad associations that provide technical information as well as trade channels. Thus the florist is most typically a designer and businessman rather than a plantsman.

Landscape Horticulture

The industry of landscape horticulture is divided into growing, maintenance, and design. The growing of plants for the landscape is referred to as the nursery business, although more broadly a nursery is the place where any young plant is grown or maintained before permanent planting. There are about 25,000 nurseries of all types in the United States, and together they occupy almost 200,000 acres. The retail value of nursery sales was about 3 billion dollars in 1977.

The nursery industry involves the production and distribution of woody and herbaceous plants and includes ornamental "bulb" crops—corms, tubers, rhizomes, and swollen roots as well as true bulbs. The production of cuttings to be grown in greenhouses or for indoor use (foliage plants), as well as the production of bedding plants, is usually considered part of floriculture, but this distinction is fading. While most nursery crops are ornamental, the nursery business also includes fruit plants and certain perennial vegetables used in home gardens—asparagus and rhubarb, for example. Christmas-tree production is also included with the nursery industry; however, though the operation is similar, the growth of seedling trees for forest plantings is usually considered along with forestry.

The nursery industry in the United States may be divided into wholesale, retail, and mail-order enterprises, although there may be some overlapping. Typical wholesale nurseries specialize in relatively few crops, dealing largely in plant propagation and supplying only retail nurseries or florists. They sell young "lining out" stock of woody material to the retail nursery, which then performs two basic functions: care of the plant until growth is complete, and resale to the public. Often, though not always, the retail nursery provides planting and maintenance service (Fig. 13-8). Many nurseries also execute the design of plantings in addition to furnishing the plants. This is analogous to the provision by some lumber yards of architectural services. This service is a complex one, since some states license landscape architects as members of a distinct profession.

Mail-order nursery companies usually grow only a small proportion of their catalogue listings, and subcontract the rest from wholesale houses. Although some of the largest and most reputable businesses in the country deal only in mail order, very often fly-by-night operators are attracted to businesses of this sort.

Ornamental trees and shrubs represent the most important of the nursery crops (ornamental shrubs are the backbone of the nursery trade); next in importance are

FIGURE 13-8. Installation and maintenance are important components of the nursery industry. *Left*, landscaping is an integral part of the home construction business. *Right*, moving large trees in winter. [Left photograph courtesy Ford Motor Co.; right photograph courtesy G. Kessler.]

fruit plants and then "bulb" crops. The most important single plant grown for outdoor cultivation is the rose. The type of plants grown by a nursery depends on location; in general, the northern nurseries provide deciduous trees and shrubs and coniferous evergreens whereas the southern nurseries provide tender broad-leaved evergreens.

The nursery industry is especially important in the Great Lakes and Middle Atlantic states, and in Florida and California. In some of these areas nursery sales (including sales of roses, shade trees, and small fruits for the home) account for nearly 40 percent of the retail trade in horticultural specialties. The nursery business is about equally divided among the production of (1) coniferous evergreens such as yew, juniper, spruce, and pine; (2) broad-leaved evergreens such as rhododendron, camellia, holly, and boxwood; (3) deciduous plants such as *Forsythia*, *Viburnum*, barberries, privets, lilacs, and flowering vines; and (4) roses.

Fields of specialization have evolved within the ornamental-shrub industry. Some firms confine their activity mostly to the production of "lining out" stock—seedlings and rooted cuttings for sale to growers for field plantings that must be tended for several years before reaching salable size. Such firms will generally offer not only the bare-root liners themselves, but also small plants started in plant bands or slightly larger ones in containers. They may also offer small grafted plants of expensive or rare cultivars. Lining out stock of a typical evergreen, such as the Pfitzer juniper (*Juniperus chinensis* 'Pfitzeriana'), is relatively inexpensive, generally several sell for a dollar; it is more economical for a grower to purchase these ready to plant in the field than to undertake the specialized propagation himself.

The field grower may, in turn, specialize in mass growing for the wholesale trade only. The field plantings are tended until they attain marketable size, having been shaped and sheared as necessary, provided with pest control, irrigated, culled, and so on. Because of the time required to produce a marketable crop, and because of appreciating labor costs, this phase of the nursery industry can easily be caught in a profit squeeze, especially if landscaping preferences change and there is no quick market for a field of mature shrubs. In contrast, wholesale growing escapes the high overhead of retail marketing in urban areas, and although many growers do raise retail stock at their nurseries they generally try to avoid the expensive merchandising required of the typical urban-area garden center. Growers are especially interested in new technology that will lessen labor requirements, and are turning to herbicidal control of weeds (as with dichlobenil) and short-cut methods for transplanting (as substitutes for hand digging and burlapping).

Most shrubs are usually balled and burlapped ("B and B") for market. The plants are dug with as much soil as possible clinging to the roots, and the root ball is then wrapped with burlap or other suitable material. Recently a trend has developed toward container-grown stock—nursery stock grown in the container (Fig. 13-9) in which it is to be sold. This allows year-round sales of plant material. Deciduous shrubs may be marketed with bare roots. There are obvious economies in not having to dig and wrap the plant carefully, thereby handling added weight and soil. Fairly satisfactory techniques have been developed for holding bare-root ornamental shrubs in viable condition until marketed. Usually this involves digging after leaf fall in autumn and storing the plants through winter in cold cellars with regulated humidity for spring sale. Unfortunately, the plants are often handled carelessly. The chief cause of loss during storage is desiccation, but respiration also plays a part when temperatures rise. But even if the plants are properly cared for during cold storage, they often suffer from warming and drying at the sales outlet. Bare-root plants ordered by mail often arrive in poor condition

FIGURE 13-9. A field of container-grown evergreens. [Photograph courtesy D. Hill Nursery Co., Dundee, Illinois.]

FIGURE 13-10. Entrance to the garden center of D. Hill Nursery Company, Dundee, Illinois.

because of warming and drying in transit. It is apparent that the great effort spent to control humidity and temperature during winter storage can be defeated if there are no cold-storage facilities at retail outlets.

In recent years a breed of "nurseryman" who is more merchandiser than plantsman has taken over much of the retail trade. Exemplifying this change are the numerous diversified garden centers that sell all sorts of horticultural merchandise and maintain display yards of shrubs, trees, and roses (Fig. 13-10). Usually these are purchased wholesale from large growers. Many such firms have associated landscaping services, and offer counsel, product, and service to the homeowner.

Turf

The turf-grass industry includes the production and maintenance of specialized grasses and other ground covers for utility, recreation, and beautification. Grass is ubiquitous in many urban and suburban facilities. In the United States, maintenance expenditures for turf run to more than 5 billion dollars, of which 70% is spent on residential lawns, 11% on highway plantings, 8% on cemeteries, and 6% on golf courses. About 10 million acres of land is tended by about 50 million homeowners in the United States, and much of it is lawn.

The two basic commodities of the turf industry are seed and sod. In the United States in 1969 seed had an estimated wholesale value of 62 million dollars; sod was valued at 100 million dollars. Associated products include fertilizer (100 million dollars), pesticides (120 million dollars), and equipment (having the greatest share—lawn mowers alone were worth 375 million dollars).

The turf-grass industry reaches a tremendous market throughout the affluent countries of the world. The maritime climate of the British Isles is particularly suited to beautiful turf. Although lawns are usually not the major landscape feature of homes as in the United States, in public areas lawns are more prominent in Great Britain than in the United States.

The major turf species in Europe is *Poa annua*. The major kinds grown in the northern two-thirds of the United States are three cool-season grasses, Kentucky bluegrass (*P. pratensis*), red fescue (*Festuca rubra*), and bentgrass (*Agrostis* species). There are a number of grass species adapted to the South: Bermuda grass (*Cynodon dactylon*), bahia grass (*Paspalum notatum*), carpet grass (*Axonopus compressus*), centipede grass (*Eremochloa ophiuroides*), *Zoysia* species, and St. Augustine grass (*Stenotaphrum secundatum*).

Seed Production

Seed production is a relatively small but essential part of the horticultural industry. The seed business involves not only a great many crops but literally thousands of cultivars and hybrids, and not only plant growing but manufacturing and processing (milling, cleaning, packaging, and storing of seeds).

The principal world areas of horticultural seed production are Holland, Denmark, England, Japan, and the United States. In the United States, the most important seed-producing areas are in the western states. California produces practically all of the flower seeds, but a good share of the vegetable seed, especially of large-seeded crops such as beans, peas, and corn, is produced in Idaho. The actual growing of plants is often contracted out to growers, although large seed houses may produce some of their own specialties directly. There are also a number of specialty growers whose production is confined to a few crops, such as pansies or sweet corn.

The seed business itself is divided into wholesale, retail, and mail-order houses. To the majority of the American public, the commission packages are the most familiar wares of the seed industry, but these represent only a small percentage of seed sales.

There is an obvious relationship between seed growing and seed improvement. The progressive seed houses have initiated the breeding of many crops, especially flowers. The Department of Agriculture and State Agricultural Experiment Stations have also supported much of the breeding programs, and, as a result, many experiment stations have sponsored seed associations. At present, horticultural crops account for only a minor portion of this activity.

Selected References

Arthur, H. B., J. P. Houck, and G. L. Beckford, 1968. *Tropical Agribusiness Structures and Adjustments—Bananas.* Boston: Division of Research, Harvard Business School. (An economic analysis of the world banana industry.)

Childers, N. F., 1976. *Modern Fruit Science: Orchard and Small Fruit Culture,* 7th ed. New Brunswick, N.J.: Horticultural Publications. (A text on temperate pomology emphasizing production practices.)

Dalrymple, D. G., 1973. *Controlled Environment Agriculture: A Global Review of Greenhouse Food Production* (Foreign Agricultural Economic Report 89). Economic

Research Service, U.S. Department of Agriculture. (A world survey of food production under glass.)

Geertz, C., 1963. *Agricultural Involution: The Process of Ecological Change in Indonesia.* Berkeley and Los Angeles: University of California Press. (See especially Chapter 2 for a comparison of shifting cultivation with sedentary agriculture.)

Gourou, P., 1966. *The Tropical World: Its Social and Economic Conditions and Its Future Status,* 4th ed. New York: Wiley. (An introduction to the tropics.)

Hanson, A. A., and F. V. Juska, (editors), 1969. *Turfgrass Science.* Madison, Wisc.: American Society of Agronomy. (A monograph on turf with each chapter contributed by a specialist.)

Janick, J., R. W. Schery, F. W. Woods, and V. W. Ruttan, 1974. *Plant Science: An Introduction to World Crops,* 2nd ed. San Francisco: W. H. Freeman and Company. (An introduction to crop science. See Chapter 15, Cropping Systems and Practices.)

Laurie, A., D. C. Kiplinger, and K. S. Nelson, 1968. *Commercial Flower Forcing,* 7th ed. New York: McGraw-Hill. (A textbook on greenhouse florist crops with emphasis on practices.)

Manshard, W., 1974. *Tropical Agriculture: A Geographical Introduction and Appraisal.* London: Longmans. (An introduction to agricultural systems of the tropics.)

Mastalerz, J. W., 1977. *The Greenhouse Environment: The Effect of Environmental Factors on Flower Crops.* New York: Wiley. (An advanced review of factors affecting greenhouse production.)

Nelson, P. V., 1978. *Greenhouse Operation and Management.* Reston, Va.: Reston Publishing. (A basic text on floriculture production in greenhouses.)

Ochse, J. J., M. J. Soule, Jr., M. J. Dijkman, and C. Wehlburn, 1961. *Tropical and Sub-tropical Agriculture.* New York: Macmillan. (A huge two-volume work dealing with tropical plantation agriculture. The first volume deals with general cultural practices, the second with crops.)

Thompson, H. C., and W. C. Kelly, 1957. *Vegetable Crops.* New York: McGraw-Hill. (An excellent text and reference for vegetable crops of temperate horticulture.)

Westwood, M. N., 1978. *Temperate-Zone Pomology.* San Francisco: W. H. Freeman and Company. (An excellent, up-to-date text on temperate pomology.)

14

Horticultural Crops

Be kind and courteous to this gentleman,
Hop in his walks and gambol in his eyes;
Feed him with apricocks and dewberries,
With purple grapes, green figs, and mulberries. . . .

SHAKESPEARE, *A Midsummer Night's Dream* (III. 1)

Horticulture ultimately reduces to the study of individual crops, the basis of the industry. Notwithstanding the diversity of horticulture, most temperate horticultural crops can be grouped into the traditional categories: fruits, vegetables, and ornamentals. However, for a world view of horticulture, additional categories are required to include beverage crops, spice and drug crops, and others.

Accurate world production figures for all horticultural crops are not available. This is due in part to inconsistent international communication about such data and in part to poor statistical records. Production statistics may be incomplete or may be computed differently in different countries. For example, some countries report only commercial production; others report total production. The lack of reliable data is especially noticeable for crops that do not enter significantly into world trade. This results in an underestimation of the highly perishable horticultural crops, even though these may be grown extensively in home or local gardens. Production figures of the major crops, shown in Table 14-1, indicate the worldwide importance of horticulture. World production statistics for selected horticultural crops are presented in Table 14-2.

FRUIT CROPS

The fruits of certain perennial plants have long been prized as a source of refreshment, for their delightful flavors and aromas, and as a nourishing food. Throughout the ages these fruits have been regarded both as luxuries and necessities, and it is the combination of utility and beauty that has earned them a special place in the relationship of people to plants:

> Sustain me with raisins, refresh me with apples;
> For I am sick with love.
> > Song of Solomon 2:5

Although there are exceptions—banana and coconut, for example—fruits are not a prime source of calories, minerals, or protein; nevertheless, considered together, fruit crops compare favorably with the world's staple agricultural crops on a tonnage basis. Orcharding is now a specialized and highly technical part of the horticultural industry.

There are a number of ways fruits can be classified. However, to the horticulturist, the basic distinction is a climatic one. Fruits are called **temperate, subtropical,** and **tropical.** Unlike vegetables, of which many tropical perennial species grow well as annuals in temperate areas (tomato, for example), long-lived fruit crops cannot be transplanted to alien conditions.

TABLE 14-1. Comparison of 1976 world production of some major crops.

Crop	Production (millions of metric tons)	
CEREALS		1,447.3
Wheat	417.5	
Rice (paddy)	345.4	
Maize	334.0	
Barley	189.7	
ROOT CROPS		557.9
Potato	287.6	
PULSES (LEGUMES)		51.5
VEGETABLES AND MELONS		311.8
FRUITS		261.4
Grape	59.2	
Citrus	50.8	
Banana	38.5	
Apple	22.2	
NUTS		3.6
VEGETABLE OILS		41.0
SUGAR (CENTRIFUGAL, RAW)		86.3
COCOA BEANS		1.4
COFFEE (GREEN)		3.7
TEA		1.6
TOBACCO		5.7
RUBBER (NATURAL)		3.5

SOURCE: *FAO Production Yearbook 1976*, Vol. 30, 1977.

TABLE 14-2. World production figures (1000 metric tons) for selected horticultural crops, 1976.

Crop	World	Africa	North America	South America	Asia°	Europe	Oceania	USSR†
FRUITS								
GRAPE								
Table	59,204	2,525	3,979	5,536	6,114	34,664	772	5,614
Wine	30,883	1,369	1,529	3,308	199	21,124	390	2,965
Raisin	913	17	196	4	471	156	69	
CITRUS								
Orange	34,091	3,080	12,404	9,134	4,892	4,034	393	154
Tangerine and clementine	7,327	452	763	669	4,319	1,086	39	
Lemon and lime	4,507	226	1,341	653	1,121	1,256	48	
Grapefruit and pumello	4,009	202	2,732	225	812	14	23	
Other citrus	909	360	77	194	271	2	5	
TROPICAL AND SUBTROPICAL FRUITS								
Banana	38,504	4,319	7,065	13,963	11,771	355	1,032	
Plantain	18,301	12,042	1,442	4,213	600		4	
Papaya	1,011	58	303	24	295		15	
Pineapple	5,499	789	1,166	982	2,431	2	129	
Mango	12,793	568	1,007	951	10,258		8	
Date	2,434	963	15	1	1,439	17		
Olive	8,527	979	83	130	1,214	6,118	3	
Avocado	1,138	57	672	364	42	2	3	
POME FRUITS								
Apple	22,262	380	3,485	882	4,077	12,959	480	
Pear	7,686	178	826	222	2,060	4,219	181	
STONE FRUITS								
Plum and prune	4,443	47	650	140	695	2,884	28	
Cherry								
Apricot	1,469	139	135	30	400	732	32	
BERRIES								
Strawberry	1,257		351	8	210	682	7	
Raspberry	160		20			137	3	
Currant	344					342	2	
CRUCIFERS								
CABBAGE	21,288	511	1,467	158	11,574	7,461	117	
CAULIFLOWER	4,052	112	163	1,513	2,112	99		

NUTS								
ALMOND	844		209		71	502	1	18
HAZELNUT	423		6		276	126		14
CHESTNUT	520			12	279	210		18
WALNUT	870	4	176	10	261	261		158
PISTACHIO	91				87	4		
COCONUT								
BEVERAGE CROPS								
COFFEE	3,653	1,235	910	1,115	350		43	
COCOA	1,387	890	89	346	25		38	
TEA	1,628	158		40	1,335		6	89
ROOTS, TUBERS, AND BULBS								
POTATO	287,554	3,965	19,611	9,204	59,075	109,649	950	85,100
SWEETPOTATO	135,855	5,065	1,282	2,699	126,168	76	565	
CASSAVA	104,952	41,939	712	32,115	29,979		208	
ONION, DRY	16,303	1,252	1,757	1,146	7,663	3,432	154	900
CARROT	6,859	317	1,127	210	2,023	3,033		
GARLIC	1,445	170	73	131	692	380	148	
LEGUMES								
BEAN, DRY	12,580	1,151	2,377	2,385	5,910	659	4	95
BEAN, GREEN	6,187	802	47	141	4,541	656		
BROAD BEAN, DRY								
PEA, DRY	13,427	276	181	114	5,430	546	81	6,800
PEA, GREEN	4,786	108	1,242	196	569	2,307	164	200
CHICK PEA	7,466	221	190	10	6,940	105		
LENTIL	1,236	142	51	35	857	60		
PEANUT (SHELL)	18,495	5,376	1,846	924	10,297	24	29	90
CUCURBITS								
WATERMELON	22,991	2,011	1,519	1,079	12,032	3,150	3,164	
MELON AND CANTALOUPE	5,627	497	757	295	2,540	1,538		
CUCUMBER	7,942	270	918	43	4,307	2,390	13	
PUMPKIN, SQUASH, AND GOURD	6,022	864	114	704	1,975	2,275	90	
SOLANACEOUS FRUITS								
CHILI PEPPER	5,617	843	654	158	1,953	2,010		
EGGPLANT	3,761	354	81	3	2,781	542		
TOMATO	40,802	4,120	8,629	2,240	9,993	11,881	239	3,700

SOURCE: *FAO Production Yearbook 1976*, Vol. 30, 1977.

*Does not include mainland China.

†Data from USSR incomplete.

There is a tremendous diversity of tropical fruits, many of which are little known. Tropical fruit plants are evergreen and extremely sensitive to low-temperature injury. Although there are fewer temperate fruits, they have received much attention. Virtually all temperate fruit crops are deciduous and require a period of cold. Subtropical crops may be either deciduous or evergreen, and they are usually able to withstand light frost; some require a period of cool weather for maximum fruiting.

Most fruits can be consumed fresh when they contain large amounts of water. They are generally extremely perishable once the climacteric has been reached. Thus the proper stage for consumption may be extremely brief and transitory. For this reason some delicious tropical fruits such as the cherimoya are virtually unknown beyond their area of production. As a result most fruits that enter commerce either ship well in the fresh state, as do citrus and apple, or adapt well to preservation through drying (date and fig), the use of preservatives (as in fruit jellies and jams), canning (pineapple), or freezing (strawberry). The advent of air transport, refrigerated storage, and new packaging techniques has recently made it possible for a tremendous variety of fruits from all parts of the world to be available fresh in the largest metropolitan centers throughout the year. Yet at the same time, in smaller cities of the United States, common fruits that were formerly plentiful at least once a year from widespread local production (raspberries, for example) have virtually disappeared and have been relegated to the status of luxury products.

Temperate Fruits

Grape

Grapes (Fig. 14-1) are the most important fruit crop, accounting for about one quarter of the fruit production of the world. Grapes are grown throughout the temperate regions, especially in warm, sunny climates with mild winters and dry periods during fruit ripening. In northern Europe, especially in Belgium and the Netherlands, there is still some production of grapes under glass. In the United States, grape production is largely located in California, although a sizable grape industry exists in the Great Lakes region.

About 78% of the world's crop is pressed into wine, 8% is consumed fresh, and 13% is dried. Specific cultivars are usually grown for each purpose; 'Thompson Seedless' ('Sultanina') can be used for all purposes, however. And, in California, the European grape, *Vitis vinifera* (Vitaceae) is grown for wine, fresh fruit, and raisins. Various combinations of soil type, climate, grape cultivar, and production technique account for the differences in wines produced throughout the world. Of the entire California crop, 59% of the grapes grown are for wine, 28% for table use, and 13% for raisins. In the past, raisins and other dried fruit made up the bulk of United States fruit exports, but this market has been decreasing.

FIGURE 14-1. Heavy fruit production in the grape. *Inset,* 'Fiesta,' a new early-ripening seedless grape, has 'Thompson Seedless' in its pedigree [Photographs courtesy USDA.]

Vitis vinifera grapes are unadapted to most areas of the United States because of their lack of resistance to native diseases and soil pests and their lack of cold-hardiness. Thus, the grape industry in the Great Lakes region has been based almost exclusively on a single cultivar, 'Concord,' which is an interspecific hybrid between *V. vinifera* and the native American fox grape, *V. labrusca.* The foxy flavor of 'Concord' is prized for nonfermented grape juice and sweet dessert wines. The grape juice is usually canned or bottled, although frozen concentrate has achieved some market acceptance.

In the United States, increasing consumption of dry or table wines (wines made from grapes with high acidity and low sugar) has resulted in a wine-grape boom that has been going on since the middle 1960s. As prices have climbed, grape acreage and wine production have increased. Much of the increase in wine production has resulted from expansion of plantings of *Vitis vinifera* grapes in California, especially such famous cultivars as 'White Riesling,' 'Chardonnay,' 'Cabernet Sauvignon,' and 'Pinot Noir.'

Because the foxy flavor of the American-type slip-skin grape cultivars derived from *Vitis labrusca* is considered undesirable for table wines, alternative wine grapes have been eagerly sought for wine production in the United States outside of California. In France, hybridization of *V. vinifera* grapes with other native American species, including *V. rupestris* and *V. lincecumii,* has long been under way. These American species were introduced to France as rootstocks because of their resistance to *Phylloxera vitifoliae,* an insect carried into France from the eastern and central United States before 1860 that threatened to wipe out the French grape industry. A series of breeding programs, carried out mainly by private French breeders (Seibel, Ravat, Seyve-Villard, Seyne, and others), has been going on for three-quarters of a century and has resulted in a new class of wine grapes known as "French hybrids."

These new grapes have not only achieved prominence in France (where they account for one-quarter of the present wine acreage) but have been found adapted to many areas of the United States, where they are now being widely planted. French hybrid cultivars adapted to the Midwest include 'Aurora' (a cross between 'Seibel 788' and 'Seibel 29') and 'De Chaunac' ('Seibel 9549').

Apple and Pear

Apple and pear, the **pome fruits,** are important tree fruits of the temperate climates. The fruits are closely related botanically, and their culture is similar. Both crops are often grown in the same orchard.

The apple, *Malus domestica* (Rosaceae), is the most important single species of tree fruit in the world. It is widely adapted and is grown in temperate regions that have a distinct cold period. Although the United States is the largest single producer, the bulk of the world's crop is grown in Europe. The majority of the European cultivars are predominantly cooking and cider apples. Cider, which is sold with an alcoholic content of about 6–8%, is quite popular in England and northern Europe. Since World War II there has been a trend away from cooking apples to dessert types, with increased emphasis on the more attractive red cultivars.

Although apples are grown throughout the United States, the largest concentrations are on the West Coast (Washington and California), in the Great Lakes area (New York and Michigan), and in the Appalachian area (Virginia, West Virginia, and Pennsylvania). The Western orchards must be irrigated, but the low humidity reduces the disease problem, and the climate is uniquely adapted for the production of fruit of high "finish" and attractiveness.

About ten cultivars account for more than 90% of the United States crop. The trend over the past sixty years has been toward a great reduction in the number of cultivars and a change to the red sports of the more popular cultivars, such as 'Delicious,' 'Rome,' 'Jonathan,' and 'McIntosh.' The small "farm orchards" have disappeared, and production has concentrated in more adapted locations. Although tree numbers have shown a gradual decline until recently, yields are going up. The export market, which has decreased from pre–World War II levels, is now stable. The net effect of the increase in population and the decrease in consumption of apples per capita has been a stabilization of the total apple consumption in the United States over the past two decades.

Production technology in the apple industry has shown great changes. In the United States there is a trend toward greater use of dwarfing rootstocks, particularly of the "semidwarf" types such as EM 2 and EM 7, but seedling rootstocks are still the most commonly used. The use of controlled-atmosphere storage makes it possible to obtain high quality apples the year around. As this technological advance has increased in use, it has brought significant changes to the pattern of apple marketing. It has already interfered with the marketing of early-summer cultivars. Probably the most significant advance in production technology has been the increased use of organic fungicides and insecticides in the spray schedules. Most Northeastern orchards use high-concentrate speed sprayers, but dusting is more common in the West.

The world's production of pear, *Pyrus communis* and other species (Rosaceae), is about a third of the world's apple production. The major production areas are located in Western Europe. More than 40% of the European crop goes into production of perry ("pear cider"); most of the rest is consumed fresh. In the United States, pears are relatively less important, and total production represents only about one-fourth that of apples. The United States industry is based largely on the 'Bartlett' cultivar, which accounts for about 80% of the total production. 'Bartlett' is used both for fresh consumption and for processing. The 'Bartlett' pear is also one of the most important cultivars in Europe, where it is known as 'Williams' or 'William Bon Crétien.' Almost 90% of the United States production is located in the states of Washington, Oregon, and California. Of the total crop, about half is processed by canning.

Although the pear is not quite as hardy as the apple, the limiting factor to pear production has been the bacterial disease **fireblight,** which has eliminated commercial pear production from the warmer humid regions of the United States. The disease had been confined to the Western Hemisphere, but it has recently appeared in England and Poland. Various breeding programs are underway in the United States and Europe to combine fireblight resistance and high fruit quality. 'Magness,' a highly fireblight-resistant cultivar introduced by the U.S. Department of Agriculture, produces fruit of extremely high quality but its commercial potential is limited by low productivity.

Stone Fruits

Species of the genus *Prunus* (Rosaceae) constitute the so-called **stone fruits.** These include the plum, peach (Fig. 14-2), cherry, and apricot. The stone fruits require a cold period to break their rest period, but they are subject to winter killing and frost injury and thus can be grown profitably only in restricted locations.

Plums and peaches are the most important of the stone fruits. There are a great number of plum cultivars and species, with a correspondingly wide range of adaptability. Plums that have a high sugar content and can be dried are known as prune-type plums, or simply as prunes. The great bulk of the world's plums are grown in Europe. The leading countries are Yugoslavia, Rumania, and Germany. In the United States, the West Coast accounts for almost the entire commercial crop, of which a large proportion is dried and sold as prunes.

The United States is the most important peach-producing country. About half

FIGURE 14-2. 'Flamecrest,' a new yellow-fleshed freestone peach produced as a shipping type for California. [Photograph courtesy USDA.]

of the crop is grown in California. Two-thirds of the California crop are clingstone peaches. These are "rubbery fleshed" peaches, and virtually all of them are processed by canning. The remaining California production, and the rest of the United States production, is devoted to freestone peaches. These have "melting flesh," and the pit more or less separates from the flesh when ripe. Probably the most famous freestone peach is the 'Elberta,' which is still used as a basis for comparison in the industry. A number of breeding programs have produced peaches that ripen over a two-month period at most locations. The early red-colored 'Redhaven' is now the most widely planted cultivar.

Nectarines are smooth-skinned peaches; this character is caused by a single recessive gene. Nectarine production, once confined almost exclusively to California, is now expanding to other peach-growing areas with the introduction of adapted cultivars.

Cherries are an important European fruit. They are relatively less popular in the United States. The European crop consists mainly of sweet cherries (*Prunus avium*), whereas more than half of the United States production consists of tart cherries (*P. cerasus*). The cultivar 'Montmorency' is the basis of the tart-cherry industry in the United States. The production of sweet cherries is concentrated on the West Coast, whereas tart cherries are grown most abundantly in the Great Lakes region.

The United States commercial production of apricots is almost completely confined to California. The bulk of the crop is processed by canning or drying.

Small Fruits

The **small fruits** grown on plants of small stature are also known as **soft** or **berry fruits.** These include the grape (usually considered separately, as here); the strawberry, *Fragaria* × *ananassa* (Rosaceae); the brambles, all *Rubus* species,

(Rosaceae), such as red and black raspberry, blackberry, dewberry, and logan-berry; currant and gooseberry. *Ribes* species (Saxifragaceae); cranberry, *Vaccinium macrocarpon* (Ericaceae); blueberry, *Vaccinium* species (Ericaceae), as in Figure 14-3; and various others. Except for grape, the most important small fruit is the strawberry (Fig. 14-4), followed by the raspberry. Although these crops are widely adaptable to other areas, the West Coast strawberry industry has expanded to the point where California accounts for more than half of the United States crop. In California the acreage devoted to this crop increases and decreases in response to price changes.

The high labor costs associated with small fruits have limited the expansion of these crops in the United States. Although production has stabilized, yields per

FIGURE 14-3. The blueberry has become an important small fruit crop, transforming many of our supposedly worthless acid soils into valuable cropland. [Photograph by J. C. Allen & Son.]

FIGURE 14-4. The strawberry is one of the most widely adapted fruit crops. Present-day cultivars descend from hybrids of *Fragaria virginiana*, native to the East Coast, and *F. chiloensis*, native to the Pacific Coasts of North America and South America. The 'Salinas' cultivar shown above is a California introduction produced by Royce Bringhurst and Victor Voth. [Photograph courtesy Royce Bringhurst.]

acre have steadily gone up with the increased care of improved cultivars and with the adoption of better cultural practices. The use of virus-free plants has increased the productivity of many small-fruit cultivars. With strawberries, irrigation has played an important role in increasing performance. Irrigation is also widely used to control damage by spring frosts. The development of the frozen-food industry and the success of harvesting machines with bramble crops indicate that future expansion of the industry is likely.

Subtropical and Tropical Fruits

Citrus Fruits

The citrus group (Rutaceae) of evergreen fruits are native to the subtropical regions of eastern Asia. They are now the major fruit crop of the subtropical climates, and they rank in commercial importance with apple and pear. Although citrus fruits can be grown in the tropics, fruit quality there is inferior, and the main centers of world production are the United States, Brazil, Spain, Japan, Italy, Mexico, and Israel. The United States is by far the largest producer, although the industry is concentrated in relatively small areas in the states of Florida, California, and Texas.

The genus *Citrus* contains a great number of edible species. Of the commercially grown types, the sweet orange, *C. sinensis*, is the most important. California orange industry is based primarily on two cultivars, 'Washington Navel' (the 'Bahia' cultivar of Brazil), which is relatively seedless, and 'Valencia' (Fig. 14-5). The Florida industry is based on a number of thin-skinned cultivars, of which the most important are 'Valencia' and 'Pineapple.' Mandarin oranges, *C. nobilis,* are typically rough skinned and have easily separated segments. (Deep orange-red mandarins are called tangerines.) Mandarins are widely popular in Japan and have recently achieved considerable importance in the United States.

FIGURE 14-5. 'Valencia' oranges in California. [Photograph by J. C. Allen & Son.]

A number of artificial hybrids within the citrus group have been produced. Among these are the tangor (tangerine × sweet orange) and the tangelos (tangerine × grapefruit). The most important of these hybrids is the 'Temple' orange, which is believed to be a tangor. The 'Temple' orange has achieved considerable importance in Florida. Grapefruits (*C. paradisi*), lemons (*C. limonia*), and limes (*C. aurantifolia*) are the other major citrus crops.

The high concentration of citrus production in the United States has led to grove management on a contract basis by cooperative or caretaking organizations. The size of individually owned groves is commonly as small as 20 acres, although the Florida enterprises tend to be larger. The great increase in the citrus industry in the United States is due to a number of factors, including low production costs (low as compared with apples, for example), efficient marketing, joint industry advertising and promotion, and the rise of citrus as an important processing crop. A major part of the Florida citrus crop is utilized as frozen concentrate.

Olive

The olive, *Olea europea* (Oleaceae), is a native to the Mediterranean region, where it has been cultivated for many centuries. World production is approximately 8.5 million tons annually, principally in Spain, Italy, Greece, Turkey, and Morocco. Most of the world's olive crop is produced for oil, but ripe (black) and immature (green) olives are pickled for table use. The California crop of 50,000 tons is used largely for pickling and canning. Olive fruits contain a bitter glycoside that must be removed by hydrolysis with sodium or potassium hydroxide. Bitter Greek olives are cured in a highly concentrated salt solution without hydrolysis.

The small, tough, evergreen tree begins production only six years after propagation, but several decades are required for full production. Although production declines after fifty years, the trees live for centuries. Biennial bearing is often a problem.

The tree is subtropical and is killed outright at 14°F (9°C), although it survives frost well. Slight winter chilling, not lower than 43–46°F (6–8°C) promotes fruiting. Because olive trees do best with low atmospheric humidity, they are typically grown in arid regions. Spacing varies with water supply.

Propagation is by cuttings and grafting. Hand harvesting is general throughout the world, although mechanical shakers are being used in California. Olives for pickling must be picked carefully to prevent bruising.

Date

The date palm, *Phoenix dactylifera* (Palmaceae), is an ancient crop plant in arid regions (Fig. 14-6). Production is centered in Asia Minor and North Africa (Iran, Iraq, Algeria, Egypt, and Morocco) with small acreages in Spain, Mexico, and the United States. Although the tree is able to withstand light frost, it requires an

FIGURE 14-6. Fruit of the date palm. [Photograph by J. C. Allen & Son.]

average temperature of 86°F (30°C) for proper fruit ripening. Low humidity during fruit maturation is essential. Because the trees require a good supply of moisture, date palms are usually planted in oases or where irrigation is possible.

The date is dioecious. Select pistillate clones are propagated through offshoots that grow from the base of the trunk. To improve fruit set, flower clusters are usually pollinated by hand with pollen from staminate trees grown in separate plantings. Pollen affects tissues of the fruit outside of the embryo sac (**metaxenia**); thus the source of pollen influences the form, size, and rate of ripening of the fruit.

Flowering may begin in three-year-old trees of precocious cultivars. The developing fruit is often enclosed in paper or net bags to protect it from birds and other pests. A mature tree may yield more than 100 pounds of fruit yearly. Harvested dates are first fumigated and then allowed to ripen to increase sugars and precipitate astringent components. The partially mature astringent fruits are eaten in the Middle East, but those that enter commerce are ripened and dried.

Fig

The fig, *Ficus carica* (Moraceae), a small deciduous tree, has been cultivated since antiquity in the eastern Mediterranean region. It was known to the ancient Egyptians as early as 2700 B.C., and numerous references to it can be found in the Old Testament and in ancient Greek and Roman writings. World production totals about 2 million tons on a fresh-weight basis. About two-thirds of the crop is dried, and much of the dried fruit is treated with sulfur dioxide to maintain a light color. Major production is in Portugal, Italy, Greece, Turkey and the Near East; U.S. production is chiefly in California. Fig trees thrive best in hot dry summers and cool moist winters. Mature trees will withstand temperatures of 14–23°F (from

$-10\,^{\circ}$C to $-5\,^{\circ}$C) if completely dormant. Uniform soil moisture is essential.

The unisexual flowers are borne inside a pear-shaped, nearly closed receptacle. The entire receptacle becomes fleshy and forms the fruit, which is called a **syconium.** Skin color may be shades of green, yellow, pink, purple, brown, or black. Figs are propagated by means of cuttings.

The many major cultivars have been divided on the basis of pollination requirements into four types: caprifigs, Smyrna figs, common figs, and San Pedro figs. **Caprifigs** are primitive cultivated types with short-styled pistillate flowers and functional staminate flowers. Most caprifigs are not edible but are grown because they harbor the *Blastophaga* fig wasp, which develops in the fruit. The winged female escapes from the fruit of the caprifig in the spring covered with pollen and seeks developing figs in order to lay her eggs. She may enter the **Smyrna** fig, which contains only pistillate flowers. The pollen she carries with her fertilizes these flowers, which allows the fruit to mature. Thus, caprifigs are essential to the cultivation of Smyrna figs, such as the large-fruited 'Calimyrna' cultivar, which will not mature if unpollinated. The wasp is prevented from laying eggs in the young synconia of the Smyrna fig because her ovipositor is too short to penetrate the long-styled flowers; the wasp either dies there or flies away. The process of growing caprifigs to achieve pollination is known as **caprification.** Fruit set can be induced with growth regulators, but such fruits are seedless and lack the characteristic grittiness of figs produced by caprification. **Common figs,** such as the 'Kadota' and 'Mission' cultivars, set fruit parthenocarpically without the stimulus of caprification and seed development. **San Pedro** types set the first crop (**brebas**) parthenocarpically from buds initiated early in the spring, but they require caprification for the second crop.

Avocado

The avocado, *Persea americana* (Lauraceae), is of New World origin, as the specific name implies. Although still not widely known in the Eastern Hemisphere, avocados have been planted commercially in South Africa and Israel. Avocado is a popular local fruit in tropical and subtropical America, and there is considerable production in California and Florida.

The fruit is extremely rich in oil—fruits of some cultivars contain as much as 30%—however, the oil content varies with location, so 'Fuerte' avocados contain about 25% in California but only 13–15% in Florida. In rural areas of Brazil, the avocado is used to make soap.

The many cultivars of avocado have been divided into three "horticultural" races as shown in Table 14-3. Fruit size varies greatly, as do other characteristics. Many cultivars are hybrids between these races—'Fuerte,' for example, is considered to be a natural Guatemalan–Mexican hybrid.

Planting usually requires more than one cultivar in order to synchronize pollen shedding and stigma receptivity. Flowers of some cultivars (in Class A) open in the morning with receptive stigmas but shed no pollen; the flower closes at noon

TABLE 14-3. Distinguishing characteristics of the cultivated races of avocado.

Characteristic	Mexican	Guatemalan	West Indian
Native habitat and elevation	Highlands of Mexico (also Ecuador, Peru, and Chile), 7900–9200 ft	Highlands of Central America (also Mexico), 2600–7900 ft	Lowlands of Central and South America, below 2600 ft
Climatic zone	Semitropical	Subtropical	Tropical
Hardiness (of dormant trees)	Slight injury at 21°F (−6°C)	Severe injury at 24°F (−4°C)	Severe injury at 28°F (−2°C)
Season of fruit maturity	Summer	Winter and spring	Summer and fall
Foliage scent	Anise odor	No anise odor	No anise odor
Fruit size	½ lb	½–5 lb	1–5 lb
Fruit stem length	Short	Long (3 inches)	Short (1–2 inches)
Skin texture and thickness	Smooth, papery (0.8 mm thick)	Rough, woody, brittle (3–6 mm thick)	Smooth, leathery (1.5–3 mm thick)
Seed size and fit	Large and tight in seed cavity	Small in proportion to fruit size and usually tight	Large and usually loose

SOURCE: Ochse, Soule, Dijkman, and Wehlburg, 1961, *Tropical and Subtropical Agriculture* (Macmillan).

and reopens the afternoon of the following day to shed pollen, at which time the stigmas are nonreceptive. Flowers of other cultivars (Class B) open with receptive stigmas in the afternoon of the first day and reopen for pollen shed the next morning or the morning of the third day.

	First day		Second day	
	A.M.	P.M.	A.M.	P.M.
Class A	♀			♂
Class B		♀	♂	

In Florida, cultivars of classes A and B are interplanted to assure adequate fruit set.

Banana

Bananas, *Musa* species (Musaceae), originated in the tropical regions of southern Asia, and are among the oldest cultivated plants (Fig. 14-7). They are true tropical fruits and require temperatures that do not fall below 50°F (10°C) or rise above 105°F (41°C). Although there are many edible species, the most important are the seedless selections of *M. sapientum*, exemplified by the cultivar 'Gros Michel.' In Asia there are many starchy types of bananas known as plantains (*M. paradisica*), which must be cooked before eating. Bananas have become one of the most important fruits of the world, and they are the best known of the tropical fruits.

The banana plant is a giant herb with an underground stem (rhizome). The above-ground portion, the **pseudostem,** is formed from a compact mass of leaf stalks through which the blossom emerges. Although each pseudostem produces only one fruiting stalk, the underground rhizome produces numerous suckers. Thus, although each pseudostem is an annual, one planting is capable of producing continuous crops.

Bananas are propagated by planting rhizome pieces or young suckers. When a fruit bunch is harvested, the pseudostem from which it is produced is cut down and a sucker from the rhizome grows to replace it. If not shortened by disease, production cycles often continue for as long as 20 years before another rhizome must be planted.

The leading banana-producing countries in the Western Hemisphere are Honduras, Costa Rica, and Panama, whereas the leading Asian producers are India, Malaysia, Taiwan, and the Philippines. The present export industry is located largely in the Caribbean countries and is controlled by a few companies, of which United Fruit is the largest. The main limitations to production in Central America have been two devastating diseases, **Panama disease** (a root wilt caused by the fungus *Fusarium oxysporum* f. *cubense*) and **sigatoka** (a leaf spot caused by the fungus *Cercospora musae*). In the past, the only economical way of controlling Panama disease was to move plantings to new wilt-free locations. Since 1963,

FIGURE 14-7. The banana plant. As the stem elongates, the blossoms unfold. The fruited stalk, or bunch, consists of many clusters of flowers called hands. Each pistillate flower develops into a fruit—the "finger." Staminate flowers are produced toward the apex end of the stem. [Photograph by J. C. Allen & Son.]

however, a new cultivar, 'Valery,' which is resistant to Panama disease, has been used to replace the susceptible 'Gros Michel.' 'Valery' is thin skinned, and must be shipped in cartons rather than on stems.

Pineapple

Pineapple, *Ananas comosus* (Bromeliaceae), is of New World origin. The fruit vaguely resembles a pine cone, and the Spanish explorers called it "Piña de las Indias." Imported to Europe and raised in greenhouses in the seventeenth and eighteenth centuries, this tropical fruit became a gourmet delicacy and a symbol of hospitality. The development of canning established pineapple as a commercial plantation crop at the close of the nineteenth century. At present, almost one-quarter of the world's pineapple comes from Hawaii, where it is grown on 75,000 acres and is the second most important crop after sugar cane. Other pineapple-producing areas include Taiwan, the Philippines, Brazil, Mexico, Puerto Rico, and parts of Africa.

The family of Bromeliaceae consists largely of epiphytes, or "air-growing" plants (Spanish-moss is an example). Although pineapple is terrestial, it has many epiphytic adaptations, including a short stem and coarse, stiff, narrow, spiny leaves arranged in the form of a rosette. The pineapple is well adapted to dry habitats; the close-fitting basal leaves form a funnel capable of holding and absorbing water. The multiple fruit is usually seedless. Therefore, propagation is from the **crown** (the leafy shoot arising from the top of the fruit), from **slips** (shoots growing below the fruit) or from **suckers** (offshoots). These propagules are remarkably resistant to desiccation.

In Hawaii, planting materials are set out under plastic mulch in strips of fumigated soil. The plants normally flower 20–24 months after planting (usually in the summer) and bear one large fruit. A number of growth regulators are used to initiate flowering out of season, and this, in turn, ensures a continuous supply of pineapple for the canning factories. After the first crop ("plant crop") is harvested, the branches yield two smaller fruits the following year (a yield referred to as the "first ratoon crop"). The plants bear four more fruits, smaller still, and after this "second ratoon crop" the field is disked and the enormous amount of plant residue is burned. Then the land is prepared and the cycle is repeated. The crop is harvested by hand in Hawaii; conveyors move the fruits through the field for quick dispatch by truck or barge to the centralized canning factory of the Dole Corporation on the island of Oahu. The Hawaiian industry is based on the 'Smooth Cayenne' cultivar, thought to have originated in French Guinea but described also at Versailles in 1841. The fruit of 'Smooth Cayenne' is tough and well suited in size and shape to the Ginaca machine, a remarkable device that trims and cores the pineapple at the rate of a hundred fruits per minute.

Pineapple by-products include bran, syrup, alcohol, citric acid, fertilizer, and fiber.

Mango

Native to Southeast Asia, the mango, *Mangifera indica* (Anacardiaceae), has been cultivated since prehistoric times in India, where it is today the most popular of any fruit. The mango is grown now in practically all tropical areas. Although it is not an important export fruit, it has very great local use, being planted widely in small groves and in backyards. The tree, magnificent and glossy-leaved, bears juicy, aromatic, one-seeded (or drupe) fruits on a terminal panicle (Fig. 14-8). The single large seed has fibers that extend into the flesh. Some select cultivars are smooth-fleshed, however, and have few noticeable fibers. The flesh has a charac-

FIGURE 14-8. 'Tommy Atkins' mango is widely planted in Florida because of its high productivity and its attractive appearance. The fruit is a bright reddish orange. [Photograph courtesy S. E. Malo.]

teristic aroma, which ranges from mild (peach-like) to very rich and strong (turpentine-like). Fruits range in color from green through yellow to bright red, and they vary in weight from $\frac{1}{4}$ to 2 pounds. Many mangos are apomictic, with adventitious embryos produced from nucellus tissue, as in citrus. Thus, mangos may breed true from seed. Often, mango trees may bear only in alternate years.

Other members of the family Anacardiaceae are the cashew (*Anacardium occidentale*), which is grown for its nut as well as for its edible pear-shaped peduncle, and the pistachio (*Pistacia vera*). It is interesting to note that poison oak and poison ivy are also members of this family. To one degree or another, all members of the family contain toxic substances that may be irritating to the skin.

Papaya

The papaya, *Carica papaya* (Caricaceae), is indigenous to tropical America but is now grown throughout the tropics. It is known by a number of other descriptive names: for example, papaw (New Zealand), mamão (Brazil), lechosa (Venezuela), and fruita-bomba (Cuba). The tree—fast growing, short-lived, and herb-like—is normally single stemmed, but occasionally lateral branches result from wounding. Fruit is borne on the axils of the large leaves that open in sequence up the stem as the tree grows. The crop is propagated by seed.

The fruit, which may be round, pyriform, or oval, ranges in weight from $\frac{1}{2}$ to 15 pounds. The skin color is orange; inside, the flesh is pale orange or light red.

Papaya has both staminate and pistillate flowers and a great array of bisexual flowers. On the basis of the flowers they bear, the trees are classified as staminate (all staminate flowers), pistillate (all pistillate flowers) or hermaphroditic (hermaphroditic and staminate). Hermaphroditic trees may be further divided into the type of hermaphroditic flowers they produce. Moreover, some papaya trees are continuously fertile and others are sterile in the summer.

Most cultivars are dioecious with staminate and pistillate plants produced in a $1:1$ ratio, but some are gynodioecious with hermaphroditic and pistillate plants. Sex determination would seem to be controlled by a single gene with three alleles M, M_1, and m, in accordance with this scheme:

$$Mm = \text{staminate}$$
$$M_1m = \text{hermaphroditic}$$
$$mm = \text{pistillate}$$

The genotypes MM, M_1M, and M_1M_1 are lethal and give rise to nonviable seeds. Dioecious cultivars produce staminate and pistillate plants in equal ratios:

$$mm\ ♀ \times Mm\ ♂ \longrightarrow \underset{\text{Staminate}}{1\ Mm}\ :\ \underset{\text{Pistillate}}{1\ mm}$$

In general only a few staminate plants are needed for pollination. Generally three seedlings per hill are allowed to grow until flowering, and staminate plants are removed as they flower. At present there is no way to distinguish the sex of the trees before flowering. In $\frac{7}{8}$ of the hills there will be at least one pistillate plant. In $\frac{1}{8}$ of the hills all of the three seedlings will be staminate, of which one is kept to provide pollination.

In Hawaii the standard papaya cultivar is 'Solo' (Fig. 14-9), a very high-quality small-fruited type composed of hermaphroditic and pistillate plants. Hermaphroditic flowers yield pyriform fruit; pistillate flowers yield round fruit that is discriminated against (apparently without reason). Seed saved from hermaphroditic fruit yields two new hermaphroditic plants for each new pistillate plant:

$$M_1 m \, \female \times M_1 m \, \male \longrightarrow \underset{\text{Lethal}}{1 \, M_1 M_1} : \underset{\text{Hermaphroditic}}{2 \, M_1 m} : \underset{\text{Pistillate}}{1 \, mm}$$

Thus, growing three seedlings per hill with subsequent thinning to one hermaphroditic plant per hill yields a crop in which only $\frac{1}{27}$ (3%) of the spaces grow the unpopular pistillate plants.

In addition to providing fresh fruit, papaya is the source of papain, a proteolytic enzyme that is used as a beer clarifier and in meat tenderizers. Papain is obtained by collecting and drying the latex exuded from scratches in the surfaces of slightly immature fruit.

FIGURE 14-9. Precocious fruiting in a hermaphroditic plant of 'Solo' papaya. [Photograph courtesy C. P. Watson.]

NUT CROPS

> I had a little nut tree, nothing would it bear
> But a silver nutmeg and a golden pear.
> The King of Spain's daughter came to visit me,
> All for the sake of my little nut tree.
>
> ENGLISH NURSERY RHYME

The edible nuts—seeds of woody plants with firm shells that separate from an inner kernel—are often considered along with fruit crops, even in nursery rhymes. However, although nuts are tree crops, their cultivation is typically more casual; many nuts are still gathered from wild stands. Nut production has been suggested as a use for marginal lands, and undoubtedly there is great potential for this through genetic improvement and better cultural practices; however, those who champion the raising of nut crops have not received the attention they deserve.

Characteristically, nuts are rich in oil; indeed, the inedible tung nut, *Aleurites fordii* (Euphorbiaceae), is grown to supply oil for paint, varnishes, and other uses. Edible nuts have been cultivated, in general, as an expensive luxury food, a special delicacy consumed often with candy or sweetmeats. Important temperate nut crops include almond, chestnut, hazelnut, pecan, pistachio, and walnut. Tropical nuts include cashew, Brazil nut, and macadamia.

Temperate Nuts

The almond, *Prunus amygdalus* (Rosaceae), is closely related to the peach. The outer flesh of the almond is astringent and tough, however, splitting as the fruit ripens to expose the pit—the almond shell. The edible sweet almond (*P. amygdalus* var. *dulcis*) is the source of confectionary nuts (Fig. 14-10); the bitter almond (*P. amygdalus* var. *amara*) contains a poison, prussic acid, and is grown only for the extraction of almond oil for use as a food flavoring. Almond crops are produced in Spain, Italy, and the United States, where California produces the largest crop.

The chestnut is important in Europe and Asia both as a livestock feed and as an edible nut. The European or Spanish chestnut, *Castanea sativa* (Fagaceae), is a semicultivated tree of southern Europe, and it produces the nut so often found "roasting by an open fire" during the Christmas season. The native North American chestnut, *C. dentata*, has been eliminated by chestnut blight. Both the Chinese chestnut, *C. mollissima*, and the Japanese chestnut, (*C. crenata*), are resistant to chestnut blight but are of inferior quality. Perhaps hybrids between *Castanea* species that are resistant to blight and those that produce nuts of high quality can restore chestnut cultivation to the United States.

Hazelnuts, various species of *Corylus* (Betulaceae), grow wild in most of the world's northern temperate areas. The cultivated types are known as filbert,

FIGURE 14-10. 'Solo,' a new almond cultivar from Israel. [Photograph courtesy P. Spiegel-Roy and J. Kochba.]

cobnuts, Barcelona nuts, and Turkish nuts. Hazelnuts are grown extensively in northern Turkey (200,000 tons—shell weight—annually), Italy (50,000 tons), and Spain (20,000 tons). United States production is mostly in Oregon (10,000 tons). The native American species *C. americana* and *C. rostrata* are pleasantly flavored but are considered too small for commercial use.

The pecan, *Carya illinoensis* (Juglandaceae), has achieved considerable commercial success in the southern United States (it is native to that area and to Mexico). Production is still largely from wild and seedling trees in Georgia and Alabama, but with selected cultivars, propagated by grafting, the industry is increasing rapidly in Louisiana, Texas, and Oklahoma. Selection favors thin shells ("papershells") and large size. Most of the crop is sold to bakers and processors, the poor-quality unshelled nuts being sold to less knowledgeable buyers—usually tourists. Efficient cracking and shelling machines can now produce a high percentage of unbroken halves.

The pistachio, *Pistacia vera* (Anacardiaceae), is a dioecious, slow-growing tree that is capable of attaining great age. The pistachio fruit has a pale red or yellow husk enclosing a small, smooth, thin, hard-shelled nut with a fine-textured green kernel. The tree is native to western Asia, and the greatest production areas are Turkey, Iran, Syria, Afghanistan, and Italy. Richly flavored, the nut is used in the United States for flavoring ice cream; red-dyed nuts are familiar at least to New Yorkers, who buy them at coin-operated nut dispensers. Commercial production in the United States is starting, though presently most of the nuts consumed in this country are imported from Turkey.

Although commercially the most important walnut with an edible nut is the Persian, Regia, or English walnut, *Juglans regia* (Juglandaceae), there are many species of walnut with edible nuts, including the black walnut (*J. nigra*) and the butternut or white walnut (*J. cinerea*). More than 100,000 tons of shelled English

FIGURE 14-11. Fruit of the English walnut, showing the outer husk and the shell-covered nut. [Photograph courtesy USDA.]

walnuts (Fig. 14-11) are produced in the United States each year, mainly in California and Oregon. By contrast, France produces 30,000 tons and Italy produces 20,000 tons. Unlike the husk of the black walnut, the husk of the Persian walnut readily separates from the shell. After being gathered, washed, and bleached, nuts are often individually labeled, when sold in the shell for table use.

Tropical Nuts

Coconut, *Cocos nucifera* (Palmaceae), which is grown throughout the tropics, is the source of **copra** (dried coconut meat that is pressed to produce oil) and **coir** (the husk fiber). Although the stately trees require 60 inches of rainfall evenly distributed throughout the year, they are easily able to withstand brackish water. They are typically found along tropical sandy shorelines. Important centers of production lie in a zone no wider than 15° of latitude. Annual world production of copra approaches 3 million tons, most of which comes from the Philippines, Indonesia, Ceylon, Malaysia, and Mexico. However, much coconut is consumed locally and does not enter into world trade.

Propagation is from seed; the young seedlings are planted about 40 or 50 to the acre. Such a planting may yield 900–1100 pounds of copra per acre, with yields per tree of up to 88 pounds. Production varies from 50 to 200 nuts per tree. Approximately 5,500 nuts will yield one metric ton of copra (of which 61.5% is oil). Trees bear at from 6 to 10 years, though full production is not reached until the trees are 15 to 20 years old. Fruits require about one year to develop. The 'Malay' tree starts to bear fruit close to the ground, but eventually these apparent dwarf trees reach great heights.

The cashew, *Anacardium occidentale* (Anacardiaceae), is native to Brazil but now grows in many tropical areas (Fig. 14-12). In temperate countries, it is well known for its high-quality edible nut, but in tropical countries the fleshy astrin-

FIGURE 14-12. Fresh cashews as harvested in Guatemala. The true fruit at the tip contains the cashew nut; the fleshy pear-shaped pedicel and receptacle can be eaten fresh or preserved. [Photograph courtesy USDA.]

gent pedicel or stalk (the cashew "apple"), which bears the nut-containing fruit, is also eaten fresh or made into juice. Moreover, the cashew shell contains a liquid—the nut-shell "oil"—that is used as a resin. Most cashews consumed in the United States are grown in India. Cashew nuts, like their pedicels, are astringent; unlike the pedicels, the nuts are toxic when raw and are roasted before export.

The large Brazil nut, *Bertholletia excelcea* (Lecythidaceae), is borne in a natural "jumble pack" in urn-like ligneous fruits on great trees of the Amazon rain forest. The nuts are gathered from natural stands, and about 30,000 tons are exported annually to the United States and Europe.

Macadamia ternifolia (Proteaceae), although native to Australia, is cultivated chiefly on the stony volcanic soils of the island of Hawaii—the "big" island—where its nuts are rapidly becoming an important crop. In the economy of the state of Hawaii, it is still a crop of minor value compared to sugar and pineapple, but it is now in third place. The spherical nut, enclosed in a fleshy husk, is extremely tough, but it cracks easily between rotating drums to yield an extremely rich kernel (70% oil) of excellent quality. The nuts fall naturally onto the smooth, stony orchard floor and are vacuumed up by special harvesting machines. In processing, the entire crop is first dried, then cracked, graded, cooked slightly in oil, salted, and vacuum-sealed in glasses or cans. Imperfect nuts are used in a number of other products, chiefly brittle and ice cream. At present, the demand for macadamia nuts outstrips the supply, and production is increasing rapidly.

BEVERAGE CROPS

Tropical plants are the source of the world's three most popular nonalcoholic beverages—coffee, tea, and cocoa. The original basis of their popularity was the "lift" provided by the caffeine (or caffeine-like alkaloids) that they contain, though their consumption with sugar and milk products also makes these products refreshing foods. As presently consumed, cocoa contains little stimulant and is more widely known for its confectionary uses. World demand has created huge industries in connection with growing, processing, and transporting the bulky materials from which these beverages are brewed. The crops represent a large proportion of the tropical world's agricultural exports; in many tropical countries, their production is the major industry.

Coffee

Coffee is made by steeping ground roasted seeds (or "beans") of *Coffea* species (Rubiaceae) in hot water. Most of the world's production is of *C. arabica* (Arabian coffee), with small quantities of *C. conephora* (Robusta coffee), *C. liberica* (Liberian coffee), *C. excelsa* (Excelsa coffee), and others. Coffee was being grown in Arabia as early as the seventh century, and it reached Europe in the sixteenth and seventeenth centuries with the opening of the East by the Dutch and English. The City of Mocha, Arabia, a city immortalized in the name still used for one type of coffee, remained the center of the trade, but the Dutch introduced *C. arabica* into Ceylon in 1658, into Java in 1699, and, about 1619, into Martinique, whence it spread into the New World.

Coffee now has its major production center in America; Brazil accounts for about one-third of the world's annual production of more than 4 million metric tons. Unlike many plantation crops, coffee tends to be produced by local entrepreneurs. Marketing is characterized by periods of glut and shortages, with corresponding fluctuations in price.

The coffee plant is a glossy-leaved shrub that flowers seasonally, usually at the onset of wet weather. The fruit ("cherry") is a two-seeded drupe. Seedlings are commonly transplanted to a density of approximately 500 plants per acre. Fields may be planted in shade, but the highest yields are achieved in full sunlight with large amounts of fertilizer. Trees are pruned to establish strong frameworks, with single or multiple-stem systems. Renewal pruning is practiced to maintain a good quantity of fruiting wood less than five years of age. Harvesting is generally done by hand, so plants are pruned to control height. Experimental machines that either shake the fruit into catch-frames or comb the limbs are now under development.

Processing requires removal of the fruit skin (exocarp), the mucilaginous flesh (mesocarp), a parchment coat (endocarp), and the delicate silverskin (seed coat). Fruit may be **dry processed:** after drying, the pulp is separated mechanically from the seed. More commonly, the fruit is **wet processed:** defective fruit is first floated off in water, then the pulp is separated from the seed by machine. Brief fermentation then removes the remaining flesh adhering to the parchment, and the beans are dried and cured (through further milling and polishing that removes parchment and silverskin). Roasting for about five minutes at 500°F (about 280°C) is the final step in either process.

Coffee requires a dry season for fruit maturation; a monsoon climate with a double dry season is preferred to obtain two crops per year. The best coffee is grown at elevations of 4000–5500 feet with rainfall of 80–120 inches and mean temperatures of 60–72°F (15–22°C).

Cacao

Although cacao, *Theobroma cacao* (Sterculiaceae), is a native of the New World, it has found its greatest production in Africa, with Ghana the major producer; Brazil ranks second. In some areas of pre-Columbian America the cacao "bean" was used as currency. The Aztecs also greatly prized a bitter, nourishing concoction made of cacao, vanilla, corn meal, and chilis. To please the Spanish palate, sugar replaced the corn meal and cinnamon replaced the chilis. The new mixture made chocolate (from the Aztec *chocolatl*) a popular beverage in Europe by the middle of the seventeenth century. Consumption increased rapidly when, in 1828, a Dutch manufacturer perfected a method for separating the fat (cocoa butter) to produce the versatile residue, cocoa powder. The Swiss invention of milk chocolate in 1876 further increased the demand for cacao.

Cacao cultivars are usually divided by "race": the Forastero has bitter purple seeds and the Criollo has sweet white or violet seeds. Hybrids between them are known as Trinitario. A mutant with white cotyledons produces a white chocolate. Football-sized fruits are borne directly on trunks and large branches (Fig. 14-13). Cacao trees are related to *Cola acuminata*, the source of cola nuts used in soft drinks.

The fruit is hand-harvested and split open with a machete. The seeds are removed and fermented to remove the sweet white pulp and to improve their flavor. After fermentation, the beans are dried, sacked, and shipped.

Propagation is commonly from seedlings. Recently improved hybrid types have been produced from selected clones. Clones may be propagated from cuttings or bud grafts. Unless a self-fertile clone is chosen, a number of clones must be planted to ensure pollination.

Cacao is a true tropical crop. Production is concentrated in lowlands in a zone

FIGURE 14-13. *Left,* a cacao tree bearing berry fruits (commonly called "pods") on the trunk and large branches. *Right,* the sweet pulp surrounding each seed can be saved and made into a jelly, but it is usually removed by fermentation. [Photographs courtesy USDA.]

that extends no more than 10° on either side of the equator. Because of its sensitivity to drought, cacao requires an even distribution of rainfall, and it is grown under the shade of a tree canopy.

Tea

Tea, an infusion produced from the dried tip leaves of *Thea sinensis* (Theaceae), is the most widely used beverage in the world. Of Chinese origin, tea was introduced to Europe by the Dutch in the seventeenth century. In England, tea replaced coffee in popularity, a preference still maintained in British Commonwealth countries.

The glossy leaves of the tea plant contain numerous glands that produce an essential oil. The terminal leaf or two are plucked and either dried immediately to produce green tea or processed by fermentation to produce black tea, the main tea of commerce in the West.

Tea can be cultivated in subtropical areas or in mountainous regions of the tropics with even distribution of rainfall. The tea plant is an evergreen shrub propagated from seeds, with some clonal propagation by cutting or bud grafting. Plants are pruned to establish a low, flat top, to keep plants within bounds, and to renew old plantings. Leguminous shade trees may be used at first, but they are removed as the plantings age. Most harvesting is done by hand, although mechanical harvesters are now being developed.

SPICE AND DRUG CROPS

> [Joseph and his brothers] . . . saw a caravan of Ishmaelites coming from Gilead with their camels bearing gum, balm, and myrrh, on their way to carry it down into Egypt.
>
> GENESIS 37:25

Plants with strong aromas and flavors, **spices** very early attracted human attention, probably at first for magical rites, spells, purification ceremonies, and embalming, but then also for fragrances and perfumes, flavorings and condiments, food preservatives, curatives, aphrodisiacs, and poisons. The distinctive odors of many of these plants (essences) are due in large part to **essential oils**, benzene or terpene derivatives that are oily, vaporizing, and flammable. Although often associated with the flower, essential oils may be found in all plant parts, usually as secretions from specialized glands. Because of these oils, the plants have been widely used with a variety of effects—some real, some fanciful. For example, clove is used for flavoring and is often tried as a remedy for stomach complaints.

Although spices and medicinal plants overlap, those whose primary value is therapeutic will be referred to as **drug crops.** Many (but not all) of the curative powers of drug crops are derived from **alkaloids,** a class of nitrogenous bases that cause profound physiological responses in animal systems (see Table 4-4, p. 109). The high value placed on spices and medicinals in the ancient world attracted traders and merchants who traversed well-established routes from sources in the East to the metropolitan centers of the West. The "spice trade" developed the first contacts between East and West.

The traffic in spices centered early in Arabia, a crossroads leading East to the Malabar coast of India by water (via the Red Sea and the Persian Gulf) and in other directions by overland caravan. However, with the rise of the West, control over the spice trade passed from Arabia to Venice and Constantinople, two cities so situated that they were effectively like bottlenecks to trade entering the Mediterranean. The prize passed next to Portugal: Although the explorations of Columbus failed to discover a competitive all-water route to the spice islands of the East, Vasco Da Gama in 1497 skirted Africa and found a direct route to Calicut, the Indian center of the land of cinnamon, ginger, and black pepper. By combining their navigational skill and daring with greed and ruthlessness, the Portuguese soon monopolized the spice trade of the East Indies until they were overwhelmed by the Dutch in the seventeenth century. Spices continued to influence the course of empire and history as power in the East Indies was contested by the French and finally (successfully) by the English. Today, the monopoly is long broken and the spices that once lured the great powers no longer have great importance in world trade, even though the mischief created in the era of spice colonialism is with us still in the form of political instability and underdevelopment. Still, spices claim their places the world over as silent partners to cooks and as essential luxuries even to jaded palates.

Today there are more than 200 species of spice and drug plants collected for use in the United States alone. Many of these are slow-growing tropical vines, shrubs, or trees—black pepper, cinnamon, clove, allspice, vanilla, nutmeg and mace; and opium, quinine, rotenone, and strychnine. Temperate condiment plants, often known as **savory herbs,** include mustard, cayenne and paprika, sage, and caraway; temperate drug plants include belladonna, digitalis, henbane, and stramonium. In the following discussion, only a few of the more important spice and drug crops are reviewed.

Cinnamon is made from the inner bark of the small, bushy cinnamon-laurel tree, *Cinnamomum zeylanicum* (Lauraceae). The bark is stripped from second-year wood, dried, and then peeled of epidermis particles. These are formed into cinnamon **quills** (hollow tubelike "pipes"), or are ground to yield powdered cinnamon. Oil of cinnamon is prepared from the leaves, fruit, or root. The highest quality cinnamon is produced in southern Ceylon. Camphor is produced from wood of the closely related *C. camphora.*

Cloves are the dried unexpanded flower buds of a small tropical tree, *Eugenia aromatica* (Myrtaceae). The Dutch at one time obtained a complete monopoly in the clove trade and confined production to a single island of the Moluccas, uprooting all other native clove trees. The French, however, introduced the clove into Mauritius in 1770, and from there production spread to Brazil, the West Indies, and Zanzibar. The ripe, bright-red flower buds, when sun dried, turn dark

531

brown. The resulting spice has a powerful fragrance and a hot, acrid taste. Oil of cloves is extracted by distillation and consists mainly of eugenol ($C_{10}H_{12}O_2$). Eugenol is the raw material for the synthesis of vanillin.

Nutmeg and mace, two fairly well-known kitchen spices, are both derived from the fruit of the nutmeg tree, *Myristica fragrans* (Myristicaceae). Nutmeg is the seed, and mace is obtained from the aril, the bright red membrane surrounding the seed. "Oil of mace" is expressed or distilled from the seed, rather than from the aril. The nutmeg tree is native to the Moluccas, but it is now cultivated also in the East and West Indies.

Black pepper, *Piper nigrum* (Piperaceae), should not be confused with chili peppers, *Capsicum* spp. (Solanaceae). Black pepper is native to India and is one of the oldest and most important of all the spices. At one time black pepper was also used as a medicine. The fruits are red when ripe, but they darken when dried to produce black "peppercorns." These are ground to produce the familiar spice in pepper shakers. Whole peppercorns are widely used in meat flavoring and preservation (of pepperoni and salami). White pepper is the kernel of the peppercorn, from which the outer skin and pulp (mesocarp) has been removed after soaking in water. The pepper plant is a perennial climbing vine usually grown on trellises. Propagation is typically by cuttings. Production is centered in India and Southeast Asia with small quantities produced in Brazil. World export ranges between 50,000 and 75,000 tons annually. The United States consumes about one-third of the world supply at the rate of about 100 grams (4 ounces) per capita.

Peppermint oil, one of the best known and most ancient of the essential oils, is a distillation product of *Mentha piperita* (Fig. 14-14), a member of the family Labiatae. Menthol, the chief ingredient, is used as a flavoring in a great number of products—chewing gum, confections, dentifrices, medicines, and tobacco products. The oil of the closely related spearmint, *M. spicata*, is used similarly. The mints are perennial herbs, well adapted to moist areas of temperate climates; oil production requires long days. Peppermint and—to a lesser extent—spearmint are cultivated on a commercial scale on muck soils in the United States (Michigan, Indiana, and the Columbia River basin of Oregon and Washington), as well as in Japan, England, and continental Europe. The plant is propagated from stolons in rows the first year (row mint); in subsequent years, the plants cover the whole field (meadow mint). The crop is mowed by machine, field dried, and steam distilled. Oil yields average about 30 pounds per acre. Verticillium wilt, one of the major diseases, has been controlled by deep plowing: great machines invert the soil to a depth of 6 feet.

Vanilla is produced from the dried, fermented seed pods of a climbing orchid *Vanilla planifolia* (Orchidaceae). The name *vanilla* is Spanish, a diminuative of *vaina*, a pod. Vanilla is indigenous to tropical America and, as noted earlier, was used by the Aztecs in compounding the beverage *chocolatl*. Vanilla beans were highly prized and were exacted as tribute. The marvelously fragrant pods yield an essential oil, which can be extracted with alcohol. The active substance vanillin

FIGURE 14-14. Peppermint (*Mentha piperita*) contains an essential oil in its foliage. The oil, which is produced when the plant is subjected to long days, is extracted for flavors and medicinals.

($C_8H_8O_3$) is now widely synthesized with other materials, but the natural flavor is greatly superior to the synthetics. The vanilla plant is grown on supports, under shade. Flowers are often hand pollinated to ensure set, and the fruits may be thinned. The plant requires a hot, moist climate. Most of the world's supply is produced in Madagascar.

Many spices are products of plants already discussed as fruits and vegetables. For example, seeds of celery, *Apium graveolens* (Umbelliferae), are used as a spice (the savory seeds are used whole or ground) or for the production of essential oil. Other species of the Umbelliferae used as spices include: anise (*Pimpinella anisum*), caraway (*Carum carvi*), coriander (*Coriandrum sativum*), cumin (*Cuminum cyminum*), fennel (*Foeniculum vulgare*), and parsley (*Petroselinum sativum*). Similarly, many *Citrus* species (lemon, grapefruit, orange, lime) are used for their essential oils. A number of crucifers are grown as spices—horseradish (*Armoracia rusticana*) and mustard (*Brassica nigra*) are the most notable examples.

VEGETABLE CROPS

The "vegetable" food plants of horticultural interest include many species and many families, of which Solanaceae, Cucurbitaceae, Cruciferae, and Leguminosae are particularly prominent. Some of the starchy crops, such as potato, sweet-potato, cassava, and yam, are important as calorie sources, and rival the cereals as nourishers of people. Many of the edible legumes (peas and beans) are important protein sources. However, many garden vegetables are classed as **protective food,** supplying vitamins and variety to the diet, and are not major sources of energy or protein.

Although many vegetable crops are merely of local interest, some have assumed tremendous economic importance. Thus, lettuce consumption in the United States is more than 6 million heads per day (300 freight cars full!), with an annual farm value of more than 400 million dollars. A few vegetables have assumed world-wide importance: The tomato, though it enters world trade only minimally as a fresh product, is produced from the tropics to the cool temperature climates and is a popular food in many cultures.

The rise of the processing industry has had a profound effect on the fortune of individual vegetable crops. Some crops, such as garden peas, have almost disappeared as fresh produce in the United States. Preferences for certain vegetables have important consequences for the industry in every part of the world. In Asia, pickled eggplants hold the place that pickled cucumbers hold in the United States. Because of the diversity of vegetables (they are much more numerous than fruit crops), the following discussion is by no means complete.

Salad Plants

Salad plants are leafy vegetables usually consumed raw, although this distinction is arbitrary. As a group, they are fast-growing cool-season plants. The major salad crops are species of three families: Compositae—lettuce (*Lactuca sativa*), endive (*Cichorium endivia*), chicory (*C. intybus*), and dandelion (*Taraxacum officinale*) (Fig. 14-15); Umbelliferae—celery (*Apium graveolens*), parsley (*Petroselinum sativum*), and chervil (*Anthriscus cerefolium*); and Cruciferae—watercress (*Nasturtium officinale*) and garden cress (*Lepidium sativum*). A number of other crucifers that may be considered as salad crops, such as Chinese cabbage, are discussed briefly with the cole crops in a subsequent section.

Lettuce is the world's most important salad crop. Lettuce appears to have been domesticated within historic times, probably derived from the wild *Lactuca serriola* native to Asia Minor. The great number of cultivars are usually classified into five types: **crisphead**, firm heads with brittle texture; **butterhead,** soft heads

FIGURE 14-15. Cultivated dandelion. [Photograph courtesy USDA.]

FIGURE 14-16. Three lettuce types: *top left,* 'Merit,' a widely used crisphead cultivar; *top right,* 'Big Boston,' the most extensively used butterhead cultivar; and *bottom,* 'Vanguard,' a distinctive winter cultivar used in desert areas that has genes from a wild lettuce, *Lactuca virosa.* [Photographs courtesy T. W. Whitaker, USDA.]

with pliable leaves and delicate flavor; **cos (romaine),** upright loaf-shaped heads with long narrow leaves; **looseleaf (bunching),** nonheading leaf lettuce; and **stem,** with enlarged edible seedstalks (Fig. 14-16). Lettuce is an annual well adapted to cool climates; temperatures that are too high result in seed-stalk elongation (bolting). Eighty percent of the United States lettuce crop is produced in Arizona and California, and production is confined to crisphead types, which are typically vacuum cooled and shipped in ice or in refrigerated transport. The butterhead, cos, and looseleaf types ship poorly and are usually grown only for local markets. Breeding programs have produced heat-tolerant strains with resistance to downy mildew, a disease that can cause severe crop losses. Mosaic, a virus disease, also a severe problem in lettuce producing areas, can be avoided by using only virus-free seed.

Although celery (Fig. 14-17) is native to marsh lands from Sweden to northern

FIGURE 14-17. Celery before and after trimming for shipment. [Photograph courtesy USDA.]

Africa and from southern Asia to the Caucasus, it is most popular in the United States and Canada, where it is the second most important salad crop after lettuce. Celery is a moisture-loving, long-seasoned annual usually grown under cool conditions; however, exposure of transplants to temperatures as low as 35–50°F (2–10°C) increases bolting in some strains. In Europe, celery is commonly blanched by covering the leaf stalks with soil or paper for about three weeks before harvest to produce white tender stalks.

Solanaceous Fruits

The family of which the potato is a member, the Solanaceae, contains three important members grown for their fruit: the tomato, eggplant, and capsicum pepper.

The tomato (*Lycopersicon esculentum*), a native of South America, was introduced into Europe before 1544 by the Spanish conquistadores. Superstitions concerning its presumed poisonous qualities discouraged its use in many European countries, with the exception of the Mediterranean regions, until the late eighteenth century. Tomatoes did not become popular in the United States until the latter part of the nineteenth century, although they had been reintroduced from Europe more than a hundred years earlier.

Although it is frost susceptible, the tomato is one of the most widely cultivated plants: it is grown from the equator to as far north as 65° latitude (Fort Norman, Canada). It is also grown largely under glass in northern Europe. In the United States, it is the most important greenhouse vegetable and ranks as the most popular home-garden food plant. The leading countries in tomato production are the United States, Italy, Egypt, and Spain; the United States accounts for about one-fifth of the world's production. Florida and California are the leading states for fresh market production. Tomatoes destined for fresh consumption are picked when pink and ripened naturally off the plant. Tomatoes can also be picked green, stored, and artificially ripened. They are then referred to as "green wraps." California is far and away the leading tomato-processing state. Tomatoes are canned whole, and as juice, puree, sauce, catsup, and paste (Fig. 14-18).

Although the tomato is a perennial plant, it is treated as an annual. The usual method of planting is by transplants, but direct seeding is increasing in popularity due to the effectiveness of chemical weed control. The mechanical harvesting of tomatoes has been developed and has resulted in great changes in the industry.

The eggplant (*Solanum melongena*) is assumed to have been domesticated in tropical India and secondarily in China. It is a staple vegetable in many oriental countries (in Japan, much of the crop is pickled). Although featured prominently in the various Mediterranean cuisines, the eggplant is only of minor importance in the United States. The round, oblong, or pear-shaped fruits vary greatly in size; their color ranges from purplish black, through light violet and white, to green. The plant is very sensitive to frost. The crop is usually started from transplants.

The capsicum peppers are native to tropical America, where they have long been an important condiment and food. "Pepper" is actually a misnomer deriving from a pungency in some species that is reminiscent of the Old World black

FIGURE 14-18. *Left,* plant breeders examine a dwarf tomato plant developed for mechanical harvesting. *Right,* 'Roma,' a paste tomato, was developed by the USDA to provide a high-solids product for processing. [Photographs courtesy Purdue University and J. C. Allen & Son.]

pepper (*Piper nigrum*, of the family Piperaceae). To avoid confusion with this totally unrelated plant, members of the genus *Capsicum* have been given the names capsicum or chili peppers. The pungent chilis were brought to Europe on Columbus's first voyage; spread first by the Portuguese, they have since gained worldwide popularity. They have become important spices in Asia and vital ingredients of such delicacies as the powerful Korean staple kimchee (a fermented cabbage relish) and Indian curry (made from roasted dried chilis mixed with tumeric, coriander, cumin, and other spices).

The capsicum peppers are an extremely variable group, and their nomenclature is confusing. Most cultivars are contained in two species: *Capsicum annuum* (annual plant, fruit borne singly) and *C. frutescens* (perennial, fruit borne in groups). *C. annuum* is unknown in the wild state. Horticulturally, capsicum peppers can be divided between "sweet" and "hot" peppers. The **sweet pepper,** also known as bell pepper and—most confusingly—as "mango" in the southern United States, is relatively nonpungent, with thick flesh; the fist-sized green fruit ripens to red or yellow. They are produced mainly in the United States. Sweet peppers are eaten either raw, cooked (often stuffed with meat), or pickled, and are a rich source of vitamin C. There are a number of cultivars and types of sweet peppers. The red "pimiento" is also used for stuffing olives or combined with processed meat or cheese. Other sweet types are dried and ground to produce the red condiment "paprika," largely in Spain and Hungary, where it is used as one of the ingredients of goulash. The **hot** or **spicy peppers** when dried are also known as "red" or "cayenne" peppers. The pungency is due to the volatile phenol capsicin ($C_{18}H_{27}NO_3$). Hot sauces, such as Tabasco, are made by pickling the pungent pulp in vinegar or brine. The largest exporter of pungent capsicum peppers is India, followed by Thailand.

Edible Legumes

The edible legumes, or pulse crops, which include many genera of the Leguminosae, have long been used as food (Fig. 14-19). There have been other uses for legumes: the expression "blackball," for example, comes from the ancient Greek and Roman practice of using beans for voting, a white bean signifying acceptance and a black bean rejection. And today, soybeans, although edible as a vegetable, are grown in the United States principally as an oil and feed crop. Edible legumes are good sources of protein, although they are deficient in the sulfur-containing amino acids, cystine and methionine, and they are especially important in Asia and South America, where they constitute a key part of the diet. In the United States, the most important of the edible legumes are the common bean and pea.

The common bean (*Phaseolus vulgaris*) is a warm-season annual quite sensitive to frost. It and the lima or butter bean (*P. lunatus*) are both native to South America, although they are now extensively grown throughout the world. In various stages of its growth the bean is edible, but various cultivars are now grown

A

B

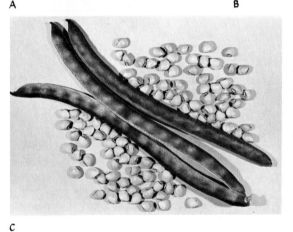

C

D

FIGURE 14-19. Some examples of edible legumes. *A*, 'Goldcrop' wax bean (*Phaseolus vulgaris*). *B*, 'Oregon Sugarpod' pea (*Pisum sativum*), which has an edible pod. *C*, 'Worthmore' southern-pea (*Vigna unguiculata*). *D*, Pigeon pea (*Cajanus cajan*) a tropical bush legume high in protein. [Photographs courtesy of All-America Selections (*A*), J. R. Baggett (*B*), J. Dan Gay (*C*), Julia F. Morton (*D*).]

to provide each type of food. The edible-podded types are called snap or stringless beans (formerly string beans!) and may be either the climbing ("pole beans") or nonclimbing ("bush beans"). Beans are also harvested when mature but still green ("shell beans") or in the dry-seeded stage. Dried beans are a substantially larger crop than snap beans and account for about ten times the acreage. Among the various types of dried beans grown in the United States are the navy, red kidney, pinto, great northern, marrow, and yellow-eye beans. The mung bean (*P. aureus*) is a small-seeded Asiatic species commonly consumed after germination, the bean sprout ubiquitous in Chinese cookery.

In contrast, the garden pea (*Pisum sativum*), also native to Asia, is a cool-season crop. It is harvested for use in either the mature green stage or the dry-seeded stage, although there are cultivars that are edible in the podded stage ("sugar peas," or "snow peas"). Almost the entire commercial green pea crop is now produced for processing. At present, these peas are one of the most popular frozen vegetables; however, the bulk of the crop is still canned. The important areas producing green peas are the northern states of Wisconsin, Washington, Minnesota, and Oregon. Dried peas, almost as large a crop as green peas, are produced mainly in Washington and Oregon.

The peanut (*Arachis hypogaea*), indigenous to Brazil, is well known in the United States in the form of peanut butter, salted table nuts, and, when roasted in the shell, as standard fare at sporting events. However, peanuts are chiefly used as an oil crop, and the by-product, the protein-rich presscake, is used as a livestock feed. Flowers are formed above ground, but after pollination they are gradually pushed beneath the ground where the shell of the peanut matures; hence the English name "groundnut."

Other edible legumes include the broad bean (*Vicia faba*), an annual with large thick pods. Native to the Old World, the broad bean is widely cultivated in Europe and North Africa. The chick-pea, also known as the gram pea or garbanzo (*Cicer arietinum*), is well adapted to arid regions and is an important crop in India, North Africa, Spain, and Portugal. The black-eyed pea, also known as southernpea or cowpea (*Vigna unguiculata*), an important fodder plant, is grown chiefly in Asia. A great number of edible legumes are popular in Asia, including lentils (*Lens esculenta*), the Congo bean or pigeon peas (*Cajanus cajan*), and hyacinth bean or lablab (*Dolichos lablab*).

Starchy Root Vegetables

The major starchy vegetable crops of the world are underground crops grown for their tubers, corms, or roots. These include potato (tuber), cassava (root), sweetpotato (root), yam (tuber), and taro (corm). Fleshy roots and underground stems are less compact than cereal grains and legume seeds; consequently they are relatively more difficult to store and transport. Root crops are high in carbohydrates and are thus prized as energy foods. However, the low concentration of protein makes them inadequate for sustenance if not supplemented with protein-rich foods.

Potato

The potato, *Solanum tuberosum* (Solanaceae), although a New World crop, now has its world center of production in Europe where it is a staple for livestock as well as for humans. It is now one of the most important food crops of the world, especially adapted to northern Europe, northern United States, and southern Canada. However, because of their bulk, potatoes have never become important in international trade. They are still cultivated as a mainstay in the Andes, where small colorful tubers are made into a flour-like product called chuño by alternately freezing and trampling them and thawing and drying them.

A temperature of 70°F (21°C) is optimum for tuber formation. At higher temperatures the increased respiration rate reduces the amount of stored carbohydrates and consequently reduces yield and quality. Tuberization is generally accelerated by short days, although some cultivars produce tubers over a wide range of photoperiods. Although adequate rainfall is essential for potatoes, excess rain has been feared because it can result in severe infestations of the late-blight disease (caused by the fungus *Phytophthora infestans*), the same disease responsible for the potato famines in Ireland in the 1840s.

Potatoes store well at 39°F (4°C). Above 50°F (10°C), sprouting is stimulated, although this can be inhibited by chemical treatment, such as with maleic hydrazide. To avoid an undesirable brown discoloration, tubers are held for a few days or weeks at 60–70°F (16–21°C) before processing. During this **reconditioning** treatment, the concentration of reducing sugars (glucose and fructose) is decreased by respiration and by conversion to starch.

In the United States, potato production is concentrated in the North—Idaho and Maine are the largest producers—although early-maturing cultivars are widely grown as fall and spring crops in southern latitudes. Even though total acreage has sharply decreased, average yields have more than doubled between 1840 and 1970. As a result, the United States production has stabilized, although consumption per capita showed a steady decrease from the 1900s until recently, when this trend was reversed. The recent increase is due in large part to the use of processed potato products, mainly as chips and other snack foods.

Cassava

Cassava, *Manihot esculenta* (Euphorbiaceae), another New World crop, is grown extensively in tropical America, but also in tropical Africa and Indonesia. The plant adapts to poor soils and casual cultivation and has become a staple in many of the poor and less-developed parts of the tropics. In the temperate areas of the world, cassava is known through a dessert pudding made from **tapioca,** heated cassava starch that agglutinates into small round pellets. Tapioca is also used for biscuits and confections. In Brazil, cassava is ground into a bland-tasting meal called **farinha** that has a coarse sawdust-like appearance. In addition to food uses, cassava starch is used for sizings, adhesives, and the production of alcohol by fermentation.

The cassava is a short-lived shrub whose tuberous roots resemble the sweet-potato. The plant is deciduous and sheds its leaves during prolonged periods of drought. The many cultivars are roughly divided into sweet and bitter types. **Sweet cassavas** have a low percentage of hydrocyanic glucoside that is generally confined to the outer layers of the tuberous root. Sweet cassavas generally have light green leaves and stems. **Bitter cassavas** have a high percentage of hydrocyanic glucoside distributed throughout the tuber. They generally have dark green or reddish leaves and stems. Bitter varieties are poisonous unless cooked, and their use is largely confined to industrial manufacture of starch, alcohol, and acetone.

Sweetpotato

The sweetpotato, *Ipomoea batatas* (Convolvulaceae), is only known in cultivation. It is one of the few New World crops that reached Polynesia before Columbus's "discovery" of America.[1] Its presence in America and Polynesia suggests pre-Columbian contacts, a theory popularized by the adventures Thor Heyerdahl wrote about in *Kon-Tiki*, but the precise route of the wandering sweetpotato remains a mystery. Easter Island, where sweetpotatoes were found by the first European settlers, may have been a possible transfer point in the trans-Pacific migration. However, some botanists do not favor this "Pacific regatta" theory and suggest that sweetpotato seed capsules, possibly attached to a floating log, may have been carried by ocean currents. Seeds will germinate after immersion in ocean water.

Sweetpotatoes are now grown throughout the tropics, principally in Africa, but major production areas also include Japan, China, United States, and New Zealand. The edible portion of this vine-like trailing perennial is the tuber-like root. The crop is typically produced as an annual from stem cuttings. Cultivars are of two flesh types: dry and mealy or moist and soft. (The moist, soft types are also known as "yams" in the southern United States.) Although yellow- and orange-fleshed cultivars are rich sources of vitamin A, many tropical cultivars have white flesh.

Yam

Although moist-fleshed cultivars of sweetpotatoes (which are dicots) are sometimes referred to as yams, the true yams are monocots of the genus *Dioscorea* (Dioscoreaceae). There are hundreds of species, many still not fully described. The most commonly cultivated species is *D. alata* (water yam or white yam), native to southeast Asia but now grown throughout the tropics with significant cultivation in West Africa, Asia, and the Caribbean (Fig. 14-20). *D. trifida* (yempi yam) is a New World species. Yams are a prestigious food in the Pacific Islands and individuals vie to provide large tubers for traditional feasts.

The yam is a dioecious perennial vine often grown on a trellis. The part of the tuber planted is the top, where the vine is attached. The crop takes 8–10 months to mature. Tubers range from the size of small potatoes to more than 100 pounds. Wild yams offer a standby food in times of food scarcity. They are also gathered as a source of steroidal sapogenins for the production of cortisone.

---

[1] Other crops known in both the New and Old Worlds include the bottle gourd (*Lagenaria siceraria*), coconut (*Cocos nucifera*), and one species of cotton (*Gossypium herbaceum*).

FIGURE 14-20. Typical tuber of 'Morado' yam (*Dioscorea alata*). [Photograph courtesy F. W. Martin.]

Taro

Taros, species of the family Araceae, produce edible corms from soft-stemmed plants that grow in shady marsh areas (Fig. 14-21). All cultivated aroids except *Xanthosoma* probably originated in Southeast Asia and Indonesia. Taros are grown commonly in the Pacific Islands as a source of food. Sweet taro, *Colocasia esculenta*, is the best-tasting species, and its leaves are also edible. The famous Hawaiian **poi,** a purplish, sticky, highly digestible food, is a fermented product made from crushed corms of sweet taro. In tropical America the plant is known as dasheen (perhaps from the French *de Chine*, "from China"); in Africa it is known as cocoyam. The wild taro, *Alocasia macrorrhiza*, is the most common species and produces corms above ground. The corms are used for hog feed because the high content of calcium oxalate crystals renders them virtually unfit for human consumption. Giant swamp taro, *Crytosperma chamissonis*, the taro of Micronesia, is the largest of the taros. After 10–15 years of growth, its corms weigh 100–200 pounds. *Xanthosoma sagittifolium* is native to the New World and is called yautias or American taro. Under the name "new cocoyam," *Xanthosoma* is spreading into Africa and Asia, where it is relatively free of pests. In New Guinea it is confusingly called taro cong, or Chinese taro.

Other Root Vegetables

In addition to those just described, there are a number of other root crops grown as vegetables throughout the world. These include the table beet, *Beta vulgaris* (Chenopodiaceae), a biennial plant whose swollen red root is eaten fresh, or is canned or pickled, and is the main ingredient of borscht, a delicious soup eaten hot or cold with sour cream. The sugar beet, an important source of sugar in temperate climates, is considered to be the same species. Carrot, *Daucus carota*

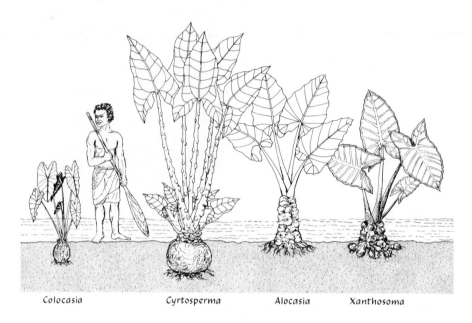

| Colocasia | Cyrtosperma | Alocasia | Xanthosoma |

FIGURE 14-21. Some edible aroids.

(Fig. 14-22), and parsnip, *Pastinaca sativa*, are well known root crops of the family Umbelliferae. Arrachacha (*Arracacia xanthorrhiza*), an interesting carrot-like vegetable, is just one of the little-known root crops with local importance in South America. Jerusalem artichoke, *Helianthus tuberosus* (Compositae), is of interest because the reserve carbohydrate in the potato-like rhizome is inulin, a polymer of fructose. Inulin, but not glucose-containing carbohydrates, can be metabolized by diabetics. Salsify, *Tragopogon possifolius* (Compositae), has fleshy roots resembling parsnip. In the family Cruciferae, radish (*Raphanus sativus*), rutabaga (*Brassica napobrassica*), and turnips (*B. rapa*) are often classed as root crops, along with horseradish (*Armoracia rusticana*), whose powerfully pungent roots are used as a condiment.

The Onion and Its Pungent Allies

The onion and the other pungent species of the genus *Allium* (leek, garlic, and chive) are among the most ancient of cultivated vegetables. The pungent odors and flavors are due to sulfur compounds, mostly *n*-propyl disulfide in onion and methylallyl disulfide in garlic. These species are all biennial bulb-forming plants of the lily family (Liliaceae). The onion, the most important of this group, is grown in all temperate climates (Fig. 14-23). The most important onion-producing countries are the United States, Japan, Rumania, Italy, and Turkey. In the

FIGURE 14-22. *Right*, a carrot field at time for harvest. *Left*, 'Imperator 58,' a popular carrot cultivar for the fresh market. [Photographs courtesy T. W. Whitaker, USDA.]

FIGURE 14-23. Onions are among the most ancient of vegetable crops. [Photograph by J. C. Allen & Son.]

United States the crop is widely grown, with California, New York, Texas, and Michigan being the chief producing states. Onions require cool temperatures during early growth and require long days and high temperatures for bulb formation. Green onions, or scallions (the nonbulbous stage of onion), are an important winter crop in the southern United States.

In the northern states onions are commonly planted from seed in the spring; in the southern states plantings are made in the fall or winter from seed, seedlings, or onion sets. The production of onion sets, which are small bulbs produced by crowding, is a specialized industry in Michigan, Illinois, and Wisconsin. Fall onions are stored for sale in the winter. A relatively small percentage of the crop is processed as flavorings for use in other products.

Cole Crops

The important vegetables of the family Cruciferae include cabbage, cauliflower, broccoli, Brussels sprouts, Chinese cabbage, kale, collards, and mustard. These are all species of the genus *Brassica*, and are known collectively as **cole crops** (Fig. 14-24). The radish (*Raphanus sativus*) is also a crucifer, but it belongs to a

FIGURE 14-24. The All-America hybrid crucifers for 1969: *top left*, 'Tokyo Cross' turnip; *top right*, 'Stonehead' cabbage; *bottom left*, 'Green Comet' broccoli; *bottom right*, 'Snow King' cauliflower.

different genus. The cole crops are cool-season, hardy plants. Except for cauli-flower and broccoli, which are annuals, they are biennials, requiring a cold treatment to flower.

The cole crops are grown widely in Europe, where they constitute the major "green" vegetables. In the United States, cabbage is the most important cole crop, and, in the past, was a common winter vegetable, since it is easily stored. The production of cabbage has sharply declined, however, reflecting a general con-sumer shift to other vegetables. About 10% of the present acreage is grown for the production of sauerkraut. Broccoli, although a minor crop compared with cab-bage, has increased in importance, and has become very popular as a frozen vegetable.

The cole crops are usually set out as transplants. In the North, they are set out in the spring; in the warmer climate of the South, they are planted in the fall or winter. Insect control is particularly important. In general, soils with a pH greater than 7 are used to control club root, a serious root disease caused by *Plasmo-diophora brassicae*. However, resistant cultivars are now commercially available. Hybrid crucifers are a recent innovation and have made possible the production of uniform plantings for mechanical harvesting.

Vine Crops

Crops of the family Cucurbitaceae, or cucurbits, are known as **vine crops** (Fig. 14-25). They include the cucumber, muskmelon, watermelon, pumpkin, squash, and the chayote. All are warm-season crops and very susceptible to cold injury.

The cucumber, *Cucumis sativus*, probably native to Asia, has been in cultiva-tion for thousands of years. In Europe it is an important greenhouse crop: the extremely long fruits, grown in the absence of pollinating insects, develop without seeds. In the United States the majority of the cucumber acreage is devoted to the production of pickling cucumbers. The crop is widely planted, but Michigan and Wisconsin lead the other states in production. Very successful mechanical har-vesting machines have been developed.

Muskmelons, *Cucumis melo*, native to Iran, are a relatively recent crop in the United States, where they are often erroneously called cantaloupes. (True canta-loupes are hard-shelled melons only rarely grown in the United States.) California, Texas, and Arizona are the chief producing states. There are many types of melons in addition to the netted, musky type. The most important of these is the late-ripening winter melon, of which the most common cultivar is the 'Honey Dew,' a white-skinned, large-fruited type.

Watermelons, *Citrullus lanatus*, grow on most United States vine acreages. Native to Africa, they have achieved their greatest popularity in the United States. The bulk of the planting is in the southern states, although there are important areas as far north as Iowa. The types range from large 30-pound fruits to small round "icebox types" weighing between 5 and 10 pounds. There are also

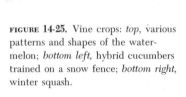

FIGURE 14-25. Vine crops: *top,* various patterns and shapes of the water-melon; *bottom left,* hybrid cucumbers trained on a snow fence; *bottom right,* winter squash.

yellow-fleshed types, but the red-fleshed type is the most popular. The seedless watermelon, a triploid fruit first created in Japan in the 1940s from crossing tetraploids and diploids, has found its greatest success in Taiwan, but it is still just a novelty in the United States.

Pumpkin and squash are native to the New World. The distinction between pumpkin and squash is a loose one. Some would use the word pumpkin only for species of *Cucurbita pepo*, referring to *C. maxima* and *C. moschata* as squash. However, the fast-growing types of *C. pepo* are commonly called summer squash. It would seem that the orange, large-fruited, smooth types will continue to be called pumpkins regardless of species. Squash represents an important food crop in South America, ranking next to maize and beans. The flowers, flesh, and seeds may be eaten. The chayote (*Sechium edule*), practically unknown in the United States, is an important South American plant. It is customarily trained on trellises.

Sweet Corn

Sweet corn, *Zea mays* (Gramineae), a sugary-seeded maize native to South America, has only been popular since the last century. Although field corn is a well-known crop in many parts of the world, the popularity of sweet corn is confined to the United States.

Sweet corn is a warm-season crop, but it is produced mainly in the northern states. This is because of the short period over which quality can be maintained in extremely hot weather. In Wisconsin, Minnesota, and Illinois, about two-thirds of the sweet corn is grown for processing, but sizable acreage for fresh-market production has built up in Florida and Texas.

The transitory nature of quality in sweet corn is due to the rapid conversion of sugar to starch. Experimental combinations of other genes, in addition to the common sugary factor, produce not only a higher sugar content but a less rapid conversion to starch. These types would no doubt extend commercial sweet-corn production to areas of hotter climates, but germination in these types is irregular.

ORNAMENTAL CROPS

Ornamentals are commonly divided on the basis of the industry into **florist crops** (flower and foliage plants) and **landscape crops** (nursery plants, including turf, ground covers, deciduous and evergreen shrubs, and trees). A tremendous number of species may be considered as ornamentals. These include many of our food plants, especially fruit trees, which are also greatly prized for their decorative value. Thus a complete list of species used as ornamentals must be encyclopedic in scope.

This available diversity notwithstanding, the commercial ornamental industry is based on relatively few species. The bulk of the United States cut-flower industry is based on less than a half-dozen flowers: chrysanthemum, rose, carnation, orchid, snapdragon, and gladiolus. One annual, the petunia, accounts for 50% of the sales of bedding plants. The turf industry is based on relatively few grasses, as was discussed in Chapter 13. Even the most popular deciduous and evergreen trees and shrubs make up a modest list: yew, juniper, boxwood, privet, forsythia, dogwood, magnolia, crabapple, oak, and maple.

Although the bulk of sales is concentrated in a manageable inventory, nevertheless the total number of species in the trade is impressive. Even more impressive are the countless cultivars and variants of each, with intense competition continually bringing out more. Often trivial variations are given the special status of an appealing name. Double naming is not uncommon. The plethora of names greatly magnifies the available variability. The registers of most ornamentals are endless; the rose fancier is bombarded with scores of new cultivars each year. In contrast, variation in vegetables and fruits is minimized at the retail level. The public is thought to be confused by cultivar names. Thus only about five or six names of apples are used, even fewer for oranges; vegetables are almost never sold by their cultivar names in the markets.

Food-crop species make up a relatively stable list, and it is very difficult to introduce new ones because of the conservatism of people's food habits. This is not true of ornamentals; apparently people are more innovative with plants they do not have to eat! One may expect that the number of species in the trade will increase as new ornamentals become introduced and domesticated. However, it is still difficult for a particular ornamental crop to become established as something other than a novelty. Recently, anthurium has increased in popularity as a cut flower as it has been made more easily obtainable through increasing air shipments from Hawaii. Proteas (Fig. 14-26) are other tropical flowers with commercial possibilities. Both anthuriums and proteas have extremely long shelf lives measured in weeks rather than days. With the trend toward suburban living, there is a greater awareness and interest in new plants, and the industry is expanding to meet this new challenge.

Chrysanthemum

The chrysanthemum, *Chrysanthemum morifolium* (Compositae), long cultivated in Asia, has become the largest cut-flower crop in the United States. It is considered a warm-temperature crop, performing best at about 65°F (18°C). The great number of cultivars are usually classified into two types: standards, or large-flowered types, and Pompons, or sprays. Pompons, small-flowered types, may be globular or daisy-like, and are largely used as fillers in floral arrangements.

The main stimulus to chrysanthemum production has been the possibility of producing continuous flowering by the manipulation of photoperiod. The plant is a perennial and will bloom for many years, but it is grown commercially as an annual crop. Chrysanthemums are grown under glass in the northern and midwestern states, under plastic in California (the center of production), and under cloth in Florida.

Rose

Roses, members of the genus *Rosa* (Rosaceae), have long been one of the major cut-flower crops. The industry is largely based on the hybrid tea rose, a perpetu-

FIGURE 14-26. Ornamental proteas: *A, Banksia bastinia; B, Banksia prionotes; C, Protea barbigera; D, Leucospermum cordifolium* (pincushion protea). [Photographs courtesy D. P. Watson (*A*) and E. Memmler (*B, C, D*).]

ally blooming, long-stemmed rose whose pedigree incorporates many species. In greenhouse culture, the plants are budded or grafted onto *Rosa manetti*, a vigorous rootstock. If properly managed, the plants bloom continuously, although the intensity of production is affected by temperature and fertility. Production can be manipulated by pruning (referred to as pinching) and temperature control to take advantage of favorable marketing periods. Producing plants are usually kept from 4 to 7 years, depending on their vigor.

The ideal growing temperature is 60°F (16°C). Commercial rose growing has been restricted, for the most part, to areas north of Lexington, Kentucky, but some establishments are found in the higher elevations of North Carolina. Evaporative cooling is being utilized to reduce daytime temperatures farther south, but, in general, areas that have night temperatures above 65°F (18°C) for extensive periods of the year are not suitable for the production of roses. Most of the rose production is carried out under glass, although there is a trend toward semipermanent construction utilizing lath and plastic, as in the San Francisco area.

The high capitalization required for rose culture has generally limited production to establishments not smaller than 10,000 square feet. The importance of novelties, as well as the better keeping qualities in the hybrid teas, has made hybridization an important part of many rose-growing establishments. The most widely grown forcing cultivar, however, is 'Better Times' originated in 1931 by Hills Brothers Nursery of Richmond, Indiana.

The production of rose plants for garden use is a very specialized operation. Most roses are budded onto *Rosa multiflora* rootstocks to achieve rapid and economical increase to meet market demands. Large-scale production of roses for the garden has tended to center in the southwest (from Texas to southern California), where the nearly year-long growing season makes rapid production possible, but it is also carried on near important consuming areas, such as northern Ohio.

Partly because of widespread promotion, and partly because of the great popular interest in the rose, the marketing of roses is nearly as important a facet of the nursery industry as is the production of other classes of shrubs. Mail-order sales from attractive catalogs that picture the roses in color account for an appreciable part of the market. Enthusiasm for new selections is heightened by an elaborate scheme for testing and selecting "All-America" roses (only a few roses receive the coveted award each year), and by a strong national rose society with headquarters in Columbus, Ohio, that encourages rose clubs and sponsors rose shows throughout the nation. Each year accredited judges compile and issue a rating sheet that ranks all available named cultivars on a numerical performance scale Thus even the neophyte can determine readily which have been the most successful named roses through the years, and purchase a 'Peace' or 'Tropicana' on the basis of authorative testing rather than extravagantly worded catalog descriptions.

The production of roses requires skilled hand labor in the budding operation and expensive field maintenance. Recently techniques have been developed to improve the distribution of roses. In the mail-order handling of bare-root roses, coated paper or plastic bags are now used to retain humidity, in contrast to the former practice of wrapping the roots in damp moss. For mass marketing through chain outlets, the stems of dormant stock are coated with wax to prevent desiccation during prolonged storage and handling, and the root system is "potted" in a moss-filled carton on which the name and a picture of the rose appear. Although much stock held dormant for long periods is not of first quality, this method of penetrating the mass market is now well established in the nursery trade.

FIGURE 14-27. The 'White Sim' carnation is a periclinal (hand-in-glove) chimeral sport of the 'Red Sim.' The internal tissues carry genetic factors for red color. The 'White Sim' carnation can be identified by flecks of red tissue in the flower. [Photograph by J. C. Allen & Son.]

Carnation

Carnations, *Dianthus caryophyllus* (Caryophyllaceae), require cool temperatures (48–50°F, or 9–10°C) and high light intensities for maximum quality production. The industry in carnations grown for cut flowers is based on perpetual-flowering types, and production proceeds throughout the year. Plantings are established from cuttings and are maintained from one to two years. Production of carnations is greatest in California and Colorado, although a sizable industry exists in Massachusetts. The production of carnations has shown a spectacular increase from 1950 to 1960, mainly as a result of expansion in the western states.

Although there are many carnation varieties in production today the 'Red Sim' and 'White Sim' cultivars are the most popular (Fig. 14-27). These cultivars and their sports are characterized by vigorous growth, heavy production, and long stems. Their chief disadvantage is a tendency toward calyx splitting, a disorder not well understood but felt to be accentuated by large differences between night and day temperatures. Recently, artificial dyeing of carnations has become popular. The dye is taken up from the cut end of the stem and is absorbed into the veins of the petals. This permits a wide variety of pastel shades to be achieved while only the white-flowered varieties need be grown.

Orchid

The culture of orchids (members of the family Orchidaceae) is a specialized greenhouse industry. The main type is the showy *Cattleya* (Fig. 14-28), but *Cymbidium* is becoming increasingly popular. Orchids are a warm-season crop with optimum growth at 65–70°F (18–21°C). Shade is necessary during the summer, but it must be removed to maintain adequate light intensity during the winter. The plants are commonly propagated by divisions. Most orchids are epiphytes, and obtain nutrients from decomposing organic matter. They are

FIGURE 14-28. The *Cattleya* orchid. [Photograph by J. C. Allen & Son.]

grown either in **osmunda fiber** (the chopped roots of a kind of fern) or in shredded white-fir bark. Since repotting, one of the costliest items of culture, is difficult with osmunda, shredded bark is more widely used by commercial growers. The plants bloom once a year. The harvested flowers are usually placed in tubes of water and are relatively long lasting.

Breeding is an important part of orchid culture, for new cultivars may become quite profitable. The time-consuming and exacting techniques involved in growing orchid plants from seed have encouraged specialized enterprises that grow seedlings on a commission basis. Successful breeding methods include hybridization of diploid and tetraploid types, since triploid orchids have proven to be extremely vigorous.

Snapdragon

Snapdragons, *Antirrhinum majus* (Scrophulariaceae), although perennial, are grown as annuals and are planted from seed. Formerly grown only under relatively cool temperatures, selection for cultivars that perform well under hot temperatures has created a year-round program for this crop. In the Midwest, snapdragons appear to fit into a profitable rotation with greenhouse forcing tomatoes. Snapdragons tolerate low light if temperatures are cool, and they can be grown during the winter when conditions are unfavorable for tomatoes.

At present, practically the entire collection of commercial cultivars consists of F_1 hybrids, which have uniformity and vigor. Artificially created tetraploid cultivars have achieved some success because of their larger flowers.

Bulbs

The "bulb" crops include such plants as tulip, hyacinth, narcissus, iris, daylily, and dahlia. Included are nonhardy bulbs for sale as indoor potted plants and summer outdoor plantings, such as amaryllis (*Hippeastrum*), anemones, various tuberous begonias, *Caladium,* cannas, *Clivia,* dahlias, freesias, gladiolus, spiderlilies, montbretias, ranunculus, *Tigrida,* tuberose, *Zephyranthes,* and others. Hardy bulbs, those left in the soil through the winter, include various crocuses (*Colchicum* and *Crocus*), fritillaries, snowdrops, hyacinths, lilies, daffodils, *Lycoris, Scilla,* tulips, irises, and others (Fig. 14-29).

Many of these plants are of Old World origin, introduced into horticulture long

FIGURE 14-29. Some hardy bulb crops: *A,* tulip; *B,* daffodil; *C,* hyacinth; *D,* bulbous iris. [Photographs courtesy Netherlands Flower Bulb Institute.]

ago, and subjected to selection and crossing through the years to yield many modern cultivars. One of the most popular is the tulip, *Tulipa* (Liliaceae). Various botanical species are grown in gardens; but tulips are especially prized in select forms of the garden tulip (which arose from crosses between thousands of cultivars representing several species). Garden tulips are roughly grouped as early tulips, breeder's tulips, cottage tulips, Darwin tulips, lily-flowered tulips, triumph tulips, Mendel tulips, parrot tulips, and so on. The garden tulips seem to have been developed first in Turkey, but were spread throughout Europe and were adopted enthusiastically by the Dutch. Holland has been the center of tulip breeding ever since the eighteenth century, when interest in the tulip was so intense that single bulbs of a select type were sometimes valued at thousands of dollars. The collapse of that "tulipomania" left economic scars for many decades. Holland remains the chief source of tulip bulbs planted in the United States; nearly 150,000,000 are imported annually. Holland has also specialized in the production of related bulbs in the lily family, and provides narcissus, crocus, and others to the United States market annually. The Dutch finance extensive promotion of their bulbs in the United States to support their market. Years of meticulous growing are required to yield a commercial tulip bulb from seed. Bulbs sent to market meet specifications as to size and quality, which assure at least one year's bloom even if the bulb (having been subjected to cold induction) is supplied nothing more than warmth and moisture. The inflorescence is already initiated, and the necessary food is

FIGURE 14-30. All-America gladiolus selections: *left to right*, 'Thunderbird,' 'Horizon,' and 'Ben-Hur.'

stored in the bulb. Under less favorable maintenance than prevails in Holland, a subsequent year's bloom may be smaller and less reliable. For featured display, such as at botanical gardens, bulbs are discarded after one year's performance although homeowners usually derive several years of satisfaction from a single bulb planting if maturation of the tulip foliage is allowed after flowering.

Gladiolus spp. (Iridaceae) have practically disappeared as a greenhouse forcing crop. Now almost all are produced in outdoor culture, mainly in Florida, California, and North Carolina. A great variety of colors and types exists (Fig. 14-30). Owing to the nature of the plant, bulb and flower production are part of the same business.

Bulbs generally grow best when planted fairly deeply (several times the vertical dimension of the bulb itself) in friable, loamy soil well supplied with phosphatic fertilizer. Autumn, a few weeks before freezing weather, is the recommended planting time for tulips, hyacinths, daffodils, and crocus. The bulbs have a chance to root and become established, and will flower in spring as soon as the soil warms. After the foliage has died back it is often recommended that the bulbs be dug up and reset, a practice that is of some value for rearranging the planting pattern and determining which bulbs are in good condition if nothing else. Some bulbs, such as daffodils and hyacinths, proliferate; digging them up provides additional stock.

Selected References

Amerine, M. A., H. W. Berg, and M. V. Cruess, 1972. *Technology of Wine Making*, 3rd ed. Westport, Conn.: Avi Publishing. (Concerned with all facets of wine manufacture.)

Arctander, S., 1960. *Perfumes and Flavor Materials of Natural Origin*. Westport, Conn.: Avi Publishing. (Comprehensive. Part One discusses processing and evaluation; Part Two, 537 raw-materials sources.)

Berrie, A. M. M., 1977. *An Introduction to the Botany of the Major Crop Plants*. London: Heyden. (A botanical description of the major crop families.)

Carpenter, P. L., T. D. Walker, and F. O. Lanphear, 1975. *Plants in the Landscape*. San Francisco: W. H. Freeman and Company. (An excellent treatment of the landscape industry and landscape plants.)

Chandler, W. H., 1950. *Deciduous Orchards*. Philadelphia: Lea & Febiger. (See next reference.)

Chandler, W. H. 1950. *Evergreen Orchards*. Philadelphia: Lea & Febiger. (The two books by Chandler are companion volumes on temperate and subtropical fruit growing.)

Cobley, L. S., 1977. *An Introduction to the Botany of Tropical Crops*, 2nd ed. (revised by W. M. Steele). London: Longmans. (A survey of the botanical features of plants grown in the tropics.)

Collins, J. L. 1960. *The Pineapple: Botany, Cultivation, and Utilization*. New York: Wiley. (An excellent treatment.)

Condit, I. J., 1947. *The Fig*. Waltham, Mass.: Chronica Botanica. (A thoroughgoing technical review of an age-old crop.)

Darrow, G. M., and others, 1966. *The Strawberry: History, Breeding and Physiology.* New York: Holt, Rinehart and Winston. (A readable review of this small fruit, covering history, breeding, and culture.)

Dawson, V. H. W., and A. Aten, 1962. *Dates—Handling, Processing and Packing.* Rome: Food and Agriculture Organization. (The United Nations takes an overall look at the date industry.)

De Hertogh, A. A., 1977. *Holland Bulb Forcer's Guide.* New York: Netherlands Flower Bulb Institute. (A grower's manual.)

Ecke, P., and N. F. Childers (editors), 1966. *Blueberry Culture.* New Brunswick, N.J.: Horticultural Publications. (A broad compilation of information on this special crop, including its botany, breeding, and growing.)

Ecke, P. and O. A. Matkin (editors), 1976. *The Poinsettia Manual.* Encinitas, Calif.: Paul Ecke Poinsettias. (A grower's manual.)

Eden, T., 1976. *Tea*, 3rd ed. London: Longmans. (Discussion of history, botany, planting, care, harvest, and processing of this important beverage crop, with references, by the Director of the Tea Research Institute in Ceylon and Africa.)

FAO Production Yearbook. Rome: Food and Agriculture Organization. (An invaluable annual compilation on world food production; published yearly.)

Graf, A. B., 1963. *Exotica 3: Pictorial Cyclopedia of Exotic Plants.* Rutherford, N.J.: Roehrs. (A comprehensive pictoral record of ornamentals, mostly tropical.)

Haarer, A. E., 1962. *Modern Coffee Production.* London: Leonard Hill. (On the history, genetics, physiology, and practical growing of coffee by an English expert with wide practical experience in East Africa.)

Harrison, S. G., G. B. Masefield, and M. Wallis, 1969. *The Oxford Book of Food Plants.* London: Oxford University Press. (A description of important food plants with remarkably beautiful illustrations by B. E. Nicholson.)

Hawkes, A. D., 1965. *Encyclopedia of Cultivated Orchids.* London: Faber and Faber. (An authoritative compilation.)

Heiser, C. B., Jr., 1969. *Nightshades: The Paradoxical Plants.* San Francisco: W. H. Freeman and Company. (A charming book on solanaceous plants.)

Holley, W. D., and R. Baker, 1963. *Carnation Production.* Dubuque, Iowa: Wm. C. Brown. (A grower's manual.)

Jones, H. A., and K. L. Mann, 1963. *Onions and Their Allies: Botany, Cultivation, and Utilization.* New York: Wiley. (A monograph on the cultivated *Allium* species.)

Kofranek, A. M., and R. A. Larson (editors), 1975. *Growing Azaleas Commercially.* Richmond: University of California Field Station. (A grower's manual.)

Langhans, R. W. (editor), 1962. *Snapdragons: A Manual of the Culture, Insects and Diseases, and Economics of Snapdragons.* Ithaca: New York State Extension Service and New York State Flower Growers Association, Inc. (A grower's manual.)

Langhans, R. W., et al., 1964. *Chrysanthemums: A Manual of the Culture, Diseases, Insects, and Economics of Chrysanthemums.* Ithaca: New York State Extension Service and New York State Flower Growers Association, Inc. (A grower's manual.)

Langhans, R. W., and D. C. Kiplinger (editors), 1967. *Easter Lilies: The Culture, Diseases, Insects, and Economics of Easter Lilies.* Ithaca: New York State Extension Service; Columbus: Ohio State University. (A grower's manual.)

Mastalerz, J. W. (editor), 1971. *Geraniums: A Manual on the Culture, Diseases, Insects, Economics, Taxonomy, and Breeding of Geraniums.* University Park: Pennsylvania Flower Growers. (A grower's manual.)

Mastalerz, J. W. (editor), 1976. *Bedding Plants.* University Park: Pennsylvania Flower Growers. (A grower's manual.)

Mastalerz, J. W., and R. W. Langhans (editors), 1969. *Roses: A Manual on the Culture, Management, Diseases, Insects, Economics, and Breeding of Greenhouse Roses.* University Park: Pennsylvania Flower Growers; Ithaca: New York State Flower Growers Association; Haslett, Michigan: Roses, Inc. (A grower's manual.)

Mortensen, E., and E. T. Bullard, 1964. *Handbook of Tropical and Subtropical Horticulture.* Washington, D.C.: United States Department of State, Agency for International Development. (A manual written for the nonspecialist.)

Nieuwhof, M., 1969. *Cole Crops.* London: Leonard Hill. (An excellent review of the production, harvesting, storage, conservation, and utilization of crops of the species *Brassica oleracea.*)

Ouden, P. den, and B. K. Bloom, 1965. *Manual of Cultivated Conifers Hardy in the Cold- and Warm-Temperate Zone.* The Hague: Martinus Nyhoff. (The complete book of conifers.)

Parry, J. W., 1969. *Spices.* New York: Chemical Publishing Company. (A two-volume work on the history and botany of spices.)

Purseglove, J. W., 1968, 1972. *Tropical Crops: Dicotyledons* (2 vols., 1968); *Monocotyledons* (2 vols., 1972). New York: Wiley. (The complete botany and culture of tropical plants; an invaluable compilation.)

Reed, C. A., and J. Davidson, 1958. *The Improved Nut Trees of North America and How to Grow Them.* New York: Devin-Adair. (Semipopular presentation of a subject seldom covered comprehensively, with references.)

Reuther, W., L. D. Batchelor, and H. J. Webber (editors), 1967, 1968, 1972. *The Citrus Industry,* rev. ed. Berkeley: University of California. (A monumental work on citrus in three volumes. Vol. 1 covers history, world distribution, botany, and varieties; Vol. II covers anatomy, physiology, mineral nutrition, seed reproduction, genetics, and growth regulators; Vol. III covers propagation, planting, weed control, soils, fertilizing, pruning, irrigating, climate, and frost control. Vol. IV, still in preparation, will cover biology and control of pests and diseases.)

Salaman, R. N., 1949. *The History and Social Influence of the Potato.* Cambridge: The University Press. (The potato throughout its history—a delightful treatment.)

Schery, R. W., 1972. *Plants for Man,* 2nd ed. Englewood Cliffs, N.J.: Prentice-Hall. (A very comprehensive text on economic botany.)

Shoemaker, J. S., 1948. *Small-fruit Culture,* 2nd ed. Philadelphia: Blakiston. (A standard reference through the years.)

Simmonds, N. W., 1959. *Bananas* (Tropical Agriculture Series). London: Longmans. (An authoritative monograph.)

Singer, R., 1961. *Mushrooms and Truffles: Botany, Cultivation, and Utilization.* New York: Wiley. (A fascinating review of the industry throughout the world.)

Singh, L. B., 1960. *The Mango: Botany, Cultivation, and Utilization.* New York: Wiley. (A treatise on one of the most popular and luscious of tropical fruits.)

Smartt, J., 1976. *Tropical Pulses*. New York: Longmans. (The botany, physiology, and culture of tropical legumes.)

Smith, J. R., 1950. *Tree Crops*. New York: Devin-Adair. (A plea for greater use of tree fruits and nuts, popularly presented but bringing diverse information into a single volume.)

Westwood, M. N., 1978. *Temperate-Zone Pomology*. San Francisco: W. H. Freeman and Company. (An excellent text on temperate pomology.)

Whitaker, T. W., and G. N. Davis, 1962. *Cucurbits: Botany, Cultivation, and Utilization*. New York: Wiley. (A comprehensive treatment of the cultivated cucurbits.)

Wood, G. A. R., 1975. *Cocoa*, 3rd ed. New York: Longmans. (A thorough review of cacao worldwide, with especial attention to its growing in different regions.)

Zucker, I., 1966. *Flowering Shrubs*. Princeton, N.J.: Van Nostrand. (A source book for ornamental plantings.)

15

Esthetics of Horticulture

Torches are made to light, jewels to wear,
Dainties to taste, fresh beauty for the use,
Herbs for their smell, and sappy plants to bear:
Things growing to themselves are growth's abuse.
Seeds spring from seed, and beauty breedeth beauty. . . .

SHAKESPEARE, *Venus and Adonis*

ESTHETIC VALUES

In addition to their utility, plants have esthetic value. Owing to particular qualities that collectively we call beauty, certain plants provide us with pleasure. Beauty is not a tangible quality that can be measured or weighed, it is a value judgment. A thing is beautiful when someone decides that it is. The artist is one who can make this judgment and communicate the experience. This judgment is a reflection of cultural tradition. People of widely different heritages will have quite different opinions about what is beautiful and what is ugly.

Experiencing visual beauty depends upon our response to things sensed visually. Although a certain amount of our perception is innate, many perceptual responses are learned. To a great extent we are aware only of what we are able to interpret. For example, upon hearing a foreign language, we do not actually perceive most of the nuances of sound and inflection until we have learned to imitate them; yet even a newborn baby is aware of a sudden loud noise and can distinguish between gentle and disapproving tones of voice. Similarly, the botanist learns to discern small differences in plants that may be all but invisible to the layman. So it is with beauty: we must learn to recognize it.

With reference to the concept of beauty, it is difficult to determine what part the innate psychological stimulation plays in the learned response. If any generalization can be made, it is that we tend to enjoy the full exercise of our perceptive facilities. Consider, for instance, the universal preferences for color, depth, and contrast for our visual experiences. Nevertheless, experiencing beauty is basically a learned response. This explains the underlying conservatism concerning beauty. We prefer what we are used to, and tend to reject the completely strange and new. Yet we learn to enjoy small, subtle differences and can be "trained" to expect them, as the automobile manufacturers and flower breeders have discovered.

If we accept that beauty is relative then it is apparent that we cannot arbitrarily define it. We cannot say absolutely whether a particular object, or arrangement of objects, is beautiful or not. We must suspend judgment until we have considered the object in relation to its beholder. Snakes, spiders, and worms are considered ugly by many people, but are considered beautiful by others. It is no coincidence that these are feared objects in our culture and that a certain fear of them is passed down by each generation. Thus, our perception of beauty is strongly affected by our emotional feelings and by our cultural attitudes toward objects. This is to say that the standards of one culture cannot in time be applied in all cases to another, for our method of evaluation—our yardstick—has been molded by the culture in which it developed. Generally, the things that have been accepted as beautiful for long periods of time, and which are more or less universally admired, have a basic simplicity and harmony of form and function. In conclusion, our concept of beauty is made up of two parts: (1) sensory stimulation and (2) our responses to this stimulation, which have strong personal and cultural components.

Most plants have an inherent capacity to stimulate visually. Their most obvious feature is their coloring; not only the brilliant hues of flowers, fruits, and (in some plants) leaves, but the muted tones of stem and bark. Green, of course, is the most common color, and it is probably more than coincidental that it is psychologically the most restful. The stimulation that plant color provides is enhanced by contrast and texture.

Also significant with respect to visual effects are the plant's structure and shape; that is, its form. Form can be seen not only in the plant as a whole but in its parts as well. The forms of plants are infinitely varied. But the same could be said of random stones, which are considerably less interesting. The perpetual interest in plants is a result of their ordered arrangements of parts, which involves symmetry, the repetition of parts on either side of an axis.

Symmetry can make any random shape an orderly one. The psychological satisfaction experienced in viewing symmetrical objects is probably due to their inherent order. Man exhibits a universal awareness of symmetry, which is not strange considering its common occurrence in biological forms (Fig. 15-1). Although all plants show some types of symmetry, the growth of many plants produces asymmetrical patterns. It is this deviation from symmetry that makes for visual interest. The basis of contemporary design is to achieve balance and harmony without the monotony of perfect symmetry.

With the possible exception of the Eskimo, human cultures have developed in plant-dominated environments. Plants provide food for people and their animals, as well as fiber, shelter, and shade. Our dependence upon plants has influenced and molded our esthetic consideration of them. We need plants, and no doubt

FIGURE 15-1. Symmetry in the rose. [Photograph by J. C. Allen & Son.]

plants have been culturally accepted as beautiful partly because they are useful. In our present American culture, in which only a relatively few people are directly involved with the growing of plants (although we still all depend on them), all of us have traditional attachments to plant material. Horticulture has a place in all our lives.

DESIGN

Design refers to the manner in which objects are artificially arranged in order to achieve a particular objective. Usually, but not always, this objective involves both a functional and a visibly pleasing arrangement. Designs are evaluated esthetically with regard to their **elements of color, texture, form,** and **line** by long-established, man-made value judgments called **design principles: balance, rhythm, emphasis,** and **harmony.** The importance attributed to each of these will vary with the objective of the design. When a design is successful, it is usually considered appropriate, functional, and beautiful.

Elements of Design

The design elements are visible features of all objects.

Color is the visual sensation produced by different wavelengths of light. Color may be described in terms of its **hue** (red, blue, yellow), **value** (light versus dark), and **intensity,** or **chroma** (its saturation or brilliance).

Texture in design refers to the visual effect of tactile surface qualities. Consider, for instance, the visual difference between burlap and silk or between the surface of a pineapple and that of a rose petal.

Form refers to the shape and structure of a three-dimensional object (sphere, cube, pyramid). However, when we view these forms in a plane, they may appear to be two-dimensional (as a circle, a square, and a triangle, respectively). In design we are concerned not only with the form of the individual objects but with the larger forms made up by their arrangement.

Line delimits shape and structure. The concept of line in design involves the means by which form guides the eye. Line becomes a one-dimensional interpretation of form. Emotional significance has been attributed to line, as shown in Figure 15-2.

Principles of Design

The principles of design (**balance, rhythm, emphasis,** and **harmony**) apply to each of the design elements as well as to their interrelations. Thus, we speak of balanced color as well as of balanced form. The artistic application of these principles is the basis of esthetic success, as measured by beauty and expressiveness.

Balance implies stability. Our eye becomes accustomed to material balance, and as a result we become uneasy about objects that appear unstable or ready to topple over. This concept is carried over to arrangements in which balance refers to the illusion of equilibrium around a vertical axis. Balance is achieved automatically by the symmetrical placement of objects around a central vertical axis. It is also achieved in nonsymmetrical arrangements (utilizing the lever principle) by the coordination of mass, distance, and space (Fig. 15-3). It should be emphasized that in design we are concerned with the illusion of balance rather than with actual physical balance.

Rhythm, in the auditory sense, refers to a pulsating beat. Similarly, rhythm in the visual context refers to the pattern of "spatial" beats that our eye follows in any arrangement of objects. Rhythm leads and directs the eye through the design. Rhythm suggests movement; design without rhythm becomes uninteresting. Its proper use makes for expression and excitement.

Emphasis in design serves to lead the eye and focus its attention on some dominant aspect of the design. By accenting and emphasizing various elements (for example, a particular form, a strong horizontal line, or a brilliant color) the separate parts of the design may be drawn together. Emphasis, properly made, coordinates the design elements and creates an orderly and simplified arrangement.

Harmony refers to the unity and completeness of the design. This quality is seldom achieved except by proper planning and organization. It relies principally on scale and proportion, the pleasing relationship of size and shape. The separate components lose their identity to become part of an idea, the basis of design.

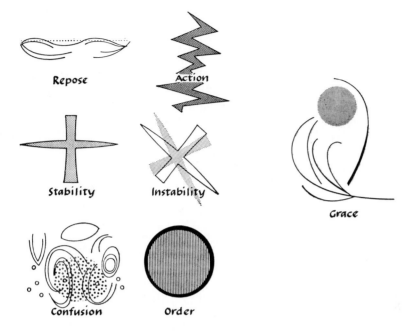

FIGURE 15-2. Line elicits emotional responses.

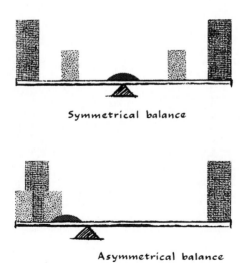

FIGURE 15-3. Graphic representations of symmetrical and asymmetrical balance.

GARDENS

The origin of the garden is rooted in the human desire to be surrounded by plants. The great variety of plants permits us to select, in addition to the useful, the' esthetically pleasing. Their pleasant fragrance, as well as their beauty, plays a role in this selection. With respect to plants that become culturally significant, as fruit trees have, it is often difficult to separate the purely functional from the esthetic.

The first gardens in recorded history were those of the ancient cultures of Egypt and China. It was in these cultures that the two opposing traditions in gardens originated, namely, **formalism** and **naturalism.**

Formalism and the Western Tradition

The Egyptian gardens (Fig. 15-4), which were developed at the edge of the deserts where the natural vegetation was sparse, represented the development of an artificial oasis. The enclosed garden is cool and leafy, typified by water and shade. The dry climate demands irrigation, which results in small, formal, orderly, arranged plantings. The garden became a human triumph, as it were, over nature.

The Egyptian garden, copied everywhere though changed by local variation in land, plants, and climate, spread to Syria, Persia, and India, and ultimately to Rome. It can be traced through the remains of Egyptian paintings, Persian rugs, and Roman frescoes. This concept of the garden as a separate outdoor room is by

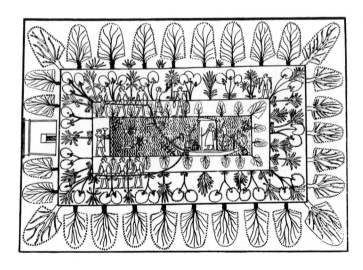

FIGURE 15-4. A portrayal of a formal Egyptian garden from a tomb at Thebes, about 1450 B.C. The garden contains doum palms, date palms, acacias, and other trees and shrubs. A statue is being towed in the lotus pool toward a pavilion. [From Singer, *History of Technology*, vol. 1, Oxford University Press, 1954.]

FIGURE 15-5. Formal French gardens of André Le Nôtre. *Left,* a view of ornamental flower plots, du midi, at Versailles; beyond them is the orangerie. *Right,* Le Chateau vu de la Gerbe, Melun. [Photographs courtesy French Government Tourist Office.]

no means outdated. The tradition has remained relatively unbroken in the Western world through the cloister gardens of the Dark Ages and the courtyard gardens of the Arab cultures to present-day patio gardens.

During the Renaissance, the grand period of the West's cultural revival, the concept of the garden was transformed from relative insignificance to a magnificent splendor befitting the age. The grounds design became the important concept, whereas the plant was treated rather impersonally as merely an architectural material. The plant was pruned, clipped, and trained to conform to the plan. Even architecture became subservient to the landscape plan, the garden engulfing and dominating the stately palace. The resultant "noble symmetry" included courtyards, terraces, statuary, staircases, cascades, and fountains. The emphasis was on long symmetrical vistas and promenades. The small, enclosed garden remained, but only within the walls of the buildings, as a component part of the grand plan. Formalism reached its peak in the age of Louis XIV. The master architectural gardens of André LeNôtre still remain unsurpassed examples of this concept of design predominant over nature (Fig. 15-5).

Naturalism and the East

Naturalism as a concept in gardens can be interpreted as an attempt to live with nature rather than to dominate it. The desired effect is the appearance of a "happy accident of nature," although it is achieved through methods fully as artificial as those of formalism. Although the separation between garden and landscape in formalism is severe, the separation in naturalism is vague and indistinct. The landscape blends into the garden. If formalism is the severe line of geometry, naturalism is the free curve.

The concept of naturalism has been traced to China, but it has reached its highest development in Japan (Fig. 15-6). It has also developed independently in the West. In the Eastern tradition, plants have symbolic significance. This concept is carried over to the arrangement of plants and miniature landscapes and to the development of the whole landscape. The fusion of Eastern naturalism and Western formalism took place in eighteenth century England, where the influence of Chinese culture coincided with a movement away from formalism. The marriage was not always a happy one. Some English gardens became interspersed with Chinese pagodas amid fake antique Gothic ruins. This idea of naturalism survives today in the use of curved walks, artificial wishing wells, and herbaceous borders (Fig. 15-7).

The contemporary trend in gardens is to develop a meaningful design for living. Freed from the confines of "naturalism" or "formalism," the modernist strives to reach esthetic expression through the capacity for abstraction. Plants and people, as in the past, make good companions. We have turned full circle to the concept of the "garden" as a vital need in our society and not merely as an esthetic mix.

FIGURE 15-6. A naturalistic Japanese garden, Rengei Temple, Kyoto. [Photograph courtesy Consulate General of Japan.]

FIGURE 15-7. A naturalistic Chinese garden in Wisconsin. The ivy-covered moon gate will seem either charming or inappropriately "cute," depending on the eye of the beholder. [Photograph courtesy Burlington Industries.]

LANDSCAPE ARCHITECTURE

Landscape architecture in its broadest sense is concerned with the relationship between people and the landscape and as such is concerned with all aspects of **land use.** The profession deals with site development, building arrangement, grading, paving, plantings and gardens, playgrounds, and pools. It is concerned with the individual home and the whole community, with parks and parkways. If the landscape architect must be first an artist, he must also be a horticulturist and a civil engineer. Although landscape architecture was in the past intimately associated with architecture, they have become—unfortunately, perhaps—distinct professions. The objectives of the landscape architect have been to integrate, functionally and esthetically, people, building, and site. The main tools at his disposal are plants and space (Fig. 15-8).

Plant Materials and Design

Although the materials of the landscape architect also include stone, mortar, and wood, the main ingredients are the plants of the horticulturist, who propagates, grows, and maintains them. Unlike the materials of the painter or sculptor, plants

569

FIGURE 15-8. The landscape architect integrates plants, structures, and space for human use. [Photograph courtesy T. D. Church.]

are not static but change seasonally and with time. The design is in living material; the composition is a growing one. These plants, along with the structural components (paving, walls, and buildings), form the elements of the landscape design. Let us consider plant material in relation to the design elements.

Color

The changing color of plant material—foliage and bark as well as flower and fruit—must be considered through the seasons in relation to the landscape. Although the material is stationary, the patterns and color contrasts are transient. Compare, for instance, the way birch trees appear in the winter with the way they appear in the spring, or compare the spring brilliance of flowering crabapple trees with their scarlet October fruit. The horticultural palette is a rich one, and variety is available for many tastes and effects. The delights of color can be planned by the allocation of space for herbaceous material (for example, bulbs, annuals, and succulents). Home landscape planning makes room for gardens to provide cut flowers for decoration within the home.

Texture

Although the variation in textural quality of plants is large, the use of plants next to structural forms (brick, stone, and wood) achieves striking contrasts—for example, trees in front of brick; grass and flagstones; ground cover against paving; or flowering stocks within stone walls. Deciduous trees and shrubs offer changing textural patterns from winter to summer.

Form

Plant material naturally duplicates most of the common forms of solid geometry as well as the more complicated and more interesting forms. Ground covers can be made to form any two-dimensional pattern. Forms can be created by massing plants. Care must be taken in planting, however, to consider plants in terms of their mature forms.

Line

Line is created, as is form, by the arrangement. The dominant vertical lines are created with trees; the horizontal lines, for the most part, by grading—using the ground itself. The use of individual "specimen" plants, by virtue of their own interesting "line," creates focal points—visual interest centers—within the landscape design.

Functional Use of Plant Materials

Terrain without plants is not a fit place for people (Fig. 15-9). The owner of a new home located on a raw piece of ground will readily testify that there are primary problems in mud and erosion, lack of privacy, and a need for protection from sun

FIGURE 15-9. A home before and after landscaping. [Courtesy Ford Motor Co., Tractor and Implement Division.]

FIGURE 15-10. Sodding a lawn. [Courtesy Ford Motor Co., Tractor and Implement Division.]

and wind. Equally dissatisfying is the uninteresting outlook onto the wounds of construction. These problems can be beautifully solved with plants.

Ground Cover

The surface of the ground may be successfully protected from erosion by covering it with plants. Living ground cover serves to disperse the force of the driving rain, but, more important, it entangles and holds the soil with roots. Grass, shrubs, trees, and vines all act as successful erosion controls, even on steeply sloping terrain. Perhaps too much cannot be said to extoll the virtues of grass sod as a landscape material (Fig. 15-10). It makes an ideal surface for recreation; it is cool and free of dust and glare. Grass, as a living floor, is beautiful, remarkably efficient, and relatively easy to maintain, although it must be cut. In areas where less maintenance is desired, such as on slopes or in inaccessible areas, vines and other spreading materials (myrtle, Japanese spurge, ivy, and so forth) make excellent ground covers. Herbaceous annual plants will also serve as surfacing materials, but they are only efficient for half the year in temperate climates.

Enclosure

Shrubby plant material that is high and dense serves to ensure privacy by restricting movement and vision. Where space is limited, plants may be used in combination with structural fences to restrict intrusion (Fig. 15-11). By screening areas with plants, the view may be controlled. Screening thus permits a planned vista. Objectional features of the landscape (garbage cans, incinerators, clothes

FIGURE 15-11. Structural materials and plants may be combined to achieve enclosure and privacy. [Photograph courtesy American Association of Nurserymen.]

lines, parking areas) may be successfully blocked from view by plant material. Screening also serves as protection from the elements. Windbreaks are important for unprotected areas in colder regions. Garden enclosures are the sides of the landscape and, as such, create the feeling of space. Depending on their composition and arrangement they can produce feelings of seclusion or grandeur.

Shade

Protection from the heat and glare of the sun becomes a vital function of plants. Shade trees are not only important to outdoor living but greatly affect the indoor temperature. Care must be taken that trees do not offer potential hazards to buildings. Well-placed trees provide a ceiling to the outdoor room. Deciduous trees further offer the advantage of shade in summer and sun in winter.

Home Landscape Design

The well-designed home landscape is planned to meet the needs and desires of the people who live there (Fig. 15-12). It is concerned with establishing a functional and esthetic relationship between building and site. It is more than the placing of shrubs around the building, although foundation planting is certainly a part of it. In the successful design the house and the surrounding area are treated as a single home unit—as "two sides of the same door." The practical plan must by its very nature be a compromise on a number of conflicting desires. This involves consid-

FIGURE 15-12. Backyards are areas children use. A circular course makes this backyard an attractive place for children to play. [Photograph courtesy T. D. Church.]

eration of the distribution of space, maintenance, and, for most people, expense.

Houses of simple form naturally subdivide a rectangular lot into areas—front, back, and side yards. The problem of home landscaping is to transform these spaces into usable areas. How many uses must be met depends upon the people who live there. However, most homes need at least three general areas: (1) for public access, (2) for service and work, and (3) for living.

The area of public access is usually the front yard. This is the portion of the house that is on view to the public, the passerby as well as the welcomed guest. It creates the setting and tone of the home, and should appear large enough to set off the house from the street (Fig. 15-13); it may need to allow for safe access by any automobile; and it should be hospitable.

The service area accommodates places for garbage and trash disposal, clothes drying, vegetable gardens, compost or mulch piles, dog houses, sandboxes, lawn mowers, storage facilities, and other untidy items. The problem is one of screening and yet maintaining accessibility. In this regard, the service area can often be separated into two locations, and a side yard usually accommodates one of them. The disappearance of back alleys and the change in garbage collections from the back of the lot to the front have complicated the once simple arrangement of service areas in home landscape planning.

To be most functional the outdoor living area should be an extension of the interior of the dwelling. This implies accessibility from the inside living area. It should be level, or should consist of at least a series of level areas, sloping only enough to provide proper drainage. This area should be sufficiently screened and sheltered to assure privacy and comfort, taking full advantage of shade and cooling breezes. All elements of this area should be planned to contribute to a total affect of pleasant and beautiful living space (Fig. 15-14).

FIGURE 15-13. A handsome lawn and a generous setback make an attractive setting for this home. [Photograph courtesy Purdue University.]

FIGURE 15-14. The outdoor living area of a well-landscaped home. [Photograph courtesy Purdue University.]

Prominent in the outdoor living area may be a patio, a paved area attached to the house, with or without a structural ceiling. Through the utilization of paving, plant enclosures, and at least one wall of the building, the patio can be made to serve as an outdoor room (Fig. 15-15). The current use of glass to separate the patio from the house helps to bring garden and house into closer harmony. A private patio that is close to (or at least on the same level as) the kitchen is ideal for dining and relaxation. It is desirable, however, to have additional access from the indoor living area. When space permits, this area may be separated from other areas used for more active recreational purposes.

The relative size of the outdoor areas is a matter of need, preference, and availability of space. Where space is limited, it would seem unnecessary to waste it on the public area. On the other hand, where space is ample, and only maintenance is the limiting factor, there is no reason to skimp. Nothing is more inviting than a generous setback for the house, but few things are as valuable as adequate living space. If large enough, the outdoor living area can be divided to include privacy for relaxation and spaciousness for more vigorous recreation.

That the landscaped home must be maintained is a horticultural truism. The completely paved lot is quite unappealing. If no-maintenance living is desired, the only real alternative is an urban apartment. However, the amount of maintenance required by a landscaped home may be modified with the plan. This entails minimizing lawn areas, annual plants, trimmed shrubs, and any plants that require special care. It involves increasing paved areas and planning for efficiency. The

FIGURE 15-15. A patio is enclosed by two sides of a house and a bamboo wall. Although there are relatively few plants, they are effectively placed. [Photograph courtesy T. D. Church.]

wise selection of hardy, easily grown plants of the proper mature size is necessary. The design of lawns to accommodate power mowers efficiently is an important consideration.

Urban Planning

Communities in the United States with populations of more than 5000 contain about 60% of the population. Ours is indeed an urban society. Yet, at the same time, our cities, sprawling at the edges and decaying in the center, appear to be committing suicide by strangulation. This is reflected in the rush, by those who can afford it, to the outskirts of the city. The suburban movement is often self-defeating, since shops and businesses soon follow, and the increased stress on already inadequate public transportation further increases congestion.

Cities need not be ugly, dirty, or congested. The problems of rehabilitating the expanding modern city can be solved by planning and subsequent action. The aim is to shape the physical development of the city in harmony with its social and economic needs. This involves land-use and transportation control, architectural design, recreation facilities, and sometimes harbor development.

It is important as a social need for cities to be beautiful, to contain fine buildings, monuments, and parks. Beauty is a positive aim, and in this respect green living plants are a necessary part of the urban pattern. This is not to say that nature must dominate the city but only that the city should develop such that nature provides contrast and relief. Just as trees are practical for the home, they are also practical for the city: to provide shade, to act as noise buffers, and to serve as windbreaks. When properly used, they contribute to the architecture and provide visual contrast with the drab colors of construction. The provision of open space in the form of parks and gardens fills a real need of the urbanite, not only for recreational use but for esthetic fulfillment (Fig. 15-16). They are not meant as an escape from the city but as an integral part of it. Finally, open green space may be considered an economic necessity. Blighted, congested cities lead to grave and expensive social problems.

Parks

> Moreover, he hath left you all his walks,
> His private arbours, and new-planted orchards
> On this side Tiber; he hath left them you,
> And to your heirs for ever—common pleasures,
> To walk abroad and recreate yourselves.
> SHAKESPEARE, *Julius Caesar* [III. 2]

Although open spaces in cities for the use of all citizens date from earliest times, the urban system of parks and playgrounds is a recent development. The industrial

FIGURE 15-16. Garden design in a backyard in New York City. Remarkable transformations are possible through imaginative planning. [Photograph courtesy J. Rose and The New York Times.]

revolution has made them a pressing need. Parks and recreational areas are classified on the basis of size and use. **Squares** and **plazas** are ornamental, restful areas, the smallest units in a park system. These are expensive, for many small units of green space create maintenance problems. **Playgrounds** (Fig. 15-17) and **athletic fields** are primarily recreational areas, and may or may not be part of a connected park system. When properly designed, golf courses and athletic fields not only accommodate recreational activities but offer all the benefits of scenic interest. **Neighborhood parks** serve areas inaccessible to large parks or are located to take advantage of local scenery. A new innovation for highly developed urban areas is the development of small green spaces popularly called "**vest pocket parks**" or "**miniparks**." These are frequently less than an acre in size and are often developed on sites formerly occupied by buildings.

Large parks (more than 100 acres), on the other hand, are designed to serve the city as a whole. They may provide facilities for such special recreational activities as horseback riding and golf, and, where water or beach is available, swimming and boating. In addition, they often include museums, memorials, zoos, and botanical gardens (which are usually horticulturally oriented). **State** and **national parks,** although serving a greater area, become important as recreational centers for nearby cities.

The way in which parks are landscaped determines their use. At one end of the spectrum is the **wild park,** which is left almost completely natural except for the creation of a few roads or trails. This limits its uses. Such parks are not suitable for large crowds unless they are unusually vast, such as national parks and forests. The **developed park** combines increased landscape development with natural effects and allows for more intensive use. Historic Central Park in New York City is an outstanding example. Finally, **formal parks** and **gardens** (Fig. 15-18) offer the most

FIGURE 15-17. A well-landscaped playground combines play activity with landscape elements. [Photograph courtesy George Patton, Landscape Architect.]

FIGURE 15-18. A formal park thoroughfare at the University of Pennsylvania that makes good use of common materials. [Photograph courtesy George Patton, Landscape Architect.]

intensive park use. Zoological and botanical parks are prime examples. When properly planned and designed, they are able to handle large crowds gracefully.

The midway atmosphere of commercial amusement parks deteriorates the landscape. These visual vulgarities are distinguished by their lack of landscaping and their general unsightliness. A new trend is the large thematic amusement park (for example, Disneyland), which is landscaped and designed to give a park effect. It is difficult to determine whether the effect of these places will be to raise the level of amusement parks in general or merely to stimulate a host of cheap imitations.

Subdivisions

The design of subdivisions for residences and businesses (such as shopping centers) is an important area for the skills of the landscape architect. The problem is to integrate great numbers of structures into a pattern that maintains function, esthetic appearance, and safety. Well-planned subdivisions preserve the character of the landscape and avoid monotonous repetition. Streets are designed for efficient traffic flow and the avoidance of congestion and dangerous conditions. The architect specified the building locations to conform with zoning regulations and he may also specify the character of the individual units. Too often poor planning or the lack of any planning creates the stereotyped development (Fig. 15-19) described by the phrase "ticky-tacky." Such neighborhoods lack individual identity and become, in effect, suburban slums.

Landscaping and Public Buildings

Public buildings include schools, hospitals, and museums in addition to national, state, and municipal buildings. (In the broadest sense, the term must also include churches.) Public buildings are often expected to be formal, even monumental, in their design and setting as compared with most commercial buildings. Representing the spirit and ideals of a municipality, they require special esthetic treatment. They should be set off so that they can be seen to best advantage. The landscaping should be spacious, dignified, and distinctive.

Schools, rural or urban, deserve special consideration—the best efforts of the community. An ideal school grounds plan could create a spacious and extensively planted park. Landscaped areas might provide adequate room for free play and supervised recreation, sports events, and outdoor ceremonies. In many cases, rural schools have greater possibilities for development than urban schools have, possessing adequate space and good soil. The development of the rural school grounds is often accomplished with community help. Natural plant materials are sometimes used to supplement nursery-grown stock.

FIGURE 15-19. Poor planning creates a suburbia ill-at-ease with the natural environment. [Photograph courtesy USDA.]

Industrial Landscaping

Factories and industrial plants are moving out of heavily congested urban areas. Many companies have begun to pay attention to esthetic considerations as part of their obligations as corporate citizens. The appearance of factories becomes an important factor in both public and labor relations. A well-landscaped industrial plant can be very attractive (Fig. 15-20).

Highways and Roadside Development

Highway design must satisfy the requirements of utility and safety, but it should also satisfy esthetic considerations. The landscaping objectives include the utilization of existing scenic advantages in the proposed routes: although bridges and pavement may need replacement eventually, scenic values can be permanently enhanced through careful planning. Details of subsequent roadside development—outlooks, picnic areas, parks, and the like—become important in planning highway routes.

Highway design should harmonize with the natural topography. Existing trees and lesser vegetation should be conserved. Plantings to control erosion, if in harmony with the natural surroundings, will accomplish a natural transition

FIGURE 15-20. A splendid example of industrial landscaping. Note the contrast created by the exciting use of water, the paving, and the plant materials. [Photograph courtesy General Motors Corp.]

between construction and landscape. To be effective, such plantings require the use of extensive land adjacent to the right of way. Zoning is essential for control and regulation of outdoor advertising and commercial structures along the highway. Safety considerations dictate roadside development that will not be monotonous but that also will not distract the attention of drivers. Interesting scenery, long sites, and gentle curves and grades all help to create these results.

Landscape objectives and engineering objectives alike include erosion control, economical maintenance, safety, sound construction, and conservation of natural beauty. That the public desires landscaped, well-designed highways is evident. Toll roads and freeways have become increasingly important in interstate travel. Tremendous amounts of tax money are spent for safe, restful, scenic driving. The **parkway**, a highway for noncommercial traffic located on a strip of parkland with limited access, only a generation ago considered an extravagent form of highway construction, is today considered an essential form.

FLORAL DESIGN

Floral design bears about the same relationship to landscape design as a string quartet does to a symphony orchestra. The principles of design are the same; only the scale is reduced. Floral design is one of the decorative arts, along with painting and sculpture.

Although arranging flowers[1] is a means of individual artistic expression, it is also the basis of commercial floriculture, which constitutes a large segment of the horticultural industry. All cut flowers are for ultimate use in some sort of arrangement. This may consist of simply placing a dozen roses in a vase, or it may involve the creation of a large floral float.

Planters

Planters, large containers for growing plants, are popular for indoor and outdoor use. Planters are well suited in outdoor courts and plazas. Depending on the season, a wide variety of plants may be used, including evergreen shrubs, bulbs, and annuals. Growing plants contribute to interior decoration just as they do to architecture. Because of the unavailability of indoor light, foliage plants having low light requirements (such as *Philodendron, Sansevieria,* and *Ficus*) are usually grown. Planters have become especially prominent as interior decor in lobbies, offices, and restaurants. The use of artificial foliage plants is also increasing, but these imitations have little esthetic appeal.

Flower Arrangement

Arrangements of flowers, plants, and plant parts have long been used for decoration. In Japan, flower arrangement, or **ikebana,** is a continuing tradition that has been an important part of cultural life for thirteen centuries. Its significance in Japanese home life was established in the fifteenth century, along with the tea ceremony. In its conception, ikebana symbolized certain philosophical and religious concepts. Today, much of the religious connotation has been lost, but the symbolism still remains a key part of the arrangement. Thus, the expression of seasonal change and the passage of time are vital parts of all arrangements, as are appropriate representations of traditional holidays and festivals. Unlike the Western concepts of floral design, the Oriental tradition emphasizes the element of line over form and color. In the classical concept, line is symbolically partitioned into the representation of heaven (vertical), earth (horizontal), and humanity (diagonal and intermediate). The chief aim is to achieve a beautiful flowing line, and to accomplish this end the most ordinary materials may be used. The concept of naturalism is used throughout. All symmetrical effects are avoided.

Flower arrangement is still an important part of Japanese life. There are many different styles and schools: **ikenobo,** classical arrangements; **rikka,** larger, ornate,

---

[1] The term *flower* is used in its broadest sense to include all decorative plants and plant parts, especially the morphological flower.

FIGURE 15-21. Bonsai, or tray culture, is an oriental art achieved through pruning and by controlled nutrition. The four examples shown are part of a gift to the United States from Japan and are on permanent display at the U.S. National Arboretum in Washington, D.C. *A*, Japanese white pine (*Pinus pentaphylla* var. *himekomatsu*), 350 years old. *B*, thorny elaeagnus (*Elaeagnus pungens*), 150 years old. *C*, azalea (*Rhododendron kiushianum*), 150 years old. *D*, Japanese beech (*Fagus crenata*), 50 years old.

upright reproductions of the landscape by means of flowers and plants; **nageire,** simple, naturalistic arrangements; and **moribana,** expressive, scenic arrangements with greater use of foliage and flowers. These schools of flower arrangement differ in opinion and conception, but the basic principles of the art are preserved in common.

There are other typically oriental types of artistic expression involving growing plants. **Bonsai,** the culture of miniature potted trees, dwarfed by pruning and controlled nutrition, is a spectacular example of the horticultural art. Living trees, some of them hundreds of years old and yet only a few feet high, are grown in containers arranged to resemble a portion of the natural landscape (Fig. 15-21).

Bonseki is the construction of a miniature landscape out of stone, sand, and living vegetation.

In Europe, flowers are readily purchased in the market and are in common use as a part of normal living. In the United States, on the other hand, the use of flowers is usually limited to special occasions such as formal dining, decoration at religious holidays, appropriate remembrances (especially birthdays, anniversaries, and Mother's day), and as "get well" gifts. The ceremonial use of flowers in weddings and funerals is the backbone of the florist business. Bouquets and wreaths are standard fare for concert artists, beauty queens, openings, and derby winners. Corsages and boutonnieres become a significant part of the costume at dances and other formal occasions.

HORTICULTURE AS RECREATION

Horticulture has long been, and will continue to be, an outlet for recreation and pleasure. Gardening is probably the true national pastime, for horticulture may be actively enjoyed at many levels. It provides either vigorous or sedentary activity that may be pursued on any scale. Both the young and strong and the aged and infirm may enjoy its respite. There is room for the innovator, the gadget-minded, the artist, and the faddist. The joys of solitude are available, as is the bustle of organizational activity. Horticulture yields the sweets of anticipation along with the bitterness of disappointment. Tangible rewards are available for a minimum of effort, and they increase in proportion to skill and persistence. All who partake of it soon acquire a keener awareness of the mysteries of life, growth, and death. The beautiful as well as the delicious are readily available to those who seek them.

Selected References

Berrall, J. S., 1966. *The Garden: An Illustrated History*. New York: Viking. (A beautiful and well-written history of gardens.)

Carpenter, P. L., T. D. Walker and T. O. Lanphear, 1975. *Plants in the Landscape*. San Francisco: W. H. Freeman and Company. (An excellent treatment of the landscape industry and landscape plants.)

Church, T. D., 1955. *Gardens Are for People*. New York: Reinhold. (Gardens and home landscaping by one of the leading designers in the United States.)

Eckbo, G., 1950. *Landscape for Living*. New York: F. W. Dodge. (An interesting and provocative analysis of landscape architecture.)

Eckbo, G., 1969. *The Landscape We See*. New York: McGraw-Hill. (Artistic principles and modern technology as they relate to landscape design.)

Eckbo, G., 1964. *Urban Landscape Design*. New York: McGraw-Hill. (Landscape design applied to specific urban situations.)

Huxley, A., 1978. *An Illustrated History of Gardening*. New York: Paddington Press. (The newest entry in the world of garden history; profusely illustrated.)

Hyams, E., 1971. *A History of Gardens and Gardening*. New York: Praeger. (A splendid history; highly recommended.)

Riester, D. W. 1959. *Design for Flower Arrangers*. Princeton, N.J.: Van Nostrand. (A beautiful book.)

Rose, J., 1958. *Creative Gardens*. New York: Reinhold. (Well-illustrated book of beautifully conceived landscapes showing unique combinations of natural and structural materials.)

Simonds, J. O., 1961. *Landscape Architecture*. New York: F. W. Dodge. (Landscape planning and site development.)

Tunnard, C., and B. Pushkarev, 1963. *Man-Made America, Chaos or Control?* New Haven, Conn.: Yale University Press. (Design problems in the urban landscape.)

Celsius–Fahrenheit Conversion Table

To convert a temperature in either Celsius or Fahrenheit to the other scale, find that temperature in the center column, and then find the equivalent temperature in the other scale in either the Celsius column to the left or in the Fahrenheit column to the right.

On the Celsius scale the temperature of melting ice is 0° and that of boiling water is 100° at normal atmospheric pressure. On the Fahrenheit scale the equivalent temperatures are 32° and 212° respectively. The formula for converting Celsius to Fahrenheit is $°F = 9/5 °C + 32$, and the formula for converting Fahrenheit to Celsius is $°C = 5/9(°F - 32)$.

C	C or F	F	C	C or F	F	C	C or F	F
−73.33	−100	−148.0	− 6.67	20	68.0	15.6	60	140.0
−70.56	− 95	−139.0	− 6.11	21	69.8	16.1	61	141.8
−67.78	− 90	−130.0	− 5.56	22	71.6	16.7	62	143.6
−65.00	− 85	−121.0	− 5.00	23	73.4	17.2	63	145.4
−62.22	− 80	−112.0	− 4.44	24	75.2	17.8	64	147.2
−59.45	− 75	−103.0	− 3.89	25	77.0	18.3	65	149.0
−56.67	− 70	− 94.0	− 3.33	26	78.8	18.9	66	150.8
−53.89	− 65	− 85.0	− 2.78	27	80.6	19.4	67	152.6
−51.11	− 60	− 76.0	− 2.22	28	82.4	20.0	68	154.4
−48.34	− 55	− 67.0	− 1.67	29	84.2	20.6	69	156.2
−45.56	− 50	− 58.0	− 1.11	30	86.0	21.1	70	158.0
−42.78	− 45	− 49.0	− 0.56	31	87.8	21.7	71	159.8
−40.0	− 40	− 40.0	0	32	89.6	22.2	72	161.6
−37.23	− 35	− 31.0				22.8	73	163.4
−34.44	− 30	− 22.0	0.56	33	91.4	23.3	74	165.2
−31.67	− 25	− 13.0	1.11	34	93.2	23.9	75	167.0
−28.89	− 20	− 4.0	1.67	35	95.0	24.4	76	168.8
−26.12	− 15	5.0	2.22	36	96.8	25.0	77	170.6
−23.33	− 10	14.0	2.78	37	98.6	25.6	78	172.4
−20.56	− 5	23.0	3.33	38	100.4	26.1	79	174.2
−17.8	0	32.0	3.89	39	102.2	26.7	80	176.0
			4.44	40	104.0	27.2	81	177.8
−17.2	1	33.8	5.00	41	105.8	27.8	82	179.6
−16.7	2	35.6	5.56	42	107.6	28.3	83	181.4
−16.1	3	37.4	6.11	43	109.4	28.9	84	183.2
−15.6	4	39.2	6.67	44	111.2	29.4	85	185.0
−15.0	5	41.0	7.22	45	113.0	30.0	86	186.8
−14.4	6	42.8	7.78	46	114.8	30.6	87	188.6
−13.9	7	44.6	8.33	47	116.6	31.1	88	190.4
−13.3	8	46.4	8.89	48	118.4	31.7	89	192.2
−12.8	9	48.2	9.44	49	120.2	32.2	90	194.0
−12.2	10	50.0	10.0	50	122.0	32.8	91	195.8
−11.7	11	51.8	10.6	51	123.8	33.3	92	197.6
−11.1	12	53.6	11.1	52	125.6	33.9	93	199.4
−10.6	13	55.4	11.7	53	127.4	34.4	94	201.2
−10.0	14	57.2	12.2	54	129.2	35.0	95	203.0
− 9.44	15	59.0	12.8	55	131.0	35.6	96	204.8
− 8.89	16	60.8	13.3	56	132.8	36.1	97	206.6
− 8.33	17	62.6	13.9	57	134.6	36.7	98	208.4
− 7.78	18	64.4	14.4	58	136.4	37.2	99	210.2
− 7.22	19	66.2	15.0	59	138.2	37.8	100	212.0

Index

Daminozide, 120. *See also* SADH
Damping-off disease, 347–348
Dandelion (*Taraxacum*), 46, 534
Dark phase (photosynthesis), 96–98
Darwin, Charles, 41
Date (*Phoenix*), 10, 11, 14, 35, 46, 172, 515–516
Datura, 109
Daucus (carrot), 48, 543–544
Daylength. *See* Photoperiodism
Daylily (*Hemerocallis*), 45, 355
Day-neutral plant, 163
Da Vinci, L., 20
DDT, 302, 303, 305
Deciduous plant, 30
Deer, as pests, 292
Deficiency symptoms, 186–189
Dehiscence, 89
Dehydration, 442
Dendrobium, 46
Deshooting, 242
Design, 563–565
  elements, 563–564
  floral, 582–585
  home, 573–577
  plant materials, 569–571
  principles, 564
Determinate plant, 388
Deuteromycetes (Fungi Imperfecti), 286
Devernalization, 167
Dewberry (*Rubus*), 513
Dewpoint, 461
Dianthus (carnation), 375, 553
Dibble, 350
Dichasium, 88
Dicotyledoneae, 43, 46–48
  structure, 57
  vasculor system, 70
Differentiation, 112–121, 149–177
Difficult-to-root cutting, 268, 360–362
Diffusion, 100
Digitalis, 531
Dill (*Anethum*), 48
Dioecism, 87, 339
Dioscorea (yam), 33, 542–543
Dioscorides, Pedanius, 13, 40, 41
Diphenylamine, 276
Diploidy, 381, 392, 397
Direct seeding, 349–350
Disbudding, 325
  chemical, 274–275
Disease
  cycle, 292–294
  nonpathogenic, 278
  pathogenic, 278–294
  signs (pathogen), 279
  symptoms, 279
Disease control
  biological, 304–305

chemical, 296–303
cultural, 295–296
genetic, 306–308
integrated, 303
legal, 295
physical, 296
physiological, 305–306
seed germination, 347–349
Distribution, 448–452
Division
  method of propagation, 353
  phyla, 42–43
DNA, 9, 59, 108, 175, 283, 329–332, 380, 413
Dodder, 281
Dolichos (hyacinth bean, lablab), 540
Dominance, 382
  hypothesis, 390–392
Dormancy
  bud, 157–158
  modification, 273
  physical, 152
  physiological, 153–15
  seed, 152–155, 309, 342–344
Dormin. *See* Abscisic acid
Double cross, 407–409
Double dormancy, 154, 344
Double fertilization, 339
Drainage, 212–213
Drug crops, 39, 530–533
Drupe, 39, 46, 89, 91
Dry fruit, 89, 91
Drying, 441–442
Dry storage (seed), 344
Duboisia, 109
Dust, 298
Dwarfing
  chemical, 273–274
  phloem disruption, 264
  rootstock, 260–262

Easter lily (*Lilium*), 45, 87, 161, 274
East Malling, rootstock, 260–261
Easy-to-root cutting, 268, 360–362
Ecolinal variation, 51
Ecology, 467
Economics
  factors in horticulture, 476–479
  of plant population, 325–326
Ecospecies, 49, 50
Ecotype, 49, 50–51
Edible plants, 39
Effective precipitation, 463
Egg, 289, 338–339
Eggplant (*Solanum*), 48, 537
Egypt
  civilization, 9
  garden, 566
Eichler, August, 42

deficiency symptom, 186–189
translocation, 101–103
Nutrition
fruit growth, 174
plant, 129, 182–194
soil cycle, 197
Nymph, 290

Obligate parasite, 280
Offset, 80, 353, 355
Offshoots, 80, 355
Oil palm (*Elaeis*), 46
Oil plants, 39
Olea (olive), 10, 13, 515
Olericulture (vegetable crops), 4
Olive (*Olea*), 10, 13, 515
Onion (*Allium*), 10, 45, 93, 167, 192, 311, 406, 415, 544–546
Oo (*Brassica*), 396
Open center (vase) tree, 246–247, 256
Open pollination, 53
Opuntia (prickly pear), 312
Orange (*Citrus*), 47, 514–515
Orcharding, 490–492
Orchid, 553–554
Orchidaceae, 45–46
Orchid family. *See* Orchidaceae
Order (taxonomy), 43, 44
Organic
acids, 105
matter, 122–123, 195–197
soil, 124
volatiles, 438–439
Organogenesis, 377
Oriental gardens, 568
Ornamentals
crops, 549–557
horticulture, 4, 5
plant, 35
production systems, 495–502
Ortet, 52
Osmosis, 100
Osmunda, 554
Ovary, 86
Overdominance hypothesis, 390–392
Ovule, 86
Oxygen, 47, 110–111, 147, 437
Ozone, 47, 147

Packaging, 430–431
Paeonia (peony), 154
Palisade cell, 83–84
Pallet box, 423, 435
Palmae, 46
Palm family. *See* Palmae
Panama disease (*Fusarium*), 519
Panicle, 45, 46, 88
Papain, 34

Papaver (poppy), 109
Papaya (*Carica*), 34, 522–523
Paprika (*Capsicum*), 538
Papyrus, 11
Parasitism, 280
Parenchyma, 63, 65–66
Park landscaping, 577–580
Parkways, 582
Parsley, (*Petroselinum*), 48, 533, 534
Parsnip (*Pastinaca*), 48, 544
Parthenocarpy, 170, 172, 270, 340
Parthenocissus (Japanese ivy), 37
Paspalum (bahia grass), 502
Pasteurization (soil), 311, 348
Pastinaca (parsnip), 48, 544
Patent (plant), 416
Pathogen, 279–292
Pathogenic RNA, 282
Patio, 576
Pausinystalia, 109
PBA, 118
Pea family. *See* Leguminosae
Pea (*Cajanus, Cicer, Pisum, Vigna*), 47, 539–540
Peach (*Prunus*), 511–512
cold requirement, 158–159
developmental history, 151
embryo culture, 344–345
genetics, 387
Peanut (*Arachis*), 47, 540
Pear (*Pyrus*), 47, 283–285, 305–306, 308, 510–511
Peat, 124, 195, 363–364
Pecan (*Carya*), 525
Pectin, 60
Pedicel, 88
Pedigree selection, 403–404
Peduncle, 88
Pelargonium (geranium), 376
Pelletierine, 109
Pelleting seed, 350
Penicillium, 439
Peony (*Paeonia*), 154
Pepo, 89, 91
Pepper
Capsicum (chili, red), 48, 537–538
Piper (black, white), 532
Peppermint (*Mentha*), 532
Perennial plant, 30, 39
Perfect flower, 87
Perianth, 85
Pericarp, 89
Pericycle, 71
Periderm, 71, 430
Perithecium, 286
Periwinkle (*Vinca*), 36
Perlite, 363–364
Permanent tissue, 63, 65–66